Lecture Notes in Computer Science 505

Edited by G. Goos and J. Hartmanis

E. H. L. Aarts J. van Leeuwen
M. Rem (Eds.)

PARLE '91 Parallel Architectures and Languages Europe

Volume I: Parallel Architectures and Algorithms
Eindhoven, The Netherlands, June 10-13, 1991
Proceedings

Springer-Verlag
Berlin Heidelberg GmbH

Series Editors

Gerhard Goos
GMD Forschungsstelle
Universität Karlsruhe
Vincenz-Priessnitz-Straße 1
W-7500 Karlsruhe, FRG

Juris Hartmanis
Department of Computer Science
Cornell University
Upson Hall
Ithaca, NY 14853, USA

Volume Editors

Emile H. L. Aarts
Philips Research Laboratories
P.O. Box 80.000, 5600 JA Eindhoven, The Netherlands

Jan van Leeuwen
Department of Computer Science, University of Utrecht
Padualaan 14, 3584 CH Utrecht, The Netherlands

Martin Rem
Eindhoven University of Technology
P.O. Box 513, 5600 MB Eindhoven, The Netherlands

CR Subject Classification (1991): C.1-4, D.1, D.3-4, F.1-3

ISBN 978-3-662-23206-4 ISBN 978-3-662-25209-3 (eBook)
DOI 10.1007/978-3-662-25209-3

Originally published by Springer-Verlag Berlin Heidelberg New York in 1991

2145/3140-543210 - Printed on acid-free paper

Preface

The innovative progress in the development of large- and small-scale parallel computing systems and their increasing availability have caused a sharp rise in interest in the scientific principles that underlie parallel computation and parallel programming. The biannual "Parallel Architectures and Languages Europe" (PARLE) conferences aim at presenting current research material on all aspects of the theory, design, and application of parallel computing systems and parallel processing. At the same time, the goal of the PARLE conferences is to provide a forum for researchers and practitioners to exchange ideas on recent developments and trends in the field of parallel computing and parallel programming. The first two conferences, PARLE '87 and PARLE '89, have succeeded in meeting this goal and made PARLE a conference that is recognized worldwide in the field of parallel computation.

PARLE '91 again offers a wealth of high-quality research material for the benefit of the scientific community. Compared to its predecessors, the scope of PARLE '91 has been broadened so as to cover the area of parallel algorithms and complexity, in addition to the central themes of parallel architectures and languages.

The proceedings of the PARLE '91 conference contain the text of all contributed papers that were selected for the programme and of the invited papers by leading experts in the field. The proceedings are presented in two volumes:

Volume I: Parallel Architectures and Algorithms

Volume II: Parallel Languages

In the Call for Papers of PARLE '91 papers were solicited describing original research in the areas of parallel architectures and systems, parallel programming, parallel languages, parallel algorithms and complexity, and applications of parallelism (including for example, neural computing). We greatly appreciate the efforts of the authors who submitted papers to this year's conference; a record number of 161 submissions were received. To maintain the high technical level of PARLE all papers went through an intensive review process. We extend our sincere thanks to the Programme Committee and to the referees who were consulted in the reviewing process, for the arduous task of selecting the programme out of these papers. Their dedication and professionalism have greatly contributed to the technical quality of the PARLE '91 conference.

No conference of this size can be organized without the contributions of many dedicated people. As for the previous PARLE conferences, the organization of the PARLE '91 conference was again handled by the Philips Research Laboratories. A special tribute in this respect is due to Fred Robert and Frank Stoots, for their skilled and efficient handling of all organizational details of the conference over the past year.

We feel that the PARLE '91 conference has once again succeeded in bringing together a wealth of material on the theme of parallel computing. We therefore hope that these proceedings will contribute to the tradition of European research on Parallel Architectures and Languages.

Eindhoven/Utrecht
June 1991

Emile H.L. Aarts
Jan van Leeuwen
Martin Rem

Scientific Programme Committee

Contents Volume I

Invited Lectures

Submitted Presentations

Contents Volume II

Invited Lectures

Submitted Presentations

List of Referees

Aerts, A.T.M.
America, P.
Anishev, P.A.
Annot, J.K.
Asveld, P.R.J.
Augusteijn, A.
Augustsson, L.
Beaudoing, B.
Beaumont, T.
Beemster, M.
Belvide, R.
Bekkers, Y.
Benveniste, M.
Best, E.
Bjornson, R.
Blanken, H.M.
Boer, F. de
Bosc, P.
Bost, B.
Bronnenberg, W.
Breu, R.
Brix, H.
Carlsson, M.
Carriero, N.
Chassin de Kergommeaux, J. de
Cheese, A.
Cherkasova, L.A.
Ciancarini, P.
Cooper, M.D.
Cortes, U.
Creveuil, C.
Dederichs, F.
Delgado-Rannauro, S.A.
Dendorfer, C.
Dennison, L.R.
Desel, J.
Dutra, J.
Eekelen, M.C.J.D. van
Ellis, J.
Esparza, J.
Feijen, W.H.J.
Feijs, L.M.G.
Fraigniaud, P.
Forshaw, M.R.B.
Fuchs, M.
Gabarró, J.
Gamané, B.
Gerth, R.
Godefroid, P.
Goltz, U.
Gonzalez, A.
Gribomont, E.P.
Gritzner, T.F.
Gupta, G.
Haaften, P.J.M. van
Hausman, B.
Helmink, L.
Hesselink, W.H.
Hofman, R.F.H.
Hofstee, H.P.
Holden, M.
Horwat, W.
Houck, C.
Houtsma, M.A.W.
Hulin, G.
Hulshof, B.J.A.
Ibañez, M.B.
Inamura, Y.
Ingels, Ph.
Jansen, P.G.
Janson, S.
Jantzen, M.
Jard, C.
Jegov, Y.

Joseph, M.
Kalker, A.A.C.M.
Kamerbeek, J.
Kaplan, S.
Karpov, Ju.G.
Kausche, A.
Kennaway, J.R.
Keen, J.
Keckler, S.W.
Kempf, P.
Khebbal, S.
Koot, J.
Korst, J.H.M.
Labarta, J.
Laborelli, L.
Langendoen, K.G.
Le Metayer, O.
Lethin, R.
Liang Liang, L.
Lukkien, J.J.
Madelaine, E.
Martin, G.R.
Milikowski, R.
Miriyala, S.
Monien, B.
Morrison, J.P.
Muller, H.L.
Murakami, M.
Mussi, Ph.
Nakata, T.
Nepomniaschy, V.A.
Nijholt, A.
O'Boyle, M.F.
Omtzigt, F.T.L.
Overeinder, B.J.
Paech, B.
Panwar, R.
Tubella, J.
Parrott, D.J.
Pazat, J.-L.
Peikenkamp, T.
Peleg, D.
Peters, J.G.
Philippe, B.J.
Plasmeijer, M.J.
Priol, T.
Quinton, P.
Raina, S.
Ratcliffe, M.J.
Raucher, R.
Raynal, M.
Remscher, R.
Renardel de Lavalette, G.R.
Roberts, G.
Roelants, D.
Rokusawa, K.
Rouwce, P.A.
Sanders, B.
Santos-Costa, V.
Schalij, F.D.
Scholten, J.
Schoute, A.L.
Sedukhin, S.
Serna, M.J.
Seznec, A.
Sijstermans, F.
Sindaha, R.
Spertus, E.
Sun, C.Z.
Stomp, F.A.
Streicher, P.
Syska, M.
Taubner, D.
Tel, G.
Thibault, O.
Watson, I.

Twist, R.A.H. van
Valk, R.
Venkatasubramanian, N.
Veron, A.
Vissers, K.A.
Vlot, M.C.
Vree, W.G.
Vries, F-J. de
Wallach, D.
Weber, R.
Wester, R.H.H.
Wei, M.
Wiedermann, J.
Wilschut, A.N.
Wolper, P.
Yang, R.
Yamasaki, S.
Zakaria, L.A.H.J.

Parallel Evaluation of Functional Programs: The $\langle \nu, G \rangle$-machine approach (Summary)

Thomas Johnsson
Department of Computer Science
Chalmers University of Technology
S-412 96 Göteborg, Sweden

For a number of years this author, together with Lennart Augustsson, have been developing fast implementations of lazy functional languages, based on graph reduction, for ordinary (sequential) computers. Our approach can be summarised very briefly as follows.

Our ideas stem from Turner's S, K, I standard combinator reduction approach [Tur79]. But instead of using a standard, fixed set of combinators, a compiler transforms the program into a new set of specialised combinators, or 'super-combinators' [Hug82]. This transformation process is called *lambda lifting* [Joh85]. Each of these super-combinators are then compiled into machine code for the machine at hand, this code implements the corresponding graph rewrite rule. In other words, the compiler constructs a specialised, machine-language coded combinator interpreter from each program. However, rather than compiling each combinator into machine code directly, we first compile them into code for an abstract machine, the *G-machine* [Joh84]. Also, rather than letting the code rewrite the graph for a combinator application into the graph of the right hand side of the combinator definition, quite a lot of improvements to this scheme is possible. The G-machine is a convenient abstraction for expressing these improved compilation schemes.

An overview of the techniques used in our compiler for Lazy ML can be found in [Aug84] and [AJ89b]. The compilation of pattern matching into efficient code is described in [Aug85]. Our method of lambda lifting is described in [Joh85], and the G-machine is described in [Joh84]. The approach to machine code generation used in the Lazy ML compiler is described in [Joh86].

Parallel computers, consisting of dozens, hundreds, or thousands of processors connected to either a shared memory or a message passing network, are now becoming available on the marketplace. Recently, we have done work on extending the G-machine techniques to perform parallel graph reduction on such computers [AJ89a], to obtain real speedup compared to the sequential implementation.

It is possible to modify the sequential G-machine into a parallel one straighforwardly, by having multiple threads of control each with one or two stacks, all of which perform graph reduction in a common graph — this is the shared memory model. Such systems have been designed and implemented by Maranget [Mar91] and [Geo89]. This is also the approach taken in the GRIP project [Jon87, JCS89].

However, there are some properties of the standard G-machine that made us want to try a different approach for a parallel implementation. Firstly, in the G-machine, when reduction of a function application starts, the arguments of the application (either in the form of a chain of binary application nodes, or a vector application node) are moved to the stack, and when reduction is finished the result is moved back into the heap by updating the root application node with the value of the function application. This seems like a lot of unnecessary data movement when the datum could have been accessed from the node in the first place (this reasoning has nothing to do with parallelism, of course). Secondly, the prospect of having to manage a cactus stack was not very appealing, we wanted something simpler.

In the machines we would like to consider, and in particular the machine we have implemented our parallel graph reducer on, the Sequent Symmetry^{TM}, a memory reference into the heap has the same cost as a reference into a stack, since they reside in the same (shared) memory. Thus the cost of moving a word to the heap while building a node is the same as pushing a word into the stack.

Thus, in the abstract machine we have designed, called the $\langle \nu, G \rangle$-machine, function applications are represented by *frame nodes*. A frame node holds the arguments of the function application and a pointer to the code for the function being applied, but in addition also contains enough space for temporaries needed for reduction of the function application. Figure 1 shows what happens when EVAL is called: the 'current point of reduction' is moved to the frame node to be evaluated, and a 'dynamic link' field is set to point back to the frame which called did the EVAL. Thus, instead of an ordinary stack we have a linked list of stack frames. In the parallel case, we have many points of reduction. For further details of the abstract machine, see [AJ89a].

Figure 1: Calling EVAL to reduce a frame node.

To be fair, the stack model has some advantages too. The spineless G-machine [BRJ88], which offers a more general and efficient tail call mechanism than the 'standard G-machine' in particular when dealing with higher order functions, requires essentially an arbitrarily big stack. However, Lester [Les89] has devised an analysis technique based on abstract interpretation, to determine the maximum size a stack might have under the 'spineless' evaluation regime. Thus it would be possible to merge the $\langle \nu, G \rangle$-machine model with the 'spineless' model of execution, by allocating a frame node of the required maximum size.

Both the stack model and the frame node model have their advantages, and it is too early to nominate an overall winner.

So far, to introduce parallelism in the LML programs the programmer has to write *spark* annotations [CJ86] in the programs explicitly. The spark annotation is advisory: if there is a processor available then it may evaluate the sparked expression, otherwise the process really needing the value will evaluate it itself.

Code generation now works rather differently from the way it was described in [AJ89a]. Code is generated by first translating the combinators into three-address form, with liberal use of temporary names. We illustrate this with the code for the combinator `f x y z = x y`, which is:

```
funstart f 3            start of f, which takes 3 args
load  t0,0(nu)          load z from frame into t1
load  t1,1(nu)          load y from frame into t2
load  t2,2(nu)          load x from frame into t3
move  t1,t4
move  t4,t5
store t5,0(nu)          store the arg y of the application into the frame
eval  t2                evaluate x, the function
move  t2,t3
do    t3,1              tail call, function is x, one arg in frame
```

The code for a combinator starts by loading all arguments into temporaries from the frame node. Then in the example above the argument y of the tail call, is moved into the current frame at the location of the last argument; the function x of the tailcall is evaluated into function form, and finally the tail call is performed with the general tail call instruction do.

This 'raw' code is then subjected to various improvement transformations; for instance, the loads and stores are moved around to minimise the number om live variables across eval. Finally, temporaries are bound to machine registers. The resulting code is:

```
funstart f 3            start of f, which takes 3 args
load  r0,2(nu)          load x from frame into register r0
eval  r0                evaluate x, the function, in r0
load  r4,1(nu)          load y from frame into register r4 ...
store r4,0(nu)          ... and store y, the arg of the application into the frame
do    r0,1              tail call, function is x, one arg in frame
```

From this code the actual machine code is generated. A notable feature of the generated code is that we have abandoned the method of coding the tag as a pointer, either to a table (as described in [Joh86]) or to code directly, as in the spineless tagless G-machine [JS89]. Instead, the tag word contains various tag bits. The reason comes from two observations: firstly, most of the time when doing eval the node is allready canonical — according to measurements 80% of the time is typical. Secondly, in most modern architectures with instruction prefetch, it is rather costly to break the sequential flow of control. We therefore implement eval with code that tests a canonical-bit in the tag field of the node to be evaluated; if canonical the next instruction is executed, and only if it is not canonical does a jump occur to code that performs the actual call to the eval routine. The call to eval is surrounded by code that stores and reloads the content of live registers. Our

implementation of the parallel $\langle\nu, G\rangle$-machine is for the Sequent Symmetry, a bus-based shared memory machine. The architecture supports up to 30 processors connected to the bus; our machine has 16 processors. This machine has some features that helps very much in the implementation of the parallel $\langle\nu, G\rangle$-machine, for instance, any cell in the memory can be used as an atomic lock.

At the moment of writing this, a new garbage collector is being tested [Röj91]. It is an improved version of the Appel-Ellis-Li garbage collector, which is an efficient real-time copying garbage collector which runs concurrently with the mutator processes. Röjemo has extended it also to collect processes which have become garbage.

Since the publication of [AJ89a] we have improved the performance somewhat due to the improved code generation method, as described briefly above. The improvement is about 25% for code purely sequential code, but for parallel programs the improvement is less than that – depending on how big a proportion of the time is spent in activities like syncronisation, task switching etc. Figure 2 shows the current speedup charts for three benchmark programs. Garbage collection time is not included in these figures.

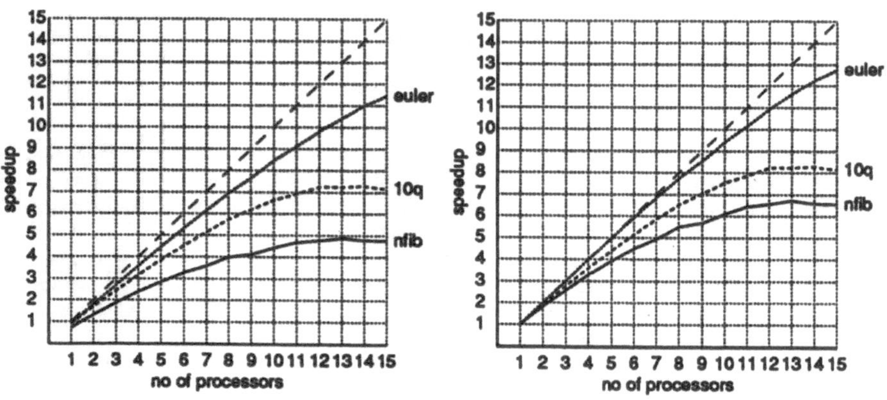

Figure 2: Speedup graphs for three benchmark programs: The left graph shows the speedup relative to one processor, the right graph shows the speedup relative to the 'standard G-machine' in the LML compiler.

References

[AJ89a] L. Augustsson and T. Johnsson. Parallel Graph Reduction with the $\langle \nu, G \rangle$-machine. In *Proceedings of the 1989 Conference on Functional Languages and Computer Architecture*, pages 202–213, London, England, 1989.

[AJ89b] L. Augustsson and T. Johnsson. The Chalmers Lazy-ML Compiler. *The Computer Journal*, 32(2):127 – 141, 1989.

[Aug84] L. Augustsson. A Compiler for Lazy ML. In *Proceedings of the 1984 ACM Symposium on Lisp and Functional Programming*, pages 218–227, Austin, 1984.

[Aug85] L. Augustsson. Compiling Pattern Matching. In *Proceedings 1985 Conference on Functional Programming Languages and Computer Architecture*, Nancy, France, 1985.

[Aug87] L. Augustsson. *Compiling Lazy Functional Languages, Part II.* PhD thesis, Dept. of Computer Science, Chalmers University of Technology, Göteborg, Sweden, November 1987.

[BRJ88] G. Burn, J. Robson, and S. Peyton Jones. The Spineless G-machine. In *Proceedings of the 1988 ACM Symposium on Lisp and Functional Programming*, Snowbird, Utah, 1988.

[CJ86] C. Clack and S.L. Peyton Jones. The Four-Stroke Reduction Engine. In *Proceedings of the 1986 ACM Conference on LISP and Functional Programming*, pages 220–232, 1986.

[Geo89] Lal George. An abstract machine for Parallel graph reduction. In *Proceedings of the 1989 Conference on Functional Programming Languages and Computer Architecture*, London, Great Britain, 1989.

[Hug82] R. J. M. Hughes. Super Combinators—A New Implementation Method for Applicative Languages. In *Proceedings of the 1982 ACM Symposium on Lisp and Functional Programming*, pages 1–10, Pittsburgh, 1982.

[JCS89] S. L. Peyton Jones, C. Clack, and J. Salkild. High-Performance Parallel Graph Reduction. In *Proceedings of PARLE'89 Parallel Architectures and Languages Europe (Vol I)*, volume LNCS 365, pages 193–206. Springer-Verlag, June 1989.

[Joh84] T. Johnsson. Efficient Compilation of Lazy Evaluation. In *Proceedings of the SIGPLAN '84 Symposium on Compiler Construction*, pages 58–69, Montreal, 1984.

[Joh85] T. Johnsson. Lambda Lifting: Transforming Programs to Recursive Equations. In *Proceedings 1985 Conference on Functional Programming Languages and Computer Architecture*, Lecture Notes in Computer Science 201, Nancy, France, 1985. Springer Verlag.

[Joh86] T. Johnsson. Code Generation from G-machine code. In *Proceedings of the workshop on Graph Reduction*, Lecture Notes in Computer Science 279, Santa Fe, September 1986. Springer Verlag.

[Joh87] T. Johnsson. *Compiling Lazy Functional Languages*. PhD thesis, Dept. of Computer Science, Chalmers University of Technology, Göteborg, Sweden, February 1987.

[Jon87] S. L. Peyton Jones. GRIP: A Parallel Graph Reduction Machine. In *Proceedings of the 1987 Conference on Functional Programming Languages and Computer Architecture*, Portland, Oregon, U.S.A, September 1987.

[JS89] S.L. Peyton Jones and Jon Salkild. The Spineless Tagless G-machine. In *Proceedings of the 1989 Conference on Functional Programming Languages and Computer Architecture*, London, Great Britain, 1989.

[Les89] David R. Lester. Stacklessness: compiling recursion for a distributed architecture. In *Proceedings of the 1989 Conference on Functional Programming Languages and Computer Architecture*, London, Great Britain, 1989.

[Mar91] Luc Maranget. GAML: a Parallel Implementation of Lazy ML. Technical report, Department of Computer Sciences, INRIA Rocquencourt, BP 105, 78153 Le Chesnay CEDEX, France, 1991.

[Röj91] Niklas Röjemo. A Concurrent Garbage Collector for the $\langle \nu, G \rangle$-machine. Technical report, Department of Computer Sciences, Chalmers University of Technology,S–412 96 Göteborg, February 1991.

[Tur79] D. A. Turner. A New Implementation Technique for Applicative Languages. *Software—Practice and Experience*, 9:31–49, 1979.

Towards a Single Model of Efficient Computation in Real Parallel Machines

Pilar de la Torre *
Department of Computer Science
University of New Hampshire
Durham, New Hampshire 03824

and

Clyde P. Kruskal
Department of Computer Science
University of Maryland
College Park, Maryland 20742.

*The research of this author was supported in part by the National Science Foundation under Grant CCR-9010445.

Abstract

We propose a model of parallel computation, the *Y-PRAM*, that allows general parallel algorithms to be designed for a wide class of the parallel models. There are two basic quantities captured by the model, which the algorithm designer must leave open as parameters: *latency* and *bandwidth.*

We design *Y-PRAM* algorithms for solving several fundamental problems: parallel prefix, sorting, sorting numbers from a bounded range, and list ranking. We show that our model predicts, reasonably accurately, the actual known performances of several basic parallel models - PRAM, hypercube, mesh, and tree - when solving these problems.

1 Introduction

There is a large variety of models of parallel computation. Typically algorithms are developed for every model individually. Some understanding of parallel algorithm design has emerged: for example, we know that parallel prefix and sorting are important basic routines. And there are now many results showing separations between and equivalences of different models. But, there is no general framework for developing algorithms that apply to a variety of models. This paper attempts to introduce such a framework.

Our basic idea is to try to find a model of parallel computation with a limited number of parameters that capture the performance of an algorithm on a large class of parallel machines. With this tool, parallel algorithms can be developed once and for all, leaving open the few parameters for customization to a given machine. Substituting the parameter values will provide some idea of how well the particular machine can execute the algorithm. Furthermore, a compiler could use the algorithm to produce code for the machine.

If the model truly reflects parallel machine performances, it can provide a simple model for proving general upper and lower bounds. Similarly, by leaving in the parameters, one can often explicitly see the bottleneck of an algorithm.

We propose the Y-PRAM model of parallel computation, which captures two basic quantities of parallel computation: *latency* and *bandwidth.* We design *Y-PRAM* algorithms for solving several fundamental problems: parallel prefix, sorting, sorting numbers from a bounded range, and list ranking. We show that our model predicts, reasonably accurately, the actual known performances of several basic parallel models - PRAM, hypercube, mesh, and tree - when solving these problems.

A model that has many of the aspects of the Y-PRAM is the X-PRAM model of

Valiant [24]. Other related work is the investigation of latency by Aggarwal, Chandra, and Snir in their LPRAM and BPRAM models [1, 2].

1.1 Performance of parallel algorithms

We are restricting consideration to tightly coupled machines. We assume that moderately sized parallel computers will be used to solve very large problems [9] i.e., problems whose sequential times are much larger than the number of processors. We follow the work of Kruskal, Rudolph, and Snir [14], who define six classes of parallel algorithms.

From our point of view, the most interesting class is **EP** (Efficient Polynomially fast, or Efficient Parallel), which are those algorithms that achieve "polynomial" reduction in running time with constant "inefficiency". In order to define these classes formally, we need some definitions.

Let $t(N)$ be the sequential time for an algorithm to solve some problem of size N. Let $T_P(N)$ be the time for P processors to solve the same problem. A problem is in **EP** if

$$T_P(N) = O(t(N)/P) + P^{O(1)}.$$

The first term on the right side guarantees that the inefficiency is at most constant for large enough problems, and the second term guarantees that the constant inefficiency obtains whenever the sequential time is polynomially larger than the machine size (for some polynomial).

1.2 Models of parallel computation

The most powerful parallel models are the synchronous PRAM's (Parallel Random Access Machines), which allow the processors to access a common memory. The weakest variant is the EREW (Exclusive Read, Exclusive Write) PRAM; it does not allow concurrent accesses to a common location. The CREW (Concurrent Read, Exclusive Write) PRAM allows concurrent reads to a common location, but not concurrent writes. The CRCW (Concurrent Read, Concurrent Write) PRAM allows concurrent reads and concurrent writes to a common location. CRCW PRAM itself has many variants including the common, arbitrary, priority, and random models.

The next most powerful model is the DCM (Direct Connection Machine). It consists of autonomous unit-cost RAM's, each with its own local memory, that communicate by message passing – there is no shared memory. (This model is also known as a Fully

Connected Direct Connection Machine [12], Module Parallel Computer [19], and Seclusive PRAM [24]).

Finally, we have the sparse networks. Here each processor are connected only to a subset of other processors. In this paper, we will concentrate on the hypercube, butterfly, shuffle-exchange, complete binary tree, 2-dimensional mesh, and 1-dimensional mesh models. For conciseness, we will refer to a complete binary tree simply as a tree, a 2-dimensional mesh simply as a mesh, and a 1-dimensional mesh as a linear array. For a fuller discussion of models than given below see [15, 11, 24].

1.3 The Y-PRAM

Before presenting the *Y-PRAM* model, we provide a few definitions. A parallel machine is *recursively decomposable into halves* if the processors can be partitioned into two groups of equal size, so that each group is itself a submachine (and therefore recursively decomposable into halves). The *latency*, $\delta(P)$, of a parallel model is the maximum time to communicate between any pair of processors (or between any processor and shared memory). Typically, it will be the diameter of a sparse machine. The *bandwidth inefficiency*, $\beta(P)$, of a parallel model is the ratio of P to the bandwidth. We will discuss these machine properties below.

Y-PRAM MODEL: The *Y-PRAM*$[\beta, \delta]$ is a recursively decomposable model consisting of $P = 2^p$ RAM processors, which operate synchronously. Furthermore,

1. It has shared memory, which is partitioned among the processors so that each processor *owns* an equal amount.

2. Any submachine can block itself off from the rest of the machine. While blocked off, the processors within the submachine interleave *periods of computation* with *periods of memory access*; the accesses may be only to the portion of memory owned by the processors within the submachine.

3. Memory accesses may not conflict, i.e., there are no concurrent accesses to a common location. Thus, a blocked off submachine acts like a little EREW PRAM, except that the cost of accessing memory depends on δ and β.

4. The time for the processors within a submachine of size S to make a total of M memory accesses is

   $$\Theta(\delta(S) + m + M\beta(S)/S),$$

   where m is the most accesses of any one processor.

In many ways the *Y-PRAM* is an especially good model of the CM* machine at Carnegie-Mellon [23] and of the Cedar machine at the University of Illinois [16]. Both machines have clusters of processors, which produces a hierarchy of memories – CM* within the context of a tree and Cedar within the context of a PRAM.

We now discuss the three properties defined above. Many parallel models are recursively decomposable into halves, including the PRAM, linear array, and hypercube models. While many parallel models are not, most come close enough. For example a 2-dimensional mesh decomposes into four meshes, each once fourth the size of the original mesh. Since this decomposition produces a constant number of copies, it will suffice for our purposes.

A ring machine does not decompose at all: proper connected subgraphs of a ring are linear arrays. However, a linear array of size P can emulate a ring of size $2P$ with only a constant loss in efficiency, by having processor i of the linear array emulate processors i and $2P-1-i$ of the ring. Thus a ring of size P can emulate a recursively decomposable ring of size $2P$.

While, a tree machine decomposes naturally into two subtrees, the original root is not in either subtree. However, every decomposition of a (sub)tree uses more than half of its processors, so there is only a constant loss of efficiency.

More problematic, are the butterfly and shuffle-exchange models. They are not recursively decomposable; in fact, no machine that has only a constant number of connections per processor and that can permute in logarithmic time is recursively decomposable [18]. However, they have many aspects of recursively decomposable machines. In particular, ascend-descend algorithms [20], which are recursive, can be implemented efficiently on them. Furthermore, since many operations take polylogarithmic time, executing them globally for the whole machine is not much slower than working locally within submachines.

Now consider the latency. For sparse networks, it will typically be the diameter of the machine, i.e., the maximum distance between any pair of processors. The DCM model has constant diameter; the hypercube, shuffle-exchange, butterfly, and tree models have diameter $\Theta(\log P)$; and a d-dimensional mesh has diameter $\Theta(P^{1/d})$. The latency bound can be used to capture pipeline delays; many machines have a high start up time to send a message.

From our point of view, the latency bound is, in some sense, irrelevant: The class **EP** is invariant under latency, as long as the latency is a most polynomial in P (i.e., $\delta(P) \leq P^{O(1)}$) [KRS]. But, without taking latency into consideration our model would predict that a linear array can solve many problems in polylogarithmic time. This seems too strange to ignore.

Our second parameter is related to the bandwidth. Consider the problem of permuting elements of an array of size N. For a DCM or sparse network, assume that each processor holds N/P elements of the array. Let $\tau(N)$ be the time it takes a machine to perform the permutation. The bandwidth of the machine is $\lim_{N\to\infty} N/\tau(N)$. The PRAM and DCM models have bandwidth $\Theta(P)$; the hypercube, shuffle-exchange, and butterfly models have bandwidth $\Theta(P/\log(P))$; the mesh has bandwidth $\Theta(\sqrt{P})$; and the tree and linear array models have bandwidth $\Theta(1)$. Bandwidth $\Theta(P)$ is optimal. We actually prefer to measure the ratio of how far from optimal a model is to make the bandwidth metric symmetric to the latency metric: The *bandwidth* bound, $\beta(P)$, is the ratio of P to the bandwidth, i.e., $\beta(P) = \lim_{N\to\infty} P\tau(N)/N$. So, for the PRAM and DCM models $\beta(P) = \Theta(1)$; for the hypercube, shuffle-exchange, and butterfly models $\beta(P) = \Theta(\log(P))$; for the mesh $\beta(P) = \Theta(\sqrt{P})$; and for the tree and linear array models $\beta(P) = \Theta(P)$. Different models obtain their bandwidths at different levels of parallelism, and the level depends on whether the routing algorithm is deterministic or probabilistic (see, for example, [21, 24] and references therein).

Although we have given here the standard latency and bandwidth bounds for the above models, these values should be considered flexible. For example, sometimes one considers PRAM's with a network between processors and memory, which have more than constant latency. Also, if wire lengths are taken into account, many models will have higher latency than indicated. Sometimes, each processor of hypercube is assumed to be able to send or receive messages on all of its wires at the same cycle. This version has bandwidth $\Theta(P)$, or $\beta(P) = \Theta(1)$.

2 Note on analysis

Sums of the form

$$\Theta(\sum_{j=1}^{\lg P} f(2^j)),$$

where the function $f(P)$ is typically constant, polylogarithmic, or polynomial in P, will often emerge in our analyses. For $f(P)$ of the form P^r, i.e., polynomial, the last term dominates, so the sum is $\Theta(P^r)$. For $f(P)$ of the form $(\log P)^s$, i.e., polylogarithmic, each term counts almost as much as the last, up to a constant factor, so the sum is $\Theta((\log P)^{s+1})$.

To keep track of these three main cases simultaneously we define

$$L[f(P)] = \frac{1}{f(P)} \sum_{j=1}^{\lg P} f(2^j).$$

Thus for $f(P)$ polynomial, $L[f(P)]$ is $\Theta(1)$, and for $f(P)$ constant or polylogarithmic, $L[f(P)]$ is $\Theta(\log P)$.

3 Parallel prefix

Given an array $A[0], \ldots, A[N-1]$, the parallel prefix problem is to compute in parallel all of the initial sums $\sum_{j=0}^{i} A[j]$ for $0 \leq i \leq N-1$. This is one of the most basic problems in parallel computation. Shared memory machines and machines with $\Theta(\log P)$ diameter can solve the problem in $\Theta(N/P + \log P)$ time. Typically, a diameter $\Theta(d)$ machine can solve the problem in $\Theta(N/P + d)$ time. A mesh, for example, requires $\Theta(N/P + \sqrt{P})$ time.

Consider the simpler problem of merely finding the sum of just P numbers. We use the standard tree algorithm:

for $j = 0, \ldots, \lg P - 1$ **do**

  for all $i = 2^j - 1, \ldots, N \bmod 2^j$ **do in parallel**

    $A[i] := A[i - 2^j] + A[i]$

At step j, the machine sends just one item within each submachine of size 2^{j+1}, so the throughput cost of the step is $\Theta(\beta(2^{j+1})/2^{j+1})$, which equals $\Theta(1)$, and the distance cost is $\Theta(\delta(2^{j+1}))$. So the total cost is

$$\Theta(L[\delta(P)]\delta(P)) + \sum_{j=0}^{\lg P-1} O(1) = \Theta(L[\delta(P)]\delta(P)).$$

Now consider, finding the sum of $N \geq P$ numbers. First, each processor locally finds the sum of N/P numbers in time $\Theta(N/P)$, reducing the problem to finding the sum of P numbers. Thus, the time to sum $N \geq P$ numbers is

$$\Theta(N/P + L[\delta(P)]\delta(P)).$$

Parallel prefix can be computed within the same time bounds, by computing the partial sums while traversing up the tree and then sending the initial sums down the tree and appropriately adding them to partial sums.

Theorem 1 *The parallel prefix problem can be solved by a P-processor Y-PRAM[β, δ] in time*

$$\Theta\left(N/P + L[\delta(P)]\delta(P)\right).$$

We now substitute typical values for $\beta(P)$ and $\delta(P)$:

SHARED MEMORY MACHINE. Setting the parameters to match a shared memory machine gives, i.e., $\beta(P) = \delta(P) = \Theta(1)$:

$$\Theta((N/P + \log(P)).$$

HYPERCUBE. Setting the parameters to match a hypercube (or butterfly or shuffle-exchange), i.e., $\beta(P) = \delta(P) = \log P$, gives:

$$\Theta(N/P + \log^2 P).$$

MESH. Setting the parameters to match a mesh, i.e., $\beta(P) = \delta(P) = \sqrt{(P)}$ gives:

$$\Theta(N/\sqrt{P}).$$

TREE. Setting the parameters to match a tree, i.e., $\beta(P) = \Theta(P)$ and $\delta(P) = \Theta(\log P)$, gives:

$$\Theta(N/P + \log^2 P).$$

Notice that our time bounds are exact for the shared memory machine and the mesh, but off by a factor of $\log P$ in the second term for the hypercube and tree. This is because our algorithm assumes that the time to send a data item within a submachine of size P is $\Theta(\log P)$, whereas the hypercube is able to send items between machines whose sizes differ by a factor of two in constant time. In other words, some logarthmic distance machines can implement ascend-descend algorithms [20] especially efficiently.

4 Merge sort

We implement here a fairly simple and straightforward sorting algorithm. The algorithm locally sorts groups of N/P items, forming P sorted lists that need to be merged into a single sorted list. It pairwise merges the lists until only a single sorted list remains. At step j, $1 \leq j \leq \log P$, the algorithm merges two sorted lists each of size $(N/P)2^{j-1}$ into a single sorted list of size $(N/P)2^j$, using all the processors of a submachine of size 2^j.

The merging can be accomplished in a variety of ways. One method is to have one processor use binary search to split the merging problem into two independent merging problems of equal size: To merge the lists $a_1 \leq a_2 \leq \cdots \leq a_m$ and $b_1 \leq b_2 \leq \cdots \leq b_n$, the algorithms finds i, j so that, $i+j = (m+n)/2$, $b_1, \ldots, b_{j-1} \leq a_i$ and $b_{j+1}, \ldots, b_n \geq a_i$, and similarly $a_1, \ldots, a_{i-1} \leq b_j$ and $a_{i+1}, \ldots, a_m \geq b_i$. The processors split into two groups of equal size and work recursively on each half until there is only one processor per merging problem. Then, the sublists can be merged sequentially.

At step j, the time to find the first splitting item is $\Theta(\delta(2^j)\log((N/P)2^j))$, and (by the previous algorithm) the time to broadcast it is $\Theta(\sum_{k=0}^{j-1}\delta(2^k))$. Notice that when we find the second two splitting items the total list size has been halved and the submachine within which a broadcast needs to be done is also halved. After each processor determines the two sublists that it needs to merge, the processor must access all of its N/P items. This process is global within the submachine of size 2^{j+1}, so it takes time $(N/P)\beta(2^j) + \delta(2^j)$. Thus, the total time for step j is

$$\Theta\left(\sum_{h=0}^{j-1}\left(\delta(2^{j-h})\log((N/P)2^{j-h}) + \sum_{k=0}^{j-h-1}\delta(2^k)\right) + (N/P)\beta(2^j) + \delta(2^j)\right).$$

Summing over all $\log P$ steps and including the initial sorting time gives

$$\Theta\left(\sum_{j=1}^{\log P}\left(\sum_{h=0}^{j-1}\left(\delta(2^{j-h})\log((N/P)2^{j-h}) + \sum_{k=0}^{j-h-1}\delta(2^k)\right) + (N/P)\beta(2^j) + \delta(2^j)\right)\right)$$
$$+\Theta\left((N/P)\log(N/P)\right).$$

This sums to

$$\Theta\left(L^3[\delta(P)]\delta(P) + L^2[\delta(P)]\delta(P)\log N + L[\beta(P)]\beta(P)N/P + (N/P)\log(N/P)\right),$$

for the parameter ranges of main interest here; namely, $N \geq 2P$ and $\beta(P)$ and $\delta(P)$ constant, polylogarithmic, or polynomial in P. This proves,

Theorem 2 *The sorting problem of size N can be solved by a P processor Y-PRAM$[\beta, \delta]$ in time*

$$\Theta\left(L^3[\delta(P)]\delta(P) + L^2[\delta(P)]\delta(P)\log N + L[\beta(P)]\beta(P)N/P + (N/P)\log(N/P)\right),$$

where $N \geq 2P$ and $\beta(P)$ and $\delta(P)$ are constant, polylogarithmic, or polynomial in P.

For comparison, we consider a few models.

SHARED MEMORY MACHINE. For parameter values matching a shared memory machine, $\beta(P) = \delta(P) = \Theta(1)$, the above result simplifies to

$$\Theta(\frac{N \log N}{P} + \log^2 P \log N).$$

This is not quite as good as the best shared memory sorting results [4], but is still efficient for $N = \Omega(P^{\log^2 P})$.

HYPERCUBE. For parameter values matching a hypercube (or butterfly or shuffle-exchange) machine, $\beta(P) = \delta(P) = \Theta(\log P)$, the above result simplifies to

$$\Theta\left(\frac{N \log N}{P} + \frac{N \log^2 P}{P} + \log^3 P \log N\right).$$

This is efficient for $N = P^{\Omega(\log P)}$, which is not as good as the best hypercube sorting results [6, 7].

It seems that no parallel merge sort can obtain extremely good performance on our model for parameters matching the hypercube: There must be $\Theta(\log P)$ merge steps after locally sorting lists of size N/P. Each merge step will permute a total of N items, and each permutation will require an overhead of $\Theta(\beta(P))$ (on average). Thus, a term of the form $\Theta(L[\beta(P)]\beta(P)N/P)$ seems inherent.

MESH. For parameter values matching a mesh machine, $\beta(P) = \delta(P) = \Theta(\sqrt{P})$, the above result simplifies to

$$\Theta(\frac{N \log N}{P} + \frac{N}{\sqrt{P}}).$$

This is optimal for a mesh: the first term accounts for the $N \log N$ comparisons that must be performed in the worst case, and the second term accounts for the data movement that must be performed. The algorithm is efficient for

$$N \geq 2^{\Omega(\sqrt{P})}.$$

TREE. For parameter values matching a tree, $\beta(P) = \Theta(P)$ and $\delta(P) = \Theta(\log P)$, the above result simplifies to

$$\Theta(\frac{N \log N}{P} + N).$$

This algorithm is also optimal for the same reasons as the mesh algorithm is optimal. Notice that it is not quite as simple as the the algorithm that sorts lists of size N/P at the leaves of the tree and pairwise merges the lists up the tree using only one processor per merge. However, our algorithm was developed in a more general context.

We could obtain slightly better results by implementing Cole's merge sorting algorithm [4] which possesses much inherent locality, but the algorithm is much more complicated.

5 Column sort

We present an alternative parallel sorting algorithm that captures the performance of the best known algorithm for the hypercube. The general idea, which is based on Leighton's column sort [17], is due to Han [10].

Column sort works by partitioning the N items to be sorted into a two dimensional matrix. The columns are sorted and the matrix is permuted (by an appropriate permutation). These two steps are executed a total of eight times, at which point the entire list is sorted. The only requirement is that the number of rows must be quadratically larger than the number of columns.

One could use this as a parallel sorting algorithm by partitioning the list to be sorted into an $N^{2/3} \times N^{1/3}$ matrix. If the number of columns ($N^{1/3}$) is at least as large as the number of processors (P), then the algorithm can be implemented directly, by assigning an equal number of columns to each processor and sorting columns sequentially. Otherwise, we assign $P/N^{1/3}$ processors per column, and execute the algorithm recursively.

Thus, the sorting time $T_P(N)$ satisfies the following recurrence:

$$T_P(N) \leq \begin{cases} 8T_{P/N^{1/3}}(N^{2/3}) + \Theta(\frac{N}{P}(\beta(P) + \delta(P))) & \text{if } P > 1 \\ O((N/P)\log(N/P)) & \text{if } P = 1. \end{cases}$$

The depth of the recursion is $\log_{3/2}((\log N)/\log(N/P))$. So, the number of sequential sorts of N/P elements is $(\log((\log N)/\log(N/P)))^{\Theta(1)}$, which implies that to total time spent sorting sequentially is

$$(\log((\log N)/\log(N/P)))^{\Theta(1)}\Theta((N/P)\log(N/P)).$$

For $(\beta(P) = \delta(P) = \Theta(\log P))$ the total time spent permuting is

$$(\log((\log N)/\log(N/P)))^{\Theta(1)}N/P.$$

Thus, for $(\beta(P) = \delta(P) = \Theta(\log P))$ the total time for sorting is

$$(\log((\log N)/\log(N/P)))^{\Theta(1)}\Theta((N/P)\log(N/P)).$$

If in addition, $N \geq P^{1+\epsilon}$, for some constant $\epsilon > 0$, the total time for sorting is

$$\Theta((N/P)\log(N)).$$

This matches the best known hypercube sorting algorithm [6].

We have proved,

Theorem 3 *Column Sort can be implemented in the P processor Y-PRAM[β, δ] to sort N elements in time*

$$\Theta((N/P)\log(N)),$$

for $\beta(P) = \delta(P) = \Theta(\log P)$, and $N = \Omega(P^{1+\epsilon})$, $\epsilon > 0$.

6 Radix sort

We implement the radix sort algorithm of [14, 22, 25] on our model. Assume we wish to sort N numbers in the range $0, \ldots, M-1$ using radix R.

One pass of radix sort does the following: Each processor locally contains N/P items. Create a $P \times R$ array. Each processor locally counts how many of its N/P items belong in each of the R buckets. This takes time $\Theta(N/P)$. Performing parallel prefix on the $P \times R$ array *by columns*, and a parallel prefix of the sequence of the total for each column, determines for each item its global location in the sorted array. By the subsection on parallel prefix, this takes time $\Theta(R + L[\delta(P)]\delta(P))$. Move the items to their determined destinations, which sorts the items by radix R. This takes time $\Theta((N/P)\beta(P) + \delta(P))$. Thus, the total time for one pass is

$$\Theta((N/P)\beta(P) + R + L[\delta(P)]\delta(P)).$$

To fully implement radix sort, we need $(\log M)/(\log R)$ passes. Thus, the total time for all of the passes is

$$\Theta(\frac{\log M}{\log R}((N/P)\beta(P) + R + L[\delta(P)]\delta(P)).$$

which is equal to

$$\Theta(\frac{\log M}{\log((N/P)\beta(P))}((N/P)\beta(P) + L[\delta(P)]\delta(P))).$$

for $R = (N/P)\beta(P)$. This proves,

Theorem 4 *A P-processor Y-PRAM[β, δ] can sort N integers in the range $0, \ldots, M-1$ in time*

$$\Theta(\frac{\log M}{\log((N/P)\beta(P))}((N/P)\beta(P) + L[\delta(P)]\delta(P))).$$

We can now substitute values for $\beta(P)$ and $\delta(P)$:

SHARED MEMORY MACHINE. Setting the parameters to match a shared memory machine gives:

$$\Theta(\frac{(\log M)}{\log(N/P)}(N/P + \log(P)).$$

This is the same result that one gets directly for the shared memory machine [14, 22, 25]

HYPERCUBE. Setting the parameters to match a hypercube gives:

$$\Theta(\frac{\log M \log P}{\log((N/P)\log(P))}(N/P + \log(P)).$$

MESH. Setting the parameters to match a mesh gives:

$$\Theta(\frac{\log M}{\log(N/\sqrt{P})}(N/\sqrt{P}).$$

For M at most polynomially larger than N, i.e. $M \leq O(N^k)$ for some constant $k > 0$, this simplifies to $\Theta(N/\sqrt{P})$, which is optimal.

TREE. Setting the parameters to match a tree gives:

$$\Theta(\frac{\log M}{\log N}N).$$

For M at most polynomially larger than N, this simplifies to $\Theta(N)$ which is optimal for the tree, although not especially interesting.

The region of interest will typically be M at most polynomially larger than N, and N at least polynomially larger than P, i.e. $N \geq P^{1+\epsilon}$ for some constant $\epsilon > 0$. In this case, our running time simplifies to

$$\Theta((N/P)\beta(P) + L[\delta(P)]\delta(P)).$$

Since routing takes $\Theta((N/P)\beta(P) + \delta(P))$ (for $N >> P$) on our model, this result shows that radix sort is essentially equivalent to routing (in this region).

7 List ranking

A problem that is fundamental for graph algorithms is list ranking: One is given a linked list of size N and wishes to determine the distance of each node from the head of the list.

We can implement Wyllie's recursive doubling algorithm [26] for $N = P$. Assume each node points to its successor via a *next* field and starts with *count* = *1*. The *next* pointer for the head field is *nil*. Each processor is assigned a node and synchronously executes the following code on its own node:

while $next \neq nil$ **do**

$count := count + count(next)$

$next := next + next(next)$

This algorithm takes $\log P$ iterations, and each iteration takes $\Theta(\beta(P) + \delta(P))$ time, so the total time is $\Theta((\beta(P) + \delta(P)) \log P)$. One could generalize this to N nodes by assigning N/P processors per node and iterating through the loop $\log N$ times. The time will then be $\Theta((\beta(P)N/P + \delta(P)) \log N)$. Notice that this is inefficient because of the $\Theta(\beta(P))$ cost of accessing all of the nodes at each iteration, and because of the the fact that there are $\log N$ iterations.

The basic idea of all efficient parallel list ranking algorithms is to compact in parallel many pairs adjacent nodes into single nodes. The difficulty is to avoid compacting node a with node b, while at the same time compacting node b with node c. After the compaction phases are completed the steps of the algorithm are unwound to "broadcast" the node ranks.

We implement the algorithm of Kruskal, Rudolph, and Snir [13] There are $O(\log(N/P))$ phases. Each phase reduces the number of nodes by at least a constant fraction until there are less than $2P$ nodes. At that point, the recursive doubling algorithm of Wyllie [26] is used finish up the list ranking.

The algorithm partitions the nodes of the list into a $P \times N/P$ array. The processors synchronously visit the columns of the array, i.e. at step j the ith processor visits the jth element of the ith row. Each processor compacts its visited node with the node's successor as long as the successor is not in the same column. This guarantees that compaction conflicts - described above - are avoided.

In order to compact within a column, the processors partition themselves into N/P groups, each of size P^2/N, one group for each column. The algorithm is then applied recursively to each column, until each column is assigned only one processor, at which point no conflicts can occur. After finishing the recursion, at most $2/3N + O(1)$ nodes remain, since if a node has not been compacted then both of its neighbors must have been compacted. The remaining nodes are packed into an array of size $2/3N + O(1)$ (or smaller). The packing is accomplished by doing a parallel prefix on the nodes with 1 assigned to *live* (not yet compacted) nodes and 0 assigned to *dead* (already compacted) nodes, then updating the pointer values for live nodes, and finally moving the live nodes (to the location indicated by the parallel prefix value). This completes the first phase. Now the whole algorithm is applied recursively to the resulting smaller list, which has $O(N/d)$ elements, $d = 2/3$.

To analyze the algorithm we let $H_P(N)$ be the time for P processors to compact N nodes. Then, by our recursive construction,

$$H_P(N) \leq \begin{cases} H_{P^2/N}(\frac{N}{P}) + \Theta(\frac{N}{P}(\beta(P) + \delta(P))) & \text{if } P > 1 \\ \Theta(N/P) & \text{if } P = 1. \end{cases}$$

Let $U_P(N)$ be the time for P processors to execute the entire list ranking algorithm on N nodes. Then, by our recursive construction,

$$U_P(N) \leq U_P(N/d) + H_P(N) + N/P + L[\beta(P)]\beta(P) + \frac{N}{P}\beta(P) + \delta(P)$$

where $N > 2P$ and $d = 3/2$. Note that computing the locations for the packing is done with a parallel prefix which takes $N/P + L[\beta(P)]\beta(P)$, and moving the elements to the computed locations takes $\frac{N}{P}\beta(P) + \delta(P)$.

Then the total time for our list ranking algorithm $T_P(N)$ is $U_P(N)$ plus the time list rank on a list with at most $2P$ nodes, $T_P(N) = U_P(N) + \Theta((\beta(P) + \delta(P)) \log P)$. We have thus proved,

Theorem 5 *The list ranking problem of size N can be solved by a P-processors Y-PRAM[β, δ] in time $T_P(N)$ such that*

$$T_P(N) \leq \begin{cases} T_P(N/d) + H_P(N) + \Theta(N/P + L[\beta(P)]\beta(P)) + \frac{N}{P}\beta(P) + \delta(P) & \text{if } N > 2P \\ \Theta((\beta(P)N/P + \delta(P)) \log N) & \text{if } N = 2P, \end{cases}$$

where $d = 2/3$, and

$$H_P(N) \leq \begin{cases} H_{P^2/N}(\frac{N}{P}) + \frac{N}{P}(\beta(P) + \delta(P)) & \text{if } P > 1 \\ O(N/P) & \text{if } P = 1. \end{cases}$$

We will not solve this recurrence in general, but rather solve some special cases of interest:

SHARED MEMORY MACHINE. Setting the parameters to match a shared memory machine gives:

$$T_P(N) = \Theta(\frac{N}{P}\frac{\log N}{\log(2N/P)}).$$

This is optimal for $N \geq P^{1+\epsilon}$ for any constant $\epsilon \geq 0$. The best parallel list ranking algorithms for a P processors EREW PRAM take $O(N/P + \log P)$ which is optimal for $N = \Omega(P \log P)$ [3, 5].

HYPERCUBE. Setting the parameters to match a hypercube gives:

$$T_P(N) = \Theta(\frac{N}{P}\frac{\log N}{\log(2N/P)}\log P) \qquad (N \geq 2P).$$

Again, this is optimal for $N \geq P^{1+\epsilon}$ for any constant $\epsilon \geq 0$. This matches the performance of the fastest known list ranking algorithm for the (strict) Hypercube, whose complexity is $O(\frac{N}{P}\frac{\log^2 N}{\log(N/P)})$ and becomes optimal for $N \geq P^{1+\epsilon}$ [6].

MESH. Setting the parameters to match a mesh gives:

$$T_P(N) = \Theta(\frac{N}{\sqrt{P}} + \sqrt{P}\log P).$$

This is optimal for $N > \sqrt{P}\log P$. For the case $N = P$, the parallel list ranking problem can be solved on a $\sqrt{N} \times \sqrt{N}$ mesh in $O(\sqrt{N})$ time, which is optimal [8].

TREE. setting the parameters to match a tree gives:

$$T_P(N) = \Theta(N + P\log P),$$

which is optimal for $N > P\log P$.

8 Conclusion

We have proposed a model of parallel computation, the Y-PRAM, that takes into account latency and bandwidth. It allows parallel algorithms to be designed independently of the parallel model. We presented *Y-PRAM* algorithms for parallel prefix, sorting, sorting numbers from a bounded range, and list ranking. It seemed to be easy to write programs for this model.

The *Y-PRAM* model seems to provide a reasonably accurate prediction of actual performance. Many *Y-PRAM* time bounds exactly match the best parallel bounds, when the parameters are set to match a particular machine. Sometimes the bounds were off by a $\log P$ factor in the second term, as in the parallel prefix algorithm for the hypercube. This discrepancy occurred because our model assumes that the time to send a data item within a submachine of size P is $\Theta(\log P)$, whereas the hypercube is able to send items between machines whose sizes differ by a factor of two in constant time. In other words, some logarthmic distance machines can implement ascend-descend algorithms [20] especially efficiently. We believe that it may unnecessarily complicate the model to try to account for this low level effect.

There are many problems left open by this work, some of which we plan to tackle in the future. For example, it would be nice to design algorithms for more problems. The *Y-PRAM* is only a first approximation; more experience designing algorithms will indicate how the model should be generalized or restricted, and also what other parameters, if any, should be included. It can be refined in a variety of ways. Many machines communicate by "randomly" sending message around the machine. In that case, a small amount of more global background traffic would not seriously degrade performance, as measured by bandwidth. Such a model would probably provide a more realistic reflection many machines including CM* and Cedar, However, it seems to be more difficult to define cleanly, and none of our current algorithms take advantage of this extra traffic.

Another variant of our model, which would also be a more realistic reflection of real machines, would be to restrict the submachines so that they can communicate only via routing, rather than allowing the full power of a shared memory. In this case, m would be maximum number of messages that any one processor sends or receives. This refinement would complicate algorithm design for questionable improvement in estimating performance. Our Y-PRAM model defines what might be called an EREW Y-PRAM. Any other PRAM model could be used for the communication within submachines.

References

[1] A. Aggarwal, A. Chandra, and M. Snir. Communication complexity of prams. Technical Report RC 14998(64644), IBM Tech. Report, 1989.

[2] A. Aggarwal, A. Chandra, and M. Snir. On communication latency of prams. Technical Report RC 14973(66882), IBM Tech. Report, 1989.

[3] R. J. Anderson and G. L. Miller. Deterministic parallel list ranking. In *Proc. 3rd AWOC*, pages 81–90, 1988.

[4] R. Cole. Parallel merge sort. In *Proc. 27th Annual IEEE Symp. on Foundations of Computer Science*, pages 511–516, 1986.

[5] R. Cole and U. Vishkin. Approximate parallel scheduling, part i: the basic technique with applications to optimal parallel list ranking in logarithmic time. *SIAM J. Comput*, 17:128–142, 1988.

[6] R. Cypher. *Efficient communication in massively parallel computers*. PhD thesis, University of Washington, 1989. Department of Computer Scienceeig.

[7] R. Cypher and C. G. Plaxton. Deterministic sorting in nearly logarithmic time on the hypercube and related computers. In *Proc. 15th Annual ACM Symp. on Theory of Computing*, pages 193–203, 1990.

[8] A. M. Gibbons and Y.N. Srikant. A class of problems efficiently solvable on mesh-connected computers including dynamic expression evaluation. *Information Processing Letters*, 32:305–311, 1989.

[9] A. Gottlieb and C. P. Kruskal. Complexity results for permuting data and other computations on parallel processors. *JACM*, 31:193–209, 1984.

[10] Y. Han. Parallel algorithms for computing linked list prefix. *J. of Parallel and Distributed Computing*, 6:537–357, 1989.

[11] R. Karp and V. Ramachandran. A survey of parallel algorithms for shared-memory machines. In *Handbook of Theoretical Computer Science*, 1988. Ed. J. van Leeuwen, North Holland, to appear.

[12] C. P. Kruskal, T. Madej, and L. Rudolph. Parallel prefix on fully connected direct connection machine. In *Proc. International Conference on Parallel Processing*, pages 278–283, 1986.

[13] C. P. Kruskal, L. Rudolph, and M. Snir. The power of parallel prefix. *IEEE Transactions on Computers*, pages 965–968, 1985.

[14] C. P. Kruskal, L. Rudolph, and M. Snir. Efficient parallel algorithms for graph problems. In *Proceedings International Conference on Parallel Processing*, pages 869–876, 1986.

[15] C. P. Kruskal, L. Rudolph, and M. Snir. A complexity theory of efficient parallel algorithms. In *Proc. 15th Annual ICALP*. Springer Verlag Lecture Notes in Computer Science, Vol. 317, pp. 333-346, July 1988. (Theoretical Computer Science, to appear 1989).

[16] D. J. Kuck, E. S. Davidson, D. H. Lawrie, and A. H. Sameh. Parallel supercomputing today and the cedar approach. *Science*, 231:967–974, 1986.

[17] F. T. Leighton. Tight bounds on the complexity of parallel sorting. *IEEE Transactions on Computers*, C-34:344–354, 1985.

[18] L. Meertens. Recurrent ultracomputers are not log n-fast. Technical Report 2, New York University, 1979. Ultracomputer.

[19] K. Mehlhorn and U. Vishkin. Randomized and deterministic simulations of prams by parallel machines with restricted granularity of parallel of parallel memories. *Acta Informatica*, 21:339–374, 1984.

[20] F. P. Preparata and J. E. Vuillemin. The cube-connected cycles: A versatile network for parallel computation. *CACM*, 24:300–309, 1981.

[21] T. Leighton B. Maggs S. Rao. Universal packet routing algorithms. In *Proc. 29th Annual IEEE Symp. on Foundations of Computer Science*, pages 256–271, 1988.

[22] J. H. Reif. An optimal parallel algorithm for integer sorting. In *Proc. 26th Annual Symp. on Foundations of Computer Science*, pages 496–504, 1985.

[23] R. J. Swan, S. H. Fuller, and D. P. Siewiorek. Cm* - a modular, multi-processor. In *Proc. AFIPS Conf.*, volume 46, pages 637–644, 1977.

[24] L. G. Valiant. General purpose parallel architectures. In *A handbook of Theoretical Computer Science*. MIT Press, 1990. J. van Leeuwen (ed.).

[25] R. A. Wagner and Y. Han. Parallel algorithms for bucket sorting and the data dependent prefix problem. In *Proc. International Conference on Parallel Processing*, pages 924–930, 1986.

[26] J. C. Wyllie. *The complexity of parallel computation.* PhD thesis, Cornell University, 1979.

Neural Computing and the GALATEA Project

Philip Treleaven

Department of Computer Science
University College London

ABSTRACT

This paper reviews the fundamentals of neural computing which includes: neural network models, neural network programming environments, and neurocomputer; specialised hardware for neural networks.

It then describes the ESPRIT II GALATEA project, and its predecessor PYGMALION, which provide the focus of neural computing research in the European Community. PYGMALION has developed a general programming environment for neural networks, and the goal of GALATEA is to build upon this environment, to produce a comprehensive neurocomputing system. This neurocomputing system will comprise: a sophisticated programming environment capable of mapping a network on to a range of conventional computers, including parallel machines; a novel general-purpose neurocomputer; and an integral silicon compiler for translating a network into VLSI *chips*.

1. Artificial Neural Networks

Neural networks[1] are a novel form of computation that attempts to mimic the functionality of the human brain, in order to solve demanding pattern processing problems. The term *neural computing* spans artificial neural networks, neural programming environments and neurocomputers, applying them to a broad class of *pattern processing* applications. These applications include: control, image processing, speech processing, inexact knowledge processing, natural-language processing, sensor processing, planning, forecasting and optimisation.

The key to neural computing is understanding the ways in which the brain uses neuronal systems for pattern processing[2]. The biological neuron basically consists of a cell body called a *soma* (Figure 1a), branching complex extensions called *dendrites*, and an *axon*, the output channel of the cell, carrying an electric signal to other cells. The axon connects to the dendrites of other cells through specialized contacts called *synapses* that can change, positively or negatively, the axon potential. The traditional view is that the neuron performs a simple threshold function - "weighted" input signals are summed; if the result exceeds a certain threshold, a signal is send out from the neuron.

The artificial neuron is made up of three sections (see Figure 1b), corresponding to the simplified model of the biological neuron: the weighted input connections, the summation function and a threshold function, that generates the unit output (usually *off* or *on*). The artificial neuron operates as a simple threshold device, depending on the state S of its input elements and the connection weights (W).

In an artificial neural network, neurons are generally configured in regular and highly interconnected topologies. Programming of a neural network, firstly involves specifying the mathematical function of the artificial neurons and their connections, and secondly involves the *training* of the network to recognise a set of patterns.

Typically the artificial neurons (called processing elements - PEs) are organised into layers with each PE in one layer having a connection to each PE in the next layer, as illustrated by Figure 2. Associated with each connection is a *weight* and with each PE is a *state* (usually *off* or *on*). Together these weights and states represent the distributed *data* of the network. The weights of a network together represent an energy surface, and their actual values determine the set of patterns recognisible by the network. During pattern recognition, each PE operates as a simple threshold device. A PE sums all the weighted inputs (multiplying the connection weight by the state of the previous layer PE) and then

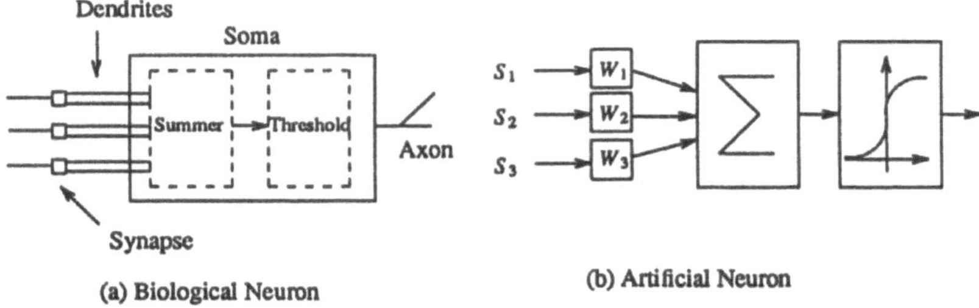

(a) Biological Neuron

(b) Artificial Neuron

Figure 1: Simplified Models of Neurons

applies a threshold function, such as a sigmoid function.

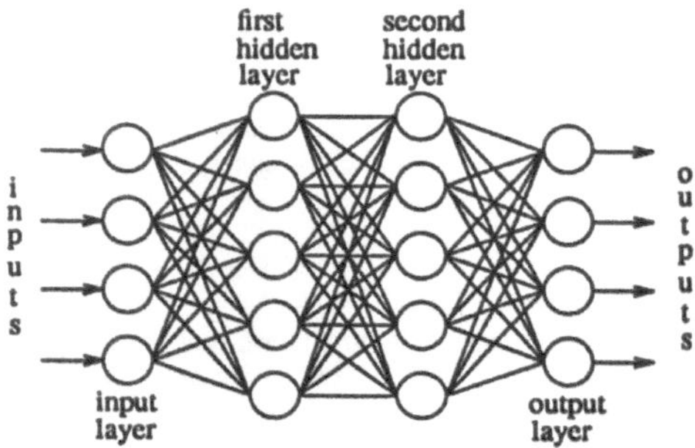

Figure 2: Multilayer Neural Network

A widely used neural network model is ***Back Propagation***[3]. A Back Propagation model is ***trained*** to recognise patterns by presenting example training pairs of input-target pattern (e.g. a hand-written *A* with a perfect *A*). The input pattern is presented and passes through the network to produce outputs. This output pattern is then compared with the "ideal" target pattern, and an ***error*** is propagated back through the network. The propagated error is used to adjust the weights of the connections. This training process is then repeated with a new training pair, and a new (hopefully smaller) error is propagated backwards. This training process is repeated with example pairs of patterns until the error is negligible, at which time the network is trained.

2. Neural Programming Environments

Interest in neural networks has stimulated the availability sophisticated software Environments for programming neural networks[3]. Environments range from commercial products from both established and Startup companies, to public-domain software, available free from university research groups.

These diverse neural programming environments share many common features. A typical neural network programming environment, as illustrated by Figure 3, might comprise:

- a ***graphic interface***, with menus and a command language: for configuring a neural network, then controlling and monitoring its execution;
- an ***algorithm library*** of common, parameterised neural network algorithms, such as Back Propagation, Hopfield, Boltzmann etc.;
- a ***high-level language***, often object-oriented, for programming or customising an algorithm or application;

- a ***network specification language***, a low-level, machine-independent, language (often based on C) defining the neural network simulation;
- ***hardware configuration*** information for defining the organisation of the target machine to run the network simulation; and
- ***translators*** for mapping the network specification language to various target machines.

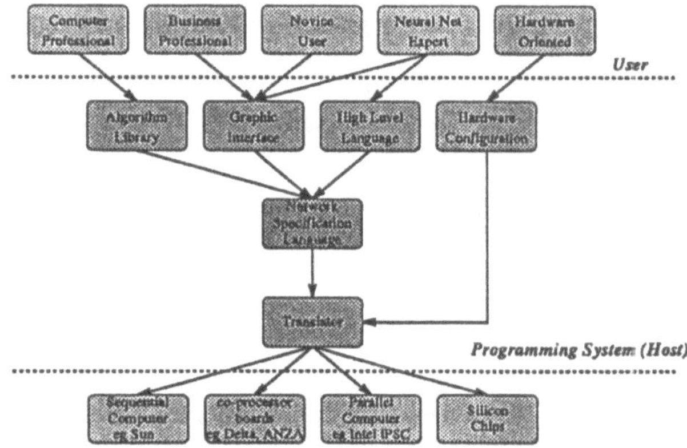

Figure 3: Typical Neural Programming Environment

Neural programming environments may be classified into three major groups: application-oriented, algorithm-oriented and general programming systems.

Applications-oriented systems
are designed for specific market domains, such as *finance* or *transportation*. These application domains form a natural subdivision. A good example is Nestor's Decision Learning System[4].

Algorithm-oriented systems
support specific neural network models. This class has two major subclasses which are: *algorithm-specific* supporting a single model, such as Back Propagation of errors, and *algorithm libraries* supporting many models coded in a common language like "C". These algorithm-oriented systems are often supplied as source code routes and therefore are easily incorporated in user applications. A popular algorithm library is OWL[5].

Programming systems
the third class, provide general "tool-kits" comprising many algorithms and programming tools, that can be used for a wide range of algorithms and applications. Programming systems can be further sub-divided into: *educational* systems for the novice user to obtain a hands-on introduction and normally support only small networks based on common algorithms; *general-purpose* systems provide comprehensive tool-kits for programming any algorithm or application; *open* systems where the user can modify any part of the system; and *hardware-oriented* systems typically supporting particular parallel computers.

As an illustration of our Taxonomy, Figure 4 lists some of the well known environments in each class.

The level of sophistication and usage of a given environment depends on the target group of users. Business professionals, interested say in financial forecasting, have little concern for the details of neural networks. For them, an applications-oriented system gives emphasis to the graphic interface and a specific, possibly proprietary, algorithm. The user merely configures a parameterised network, applies the data and monitors the network's execution.

Class	Category	Organisation	System
Application-Oriented	Financial	Nestor	Decision Learning System (DLS)
	Transportation	BehavHeuristics	Airline Marketing Tactician (AMT)
		Excalibur	Savvy
Algorithm-Oriented	Algorithm-Specific	Cal. Scient. Softw.	BrainMaker
	Algorithm Libraries	Olmstead & Watkins	Owl
		Mimetics	Galatea C-library
Programming Systems	Educational Systems	UCSD	PDP
		NeuralWare	Explorer
	General-Purpose Systems	SAIC	ANSpec
		HNC	Anza/Axon
		NeuralWare	NeuralWorks Professional II
	Open Systems	Lucid	Plexi
		UCL	Pygmalion
	Hardware-Oriented Systems	Oregon Grad. Institute	Anne

Figure 4: Neural Network Programming Environments

For the novice user wishing to learn about neural networks there are many good educational systems that allow small parameterised networks to be configured, and their execution monitored. These systems typically have an easy to use graphic interface, a library of a few common networks, such as Hopfield, Back Propagation etc. and simple demonstration applications. Essentially these environments are "skeleton" general-purpose programming systems.

For the computing professional wishing to incorporate a common neural network algorithm in an application, there are source code algorithm libraries available in C and LISP. These libraries, provide an optional rudimentary graphic interface, and any additional programming is done in C or LISP. This is a straight-forward, and popular, way of incorporating neural networks into conventional applications.

For neural network experts who need to program their own algorithms and applications there are many general-purpose programming systems. Besides providing a comprehensive graphic interface and algorithm library, they also usually provide a high level language with specialised data structures, classes and functions. These languages are usually object-oriented, possibly based on C++. Many of these general-purpose systems are also "open", allowing the programmer to modify any part of the environment.

Finally, an increasing number of users are interested in mapping neural networks on to specific hardware, such as parallel computers like Transputers, or even into silicon. Hardware-oriented programming systems emphasis the hardware configuration component, specifying the structure of the target hardware, and the translator which uses this configuration information to map the neural specification language on to the hardware. Certain hardware-oriented systems are little more than parameterised code for running a specific neural network. Others are translators for mapping the network specification language to a parallel machine or to a silicon chip.

3. Neurocomputers

The computational demands of neural networks have also stimulated the development of specialised hardware, referred to as *neurocomputers*, to speed up execution. There are two basic classes of neurocomputer:

- **General-Purpose Neurocomputers** - generalised, programmable, neural computers for emulating a range of neural network models, thus providing a framework for executing neural models in much the same way as traditional computers.
- **Special-Purpose Neurocomputers** - that are dedicated hardware implementations of a specific neural network model.

The essential difference between them is whether the neurocomputer is programmable, hence general-purpose and capable of supporting a range of neural network models, or is special-purpose, implementing a dedicated neural network.

General-purpose neurocomputers subdivide into *commercial co-processors* boards and *parallel processor arrays*[6]. Commercial co-processors are typically floating-point or signal processing accelerator boards, usually supplied with a large memory (e.g. 4M byte), that plug into the backplane of an IBM PC, or interface to a SUN Workstation. Parallel processor arrays are cellular arrays, composed of a large number of primitive processing units, connected in a regular and usually restricted topology. Their general architecture is shown in Figure 5. The structure resembles a parallel array processor, comprising identical processors connected through a parallel broadcast bus, where each physical unit executes a section of the "virtual" network. To program the neurocomputer, the virtual PEs are partitioned across the local memories of the physical processors. Updating a virtual PE implies broadcasting the update through the bus. Units that need access to that information accept and store the update in their system state memory.

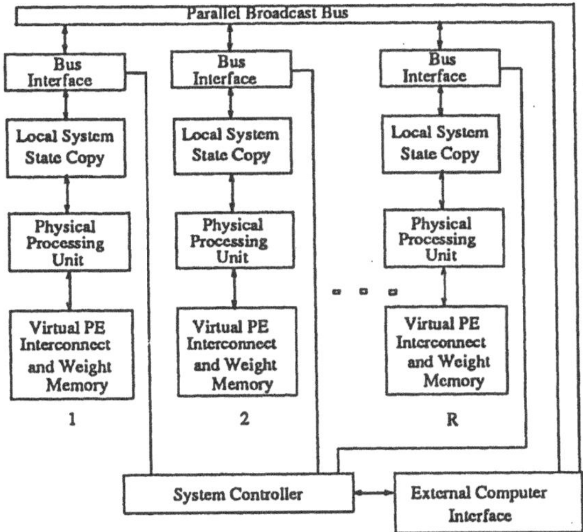

Figure 5: General-Purpose Virtual Neurocomputer Architecture (from Hecht-Nielsen)[7]

These two categories of general-purpose systems differ basically in the number and complexity of the physical processing units employed. Parallel processor arrays aim to attain high performance and real parallelism by (mainly) increasing the number of the implemented processing units, while co-processors attempt to improve performance by strengthening the processing/storage capabilities of standard microprocessors.

Hardware accelerators, such as the HNC ANZA[8], the SAIC SIGMA-1[9] and, from Japan, the NEC Neuro-07[10], have allowed neural network experiments to be carried out over 100 times faster than the usual simulators of neural network models. They are programmable and can implement large networks of virtual processing elements with a limited number of hardware implemented processors. Processors are usually implemented by means of an industry-standard signal processing chip or microprocessors such as the MC68020 (and its MC68881 floating point co-processor), interconnected through a standard parallel broadcast bus, such as the VME bus. Physical processors and interconnections are multiplexed across a large number of virtual processing elements and virtual interconnections, so demanding large memories to represent them. Performance comparisons between these products are based on capacity, meaning the maximum size of neural network, and speed, the time to process a neural network. Speed is usually expressed in network updates per second, for both the training phase and the recall phase. For instance, the ANZA Plus supports 1M PEs with 1M interconnections, and is capable of 1.5M connection updates per second during training and 6M updates during recall.

Several examples of very high performance special-purpose neurocomputers have been successfully built[6]. They are, typically, analog electronic implementations of neural networks and usually employ structures that resembles the simplified model of the neuron.

In a simple model of the neural function, the neuron state is given by:

$$s = T(\Sigma\ S.W - \theta)$$

The circuit of Figure 6a is a straightforward electronic implementation of the above equation[11]. Wires replace the input structure (dendrites) and the output (axon); the conductances (1/R) model the synaptic connections between neurons; and the amplifier models the cell body by executing the threshold function. Inputs appear as voltages to the summing wire. Using these electronic neurons, a neural network is implemented as a crossbar representation (Figure 6b). Outputs (vertical lines) are connected through resistors to the horizontal summing wires, which represent the weighted sum of the output signals of other PEs. Inverted input lines are provided when both positive and negative values are required.

For many neural network applications, however, it is necessary to change the synapses values[12]. To implement programmable connections, resistors can be replaced by a "synapse circuit", capable of providing different types of connections: excitatory, inhibitory and disabled.

In the design of special-purpose chips, the coupling network occupies most of the chip area. This is due to the difference between the silicon areas required by the PEs (containing many transistors) and by the wires (plus spacing).

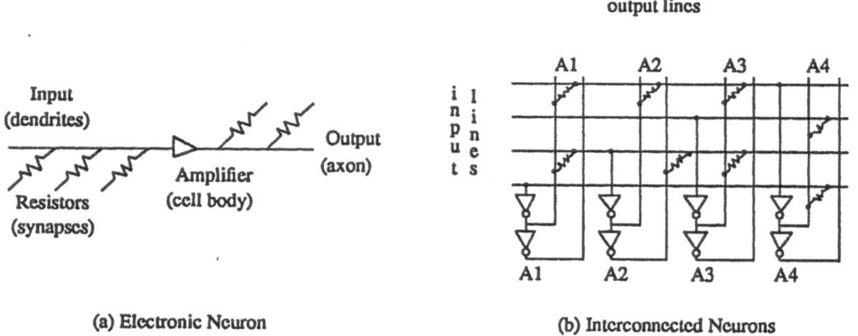

(a) Electronic Neuron

(b) Interconnected Neurons

Figure 6: Special-Purpose Neurocomputer

Special-purpose neurocomputers are naturally subdivided by their implementation technology into: *silicon processors*, *optical processors* and *molecular processors*. The approach for special-purpose neurocomputer architectures is to directly implement a specific neural network model in hardware to produce a very high performance system. Any neural network model could theoretically be chosen, although currently the Kohonen or the Hopfield associative memory model[3] are typically favoured, because of their simplicity. Most silicon implementations use an analog electronic neuron model, as described above.

Further details of neural networks, neural programming environments and neurocomputers are given in the references[6].

4. The GALATEA Project

In the European Community, the focus of neural computing research is the ESPRIT II GALATEA Project, and its predecessor PYGMALION. PYGMALION and GALATEA broadly aim to promote the application of neural networks by European industry, and to develop European "standard" computational tools for programming and simulation of neural networks. The projects, as shown by Figure 7, bring together many of the leading neural computing research groups from European industry, research institutes and universities.

Partner	Laboratory	Role
Thomson-CSF	Division Outils Informatiques, Paris	*Prime Contractor* Manager integration workpackage OCR application
Philips	Lab. d'Electronique et de Physique Appliquee,Paris	Manager image processing application neurocomputer
Siemens AG	Central Research laboratories	Manager hardware workpackage neurocomputer programming environment
Mimetics SA		Manager silicon compilation Manager OCR application programming environment
UCL	University College London	Manager software workpackage programming environment silicon compilation
SGS-Thomson	Microelectronics Srl	silicon compilation
INPG	Institut National Polytechnique de Grenoble	silicon compilation
IS	Informatica Sistemi Spa	image processing application
CRAM	Consorsio per la Ricerca in Agricultra nel Mezzogiomo	image processing application
INESC	Instituto de Engenharia Sistemas e Computadores	algorithm library
CTI	Computer Technology Institute, Patras	parallel processing

Figure 7: Partners in the GALATEA Project

PYGMALION[13] has produced a rudimentary environment for programming neural networks. Its design philosophy is threefold. Firstly, to provide a general environment for neural networks with the same facilities (e.g. graphic monitor, algorithm library etc.) and functionality as commercial systems such as those of Hecht-Nielsen, NeuralWare and SAIC. Secondly, to provide a rudimentary "platform" - that can be easily extended and interfaced to other tools. For this reason the core of the environment is ***X-windows***, ***C*** and ***C++***; running on a colour workstation. Thirdly, to provide "portable" neural network applications, so that trained and partially trained networks can be easily moved from machine to machine. For this reason the (partially) trained neural network applications are specified in a subset of *C*; essentially a *C* data structure.

The environment comprises 5 major parts:

- **Graphic Monitor**, the graphical software environment for controlling the execution and monitoring of a neural network application simulation.
- **Algorithm Library**, the parameterised library of common neural networks, written in the high level language *N*.
- **High Level Language** ***N***, the object-oriented neural programming language based on ***C++***.
- **Intermediate Level Language** ***nC***, the low level machine independent network specification language, which is a subset of ***C***.
- **Compilers** to the target UNIX-based workstations and parallel Transputer-based machines.

PYGMALION embodies a hierarchical data structure encompassing: system, network, layer, cluster, plus neuron and synapse. Network information is encoded in the *C* data structure **system**, as shown in Figure 8.

```
typedef struct {
        int n_rules;                    /* # of rules */
        rule_type *rules;               /* list of rules */
        int n_parameters;               /* # of user parameters */
        par_type *parameters;           /* list of user_parameter */
             (...)                      /* system variables (e.g. int clusters;) */
             (...)                      /* lower level elements (e.g. cluster_type *cluster;) */
} (...)_type;                           /* name of the level (e.g. layer) */
```

Figure 8: *nC* Data Structure for Network Information

The Graphic Monitor sits on top of the *nC* data structure and displays its contents, with a window corresponding to each level in the data structure. There are two types of window: **Top Window**, providing facilities for controlling the simulation and displaying status information plus the program text of the neural network; and **Level Windows**, providing control facilities, and displaying status information and a graphic representation, for each specific level. A window has three areas: commands, graphics/text, and parameters (i.e. status). A command area, comprises a series of button boxes (i.e. labels), each associated with a command or a "pull-down" menu. For example, at the top level, the boxes are create, start, pause, resume, input/output and quit. A graphic area presents a graphical view of the specific level and its pattern of intra-connectivity. Lastly, a parameter area displays status information in a textual form.

The PYGMALION programming environment is available free from University College London, and has been distributed to over 250 organisations.

The goal of GALATEA is to build upon this environment, to produce a "general-purpose" neurocomputing system. We visualise a general-purpose neurocomputing system as being a fully integrated environment of software and hardware components for the development and implementation of artificial neural networks. This neurocomputing system (see Figure 9) will comprise: a sophisticated programming environment capable of mapping a network on to a range of conventional computers, including parallel machines; a novel general-purpose neurocomputer; and an integral silicon compiler for translating a network into VLSI *chips*.

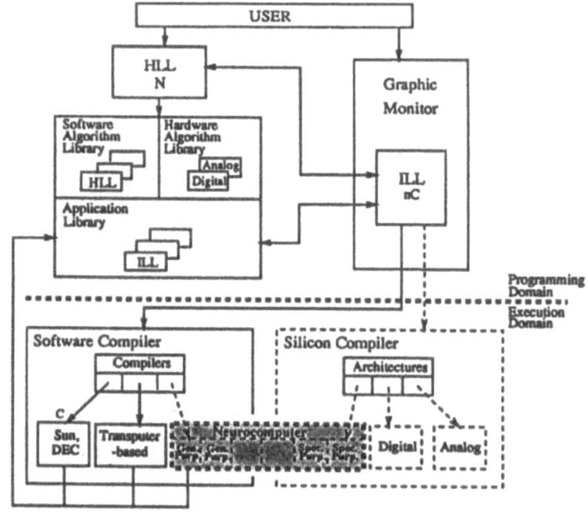

Figure 9: The GALATEA Neural Computing System

The programming environment will be similar to PYGMALION, and execute on a conventional workstation. Once programmed, a neural application can be compiled into binary (i.e. software), or silicon (i.e. hardware) for execution:

- ***workstation simulation*** - a binary representation could be executed on the workstation itself (e.g. SUN, DEC).
- ***parallel simulation*** - the binary could be down-loaded to a conventional parallel computer connected to the workstation (e.g. a Transputer accelerator).
- ***neurocomputer emulation*** - the binary could be down-loaded to general-purpose sub-systems used in a specialised neurocomputer connected to the workstation.
- ***silicon chip*** - a silicon representation in a special-purpose integrated circuit (digital or analog).
- ***special-purpose sub-system*** - special-purpose silicon chips mounted in a sub-system for use in the neurocomputer.

In addition, applications composed of multiple networks may combine the binary or silicon execution strategies specified above.

In conclusion, GALATEA aims to develop a fully integrated, general-purpose neurocomputing environment, covering all possible aspects of neural network application in real-world tasks and research. A key component of the environment is the neurocomputer, viewed as a heterogeneous system containing a number of general-purpose programmable sub-systems. It could also include one or more special-purpose, domain-specific sub-systems.

References

1. I. Aleksander and H. Morton, "Introduction to Neural Computing," in *North Oxford Press*, 1990.
2. G. Shepherd, "Synaptic Organisation of the Brain," *John Wiley & Sons*.
3. P.D. Wasserman, "Neural Computing: theory & practice," in *Van Nostrand Reinhold*, 1989.
4. Nestor, "Nestor Development System User's Guide," *Nestor Inc.*, 1988.
5. Olmsted & Watkins, "Neural Network Library," *Olmsted & Watkins Inc*, 1988.
6. P.C. Treleaven, M. Pacheco, and M. Vellasco, "VLSI Architectures for Neural Networks," *IEEE Micro*, vol. 9, no. 6, pp. 8-27, December 1989.
7. R. Hecht-Nielsen, "Performance Limits of Optical, Electro-Optical, and Electronic Neurocomputers," *Optical and Hybrid Computing SPIE*, vol. 634, pp. 277-306, 1986.
8. Hecht Nielsen, "Hecht-Nielsen Neurocomputing ANZA and AXON," *Hecht-Nielsen Neurocomputing*, 1988.
9. SAIC, "DELTA/SIGMA/ANSim, editorial," *Neurocomputers*, vol. 2, no. 1, 1988.
10. NEC, "Neuro-07 (in Japanese)," *Nikkan Kogyo*, February 1989.
11. L.D. Jackel, H.P. Graf, and R.E. Howard, "Electronic neural network chips," *Applied Optics*, vol. 26, pp. 5077-5080, December 1987.
12. H.P. Graf, L.D. Jackel, and W. Hubbard, "VLSI Implementation of a Neural Network Model," *IEEE COMPUTER*, pp. 41-49, March 1988.
13. Angeniol B., "Pygmalion: ESPRIT II Project 2059, Neurocomputing," *IEEE Micro*, vol. 10, no. 6, pp. 28-32, December 1990.

A Novel High-Speed Memory Organization for Fine-Grain Multi-Thread Computing

Herbert H.J. Hum
Centre de recherche informatique
de Montréal
3744 Jean Brillant, Bureau 500
Montreal, Canada, H3T 1P1.
and
McGill University

Guang R. Gao
McGill University
School of Computer Science
McConnell Engineering Building
3480 University St.
Montreal, Canada, H3A 2A7.

Abstract

In this paper, we propose a novel organization of high-speed memories, known as the *register-cache*, for a multi-threaded architecture. As the term suggests, it is organized both as a register file and a cache. Viewed from the execution unit, its contents are addressable similar to ordinary CPU registers using relatively short addresses. From the main memory perspective, it is content addressable, i.e., its contents are tagged just as in conventional caches. The register allocation for the register-cache is adaptively performed at runtime, resulting in a dynamically allocated register file.

A program is compiled into a number of instruction threads called *super-actors*. A super-actor becomes ready for execution only when its input data are physically residing in the register-cache and space is reserved in the register-cache to store its result. Therefore, the execution unit will never stall or 'freeze' when accessing instruction or data. Another advantage is that since registers are dynamically assigned at runtime, register allocation difficulties at compile-time, e.g., allocating registers for subscripted variables of large arrays, can be avoided. Architectural support for overlapping executions of super-actors and main memory operations are provided so that the available concurrency in the underlying machine can be better utilized. The preliminary simulation results seem to be very encouraging: with software pipelined loops, a register-cache of moderate size can keep the execution unit usefully busy.

1 Introduction

The current microelectronics technology is passing the mark of a million transistors per microprocessor chip, and computer architects are facing the increasing challenge of ULSI – *ultra large scale integration* – technology, which may boast the capability of 50—100 million transistors on a chip by the year 2000[10]! One direction to utilize such enormous hardware parallelism is to increase significantly the architectural support for fine-grain parallelism. Examples include superscalar machines which can issue multiple instructions per cycle like the Intel i860 and the IBM RS/6000, superpipelined machines which use deep instruction pipelining like the CDC-7600, or a combination of both[14].

However, conventional single-instruction-stream processors have inherent limitations in fully exploiting instruction level concurrency. This is due to the fact that a processor equipped with only the mechanism of executing a totally ordered instruction stream, lacks the capacity of tolerating long and unpredictable memory and communication latencies – latencies which are unavoidable in a multiprocessing system [4]. An alternative approach is to directly support multiple instruction threads at the processor architecture level, the so-called *multi-threaded architectures*. Multi-threaded architectures have the potential to keep the processor pipelines usefully busy by rapidly switching between threads on long-latency operations. Research on multi-threaded architectures can be found in [1, 15, 17, 13].

For a pipelined multi-threaded architecture, including the target architecture in our research, the ability to exploit the principle of locality (temporal and spatial) is both important and challenging. In conventional modern RISC architectures, the reduction in memory latencies is achieved by providing (explicit) programmable registers and (implicit) high-speed caches. A small number of programmable registers alone can only provide a partial solution – two reasons for this are: (1) the register allocation for subscript variables of array (vector) data is difficult. In fact, "most compilers fail to recognize even the simplest opportunities for re-use of subscripted variables" [7]. (2) Increasing programmable registers will increase the "context" of a thread. If the registers are to be shared between different threads, a large context may present a high context switching overhead.

For the conventional cache solutions, we point out the following important limitations: 1) the published high hit ratios were reported mostly on non-scientific benchmark programs. The effectiveness of a cache for scientific applications where large arrays (vectors) of data are accessed in the computation is less than satisfactory [6]. 2) When a cache miss occurs, the instruction pipeline usually stalls or freezes, causing considerable performance degradation [12]. This degradation will become more severe as the mismatch in processor speed and memory access times continues to grow, as witnessed in new generations of processors. 3) The fact that a conventional cache is transparent to the programmers (compilers) makes performance improvements by optimizing compilers difficult. 4) Lastly, the conventional cache memory is not designed to accomodate multi-threaded architectures. Frequent switchings between instruction threads have a negative impact on the locality of reference. Moreover, multiple active contexts contend for limited cache space, thus further eroding the benefits of the cache because of unwanted cache line replacements. Some notable examples of multi-threaded architectures have rejected the use of caches, such as in the Horizon[17].

The target architecture studied in this paper is the Super-Actor Machine (SAM) — a multi-threaded architecture based on an hybrid dataflow and von Neumann evaluation model. A number of instructions can be issued simultaneously in the SAM so that effective overlapping of floating point ALU operations with other operations can result in a higher floating point performance. This processing model demands low latency and high throughput from its memory subsystem. In this paper, we propose a novel organization of high-speed memories, known as the *register-cache.* As the term suggests, it is organized both as a register file and a cache. Viewed from the execution unit, its contents are addressable similar to ordinary CPU registers using relatively short addresses. From the main memory perspective, it is content addressable, i.e., its contents are tagged just as in conventional caches. The register allocation for the register-cache is adaptively performed at runtime, resulting in a dynamically allocated register file.

In our execution model, a program is compiled into a number of instruction threads called *super-actors.* A super-actor becomes ready for execution only when: 1) the data dependence is satisfied, i.e., all its input data are generated and its result data from the previous activation, if any, have been used; and 2) space locality is satisfied, i.e., its input data are physically residing in the register-cache and space is reserved there to store its result. The first condition is similar to the so-called *firing rule* in a traditional dataflow machine, however each scheduling quantum in the SAM is an instruction thread instead of one instruction. The second condition, a more radical feature unique to the SAM architecture, ensures that an enabled super-actor can be scheduled for execution only when all memory accesses of its instructions are guaranteed to be in the high-speed buffer memory. Therefore, the execution unit will never freeze when accessing instructions or data. This eliminates one main source of pipeline performance degradation.

To study the multi-threaded capability of the SAM, and indirectly, the effectiveness of the register-cache, detailed simulations were performed. The simulation results are very encouraging: with software pipelined loops, a register-cache of moderate size can keep many threads in concurrent execution and effectively hides the local memory latencies and the latencies associated with fine-grain synchronization support.

In the next section, the Super-Actor Machine's abstract model will be described. Section 3 will provide an overview of the Super-Actor Machine and section 4 will describe the architecture of the register-cache along with the 'check-in process' – the process of ensuring that the required data are present in the register-cache. Section 5 will examine the effects of the register-cache on multi-threaded computing in the Super-Actor Machine via simulation results and analysis, and discussions of related work will ensue. Finally, conclusions are drawn in section 6.

2 The Abstract Model of the Super-Actor Machine

In a dataflow graph, each individual instruction ('actor') is the basic unit of work and scheduling quantum for the underlying machine, and fine-grain synchronization is performed to schedule each instruction. However, some actors can be logically grouped into *threads* so that the cost of synchronization can be reduced by performing the synchronizations only among the threads, while actors within a group can be scheduled via the conventional technique of sequencing with a program counter – à la von Neumann. Grouping instructions into threads and sequentially executing the instructions within the threads while performing dataflow-like fine-grain synchronizations at the thread level is referred to as the *hybrid dataflow/von Neumann* model of computation. It belongs to a

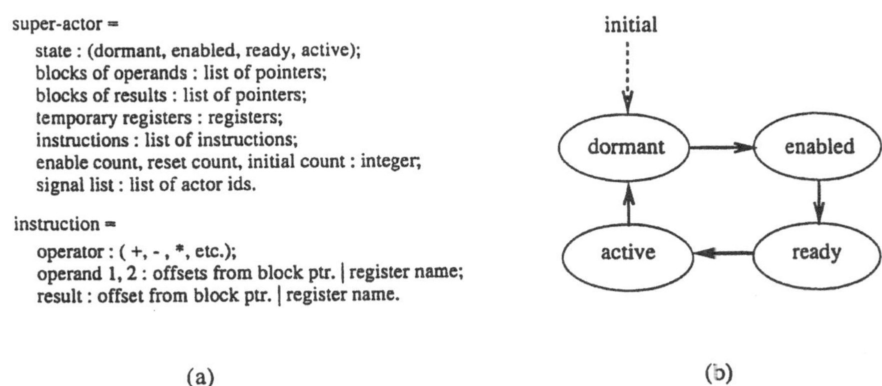

Figure 1: Definition and States of a super-actor.

subclass of the so-called *multi-threaded* architectures [11].

In our work, we call aggregates of one or more actors, *super-actors*. From this point onward, individual actors in the pure dataflow terminology will be referred to as instructions and a thread of one or more of these instructions will be called super-actors, or actors for short. The *attributes* of a super-actor are illustrated in 1 (a), and the operational semantics of a super-actor are defined by the state transition diagram in figure 1 (b).

A super-actor has 4 states, and typically, it goes through the following state transitions: 1) A super-actor is in its *dormant* state when it is waiting on its neigbouring actors to signal to it that it can be enabled. 2) For an *enabled* super-actor to make a transition into the ready state, the following must be prepared: all of the memory blocks containing its input values must be in fast memory and a block of fast memory must be reserved for its result values. 3) A *ready* super-actor enters the active state when it is assigned an available physical domain (context) for it to be executed. 4) Instructions in an *active* super-actor can be scheduled for execution (to be explained later). After execution, an active super-actor will signal its completion to all of the actors requiring notification that it has been executed, and re-enter the dormant state.

One important feature of the architecture model is the *atomicity* of the super-actor activation. Once a super-actor becomes active, it will be executed *atomically* until its completion without the possibility of suspension, i.e., all instructions in the super-actor perform operations entirely *local* to the execution unit, causing no external transactions. Furthermore, it requires no synchronizations with other super-actors during its execution.

Since super-actors are processed atomically, scheduling them based on the data-driven principle will ensure that the data dependencies among the super-actors are satisfied, thus the determinancy of the dataflow computation model is retained, where a node in the super-actor machine model is an instruction thread. A more radical feature unique to the model requires that an enabled super-actor be scheduled for execution only when all memory accesses of its instructions are guaranteed to be in the high-speed buffer memory. Thus, not only do super-actors decrease the synchronization cost, but they also offer the opportunity to exploit the locality of reference so as to minimize the latencies in memory accesses in the execution system.

Two types of super-actors have been investigated in this paper. In a *sequential super-actor*, the data dependencies between the instructions requires that they be sequentially executed. A sequential super-actor may contain conditional branch instructions which jump to another instruction within the same super-actor. We call these instructions, *short branches.* Conditional instructions which fork multiple super-actors or alter the stream of evaluation of super-actors are restricted to being tail-instructions since the scheduling of super-actors are performed in a separate unit from the execution unit. The second type of super-actor is called a *parallel super-actor*, a special case of sequential super-actors where the instructions are data independent, i.e., instructions within a parallel super-actor does not depend on any results produced by any other instruction within the super-actor. Instructions in parallel super-actors can be executed every pipe beat.

Instructions with long and unpredictable latencies are excluded from ordinary super-actors. These instructions include non local memory access operations, explicit 'send' and 'receive' instructions which perform inter-PE communications, etc. A long-latency instruction is grouped by itself and the actor containing it is called a *long-latency actor* ('L-actor' for short). L-actors will be handled by a dedicated unit.

Furthermore, instructions which modify the memory addresses of the lines a super-actor operates on (its operand or result lines) should be grouped separately into aggregates called *support-actors.* (Instructions which modify memory addresses are used to realize address computations in the underlying machine.) The separation of these instructions from super-actors will be explained when we discuss the architecture of the machine.

3 The Super-Actor Machine

The Super-Actor Machine is to be a multi-processor system consisting of multiple processing elements linked together by some interconnection network. Memories are distributed to each processor in the machine, and the aggregation of these memories present a global address space which is shared among all processors. In this paper, we will concentrate our discussions on one processing element. Due to space limitations, only a brief overview of the Super-Actor Machine will appear below, nonetheless we hope it provides enough background information for understanding the organization of the register-cache (R-cache).

A processing element of the Super-Actor Machine has 5 basic components: the Super-actor Execution Unit (SEU), the Actor Preparation Unit (APU) which has an adjoining support-actor execution pipe, the Actor Scheduling Unit (ASU), the L-actor Execution Unit (LEU), and the local main memory (cf. fig. 2).

The structure of the execution unit is shown in figure 3. In this section we discuss the smooth execution pipeline and the collection of physical contexts realized by multiple sets of registers. The organization of the register-caches (*i-R-cache* and *d-R-cache*) is left for a later section.

The architecture of the *smooth* execution pipeline is like any standard instruction

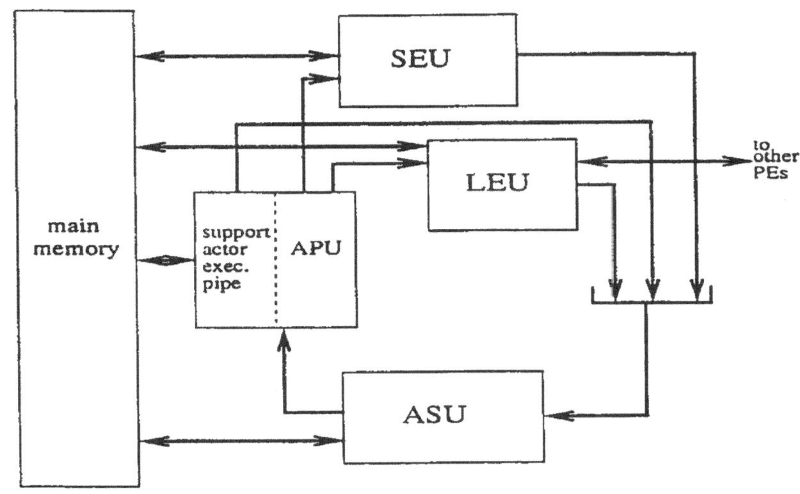

SEU = Super-actor Execution Unit
LEU = Long-latency actor Execution Unit
APU = Actor Preparation Unit
ASU = Actor Scheduling Unit

Figure 2: A Processing Element of the Super-Actor Machine.

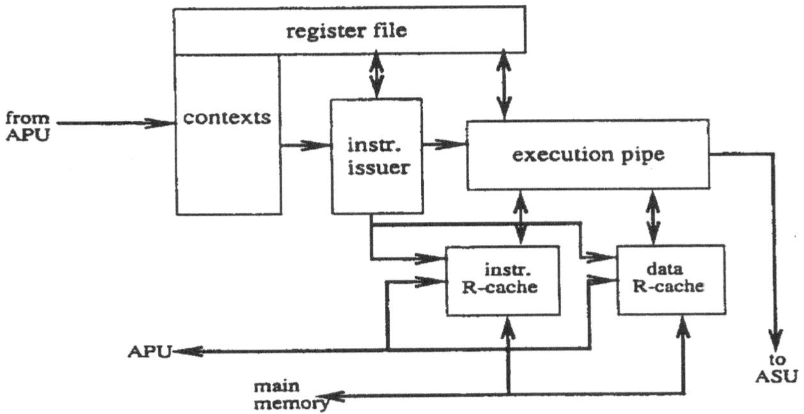

Figure 3: The Super-Actor Execution Unit.

processing pipeline except that the ALU stage is made up of sub-pipes which can handle integer and floating point operations. The other stages are the standard ones like the instruction fetch, operand fetch, etc. The aim of the smooth pipeline is to initiate an independent instruction at a sustained rate of one instruction per cycle, thus the pipeline has the following features: (1) it is *clean* or free of *structural hazards*, and (2) all stages in the pipeline have a uniform and fixed processing time for all types of instructions.

A physical *context*, realized by a set of registers, will be assigned to each super-actor when it becomes active, and will be returned to the pool of free physical contexts when it leaves the execution unit. The purposes of the set of registers are to store information of a super-actor and to be temporary scratch-pad registers for an active sequential super-actor. Values in the registers are not retained after the activation of the super-actor and cannot be used by other super-actors.

All contexts share an *instruction issuer*. The issuer chooses a ready context, increments its counter value and sends the instruction into the execution pipe. An *activation id* is associated with each context and is sent along with each instruction when it enters the execution pipe so that the proper set of registers are used. The issuer is also responsible for sending a 'decrement-reserve-counter' signal to the R-caches when a super-actor exits the SEU (The purpose of this will be explained later). If a context is assigned to a parallel super-actor, it can be ready every machine cycle. Otherwise, the instructions in the super-actor must be executed sequentially and the context must wait for a signal from the execution pipe before it can progress.

Attached to the APU is a simple RISC pipeline which is responsible for processing instructions within a support-actor. The only instructions which the pipeline can process are loads and stores, and integer add and multiply since the sole purpose of support-actors are to perform address calculations, e.g., array indexing, etc. The reason for processing address calculations in the APU is that the R-cache loader must access those calculated addresses, thus the attached RISC pipeline. The LEU is responsible for fetching the instruction and necessary operands from main memory, and processing the long-latency instructions.

Upon completion of an active super-actor, the SEU sends a done signal to the ASU indicating that that super-actor has been executed. The ASU, in turn, processes the signals and decrements the associated enable count of actors. When an actor is enabled (its enable count is zero), its enable count is reset and the enabled actor along with its attributes is sent to the APU. There the enabled actors are enqueued for entry to either the SEU, the LEU, or the support-actor execution pipe. The structure of the ASU and the handling of the signals is similar to the instruction scheduling unit as described in [9].

4 The Register-Cache Architecture

The R-caches are organized both as a register file and a cache. Viewed from the execution unit (SEU), its contents are directly accessible using relatively short addresses; a process similar to the addressing of general registers in conventional CPUs. Moreover, from the APU's perspective, it is content addressable, i.e., its contents are tagged just as in conventional caches (cf. fig. 4). To make effective use of all R-cache lines, the APU will

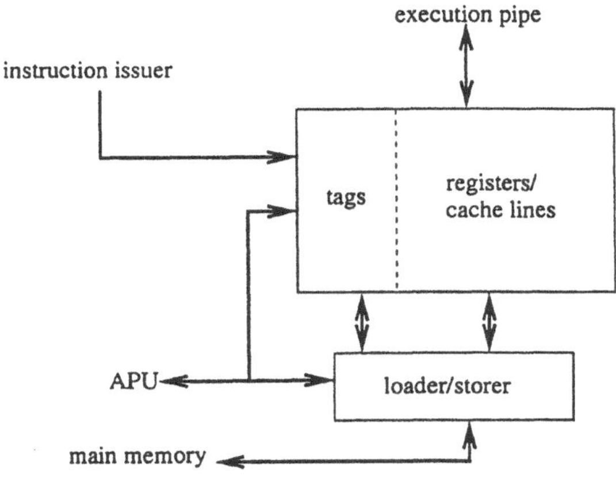

Figure 4: A register-cache.

see a fully-associative cache.

The R-cache retains the transparency feature of conventional caches in the sense that it is not visible to the programmers or compilers; thus, no register allocation by the compiler is required for the R-cache. The allocation of an R-cache line for a block of memory is done entirely at runtime and is performed using cache update and replacement algorithms. Once this is done, the R-cache locations within a line can be accessed by the SEU directly using short addresses, just as if they were general registers (cf. fig. 5). This binding process is called *registering*.

4.1 The Check-In Process for the R-Caches

The APU consists, among others, a queue for enabled parallel and sequential super-actors (PSA/SSA) and a queue for *ready* super-actors (cf. fig. 6).

The APU also contains an R-cache loader which is responsible for *checking-in* enabled super-actors, i.e., ensuring that all the necessary data for the operation of the super-actor (SA) is in the R-cache and that space is reserved in R-cache for its results. The checking-in algorithm is shown in figure 7. An analogy of this check-in process can be found at the ticket counter in an airport. Before boarding the airplane (execution unit), the tour group (super-actor) must first receive their boarding passes indicating that a block of seats are reserved (set of cache lines L it requires is in place). The assignment of seats within the block (relative locations of operands and results) to each member (instruction) of the tour group can be done statically by the tour group manager (at compile-time). However, the final row numbers are assigned dynamically prior to the departure (the processing of the super-actor) during the check-in time. The key (and a divergence from the analogy) is that the SAM architecture can overlap the check-in process of super-actors with the execution of other ready super-actors.

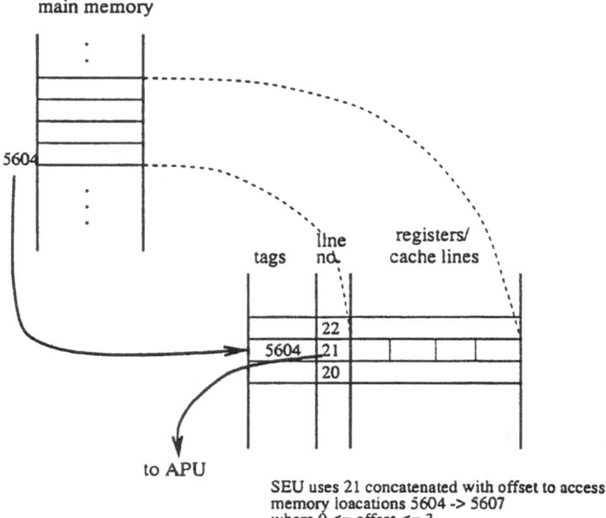

Figure 5: The registering process.

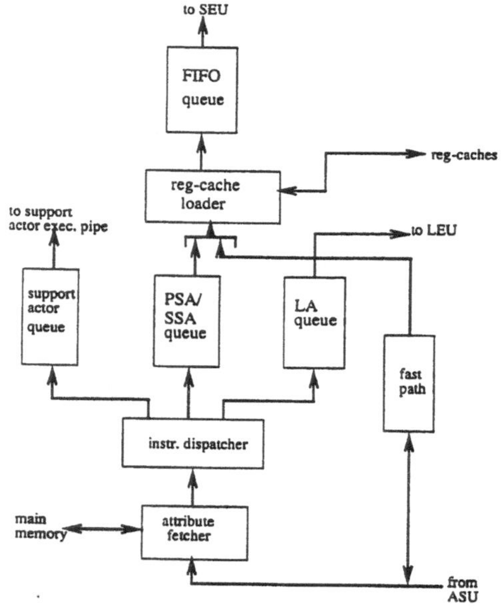

Figure 6: The APU

```
algorithm check-in
   do forever
      if PSA/SSA ready queue has an empty space then
         take next available SA from PSA/SSA enabled queue,
         fetch and/or calculate addr. ptrs. to operand and result line(s), if needed,
         send addr. ptrs. to d-R-cache and receive R-cache line nos.,
         send head-instruction addr. to i-R-cache and receive R-cache line no.,
         package R-cache line nos. with SA packet and put it in PSA/SSA
            ready queue,
      endif
   enddo.
```

Figure 7: The check-in algorithm.

Now let us elaborate on the algorithm. The addresses for the operand and result lines will need to be calculated if they are offset values from the base address. If they are pointer values, then the memory location must be calculated ($ptr + baseaddress$) and the address fetched from the data cache which is shared with the support-actor execution pipe. Otherwise, the values are absolute addresses and are sent to the d-R-cache without modification. Once the addresses are sent to the d-R-cache, the R-cache will return a line number for each address sent. When the loader has received all the line numbers from the two R-caches, it will send the super-actor, consisting of its id, base address, length, instruction and data R-cache line numbers, to the ready PSA/SSA queue. There, it will wait till a context is free in the SEU.

The algorithm describing the operation of the data R-cache is shown in figure 8. The registering process begins when a memory address is sent to the R-cache from the APU. Read-in requests are issued for operand lines and reserve requests are issued for result lines. After the registering process, we say that the instruction is *checked-in*. If the R-cache is full, the Least-Recently-Used (LRU) cache replacement policy is used on lines which are no longer needed (i.e., lines with reserved counters equal to zero) to find a replacement line. The LRU algorithm uses the "age counters" to decide which line to replace. Note that the age counters are only updated by requests from the APU, not by accesses from the SEU.

A decrement-reserved-counter signal along with the line numbers are sent from the instruction issuer in the SEU when a super-actor exits the SEU. Forced write-backs are used to handle super-actors passing their results to long-latency actors, because the LEU does not access the R-cache. Mandatory read-ins are necessary in the d-R-cache because operand lines of a super-actor which was written by a long-latency actor must be brought in since the the LEU can only write into main memory.

The i-R-cache check-in algorithm is similar except simpler due to the fact that it is read only.

```
algorithm d-R-cache
  in parallel do forever
    if read or write request from SEU then service it,
    if decrement-reserved-counter signal from SEU then service it,
    if forced write-back from SEU then
      write corresponding line to memory and send acknowledgement back
        to SEU,
    /* the registering process */
    if mandatory read-in request from SPU then
      use LRU replacement policy on not needed lines if there is no free line
        and write back dirty line if necessary,
      read in from memory, update reserved counter, and send assigned line no. to SPU,
      update age counter of not needed lines,
    else if read-in request from SPU then
      check, in parallel, tags for requested line,
      if there then update reserved counter and send line no. to SPU
      else do as in mandatory read in,
    else if reserve request from SPU then
      check, in parallel, tags for requested line,
      if there then update reserved counter and send line no. to SPU
      else use LRU replacement policy on not needed lines if there is no free line and
               write back dirty line if necessary,
           send assigned line no. to SPU,
      update age counter of not needed lines,
    endif
  enddo.
```

Figure 8: The algorithm for the data register cache.

4.2 The Size of the Register-Cache

For the tandem of the APU and R-cache (i-R-cache or d-R-cache) to function correctly, i.e., the SEU is guaranteed that its operations will always find their value in the register-cache, there is a minimum number of required R-cache lines. The minimum number of lines is $(J + K) \times L$ for $(J + K)$ active super-actors. J is the number of slots in the PSA/SSA ready queue, K is the maximum number of allowable super-actors in the SEU, and L is the maximum number of register-cache lines allocated to a super-actor. The use of the 'reserved counters' in the R-cache guarantees that the reserved or read-in register-cache lines of the $(J + K)$ active super-actors will not be replaced until the super-actor that requested it exits the SEU.

4.3 A Bypass Path to Avoid Unnecessary R-Cache Probing

The bypass path called a *fast-path* is used to avoid unneccessary probing of the R-caches. For super-actors in loop constructs which are enabled every time the loop iterates, the lines they use might still be in the register-caches when they are enabled. These super-actors can be tagged by the compiler as possible *fast-path candidates* so that when they are enabled, a small cache (called an *actor-cache*) containing the recently fired super-actors can be checked for its presence. To ensure that the super-actors present in the actor-cache would still have their lines in the R-caches, the cache should only have $J + K$ entries. If an enabled super-actor has its entry in the actor-cache, then the cache line numbers which it used previously are retrieved, the lines reserved, and the super-actor enters its ready state immediately. But if that instance of a super-actor is not present in the actor-cache, it will be sent back to the regular path where its other attributes can be fetched and the R-caches probed.

5 Performance Effects of the Register-Cache

To investigate the effects of the register-cache on multi-thread computing, a detailed simulator was written in Common Lisp with Flavors. The architectural simulator models the Super-Actor Machine down to the machine cycle level, i.e., for each object modelling a particular task, a processing time of x machine cycles was assigned for which the authors believe is attainable with today's device technology.

5.1 The Simulated Architecture

In the simulations, some of the following design parameters were arrived at arbitrarily while others were based on rough calculations of the requirements for efficient processing.

The local main memory is made up of 16 banks with access times of 6 machine cycles and a memory controller regulating access to them. Addresses are interleaved amongst the banks and each bank can service a request independently from the others. FIFO queues are used to smooth out the throughput rates between the ASU, APU, SEU and LEU.

The signal processor and the enable controller of the ASU are pipelined functional blocks with a pipe beat of 1 machine cyle. A 1K word 4-way set-associative cache with a line size of 8 words is used to buffer requests for signal lists from the signal processor to the main memory. The enable controller goes through a 512 word 4-way set-associative cache with a line size of 4 words. (Note that an enable count is only 4 bits, so 1 word contains the enable counts of 8 actors.)

There are 16 available physical contexts in the SEU, so there are 16 sets of registers where each set contains 8 32-bit registers. We chose the value 16 because the execution pipe has 10 stages through its longest path, and if the contexts contained only sequential super-actors, then we would require a minimum of 10 active super-actors to keep the pipe fully busy. The execution pipeline has a pipe beat of 1 machine cyle and the instruction issuer is also pipelined with a cycle time of 1 machine cycle. The floating point add, multiply and approximate reciprocal pipes are six stages long with a pipe beat of 1 cycle, and the integer pipe is one stage long. Fetch and stores from the register set or register-cache each take 1 cycle. The i-R-cache is 1K words with 16 words per line. This implies that a super-actor can only contain a maximum of 16 instructions. It has been found that the average grouping of dataflow actors are of size 4 [3], so 16 should be plenty[1]. The d-R-cache is also 1K words with 4 words per line and the path to main memory is 4 words wide. Each super-actor is allowed a maximum of 4 lines.

The fetching of actor attributes in the APU goes through a 1K word 4-way set-associatve cache with a line size of 8 words. The R-cache loader can load a R-cache line in a maximum of 8 cycles (one to form the address, 6 for the access and one to load). Requests to the i-R-cache and d-R-cache are performed in parallel. The actor-cache in the fast-path has space for 63 entries.

The support-actor execution pipe is a basic RISC pipe with a pipe beat of 1 cycle. The i-cache and d-cache are both 1K words 4-way set-associative with a line size of 4 words. The LEU was not modelled since the preliminary experiments were only used to investigate the impact of the register-caches in a PE of the Super-Actor Machine.

5.2 The Test Programs

Work is currently under way in producing an assembler for SAMAL (Super-Actor Machine Assembly Language) and in generating SAMAL code from the program graph form of the Id compiler[18]. For this study, we have hand-coded three small benchmark programs.

The three benchmark programs are: SAXPY, SAXPBYPC, and Lawrence Livermore Loop 1. SAXPBYPC is the same FOR-loop construct as SAXPY except the expression is $a * X[i] + b * Y[i] + c$ instead of $a * X[i] + Y[i]$. For all three benchmarks, the loops were unrolled four times so that parallel super-actors can be formed by aggregating four identical operators in the loop[2]. The index sequencing is handled by a sequential super-actor which can either trigger super-actors in the loop body or exit when it is finished. The

[1]This does not mean that the rest of the cache line goes to waste. In fact, other super-actors can share the same i-cache-line, the only requirement being that an assembler or compiler must handle the arrangement of instructions into contiguous blocks which are aligned on 16-word boundaries.

[2]Other techniques for grouping instructions into super-actors are possible, but they are not investigated in this paper.

Benchmark	1 Loop		4 Loops		
	execution time (cycles)	SEU utilization	execution time (cycles)	SEU utilization	Speedup
SAXPY	30799	13%	8621	46%	3.6
SAXPBYPC	33527	19%	10797	59%	3.1
Loop1	32151	23%	13442	56%	2.4

Table 1: Results for SAXPY, SAXPBYPC and Loop1.

arrays which the loops process are stored locally in main memory and support-actors are used to perform address calculations for the super-actors in the loop body. The indexing super-actor has 6 instructions while the parallel super-actors in the loop body have 4 instructions each. The support-actors have an average of 7 instructions each.

Dataflow software pipelining[8] was utilized to increase the amount of exposed parallelism in the programs. With dataflow software pipelining, a code body for SAXPBYPC was reconstructed with 4 stages, thus handling 2 simultaneous iterations and exposing more parallelism, while a code body for Loop1 was reconstructed with 5 stages. However, SAXPY could not really benefit from software pipelining due to its small loop body – only 2 stages were produced when it was software pipelined.

5.3 Simulation Results

Two versions of each program were written: one which only had one loop that iterated from element 1 to 1200, and the other version with 4 simultaneous loops, i.e., each loop was invoked in parallel where a loop iterated through 300 elements. The results are shown in table 1. The reader should note that several operations can be issued each cycle besides an ALU operation in the SEU, and the utilization rate shown in table 1 does not reflect the processing of those other operations. The speedup factor was calculated by dividing the execution time for the 1-loop version by that of the 4-loop version. From the table, one can conclude that more parallelism results in a higher utilization of the SEU and the more parallelism the compiler exposes in a program, the faster the execution due to the opportunity of overlapping memory loads to the R-cache with the processing in the SEU. However, there appears to be a limit as to how much exposed parallelism the current configuration of the Super-Actor Machine can exploit, as shown by the smaller speedup for Loop1 as opposed to SAXPY and SAXPBYPC, and by the leveling off of the SEU utilization.

On closer monitoring of the simulator, we realized that the register-cache loader had become a bottleneck because of the increased number of ready super-actors which must wait to enter the loader. To overcome this bottleneck for memory-intensive programs, the loading phase could be sped up by pipelining the R-cache loads or a load request can bring in a larger memory block. We decided to investigate the latter since it is less costly in terms of hardware. In the following experiment, the d-R-cache line is increased to 8

	8-word			16-word		
loop(s)	execution time (cycles)	SEU util.	Speedup	execution time (cycles)	SEU util.	Speedup
1	15759	19%	1.0	8845	30%	1.0
2	7981	38%	2.0	5321	50%	1.7
4	4897	62%	3.2	3690	73%	2.4
8	4005	76%	3.9	3422	80%	2.6

Table 2: SAXPY for 8-word and 16-word long d-R-cache line.

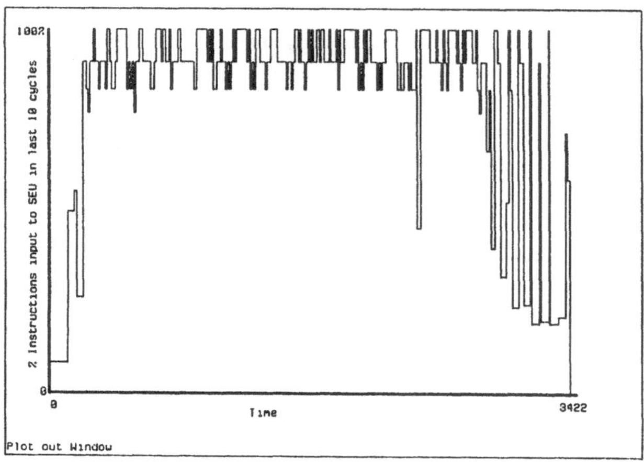

Figure 9: SEU utilization profile for the 8-loop, 16-word wide version of SAXPY.

words and 16 words[3], so the R-cache size increases accordingly. The i-R-cache organization remains the same since the real backlog is caused by the loading of array elements into the d-R-cache. To exploit the larger d-R-cache lines, SAXPY was re-written with an unrolling factor of 8 and 16 so that a parallel super-actor has 8 and 16 instructions respectively. Table 2 shows the results of this experiment where the loops iterated from 1 to 1152. The numbers in the speedup column is relative to the 1-loop case. As one may notice, the SEU utilization has increased significantly for the 16-word R-cache line case as compared to the 4-word R-cache case. Indeed, the bandwidth from R-cache to main memory is a major factor in keeping the SEU usefully busy. In fact, for the 8 simultaneous loop and 16-word R-cache line case, the SEU can be kept at about 90% busy when the start-up and wind-down phases are discounted (fig. 9).

[3]A 16-word cache line is not impossible, the IBM RS6000 has a 128 byte cache line.

5.4 Discussions and Related Work

Local memory latencies and the latencies associated with fine-grain synchronization have always been a challenge to dataflow architectures. The above simulation results show that with enough exposed instruction-level parallelism, these two sources of latencies can be effectively hidden via the actor-preparation unit and R-cache mechanism of the SAM. However, further research remains in addressing the efficiency issues and the multitude of tradeoffs present in the Super-Actor Machine.

In the SAM, a number of supporting operations can be issued at each cycle while the SEU is issuing an ALU operation. These operations may include: (1) an integer operation (+/-) in the APU for address calculation, (2) a memory load/store operation (integer or floating-point), (3) a 'fork' or 'join' operation in the ASU[4], and (4) some long-latency operation in the LEU. Moreover, the execution pipes in the SEU are fully pipelined. In this manner, one processing element of the SAM can be considered a superscalar, super-pipelined machine as defined in [14]. However, it really belongs in a superset of this class due to its support for partial order execution and the issuing of multiple instructions from multiple streams. (A superscalar machine issues multiple instructions around one program counter, i.e. from a single stream.) To roughly guage the efficiency of the SAM in hiding local memory latencies and fine-grain synchronization costs, we examined the performance of SAXPY on the IBM RS6000/530 - a superscalar, somewhat superpipelined machine. The sustained performance is about 1 floating-point operation per clock cycle[5] when the combined floating-point multiply add instruction is used. In the SAM, a 0.76 floating-point operation per cycle can be obtained on SAXPY (the 8-loop, 16-word version). Note that the floating-point pipe on the SAM is 6 stages deep whereas the RS6000 only has a 3-stage floating pipe. This crude comparison has given us great hope in that a fine-grain multi-threaded architecture can indeed hide the overhead associated with fine-grain synchronization.

Some multi-threaded architecture researchers are also investigating high-speed memory between main memory and the execution unit. In particular, Nikhil and Arvind (P-RISC)[15], Agarwal et al (APRIL)[1], and Iannucci (EMPIRE)[13]. In P-RISC as in EMPIRE, the key to exploiting temporal and spatial locality is to continuously process all the threads within an active frame (the frame contains threads of a code block) until there are no more. A set of active frames would be kept in a high speed memory to minimize the local memory latency. EMPIRE is more unique in that a frame must be in fast memory before any of its threads are picked for execution. These architectures have addressed the local memory latencies, however, their fine-grain synchronization support mechanism is within the execution unit, thus those associated costs cannot be totally hidden. In APRIL, a regular cache is put between main memory and a RISC processor. For its cache to be effective, a small amount of active threads can only be supported, otherwise cache interference can have deleterious effects. Since only a limited number of active threads are loaded into fast memory at any given time, the ominous possibility of constantly loading and unloading active threads exists. Furthermore, the fine-grain synchronization support is also embedded within the execution unit. With all RISC implementations, the question

---

[4]We note that the ASU performs fork/join operations implicitly through signal processing functions, while some other multi-threaded architectures may execute explicit machine instructions [16].

of floating-point performance remains.

Thus far, we have only mentioned architectures with non-superscalar processing elements. Burton Smith's Tera computer[2], however, is a multi-threaded superscalar architecture. Again, his architecture does not rely on cache memory, instead he utilizes a huge register set file. It will be very interesting to quantitatively compare the SAM with the above mentioned architectures.

6 Conclusion

The SAM architecture has put a strong demand on its memory organization in terms of low latency and high throughput requirements. In this paper, we have presented the organization of the register-cache as a high-speed buffer memory to meet such a demand. It ensures a low and fixed latency to support the highly pipelined instruction processing capability in the execution unit. Architectural support for overlapping the super-actor execution and main memory operations are introduced to provide high memory bandwidth. An adaptive "check-in" process which dynamically registers and binds the keys to access register-cache lines helps to avoid some of the difficulties of register allocation at compile-time, in particular, for subscript variables of large arrays containing floating-point numbers. The preliminary simulation results have provided evidence of the effectiveness of the register-cache in hiding local memory latencies and the latencies associated with supporting fine-grain synchronization.

7 Acknowledgment

We would like to thank the Natural Sciences and Engineering Research Council for their support. We are very grateful to the members of the ACAPS group for their interesting discussions on this subject, in particular, Erik Altman, Russ Olsen, Philip Wong, and Kevin Theobald. Finally, we would like to thank Dr. J.B. Dennis for his comments on an earlier draft of this paper.

References

[1] A. Agarwal, B. H. Lim, D. Kranz, and J. Kubiatowicz. APRIL: A processor architecture for multiprocessing. In *Proceedings of the 17th International Symposium on Computer Architecture*, pages 104–114, 1990.

[2] R. Alverson et al. The Tera computer system. In *Proc. of the 1990 Int'l. Conf. on Supercomputing*, 1990.

[3] Arvind. Personal communication, 1990.

[4] Arvind and R. A. Iannucci. Two fundamental issues in multiprocessing. Computation Structures Group Memo 226, Laboratory for Computer Science, MIT, 1987.

[5] R. Bell. IBM RISC system/6000 preformance tuning for numerically intensive FORTRAN and C programs. Technical Report GG24-3611, IBM Int'l. Technical Support Center, Aug. 1990.

[6] D. Callahan and A. Porterfield. Data cache performance of supercomputer applications. In *Proc. of the Supercomputing '90 Conference*, pages 564—572, New York, New York, 1990.

[7] David Callahan, Steve Carr, and Ken Kennedy. Improving register allocation for subscripted variables. *Proceedings of the SIGPLAN '90 Conference on Programming Language Design and Implementation*, June 1990. White Plains, NY.

[8] G. R. Gao, H. H. J. Hum, and Y. B. Wong. Towards efficient fine-grain software pipelining. In *Proceedings of the ACM International Conference on Supercomputing*, Amsterdam, Netherlands, June 1990.

[9] G. R. Gao, R. Tio, and H. J. Hum. Design of an efficient dataflow architecture without dataflow. In *Proceedings of the International Conference on Fifth-Generation Computers*, pages 861–868, Tokyo, Japan, December 1988.

[10] P.P. Gelsinger et al. Microprocessors circa 2000. *IEEE Spectrum*, pages 43—47, Oct. 1989.

[11] R. H. Halstead Jr and T. Fujita. MASA: A multithreaded processor architecture for parallel symbolic computing. In *Proceedings of the 15th Annual International Symposium on Computer Architecture*, pages 443–451, 1988.

[12] J.L. Hennessy and D.A. Patterson. *Computer Architecture: A Quantitative Approach.* Morgain Kaufman Publishers Inc., San Mateo, CA, 1990.

[13] R. A. Iannucci. Toward a dataflow/von Neumann hybrid architecture. In *Proceedings of the 15th Annual International Symposium on Computer Architecture*, pages 131–140, 1988.

[14] N.P. Jouppi and D.W. Wall. Available instruction-level parallelism for superscalar and superpipelined machines. In *Third Int'l. Conf. on Arch. Support for Prog. Lang. and Operating Sys.*, pages 272—282, 1988.

[15] R. Nikhil and Arvind. Can dataflow subsume von Neumann computing? In *Proceedings of the 16th International Symposium on Computer Architecture*, pages 262–272, Israel, 1989.

[16] R. S. Nikhil and Arvind. Id: A language with implicit parallelism. Computation Structures Group Memo 305, Laboratory for Computer Science, MIT, 1990.

[17] M.R. Thistle and B.J. Smith. A processor architecture for Horizon. In *Proc. of the Supercomputing Conference '88*, Florida, 1988.

[18] K. R. Traub. Sequential implementation of lenient programming languages. Technical Report MIT/LCS/TR-417, Laboratory for Computer Science, MIT, 1988.

Evaluation of Futurebus hierarchical caching

K.G. Langendoen H.L. Muller
L.O. Hertzberger
University of Amsterdam,
Kruislaan 403, 1098 SJ Amsterdam, The Netherlands
koen@fwi.uva.nl

Abstract

This paper presents a simulation model for hierarchically structured multiprocessors based on the Futurebus+. The model simulates the behaviour of the buses and caches at the level of individual memory references. These memory references are generated by a set of "stochastical processes" which are based on measured statistics of actual programs. The model is validated with published trace driven simulations of single and two level cache systems.

We have used the model in some experiments to study the performance effects of cache parameters in various multilevel cache hierarchies. We conclude that a two level hierarchy of caches is attractive for those applications that cause a lot of bus traffic. The parallel application of our benchmark, which heavily uses shared data, showed a performance increase of 44% when a flat bus was replaced by a two-level hierarchy. Finally we observed that 99% of the total of bus transactions in all simulations used only 5% of the Futurebus+ cache-coherency protocol. We conclude that many of the optimizations in the protocol only increase complexity without a clear performance benefit.

1 Introduction

The current generation of bus based shared-memory multiprocessors is equipped with caches for two reasons. First the performance of modern processors critically depends on the usage of caches, and secondly caches reduce bus traffic and therefore bus contention in a multiprocessor system. The caches run a cache-coherency protocol to keep the data values consistent. Many of these protocols require the caches to snoop (monitor) all traffic on the memory bus and take appropriate action when they have the requested data, they depend on the broadcast capability of the memory bus. Since this severely limits the number of processors in a machine coherency protocols have been designed to operate in systems with a *hierarchy* of buses. Caches located between two levels snoop traffic at both sides and pass requests and responses on to the other level when necessary.

The Futurebus+[1] [Futurebus89] is an industry standard bus definition for scalable

[1]Note, in the sequel we omit the '+' for readability.

shared memory multiprocessors. It is the definition of a high performance multiprocessor system bus and includes a hierarchical cache coherency protocol. Because the Futurebus is targeted as a general industry standard for several generations of computer systems, it is defined in architecture, processor, and technology independent terms. For example the cache protocol is described in terms of lines and sets. As a consequence the Futurebus specification has many degrees of freedom (transfer rate, cache size, etc) which have to be fixed for an actual implementation. In general it is quite difficult to find the "optimal" set of parameter values because many of the relations between the parameters are unclear. Since experimenting with real hardware is expensive we need a performance model to evaluate various design alternatives.

Despite the great interest in caches, hence the volume of literature, we were not able to find a suitable multiprocessor model which includes both hierarchical and snooping caches. Many different snooping protocols for single bus systems have been described and compared, see for example [Archibald86] and [Eggers89]. Hierarchical cache memories have been studied in the context of fast single processor system; [Bugge90] is an interesting example because it uses the Futurebus. Some analytical models of hierarchical snooping cache systems do exist, but either these models are too restrictive or the parameters are too high level. For example, in [Vernon89] one has to specify the probability that a cache miss at some level can be satisfied by a neighboring cache at the same level. We do not want to specify such a parameter, but rather derive it from results obtained with the model.

The simulation model presented in this paper is based on the Futurebus specification and can model a wide range of hierarchically structured multiprocessors. The model is heavily parameterized so it can be used as an experimentation tool to gain insight in the behaviour and capacities of the Futurebus protocol. To capture the effects of some low level parameters the simulator operates at the level of individual bus cycles and records the complete status information of cached data. Although this requires large amounts of memory and computation the model is still capable of simulating an eight 20 MIPS processor system; a simulation run representing 50 millisecond takes 1 hour on a SUN-4.

The remainder of the paper starts with a detailed description of the simulation model in section 2. Next, some experiments to validate the model with published measurements are described in section 3. In section 4, we discuss some experiments. We determine the optimal line sizes of caches in a simple hierarchical system, and study the effects of various degrees of associativity. Then we fix the line size and associativity, and determine the performance effects of different multiprocessor topologies. The bus hierarchy varies from a flat bus up to a three level bus system. The conclusions about these experiments are summarized in section 5.

2 Model

The multiprocessor model covers the complete trajectory from application down to bus cycles and consists of two parts as depicted in figure 1. The upper layer models applications running on multiple processors and generates a sequence of memory read/write requests (an address trace), which is serviced by the lower layer. This memory layer models a hierarchy of caches and Futurebuses.

We have imposed some restrictions on the application-memory interface to keep the

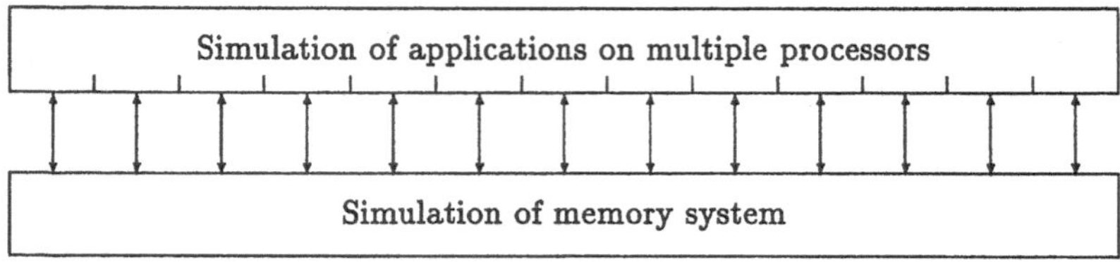

Figure 1: Simulation layering.

simulation model feasible:

- Only the addresses of memory locations are considered, not their values. This excludes the interpretation of the applications at assembly level, but the amount of memory required to store the contents of all simulated memory and caches of a multiprocessor would simply exceed the capacity of our computer systems. Besides, interpretation of assembly code is also unfeasible in this case because of computational demands.
- The interface is limited to simple read/write requests, no I/O instructions or read-modify-write cycles are considered.
- We do not consider the mapping of virtual to physical addresses. Instead we assume that processors directly generate physical addresses and that all applications fit into main memory: there are no page faults.

These limitations make it feasible to exactly simulate the Futurebus cache coherency protocol in the memory layer. The simulator only has to record the status information of each cache line. The applications, on the contrary, do have to be modeled as simple processes generating memory references and may not depend on the result values. This is elaborated in the following sections.

2.1 Application layer

The application layer is responsible for generating memory references to drive the memory layer and models both applications and processors. Since the memory layer only handles memory addresses, the application layer just has to generate an address trace. Although appealing, the usage of real multiprocessor traces of existing programs has a few drawbacks. First the traces will be huge because of the number of processors and the amount of references needed to avoid the long-term cold start effects of second level caches. The measurement and handling of such large traces is a research topic in its own right. Secondly a multiprocessor address trace lacks flexibility because it is only valid for one specific multiprocessor configuration.

To avoid the drawbacks of real multiprocessor traces we use the alternative of stochastical simulation. Applications are modeled as simple processes that generate 'random' addresses. To capture the locality in access patterns of real world programs, a model is used which distinguishes instruction, stack, and data access, each with their typical locality, see section 2.1.1. By tuning a few parameters, like grainsize and data access rate, a diversity of applications can be specified. The major drawback of stochastical simulation is its inaccuracy. Therefor we have validated the model with the experiments described in section 3.

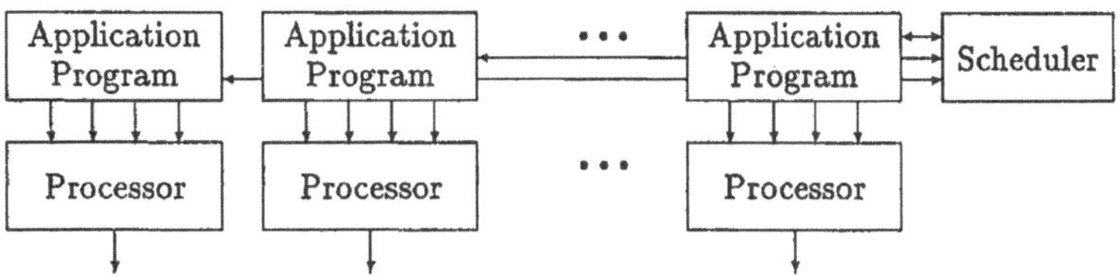

Figure 2: Structure of application layer.

The application layer of figure 1 is detailed in figure 2. Each application process generates an address trace based on measurements of a specific real world application. The trace is fed to a processor filter, which tailors it to a certain type of processor. This enables us to model processor speed, instruction density, register banks (by masking off some of the stack accesses), memory pipelines, etcetera. The scheduler maintains a FIFO queue of ready applications and allocates them to processors on context switches. Note that during a single simulation run, a program can be scheduled in on a different processor from where it was scheduled out last time. Like in a real machine, more than one instance of an application can run at the same time.

2.1.1 Application model

An application is modeled as a simple loop process. Each iteration generates an instruction fetch and possibly some stack and data references. We distinguish three types of instructions to take care of locality in the instruction stream:

1. Jump instructions. The next instruction is an instruction in the neighborhood of this instruction. The jump target is randomly selected from a normal distribution. Both the average distance of jumps and the probability of forward jumps are parameters of the application. It turns out that a normal distribution reasonably matches the real world jump behaviour.

2. Call/return subroutine instructions. The program control transfers to one of the subroutine entry points. The selection of an entry point is based on an exponential distribution to model the typical runtime preference of a program. The subroutines themselves are uniformly distributed over the text segment.

3. Arithmetic instructions. This category includes all other instructions and the program simply continues with the next instruction.

The text segment is shared by all instances of the same application. Besides instruction fetches, an application generates data references as well. An application contains three different data segments with their own locality:

Stack. The application maintains a stackpointer, accesses to the stack are in the neighborhood of this stackpointer. The stackpointer is updated when a call- or return-subroutine instruction is simulated. The average size of the stack frame, and the chance of doing a stack access are application specific parameters.

Each application has its own private stack segment.

Private data. When an instruction references this segment the memory address is selected from a normal distribution around the previous reference. This simple scheme is not very realistic and we plan to improve upon it by using a set of datapointers analogous to the subroutine entry points in the text segment. That would be a simplified version of the model described in [Archibald86], who uses an exponential distribution over a LRU-chain of data references.

Each application has its own private data segment.

Shared data. The address selection is the same as for the private data segment, but this segment is shared with all instances of that application. Different types of shared applications can be modeled by varying the access rate and load/store ratio of this segment.

For each of the segments a pair of parameters specifies the access rate and load/store ratio of the data references. After a sequence of instructions an application process will do a context switch to make a system call. The scheduler puts the application at the end of the ready list, runs a kernel job for a short time, and allocates a new application to the idle processor. The context switch interval is a model parameter and controls the grainsize of the application.

2.1.2 Processor model

The task of the processor model is to adapt the traces coming from the application to a specific processor. RISC and CISC architectures show their own particularities in the address trace. RISCs tend to do more instruction accesses at a higher clock frequency and have larger code sizes, while CISCs do more data accesses, because of less register usage. The following parameters are used to transform the universal application trace into a processor trace:

Speed. The instruction fetches of the application trace are fed to the memory layer at the MIPS rate of the specific processor. Note that the MIPS rate influences the stack and data references as well.

Instruction size, Instruction power. A CISC uses fewer instructions than a RISC to perform the same application and instructions are coded more dense as well. Hence, the average jump distance and text size of an application have to be scaled accordingly. Therefore the processor model includes two scaling parameters: instruction power and instruction density. These can not be merged into one parameter because they scale differently: The density operates on all instructions, whereas the instruction power only affects the arithmetic instructions because the absolute number of jump instructions doesn't vary between RISCs and CISCs.

Register usage. The usage of register windows, large register banks, and compiler optimizations leads to reduced data traffic. Especially the accesses to the stack segment are greatly reduced. The processor model accounts for this effect by masking off some of the stack references from the application trace, they are simply discarded. The register-usage parameter specifies the percentage of masked references.

Context size. The size of the data that should be saved and restored on context switches. This is related to the number of registers.

2.2 Memory layer

The memory layer of figure 1 simulates buses, caches, and shared memory at bus cycle level. To exactly simulate the behaviour of the Futurebus cache coherency protocol, the simulator maintains the tags and states of all cache lines. The simulator is written in Pearl [Muller90], an object oriented language for architectural simulation and evaluation. By modeling the buses, caches, and memory as individual objects it is easy to specify arbitrary multiprocessor configurations.

Some examples of multiprocessor cache hierarchies are shown in figure 3. The top level caches are connected to the address generators found in the application layer. In the remainder of the paper we often name caches by their level number. We start counting at the processor, which is connected to a first-level or primary cache. When referring to other caches in the hierarchy relative to a specific cache, we use downwards to denote caches closer to memory, and upwards for the ones closer to the processors. Likewise we refer to buses.

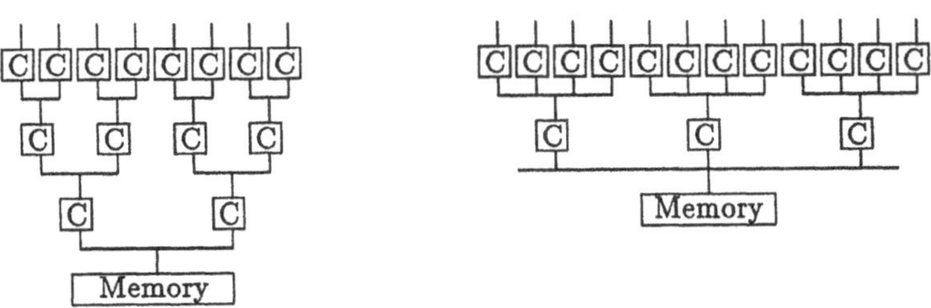

Figure 3: Two example memory hierarchies.

To simplify the simulator all caches in the memory hierarchy have to use the same line size. Furthermore all data transport in the memory layer occurs as line transfers. The performance deviation in case of a write back of a single word will be quite small because of the large setup time compared to the data transmission speed.

2.2.1 Bus model

In general bus specifications define the electrical behaviour of the bus, and the protocol how to drive the bus wires to perform basic actions like read and write. The Futurebus specification, however, also includes a cache coherency protocol, and even a complete message protocol. In this article, we ignore the message protocol and the electrical layer. We concentrate on the transaction layer and the cache coherency protocol. The bus model implements the transaction layer, it handles bus requests from caches in FIFO order and broadcasts the transactions to all other devices connected to the bus. The speed of the bus transactions is specified with two parameters: the setup (arbitration) time and the transmission speed. The cache coherency protocol is implemented by the cache model.

2.2.2 Cache model

The Futurebus specifies a MOESI-like[Sweazy86] cache coherency protocol for hierarchical caches. Only three states are used in the Futurebus protocol: *Exclusive* (modified), *Shared* (unmodified), and *Invalid*. The protocol is described with four basic bus transactions: `read_shared`, `read_modify`, `write`, and `invalidate`. These actions are for reading (shared) data, for reading/writing exclusive data, for writing modified data back and for invalidating a cache line. In normal operation, when there are no cache coherency conflicts, reads and writes of a bus are handled by the cache or memory downwards in the hierarchy. However, when a cache monitors on the downwards bus that a neighbouring cache wants to access data it owns exclusively then it intervenes in the bus transaction and passes the data to the requesting cache. An example of the cache protocol is shown in figure 4: cache `A` tries to read data that is exclusively owned by `B`. Cache `B` intervenes in the bus transaction and supplies the data, thereby preventing the memory to answer the read request (with outdated data).

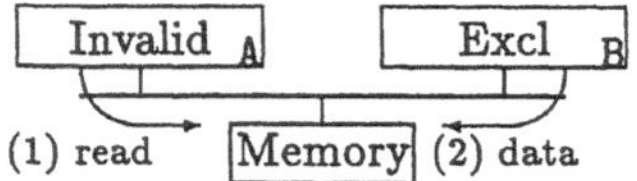

Figure 4: An example of an intervening cache.

In a multi-level hierarchical system, it frequently happens that a cache has to intervene a bus request, but cannot immediately answer the request because it has to get the data from another bus segment. To prevent the first bus from being blocked for a long time, the Futurebus introduces the concept of a *split*. The *split* of a bus transaction causes the requester of the bus action to suspend itself and release the bus for other transactions. The splitting cache issues a response transfer as soon as the answer is available. The originator of the request catches this response, and continues with the original transaction. In the meantime, both the cache and the bus are free to be used for requests of other cache lines.

Figure 5 shows two example cases of a split transaction. In the first simple case, cache `B` issues a split because it does not have the data requested by `A` and repeats the read on the lower bus. In the second case, cache `C` tries to read a value. Cache `D` intervenes this read because it registered that some cache up in the hierarchy has the actual data. It splits the read and issues a read transfer on the upper bus where cache `E` will respond because it has the data. The data is then propagated downwards by cache `D` which issues a response transfer on the bus connected with cache `C`. Cache `C` catches the response and finishes its pending read request by supplying the data to the upper bus.

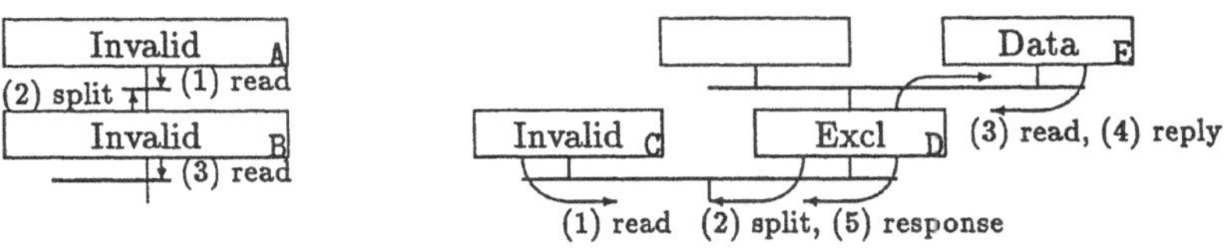

Figure 5: Two examples of *splits*.

The Futurebus protocol does not specify which cache line to replace when a set is full, nor what actions to take during the replacement. Things get really hairy when neighboring caches start referencing either the line selected for replacement, or the line that caused the replacement. Our cache model randomly selects a line for replacement and carefully records the status of both lines.

The cache simulator is implemented as a state machine that reacts to bus transfers. A cache snoops all bus cycles on both buses and has to be capable of handling two transactions in parallel, so the transition table becomes quite large. In practice, however, only a small fraction of the table is used; specifying 252 entries out of the 8620 suffices to run all simulations described in the following sections. The transition-table usage is discussed further in section 4.4.

2.2.3 Memory model

The task of the memory model is easy: Just listen to the bus and service read/write requests unless one of the caches on the bus intervenes. The memory is characterized by its service time. Since we do not store data values, the size of the memory is not relevant.

3 Validation

As noted, the stochastical application model, described in section 2.1.1, introduces inaccuracies. Especially the locality in data and text references is a potentially weak point. To verify the stochastical application model we made simulation-runs to compare the model results with experiments described in literature. The published results have been derived from real world address traces, and cover important aspects of our hierarchical multiprocessor model:

- [Hennessy90, Hill87]: A uniprocessor system with one single level cache. The trace is derived from a UNIX environment.
- [Bugge90]: A uniprocessor system with a two-level cache hierarchy. The trace is derived from some data manipulation programs running under SINTRAN III.

Running the same experiments with our simulation model required the specification of the model parameters of section 2. The articles provide values for many of the memory-layer parameters, but none for the application parameters since these are hidden in the address traces. The next section discusses the application benchmark we have used as workload in various experiments. The two sections thereafter present the validation results.

3.1 Benchmark

To get realistic application parameters we have taken values from [Hennessy90], which lists some program specific parameters like instruction usage distribution and percentage of data references. We did additional measurements of UNIX programs to obtain the remaining parameters. This included parameters like size of text/data segment, number of procedures, and size of stackframe. To better quantify the effect of sharing on system performance, we have defined two job mixes:

UNIX mix: A set of jobs consisting of multiple editors (vi), TeX formatters and C-compilers together with a UNIX kernel. All instances (jobs) of one application program share one text segment, but have their private stack and data segments. The UNIX kernel is a small process duplicated on each processor, which shares both text and data segments.

Fine grained parallel program: A set of identical jobs that generate many read/write references to a small shared data segment. The grainsize is controlled by setting the context-switch rate (which is typically much higher than for the UNIX jobs) and by setting the access rate to the shared data. Since locality and access rate of shared data is essential to performance, we took parameters (10Kbyte segment, one access per 10 instructions) which give comparable write-broadcast ratios as reported in [Eggers89]. In this way, our parameters are in the right order of magnitude, but we emphasize that real parallel programs may behave quite different.

In the future we would like to include the concept of locking in our simulator, to model synchronization phases in a fine grain application. At this moment the memory layer can not support locks because it handles only addresses of memory locations, not their contents. We foresee to include a special lock manager which keeps the lock value. A lock operation in an application will first issue a read/write to the memory layer to get access to the lock, and then get/set the lock by consulting the lock manager.

3.2 Single processor, one-level cache validation.

This experiment is described in [Hennessy90, Hill87]. It is a trace driven simulation of a VAX processor with a single cache, running a multi programming workload. The cache size and associativity are variable. The study of [Hill87] has decomposed the cache miss rate into three fractions, but we will only use the reported total miss rates. We have run the same experiments on our simulator with the UNIX-mix benchmark as workload.

The data in figure 6 shows that our simulation model compares well to the measurements of [Hill87]. The miss rates for small caches are a bit too high, and the miss rates for larger caches are a bit too low. For caches with increased associativity the miss rates show similar results with a slightly higher deviation for large cache sizes.

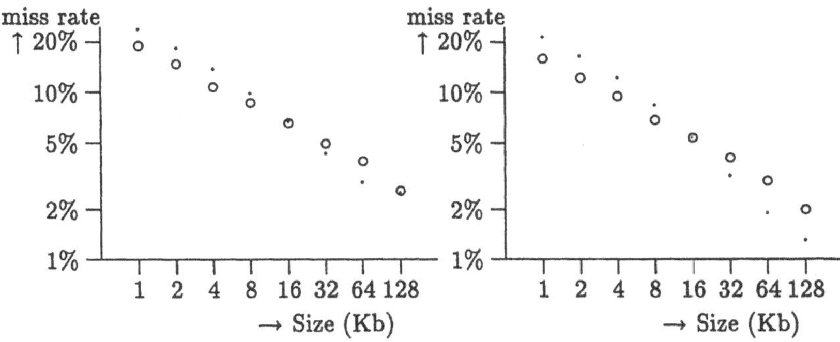

Figure 6: Miss rates measured by [Hill] (circles) and simulated miss rates (dots). A direct mapped cache (left) and a 2-way associative cache.

3.3 Single processor, two-level cache validation.

This experiment considers a uniprocessor with a small fast primary cache, followed by a large secondary. cache. [Bugge90] reports the miss rates of the secondary cache with various parameter settings of line size, associativity, and total cache size. The primary cache is fixed as a 128Kb direct-mapped cache with 16 byte lines. To deal with the effects of cold start misses in the address trace, that paper contains three miss rates: a *worst*, *best*, and *estimate* case. The worst case assumes that cold start misses are indeed real misses, whereas the best case counts those misses as hits. The estimated miss rate simply ignores cold start misses (nor miss, nor hit).

We did the same experiment using our stochastical address-trace synthesizer. Since our model is limited to equal line sizes in all levels of the hierarchy, and [Bugge90] uses a 16 byte line for the primary cache, we only report miss rates for the cases with a 16 byte line size (this is different from the Futurebus standard of 64 bytes). To reduce the effects of cold start misses we report the miss rate over the last 30% of the synthesized address trace. The trace stems from a mix of UNIX jobs (see section 3.1) and contains 46M references to the primary cache, which has a miss rate of 3%.

The results in table 1 show that the miss rates of our simulation model are close to the figures reported in [Bugge90]. The simulated figures of the 8 Mbyte secondary cache are inaccurate because the cache did not reach a steady state before the end of the simulation run[2]. The 8 Mbyte figures of Bugge are also distorted by cold start effects, as can be seen from the relatively large difference between the worst and best miss rates. Our optimistic 1 Mbyte results are presumably caused by the usage of shared text segments in the UNIX mix. We observed that the miss rates, especially those for large caches, are quite sensitive to the exact configuration of the workload (number of applications, context switch rate). The simulated miss rates of the caches up to 4 Mbyte however, show the same trend as the miss rates of Bugge: the influence of the cache size is larger than the effects of the associativity. This indicates that our stochastical application model has the same long-term locality behaviour.

[2]Longer simulation runs are underway to improve the 8 Mbyte figures.

		[Bugge90]			Simulation
Size	Associativity	Worst	Est	Best	model
1M	2-way	22.6	22.2	22.1	20.9
	4-way	18.6	18.2	18.1	17.9
	8-way	16.4	16.0	15.9	15.5
2M	2-way	11.6	10.7	10.6	12.0
	4-way	9.0	8.1	8.0	9.4
	8-way	8.0	7.1	7.0	9.4
4M	2-way	7.3	5.6	5.5	7.3
	4-way	6.0	4.2	4.2	6.2
	8-way	5.9	4.0	4.0	5.8
8M	2-way	5.0	2.0	1.9	5.9
	4-way	4.6	1.3	1.3	5.4
	8-way	4.4	1.0	1.0	5.3

Table 1: Miss rate of secondary cache (in %); [Bugge90] and model values

4 Simulation results

We have used our simulation model to study the performance effects of different multiprocessor topologies. The experiments include a flat, a two-level, and a a three level deep bus hierarchy with a constant number of processors. Before running the experiments we had to determine some reasonable parameter values for the associativity and line size of the caches. Reasonable values for these parameters are determined in the experiments reported in the next two subsections. First we determine the optimal associativity of the caches with a fixed linesize of 64 bytes. Given this associativity we determine the best linesize.

Throughout the remainder of the paper we use the instruction execution rate, expressed in MIPS, as measure of system performance. All the experiments have used the two benchmarks from section 3.1 that consists of a mix of UNIX jobs and a fine grained parallel program. The simulated processors are single-cycle processors running at 20 MIPS. The parameters of the Futurebuses are set to 110 ns arbitration time and a transfer rate of 700 Mb/s. All experiments run for 500 (simulated) milliseconds.

4.1 Associativity

The associativity in a first-level cache has a known important effect on the miss rate, see for example [Hennessy90]. What associativity should we pick for lower level caches in a hierarchical multiprocessor? A common line of reasoning is that low-level caches should be at least large enough to hold all lines of the caches upward in the hierarchy (inclusion property, [Baer88]). Otherwise the high level caches will have to compete with neighboring caches for space in the low-level cache, which causes the low-level caches to frequently invalidate upward copies to service requests that hit a full set. The performance of high-level caches will decrease because they have to invalidate useful lines, which results in a low hit-rate. In [Baer88] it is proved that to enforce the inclusion property, the associativity of a low-level cache memory should be at least the sum of the associativity of its upward connected caches. Then one set in the low-level cache can hold all lines in the primary caches which fall into that set. Since the paper does not quantify the performance effects of obeying the inclusion property, we performed an experiment with various associativities around the inclusion value.

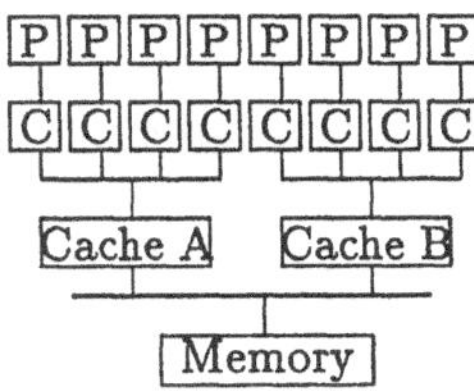

Figure 7: The memory hierarchy used in the associativity and line size experiments

The experiment uses the architecture depicted in figure 7. The first-level caches have a size of 64 Kbyte, and are 2 or 4-way associative. The line size is 64 bytes according to the Futurebus specification. The size of caches A and B is fixed at 2 Mbytes, while the associativity ranges from 4 to 64. The inclusion property requires at least an associativity of 8 respectively 16 for the second-level caches.

second-level	2-way			4-way		
associativity	MIPS	missrate	bus-util	MIPS	missrate	bus-util
4	13.4	3.2%	65%	13.6	3.0%	64%
8	13.3	3.3%	66%	13.8	3.0%	63%
16	13.5	3.2%	65%	13.7	3.0%	64%
32	13.5	3.2%	65%	13.7	3.0%	63%
64	13.5	3.1%	65%	13.7	3.0%	64%

Table 2: The MIPS and miss rates of 2-way and 4-way first-level caches for varying associativity of the second-level cache.

The results of the UNIX mix in table 2 show that the performance is hardly influenced by the associativity of the second level cache. This unexpected result is probably caused by the size of the second-level cache. The large number of sets compared to the primary caches effectively increases the associativity of the second-level cache. Possibly colliding lines between different primary caches usually fall into different second-level sets because of the spatial locality in the applications. Higher associativity does decrease the miss-rate of the second-level cache (as observed in section 3.3), but the overall performance is dictated by the miss rates of the first-level caches. Only architectures with saturated busses will benefit from high associative second-level caches. In the remaining experiments we use the lowest possible associativity that satisfies the inclusion property.

4.2 Line size

In this experiment, the line size is varied to find an "optimal" value. We have used the same architecture as in the previous experiment, drawn in figure 7. The associativity of the first level caches is set to 2, and the second level caches have a fixed associativity of 8. The second- and first-level caches use the same line size, which is varied between 32 and 1024. The results of the two benchmark applications are shown in figure 8.

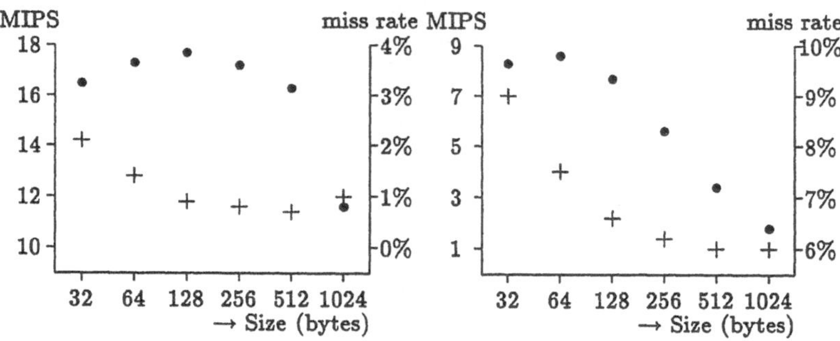

Figure 8: Processing power (dots) and miss rates (+) of various line sizes in the case of a UNIX application (left figure) and a parallel application (right figure).

The data in figure 8 shows that the line size has different effects on the system performance and the second-level cache miss rate. For a UNIX mix of programs, the optimal performance is reached with a 128 bytes line, whereas a 512 bytes line yields the lowest miss rate. Apparently, the decreased miss rate does not outweigh the increased miss penalty. The parallel application results are similar to the UNIX mix, only the optimal performance is reached with a smaller line size of 64 bytes. This optimum depends on the way the shared data is used. As stated in [Eggers89] applications can be specifically coded for a certain line size. The 64 bytes line size of the Futurebus seems a reasonable choice, and we used it in the rest of the experiments.

4.3 Different architectures

This experiment studies the performance differences of comparable multiprocessor systems with different cache hierarchies. We have defined several architectures consisting of 8 processors and a comparable amount of cache memory. These are the architectures labeled a through d in figure 9. The architectures e and f have been added to the experiment when we observed a low bus utilization for the UNIX-workload. The line size is 64 bytes in all architectures. The first-level caches are 2-way associative, while the other caches have the associativity enforced by the inclusion property. For example, the third-level caches in figure 9-d are 8-way associative.

The performance of the multiprocessors is listed in table 3 and shows a large difference between the two benchmark programs. The UNIX-mix performs best on an architecture with a single bus, whereas the parallel application prefers a two-level hierarchy.

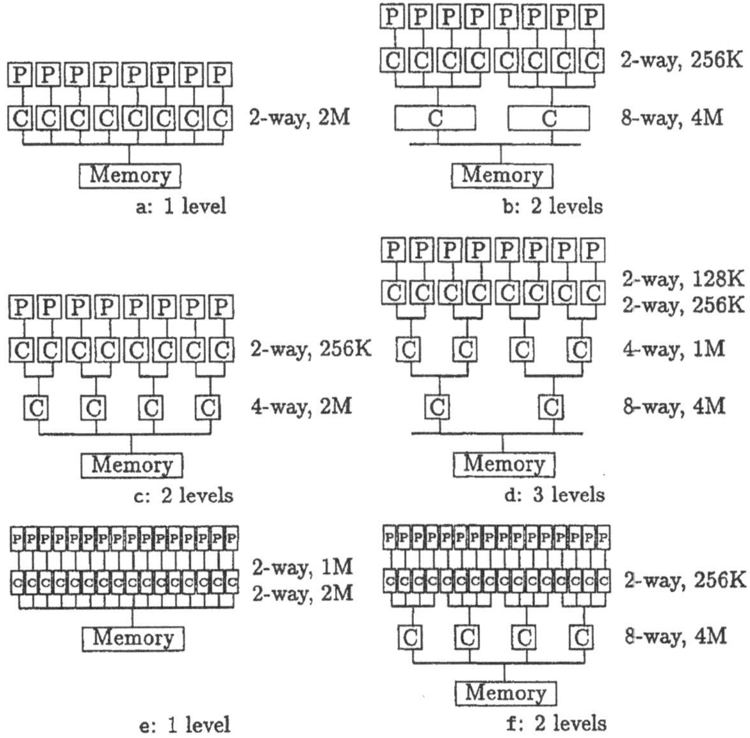

Figure 9: Hierarchical architectures, a-f. The associativity and size of the caches are denoted at the right hand side of the caches. Architectures d and e are evaluated with two different sizes.

The performance of the UNIX-mix largely depends on the effectiveness of the first level cache since jobs are independent and do not share writable data. A cached line of one job will *never* be invalidated by a job running on another processor; the cache line is replaced when the job itself causes a cache miss in a full set. The big caches of architecture-a are well suited for the UNIX-mix and only cause a moderate load on the main-memory bus, therefore the UNIX-mix does not benefit from a hierarchy which potentially decreases the effects of bus contention. In fact the performance degrades because of the smaller primary caches (increased miss rate) combined with a higher miss penalty. The results of doubling the number of processors to 16 are shown in table 4. This number of processors saturates the bus and favors a two-level hierarchy. The memory bus is still the performance bottleneck, larger second-level caches may pay off.

The bus utilization figures show that the parallel application generates more bus traffic than the UNIX-mix to handle the updates of shared data. All hierarchical architectures succeed in diminishing bus contention and substantially increase the overall performance,

	total cache size	UNIX mix					parallel application				
		MIPS	miss-rate	bus util (%)			MIPS	miss-rate	bus util (%)		
				1st	2nd	3rd			1st	2nd	3rd
a	16Mb	16.4	1.6%	69	-	-	5.0	7.0%	99	-	-
b	10Mb	15.0	2.2%	59	60	-	7.2	7.2%	93	87	-
c	10Mb	14.7	2.2%	32	70	-	6.4	7.2%	52	99	-
d	13Mb	13.5	2.7%	34	50	59	6.6	7.5%	55	83	83
	14Mb	13.8	2.3%	33	50	57	6.7	7.2%	54	83	82

Table 3: Performance of the 8 processor architectures (a-d).

architecture	total		miss	bus utilization	
architecture	cache size	MIPS	rate	1_{st}	2nd
e	16Mb	8.8	1.9%	99%	-
	32Mb	9.5	1.8%	99%	-
f	20Mb	11.9	2.3%	56%	98%

Table 4: Performance of the 16 processor e and f architectures for the UNIX work-load.

up to 44%. Note the remarkable difference between the two-level hierarchies b and c. The configuration with two large second-level caches has a clear advantage over four smaller caches, even the three level deep architecture d outperforms architecture c. This is a consequence of the saturation of the memory bus (99%) in architecture c, which is most likely due to sharing in the low level caches.

4.4 The usage of the transition table

During the simulation runs we did not only measure performance parameters like miss rates and bus utilization, but also the usage of the Futurebus cache coherency protocol. The protocol has been implemented as a state machine, and the measurements show that 90% of the transitions are executed in just 17 states. As to be expected, these states correspond to cache hits on reads/writes and snooping hits. Only 38 states cover 99%, while the remaining 1% requires another 214 states. These 252 states out of a total of 8620 sufficed to run all simulation experiments described in this paper, which issued a total of 3.10^9 instructions. Unfortunately we had already specified over 300 states before trying to run the simulator.

5 Discussion and Conclusions

This paper has reported on the evaluation of hierarchical multiprocessor architectures based on the Futurebus. In section 2 a simulation model was presented to study the performance effects of various system parameters. The experiments include various line sizes, associativities, and hierarchical cache configurations.

The model does not use multiprocessor address traces as a workload, but contains a set of stochastical processes instead. These processes generate memory references based on measured statistics of actual programs. Throughout the experiments we have used a benchmark of two different synthesized workloads: a multiprogrammed set of UNIX applications, and a fine grained parallel program which uses shared data. We have run some simulations to validate the model with published trace-driven results.

The first experiments with the model studied the effects of line size and associativity of caches in a hierarchical multiprocessor architecture. The simulation results show that a line size of 64 bytes, as chosen by the Futurebus, will give good performance for both the UNIX and parallel applications. Higher associativity of (large) caches down in the hierarchy (close to memory) will decrease the low-level cache miss rates which only has an effect on the overall performance when the lower busses are saturated. In general the size of the low-level caches is more important than the associativity.

The benchmark was run on various multiprocessor architectures, see figure 9, to study the performance effects of various hierarchical topologies. It shows that a hierarchy of caches decreases the bus utilization, but raises the miss penalty. The performance of the architecture critically depends on the type of program. For the UNIX applications, which do not share writable data, a flat hierarchy is optimal up to at least 8 processors. A multilevel cache hierarchy is beneficial for a large number of processors, the break-even point lies between 8 and 16. Our fine-grained parallel program benefits from a hierarchy even with a small number of processors. With 8 processors we noted a performance difference between a flat and a two-level hierarchy of 44%. It is difficult to generalize this result because the performance benefits of a cache hierarchy for parallel programs heavily depends on the locality in the application. A hierarchical cache system only succeeds in diminishing bus contention if most data sharing is between adjacent processors in the tree. To better quantify the effects of data sharing on system performance we will improve our synthesized application model to include synchronization patterns, see section 2.1.1.

The usage of the future bus transition table is quite remarkable: 38 states covered 99% of the transactions, 252 covered all transactions. The definition of the bus contains a lot of small optimizations to save some bus cycles. Many of these optimizations were rarely executed, or never at all. We expect that only systems with much rapidly changing shared data will exercise these optimizations. Therefore we question the value of these optimizations since they do increase the complexity without a clear performance benefit.

The simulation tool has shown to be useful in evaluating the performance of hierarchical multiprocessors. Additional work will have to be performed to obtain better understanding about the performance effects of locality in fine grained parallelism programs, and its impact on the stochastical application model.

6 Acknowledgements

We like to thank E. Odijk for his stimulation to start work on the design of the Pearl/Oyster simulation system, which has been developed in the PRISMA project. Marius Schoorel of ACE provided us with the Future-bus problem. Furthermore we thank Rutger Hofman, Pieter Hartel and Wim Vree for reading and commenting on a draft version of this paper.

References

[Archibald86] J. Archibald and J. Baer, "*Cache Coherence Protocols: Evaluation Using a Multiprocessor Simulation Model*", ACM Transactions on Computer Systems, Vol 4, No. 4, November 1986, pp 273-298.

[Baer88] J. Baer and W. Wang, "*On the inclusion properties for multi-level cache hierarchies*", Proceedings of the 15th Annual International Symposium on Computer Architecture, 1988, pp 73-80.

[Bugge90] H.O. Bugge, E.H. Kristiansen and B.O. Bakka, "*Trace-driven simulations for a two-level cache design in open bus systems*", Proceedings of the 17th Annual Int. Symposium on Computer Architecture, 1990, pp 250-259.

[Eggers89] S.J. Eggers and R.H. Katz, "*Evaluating the performance of four snooping cache coherency protocols*", Proceedings of the 16th Annual International Symposium on Computer Architecture, 1989, pp 2-15.

[Futurebus89] Futurebus+, Logical Layer Specifications, Draft 8.1, P896.1 Working Group of the IEEE Computer Society, December 1989.

[Hennessy90] J.L. Hennessy and D.A. Patterson, "*Computer architecture: a quantitative approach*", Morgan Kaufmann Publishers, Palo Alto, California, 1990.

[Hill87] M.D. Hill, "*Aspects of Cache Memory and Instruction Buffer Performance*", Ph.D. Thesis, Univ. of California at Berkeley Computer Science Division, Tech. Rep. UCB/CSD 87/381, November 1987.

[Muller90] H.L. Muller, "*Evaluation of a communication architecture by means of simulation*", Proceedings of the PRISMA workshop on Parallel Database Systems, Noordwijk, The Netherlands, September 24-26, 1990.

[Sweazy86] P. Sweazy and A.J. Smith, "*A class of compatible cache consistency protocols and their support by the IEEE Futurebus*", Proceedings of the 13th Annual International Symposium on Computer Architecture, 1986, pp 414-423.

[Vernon89] M.K. Vernon, R. Jog and G.S. Sohi, "*Performance Analysis of Hierarchical Cache-Consistent Multiprocessors*", Performance evaluation 9, 1989, pp 287-302.

Efficient Global Computations on a Processor Network with Programmable Logic *

Jean Marie Filloque

LIBr-ENST Bretagne, Kernevent-Plouzane, 29285 Brest, FRANCE

Eric Gautrin

IRISA-INRIA, Campus de Beaulieu, 35042 Rennes Cedex, FRANCE

Bernard Pottier

LIBr-UBO, UFR Sciences, av. Le Gorgeu, 29287 Brest, FRANCE

Abstract

A new parallel MIMD architecture is described each node of which is tightly coupled to a global programmable logic layer. This layer gives local acceleration to the node processors by massive micro-grain parallelism. It also provides fast computation services to distributed algorithms by synthesis of global dedicated units operating directly on node operands. As a result, fine approximations of global states become transparently visible in each node, in contrast with usual difficulties and delays in sharing and computing control data.

This point is emphasized by the description of two parallel virtual time mechanisms. The first study involves increasing virtual clocks, and the second one takes into account time counter overflows in a time warp environment. Implementations are based on global systolic networks, fed by the array of local operands and controlled by a small automaton. Thus, global states are handled for each cycle of the mechanism, and results become visible after one pipeline delay with no cost for the accelerated parallel machine.

To summarize general characteristics of this architecture are: general purpose, reconfigurability, cheapness, extensibility.

Introduction

Conventional MIMD architectures are usually built using Von Neumann processors, communication links, or external addressing facilities. Machines belong to the *shared memory class* or the *distributed memory class*, depending on the way the nodes communicate. Algorithms must take care of global computation states for various purposes: termination or deadlock detections, calculation convergences, minimum of local stamps, ...

*This work is supported by *Région Bretagne* and *Municipalité de Brest*. The Armen machine implementation is supported by *ANVAR*.

Shared memory multiprocessors can use shared variables to compute these conditions. Distributed memory computers must implement periodic visits of the whole network to calculate global flags or values. In each case, control operations are merged with computation operations. Periodic accesses to shared variables from the whole network, processor and communication links preemption and network delays raise barriers to the repetition of such operations.

Thus, it appears that MIMD machines cannot execute simultaneously control and computation operations. Moreover, it is evident that the computation bandwidth is dependent on the frequency of control: MIMD machines cannot provide adequate global state visibility.

As an example in the distributed simulation field, the *Time Warp* algorithm requires that the minimum of local *Time Stamps* will be available to garbage memories. Bellenot[2] has reported that such an operation takes about 150ms on a 32 nodes hypercube with the application being stopped.

Beside the strong interest for fast and flexible interconnection networks, a very attractive hardware support in an MIMD machine could be some global unit receiving data from node interfaces and sending computation results back to these nodes. To cover large fields of applications, it is necessary to define such a unit with a technology allowing deep modifications of its behavior. Recent advances in reconfigurable logic technology have given the opportunity to investigate all kinds of global hardware supports to accelerate control and computation in parallel architectures.

This paper presents the architectural concept of *global reconfigurable coprocessors* for MIMD machines. For this purpose, local *reconfigurable logic sockets* are added to each node and connected together to build a *linear logic layer*. The topology of the parallel machine does not need to be specified, but there is a requirement for some stable primary communication services. To get additional hardware support, the operating system must synthesize services into the logic layer. This task is achieved by sending first configuration specifications to each node, and then writing them into the configuration memory of the local socket. Delays for this last task are currently from 0,1s to 1s, and this process can be repeated and interleaved with execution.

The addition of a reconfigurable logic layer to an MIMD machine has two strong advantages with respect to technology and architecture. First, reconfigurable logic *is* an integration technology and allows very efficient circuits to be synthesized and used. Second, the coprocessor has a strategic position in the MIMD machine. It is strongly tied to each node of the machine, but conserves properties of a dedicated centralized functional unit. It can improve intensive computations as a local accelerator, or distributed computations as a global coprocessor.

The objectives of this paper are twofold:

1. the configurable layer use is illustrated and demonstrated by a description of two Global Virtual Time coprocessors for distributed algorithm support. It is shown by these simple examples that inefficient software tasks can be improved in a smart way by the reconfigurable layer.

2. an original algorithm is proposed where a global controller is synthesized to synchronize the nodes periodically. The period is an application dependent tunable constant.

The paper is organized as follows:

- The first part is a short description of architecture principles. The general coprocessor status is emphasized by the notion of dedicated synthesized architecture.
- In the second part we introduce some notions from the distributed simulation field, and describe an implementation of global virtual time computation on the configurable layer.
    - A minimum approach is first investigated without any attempt to manage the time counter overflow. The synthesized service can be used to prevent mutual drift between the logical local clocks.
    - A second approach is virtual time management processes on user-specified time slice boundaries.

Information relative to a practical implementation and the whole project is given, and we conclude by general considerations and fields of application.

1 Accelerated parallel architecture

1.1 Node architecture

The proposed architecture principle involves a general purpose parallel machine with a shared or distributed memory, and a complementary global synthesized coprocessor. Figure 1 shows a node with a processor, local memory and a configurable socket. The socket interconnection has a ring topology.

The socket can be implemented with a large commercial reconfigurable logic array providing at least three data ports to the local system bus and the two adjacent sockets. The local interface of the socket is connected to the processor interrupt and arbitration signals, as well as to local memory control lines. Access to the configuration memory of the socket is mapped into the processor address space, and normal memory processor transactions are passed to the socket logic to be internally interpreted. Therefore the socket can be seen as a second processor rather than a slave unit.

1.2 Synthesized coprocessors properties

Coprocessors can operate on control information, instructions and data[15]. They can be synthesized to yield three distinct classes of computation:

- local coprocessing, examples of which are : instruction set emulations, data intensive algorithms, support for heavily used functions.
- massive parallel computations on the reconfigurable layer which may be used as a large operator controlled and fed by the parallel machine. Two fields of application are systolic signal processing and cellular automata.
- architectural support for global control of the parallel machine. Expected applications are load balancing, fast termination detection, global synchronization and virtual clock support.

The two following paragraphs describe the architecture's design and properties of the reconfigurable logic technology respectively.

Sockets are tightly coupled to node processors and embedded into the reconfigurable layer, thus providing local interfaces to global synthesized operators. Data are directly

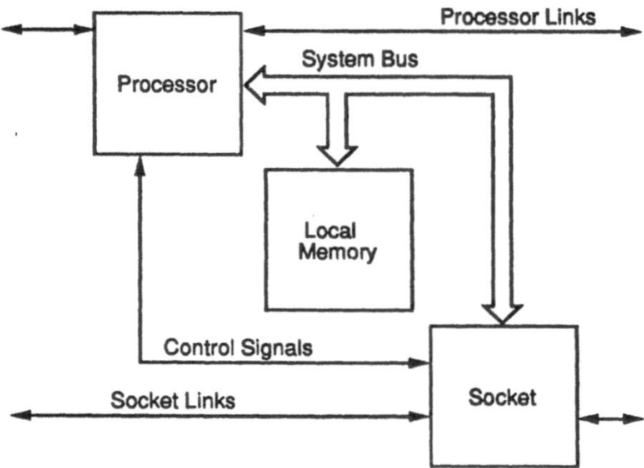

Figure 1: Node Architecture

used as operands for calculations, avoiding the very heavy control process of an MIMD machine, where objects must be carried from node to node to be processed on Von Neumann processors. Therefore there is no longer a bottleneck from bus or network contentions, and there are no prohibitive delays from transport layers. Global intensive computations can be achieved on networks of self-synchronized operators, possibly controlled by a small automaton on a special node. This property applies either to highly regular algorithms or to the control support of irregular applications. This last point has a predecessor in the *Fetch-And-Op* primitive operation of the Ultracomputer [4], where a dedicated network provides a global service with implicit mutual exclusion. Another accurate comparison is the systolic architecture, which is known to minimize regular application control by allowing fast communication between interconnected processors. In a similar way, reconfigurable coprocessors provide tight coupling between nodes and global services, thereby greatly improving control and synchronization for irregular distributed applications.

Von Neumann processors use fixed size general operative units, sequence test and execution, reject constants into data or program space. Accessing data involves the use of memory tables or register files. In contrast to these processors, a synthesized operative unit matches the operand size, implements test and execution in parallel and integrates temporarily stable data into the operators. Memory tables can be mapped into internal trees with very fast access time. As a result, synthesized logic efficiently implements massive micro-grain parallelism. Previous work from various authors has taken advantage of these properties to allow considerable speedups for many applications like image processing, encryption, data compression, long integer arithmetics [1, 8, 14, 18]. The proposed architecture will obviously benefit from the technology, enlarging node fields of application.

The following section shows the need for development activities to build synthesized dedicated architectures on very general purpose hardware.

1.3 Coprocessor development model

Considering a conventional working application we can distinguish three logical components:

3	Application software
2	System support
1	Machine hardware

Each layer in the machine brings services and constraints to the upper layers. As a result application software must deal with the characteristics of underlying components. A common alternative to general purpose machines are specialized ones with the following problems: (i) small market segments involve higher costs, (ii) hardware and software investments are more difficult to preserve. The programmable logic layer architecture introduces an additional flexible component into the usual decomposition, allowing temporarily specialized machines to be synthesized on general hardware :

4	Application software
3	System support
2	Programmable logic layer
1	Machine hardware

Such a machine inherits properties of the conventional initial hardware, because of the transparency of *layer 2*. The behavior of a specialization is similar to the addition of optional arithmetic or dedicated[1] coprocessors to existent machines. It becomes possible to obtain efficient dedicated services by rejecting some difficult points of software implementation into programmable logic. Another point of interest is that new applications are more independent of technology : a specific configuration is defined to match the problem exactly, and the influence of processor integration advance is minimized.

Layer 2 definitions come from creations of *configuration files*. This is currently a CAD activity, similar to peripheral driver writing at operating system level, but there is room to design more dynamic schemes. The next section will show two virtual time services for distributed algorithms. It is envisioned that such services can be part of independent resource libraries to be released for application developments. Speed-ups and additional supports are two benefits from the proposed architecture on the quantitative and qualitative sides.

2 A first example of a global virtual time service

Introduction

Distributed systems with pure message passing communication usually use *logical clocks* to timestamp events and messages used to bring them from one process to another. Lamport has shown in [12] that it is possible to construct a total order over their occurrences by using strictly increasing counters, incremented on each emission and updated at reception. So, throughout the system, reception always occurs *after* emission.

In the distributed simulation domain, the problem of virtual clocks (another name for logical clocks) is a little bit more complicated because these clocks are not completely unrelated. Jefferson [6] has proposed the paradigm of *virtual time* that coordinates execution with an imaginary virtual clock. *Virtual time* represents global information and each

[1]An example is the coprocessor board for parallel simulation proposed in [3]

site can have only an approximation of it. *Virtual time* can be implemented with either a pessimistic or an optimistic approach. With the first, a process on a site can safely increase its local clock only if it is sure that it will receive no message in its past. The respect of this *causality* constraint may lead to deadlock. This approach is presented in [13]. It consists in avoiding or resolving deadlocks. Local virtual clocks, as well as *virtual time* never decrease. The second approach assures only the growth of *virtual time* but not of local clocks. So it allows rollbacks in the past to occur on a site. This is described in [5, 6].

It is, a priori, impossible to have a consistent view of global state and time in such an environment without a shared memory and a common clock. So, processes must content themselves with a best possible approximation of this global information. The construction of a global time approximation is proposed by several authors. It consists in a steady evaluation of a lower bound of all the local clocks in the network. This approximation is used to prevent mutual drift between logical clocks like in [16], to update queues and to avoid memory saturation in time warp systems like in [5, 17], to estimate load ratio of processors for load balancing... This type of computation is suitable for implementation in the programmable layer of the machine, and the following sections describe two applications used to support this assertion.

2.1 A global computation for increasing virtual time

The goal of this section is to emphasize the use of basic mechanisms to build coprocessors. This presentation is driven by the example of a global computation for increasing virtual time. A coprocessor is synthesized to swiftly calculate a lower bound of all *Local Virtual Time* with a circulating token and so, to deliver either this bound (a fine approximation of the global virtual time) or an upper limit for message emission timestamps to each node . This limit may also be evaluated in the programmable layer. It is a simple addition with a constant.

For the sake of simplicity, this first proposal does not manage time counter overflows. The coprocessor must compute a *Global Virtual Time* as the minimum of each node *Local Virtual Time.* For the following, let us define *GVT* to be an evaluated *Global Virtual Time*, and *LVT* to be a node *Local Virtual Time.*

2.1.1 Coprocessor Architecture

To achieve global evaluation, the coprocessor will receive local data, like node LVTs, and send back results, like a GVT. Transparency of coprocessor parallel services is given by *asynchronous channels* with nodes. These channels are implemented with double-register directional mechanisms connecting the coprocessor to a node. The coprocessor periodically reads or writes channels while node processors execute less intensive write or read operations respectively.

An asynchronous channel from coprocessor to node works as follows. The coprocessor is always allowed to write its own register, and the node to read its own register. Data are transferred from a coprocessor register to a node register when the node does not execute a read operation. In our example (see figure 2), the interface consists of two asynchronous channels:

- **LVT**: from the node to the coprocessor;
- **GVT**: from the coprocessor to the node.

To obtain fast computation cycles, the coprocessor has a pipeline topology in which one stage is associated to one socket. Partial results between stages are embodied in

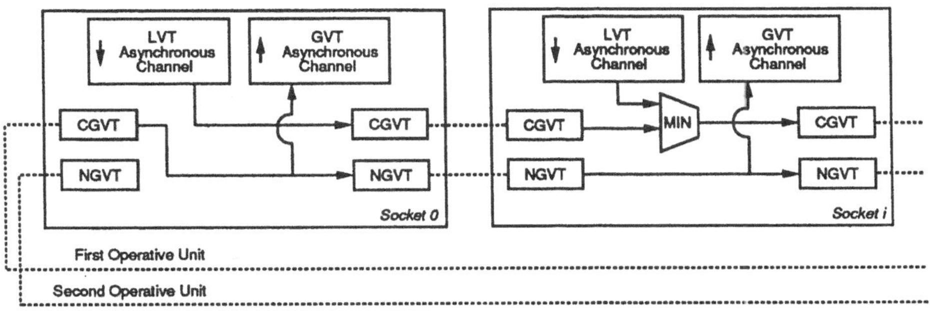

Figure 2: Socket Internal Configuration

so-called *tokens*. Token communications are asynchronous. After completion of its task, a socket writes a modified token to its righthand neighbor.

The coprocessor pipeline architecture is composed of two parts:

- A *large operative unit* distributed across every socket. This unit executes a systolic computation on an array of values from asynchronous channels. Results are fed back to the pipeline head.
- A *control unit* implemented at the pipeline head. This controller is in charge of the initialization of the operative unit, and the token generation. It also receives computation results from the operative part.

In practical implementation, the control unit and the first operative unit stage can be merged on the same socket. Furthermore, the control unit automaton can drive several operative units.

2.1.2 Coprocessor Service

The coprocessor service is defined by two successive global operations:

$NewGVT := min_{i=1..n}(LVT_i)$, where LVT_i is defined to be LVT of node i.
$GVT_i := NewGVT \; for \; i = 1..n$ where GVT_i is defined to be GVT of node i.

Each operation is implemented by an operative unit. The first one completes the systolic computation of the minimum by passing the $NewGVT$ result to the control unit. The second operative unit broadcasts the $NewGVT$ value to every node.
Partial results of the two operative units are embodied in a single token:

Computed Global Virtual Time (CGVT) : partial $NewGVT$ value.
New Global Virtual Time (NGVT) : broadcasted $NewGVT$ value.

The control unit and the first operative unit stages are merged, as shown in *socket 0* of figure 2. It can be seen that the first operative unit is initialized with the LVT value of its socket asynchronous channel at each pipeline cycle. The control unit also feeds the input of the broadcast unit with the output of the minimum computation.

This figure also shows, in the block at socket i , the parallelism between the local minimum and broadcast operations which overlap with one pipeline latency.

3 A hardware service for Time Warp Simulation

This section gives a brief description of *Time Warp* principles as defined in [5] and shows the interest of knowing *Global Virtual Time.*

In the *Time Warp*, all processes are independent and there is no constraint on their asynchronous evolution. Each message is timestamped with the addressed simulation time, t_r. If the LVT of the receiver is already higher than t_r when reception occurs, then the process must roll back to time t_r, and must undo actions between t_r and LVT. All the messages it has sent must be unsent using *anti-messages.* The roll-back mechanism imposes that processes retain the history of states and lists of all messages sent and received. Maintaining all previous information obviously requires an unbounded amount of memory. It has been proved that there is a lower bound on virtual time which the system will never roll-back to [17]. Knowing this lower bound, it is possible to forget older information. This time is called *Global Virtual Time* and is defined as $GVT = min(LVT_i, t_{ri})$, where t_{ri} is the timestamp of message not yet received. GVT must be computed regularly and generally freezes the simulation progress for *one network diffusion time at least*[17]. Notice that LVT has not the same signification as in the previous section: here, LVT is the minimum of all timestamps of one node.

3.1 Algorithm presentation

To implement the Time Warp Simulation, the nodes need to know the Global Virtual Time. Our previous example presents two restrictions: first, there is no provision for a roll back mechanism; secondly, it does not consider time counter overflows. In this section we present a practical solution taking into account these restrictions.

Instead of computing the GVT, this approach tracks a condition where all nodes have overtaken an LVT bound. When this condition is verified, an approximation of GVT has occurred, and then memory garbage collection is possible. To minimize simulation process freezing on memory saturation, the application must tune the GVT progress intervals to deal with node memory capacities and application characteristics. For the sake of simplicity, the intervals between bounds are equal and the same for all processors.

The configurable coprocessor will compute this condition of a global bound overtake.

3.2 Node Message Passing

On message reception, an advanced process can roll back its LVT to a time less than the next bound to overtake. The computation of the condition must take care of unreceived messages. This problem is solved by message acknowledgment.

Each node is supposed to have its current simulation time, and two queues for input and output messages. The current LVT is deduced from the minimum of all time stamps on the node including messages in the input and output queues [5]. Message deletion from an output queue requires an acknowledgment from the communication service to ensure the visibility of the minimum LVT on the coprocessor.Thus, if the bound is not overtaken, it guarantees that there is at least one node which discards this state.

Message passing from node A to B must respect the following protocol:

Node A sends a message from its output queue;
Node B receives this message then places it in its input queue;
Node B computes its new LVT;
Node B sends an acknowledgment to Node A;
Node A receives the acknowledgment;
Node A deletes the message from its output queue;
Node A computes its new LVT.

3.3 Global Condition Computation

A global computation is an operation on an array of values distributed on every socket. This operation can not be instantaneous because of propagation delays. In the previous proposition, local virtual clocks are strictly increasing. So, GVT is evaluated in a systolic way.

In Time Warp Simulation, the condition of a global bound overtake can be expressed as follows. Let us define GO as the Global Overtake condition, and NB the next bound to overtake.

$$GO = min(LVT_{i\ i=1..N}) > NB$$

This expression could be calculated in a systolic way by a distributed operative unit, where PO is defined as a Partial Overtake :

$$PO_0 = true;$$
$$PO_i = PO_{i-1} \ and \ (LVT_i > NB);$$
$$GO = PO_N;$$

Note that the comparison $(LVT_i > NB)$ can be carried out by the node. Only the boolean result is discarded to the socket through a flag O (*Overtake*). So, the operative unit computes the boolean product of the flags O in a systolic way.

With the roll back effect, local virtual clocks are not strictly increasing. A simple systolic computation can provide an erroneous result. The following sequence on message passing from node i to node j illustrates this problem:

Initial conditions: $i > j$; $LVT_j > NB$; $LVT_i < NB$;

1. Computation of PO_j;
2. Node j receives a message from node i producing a roll back, and decreases LVT_j such $LVT_j < NB$;
3. Node i, receiving the acknowledgment from node j, updates LVT_i such $LVT_i > NB$;
4. Computation of PO_i.

In this example, the operative unit delivers a true GO value to the control unit, but $LVT_j < NB$. Note that at least one of the N following GO will discard a false value.

Property: If N consecutive true GO values are received, the condition of a global bound overtake is true.

Proof: A proof by contradiction can be given. Suppose $W(c)$ is true and $\exists\ O_i(c)$ false with i in $1..N$.

Note $O_i(c)$ the boolean value of flag O of node i at pipeline cycle c.

Then:

$$GO(c) = \bigwedge_{i=1}^{N} O_i(c - N + i - 1)$$

And $W(c)$ which is equal to

$$\bigwedge_{\gamma=1}^{N} GO(c - \gamma)$$

can be rewritten as :

$$W(c) = \bigwedge_{\gamma=1}^{N} \bigwedge_{i=1}^{N} O_i(c - \gamma - N + i - 1)$$

so

$$\neg O_i(c) \Rightarrow \exists \neg O_j(c_\delta) \mid c_\delta \in]c - N, c]$$

which is in contradiction with $W(c)$.

To detect this condition two implementations are proposed: when detecting a first true GO, the control unit can either push a marker CO (*Confirm Overtake*) into the pipeline through an operative unit and wait for its return, or count the pipeline cycles to ensure a total dump. The first solution is chosen for the sake of simplicity.

Systolic arrays cannot exactly implement a computation over the array of operands because of the technological depth limitation of reconfigurable sockets. Each token does not operate on simultaneous sample when circulating in the operator. Therefore it is necessary to observe full pipeline results to get accurate conclusions about what has occurred one pipeline delay before.

3.4 Coprocessor Behavior

The general behavior of the coprocessor can be described in three stages:

1. Tracking N consecutive true GO values;
2. Broadcasting the condition of the bound overtake to every node;
3. Waiting for a global acknowledgment from every node.

To implement the second stage, the control unit can push a marker W (*Wave*) through an operative unit and wait for it to come back.

Assuming that each node acknowledges the coprocessor through a flag A, the third stage can be implemented with an operative unit which calculates the boolean product of all the flags in a systolic way. Note that this computation is carried out in a single systolic pass.

In conclusion, the coprocessor consists of four operative units:

PO : computes the boolean product of flags O in a systolic way.
CO : pushes the marker CO through the pipeline.
W : broadcasts the condition of the global bound overtake.
PA : computes the boolean product of flags A in a systolic way.

and a control unit feeding the pipeline with tokens. Figure 3 illustrates the control unit automaton. The transition conditions are flag values from input tokens.

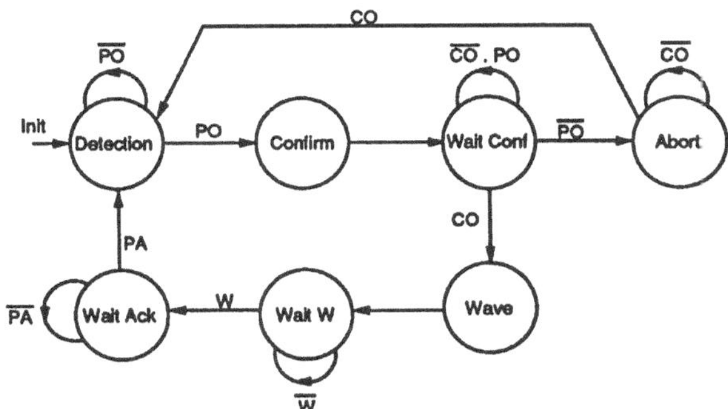

Figure 3: Node 0 Automaton

The token structure is:

PO : the partial boolean product of previous node flags O.
CO : a boolean marker to indicate the *Confirm* condition.
W : a boolean marker to indicate the *Wave* condition.
PA : the partial boolean product of previous node flags A.

The control unit can be implemented in the socket 0. The values for the output token of the control unit are deduced from the automata state:

Detection	: (PO = node.O,	CO = false,	W = false,	PA = node.A)
Confirm	: (PO = node.O,	CO = true,	W = false,	PA = node.A)
Wait Conf	: (PO = node.O,	CO = false,	W = false,	PA = node.A)
Abort	: (PO = node.O,	CO = false,	W = false,	PA = node.A)
Wave	: (PO = node.O,	CO = false,	W = true,	PA = node.A)
Wait W	: (PO = node.O,	CO = false,	W = false,	PA = node.A)
Wait Ack	: (PO = node.O,	CO = false,	W = false,	PA = node.A)

The operative units are distributed on each socket executing the following operations on the tokens:

tokenOut.PO = tokenIn.PO *and* flag O
tokenOut.CO = tokenIn.CO
tokenOut.W = tokenIn.W
tokenOut.PA = tokenIn.PA *and* flag A
if tokenIn.W *then* reset flag A

This second example illustrates the use of the coprocessor to compute global conditions in a systolic way by a simple pass, or a pipeline dump. Moreover, the coprocessor controls and sequences actions over the whole network, like broadcasting a condition to every node. Implementation obviously requires very few logic resources, giving way to additive functionalities.

4 Further work : the Armen project

An implementation of the reconfigurable logic layer parallel architecture is currently being built by the LIBr[2]. An MIMD experimental machine called *ArMen* has been designed to investigate most of the capabilities of the architecture. An INMOS T800 has been chosen as the processor node and a Xilinx 3090 LCA [19] as the reconfigurable socket. This leads to small and affordable modules where the socket can operate on addresses, instructions and data from the 32-bit processor multiplexed bus. This is not the most powerful design one could create today, but it is sufficient as a test vehicle. On the other hand, no commercial processor exactly matches our requirements, and there remains a real problem in that we cannot experiment on processor to processor exchanges with the first machine.

Another goal of the ArMen project is to build a software environment for the architecture. Applications can be either specific or general. In the first class, signal processing with use of generic tools, fixed global services like virtual clock support and cellular automaton are considered. Support for these applications can be currently designed as parallel programs and configuration file libraries. The second class of applications is a challenge involving the production of High Level Language development tools for coprocessor synthesis. Given a coprocessor model involving a *node 0* automaton, regular pipelined operative parts and standard interfaces into the node, it is expected that coprocessor generations could be considerably facilitated.

We have shown that the proposed architecture with its accelerated network layer is able to compute global information all over the system with low time cost[3]. It is of obvious interest for the efficient implementation of many applications requiring multiprocessor computation like large logic simulations, signal or image processing, etc...

More generally, one can take advantage of the active communication layer to implement every algorithm requiring global knowledge computation. This can be done very simply on the hypothesis of always empty communication channels (of the application layer). Another use of the active layer can be found in the out-of-band communication between sites. Urgent messages can be routed via this layer by a token containing data and destination. The token can be used either for point-to-point communication, partial diffusion (with an associated list) or complete diffusion.

These points are currently being studied in the framework of distributed discrete event simulators.

5 Conclusion

The proposed architecture complements current parallel designs on many levels.

A first important property of configurable logic is its ability to synthesize small dataflow sequenced operators, and thus to increase the level of parallelism within the nodes.

The connection of adjacent logic arrays provides a global programmable logic resource, on which very large operative parts with arrays of input/output ports are implemented. These ports handle the whole state of the machine repeatedly by feeding systolic arrays with it. We have shown some internal points of the coprocessors, with *global operators* which are small automata controlling systolic linear parts, as well as asynchronous channels and interrupt waves to interfere with node behavior. These tools are useful in computing global resources, or controlling the whole network behavior.

---

[2]Laboratoire d'Informatique de Brest is a common structure to *Université de Bretagne Occidentale* and *Ecole Nationale Supérieure des Télécommunications de Bretagne*

[3]pipeline delays are in the order of 50ns

Accessing global conditions over a distributed system has often been considered to require heavy local computation and communication or synchronization tasks. Pure distributed implementations can fail[7] because of the inefficiency of these mechanisms: communications and local computations are involved in calculating results which must be dispatched back to the nodes. It is expected that the logic layer architecture will encourage the use of efficient global services within MIMD machines for distributed systems, languages or algorithms.

References

[1] P.Bertin, D.Roncin, J.Vuillemin, "Introduction to programmable active memories", in *Systolic Array Processors*, Prentice Hall, pp. 301, 1989.

[2] S. Bellenot, "Global Virtual Time Algorithms", in *Proc. of the SCS multiconference on Distributed Simulation*, San Diego, Californie, pp. 122, 19 January 1990.

[3] C.Buzzell, M.J.Robb, R.Fujimoto, "Modular VME rollback approach for Time Warp", in *Proc. of the SCS multiconference on Distributed Simulation*, San Diego, Californie, pp. 153, 19 January 1990.

[4] A.Gottlieb and al., "The NYU Ultracomputer — Designing an MIMD shared memory parallel computer", in *Proc. International Conference on Computer Architecture*, ACM, pp. 175. 1982.

[5] D.Jefferson, H.Sowizral, "Fast concurrent simulation using the time warp mechanism", in *Proc. of the SCS Conference on Distributed Simulation*, San Diego, pp. 63-69, Jan. 1985.

[6] D.Jefferson, "Virtual Time", *Transactions on Programming, Languages and Systems*, ACM, vol. 7, no. 3, pp. 404, 1985.

[7] R.Fujimoto, J-J.Tsai, G.Gopalakrishnan, "Design and performance of special purpose hardware for Time Warp", in *Proc. of International Symposium on Computer Architecture*, IEEE, pp. 401, 1988.

[8] T.Kean, J.Gray, "Configurable hardware: two case studies of micro-grain calculation", in *Systolic Array Processors*, Prentice Hall, pp. 310, 1989.

[9] B.W. Lampson, K.A.Pier, "A Processor for a High-Performance Personal Computer", in *Proc. of Computer Architecture Symposium*, IEEE-ACM, pp. 146, 1980.

[10] H.T.Kung, "Why systolic architectures?", *IEEE Computer*, vol. 15, no. 1, pp. 37, 1982.

[11] H.T.Kung, "Network-based multicomputers : redefining high performance computing in the 1990s", in *Proc. of the Decennial Caltech Conference on VLSI*, The MIT Press, pp. 49, 1989.

[12] L.Lamport, "Time, Clocks, and the Ordering of Events in a Distributed System", *Communications of the ACM*, vol. 21, no. 7, pp.558, 1978.

[13] J.Misra, "Distributed Discrete Event Simulation", *Computing Surveys*, vol. 18, no. 1, pp. 39, March 1986.

[14] B.Pottier, D.Lavenier, "High rate sigma filtering, feasibility studies on processor networks", in *Proc. of IFIP Workshop "Parallel architectures on Silicon"*, INP Grenoble, pp. 182, 1989.

[15] B.Pottier, "Machines parallèles à accélérateurs reconfigurables", *Thèse de l'Université de Rennes 1*, Dec. 1990.

[16] M. Raynal, A distributed algorithm to prevent mutual drift between N logical clocks, *Information Processing Letters*, vol. 24, no. 3, pp. 199-202, Feb 1987.

[17] B. Samadi, "Distributed Simulation, Algorithms and Performance Analysis", PhD. Num 8513157, University of California, Los Angeles, pp. 35-64, 1985.

[18] J.Viitanen, T.Kean, "Image pattern recognition using configurable Logic Cell Arrays", in *Proc. of Computer Graphic International '89*, Springer-Verlag, pp. 355, 1989.

[19] Xilinx, The Programmable Gate Array Data Book, *Xilinx*, San Jose, 1990.

POMP* or How to design a massively parallel machine with small developments

Philippe Hoogvorst Ronan Keryell Philippe Matherat
Nicolas Paris

—

Laboratoire d'Informatique de l'Ecole Normale Supérieure
URA 1327 CNRS
45 rue d'Ulm, 75005 PARIS
Tel: (+ 33 1) 43.26.58.85, fax: (+ 33 1) 46.34.05.31.
E.mail: ...@dmi.ens.fr
...@frulm63.bitnet

—

Abstract

The design of a SIMD machine is usually complex because it leads to developping an efficient Processing Element and to writing all the softwares required by the chip and the control of the machine. We propose a different approach by using an efficient 32-bit off-the-shelf processor with its software environment (compiler and assembler) and a programmable gate array for the network. It limits the development to the minimum and leads to a rather general SIMD cluster built with off-the-shelf chips which can be considered as a SIMD transputer.

1 Motivation

In this article we propose a methodology for the development of a SIMD machine. The philosophy of the development consists in minimizing the development effort. The excessive complexity of parallel machines is probably the major cause of failure in academic projects. The first quality of a machine is its existence at the end of a project. In this article, we show that it is possible to develop a coarse-grain SIMD machine that offers good performance with very little effort on both hardware and software aspects.

Even if the specificity of this article is to show development reduction, it is important to explain why we have decided to develop this kind of machine. Our main field of interest is image synthesis.

**un Petit Ordinateur Massivement Parallèle*: a small massively parallel computer. Project supported by the French Ministry of Research and Technology, in collaboration with Thomson Digital Image.

2 Why choose SIMD for image synthesis?

Commercial machines for image synthesis are often very specialized with dedicated hardware to speed up the computation of a single algorithm [JHH80,AJ88]. This specialization is the major drawback of this approach. Machines become rapidly obsolete because new rendering algorithms require ever hardware. Only large companies are able to invest large amounts of money and man-power to develop custom machines that will be obsolete in a few months' time. We propose POMP as a non-specialized architecture (with no hardware dedicated to any special algorithm), which is a step beyond the other alternatives (partially non-specialized) proposed in [KV90,FPE*89].

We have to balance the loss of power due to this non-specialization by a massively parallel approach (up to 256 32-bit processors, in fact 8,192 bits of data-paths). This massively parallel organization prohibits the organization in a multiprocessor with shared memory. Each processor has its own local memory and an interconnection network enables data interchange.

This class of architecture contains 2 major subclasses:

- The MIMD machines. Each processor runs its own program on its own data.
- The SIMD machines. Each processor executes the same instruction at the same time on its own data. We do not need a program memory for each processor.

In the graphic pipeline, the last stage is rasterization, which requires most of the computation. A SIMD structure offers the best performance on these computation. [FP81] introduces the concept of *smart memory* which are a set of SIMD memory-PE[1] clusters.

The POMP project tries to generalize this *smart memory* concept to all the algorithms of the whole pipeline. We need for each cluster a general purpose PE, which is able to handle 32-bit integer numbers, floating point numbers, pointer data types, etc.

We also prefer the SIMD structure because a lot of synchronization problems are avoided [BCJ89] and a high MFlop/dm^3 ratio can be reached [BDW85]. Furthermore we can build a very simple programming model which enables to develop debugging environments.

3 The basis of the machine: the programming model

The efficiency and the programmability are the final targets of computer designing. The relationship between hardware and software is the main problem. Most of our choices for the architecture of POMP are consequences of the programming model.

3.1 The programming model

Variables belong to two classes:

- scalar variables (for standard calculation and flow control),

[1]Processing Element.

- parallel variables (also called vectors).

An n-PE SIMD machine is able to simultaneously perform the same operation on a vector of size n. Some SIMD programming environments try to hide the number of processors behind the concept of virtual processors (for instance the Connection Machine). The size of massively parallel variables is assumed to be larger than the number of PEs. Each physical processor emulates one or more virtual processors (*vp*). Vectors are not broken into individual elements but into smaller arrays equally distributed over the PEs.

In a typical massively parallel application, vectors of different sizes are required and need to interact. The vectors must be partitioned into classes called *vp-set* for the CM and *collection* for POMPC. Each *collection* corresponds to one set of virtual processors.

The size is the first attribute shared by the vectors of a same *collection*. The other attributes of the *collection* are:

- the activity. This vector of boolean elements (also called *context*) is the mask which indicates which elements of the vectors of the collection are active.
- the topologic organization. These information describes the topologic relative organization of the virtual processors and the mapping of these virtual processors on the PEs.

3.2 The POMPC language

A detailed description of this language can be found in [Par90]. This model has led to designing of a programming language which is called POMPC. POMPC must be considered as a symbolic macroassembler for SIMD machine as is C for general computers. This language is the direct translation of a programming model and emphasizes the SIMD aspect of the machine: we do not provide an autovectorizing language which hides the structure of the machine from the programmer. This kind of higher level languages can be implemented over POMPC.

Most of the SIMD machines provide this kind of basic language and a taxonomy of many SIMD languages and machines can be found in [Tuc90]. POMPC has been inspired by the previous version of C* [Tmc87, pages 35–41] and is rather similar to the new version of this language. MPL [Chr90] and MultiC [Mlc90] are also alike (without the collection mechanism).

To implement this model, we must define at any time what the different processors (the scalar one and the different virtual SIMD machines) are doing. As only the scalar processor has the control over the program flow and as the PEs are slaves, the best way to express this dependency is to include the instruction for the PE into the sequential program of the scalar processor (it leads to the definition of a very simple and very convenient controller explained in the next section).

As we expect to write the addition of 2 vectors like the addition of 2 scalars, the major problem is to determine from the source file the location of each calculation. Typechecking on expressions and statements provides these informations.

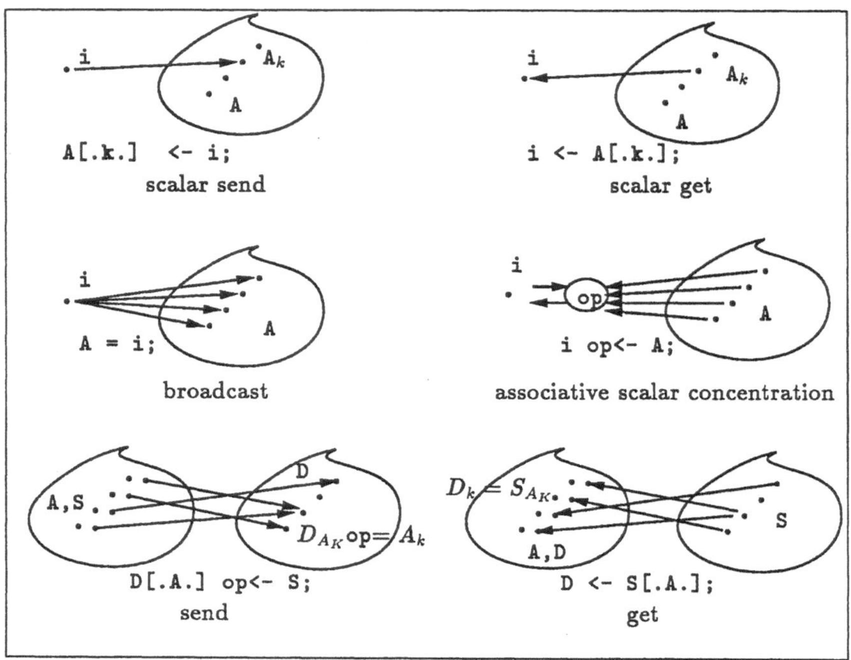

Figure 1: The communications used by POMPC.

POMPC is an extension of the Kernighan & Ritchie C [KR78]. The extensions are as follows:

- It is possible to define collections.
- Each variable can be either a scalar (like in C), a vector of the particular `processor` collection (one datum per PE) or a vector belonging to another collection. Each vector is declared as a member of a collection. Thus it is possible to associate a collection with each vectorial statement or each vectorial expression.
- The `where/elsewhere` operators allow to change the activity of a collection, during the execution of a block. This activity is modified according to the value of a boolean vector of the collection. The `where` statement is the equivalent of the `if` statement except that the block is always executed (it mays contains scalar statements or statements concerning other collections). Every other flow-control statements (even `break`, `continue` and `return` but not `goto`) has been translated for a parallel usage.
- Communications are required to perform non-local interactions. Most of the communications are expressed in the syntax of the language because they require only

```
/************************************************************************
  Mapping a picture on a scrambled surface :
      The view of an underwater chessboard under a dripping tap
************************************************************************/
#include "pompc.h"                    /* pompc standard include file */
#include "pc_math.h"                  /* pompc math include file */
collection [256,256] pixel;           /* pixel is a 2D 256 x 256 collection */

pixel chessboard()                    /* returns a chessboard picture */
{
    pixel x,y;                        /* x,y : the local coordinates */
    x = pc_coord(0);y = pc_coord(1);
    where((x & 16) ^ (y & 16)) return 255;
    elsewhere return 0;
}

main()
{
    pixel char color,picture;         /* two pictures */
    pixel int x0,y0;                  /* local coordinates */
    pixel int u,v;                    /* mapping coordinates */
    int time,screen;                  /* current time and screen number */

    screen = gr_open_graphic();       /* gets a window where to display the movie */
    gr_set_cmap(screen,0,0,0,-1);     /*      sets a standard color map table... */
    color = chessboard();             /* gets the picture of a chessboard */
    x0 = pc_coord(0) - 128;           /* x0,y0 : coordinate system from */
    y0 = pc_coord(1) - 128;           /*      the center of the chessboard */
    for(time=0;;time++) {             /* and let's go forever... */
    {
        pixel double X,Y,d,d1,phi;
        X = x0;Y = y0;
        d = pc_sqrt(X*X+Y*Y);         /* d : distance from the origin */
        phi = time - d/16.0;          /* phi : phase delay */
        where(phi < 0) phi = 0;       /* drop touches the surface at phi=0 */
        d1 = 1 + 8*pc_sin(phi)/d;     /* d1 : new distance from center */
        u = X * d1 + 128;             /* coordinates where to get the */
        v = Y * d1 + 128;             /*      color of the local pixel */
    }
    picture <- [u,v]color;            /* global indirection */
    gr_flash(screen,&picture,0,0,1,1); /* displays the result */
    }
}
```

Table 1: Example of a POMPC program.

standard network specificities, the rest being carried out by library functions.

Figure 1 summarizes the different syntactical constructions for POMPC communications. The first 4 types of communications are interactions between scalar variables and vectorials ones. The 2 last types are interactions between collections. The `[.  .]` operator specifies transformations on the rank of the elements. When a communication may send more than one datum on a given element, a accumulative operator can be specified to accumulate different data in the resulting element. Accumulative operators are addition, subtraction, multiplication, bit-wise and, bit-wise or, exclusive-or, minimum and maximum.

Table 1 shows an example of the POMPC language.

This program computes a chessboard picture (the `chessboard` function) and distorts it (as the deformation of a water surface under a water drip) according to a mapping achieved by a *get*.

4 Architecture of POMP

4.1 Processor designing: a necessary evil?

The first choice during the design of a SIMD machine is the size of the PEs. In fact, this is the first choice because everyone considers that PEs are necessarily custom-made and that we can freely choose the width of the datapath of the PEs.

For some very special applications (mostly image processing), it is interesting to choose 1-bit PE because of the size of the data (from 1 to 8 bits). In order to be efficient these machines require full-custom processors [Gap84]: classical sequential processors are not adapted for this computation, because of the inadequation of the width of the 32-bit processors to 1-bit and 8-bit data.

In the other fields of application for the SIMD (like ours), the required data sizes are more conventional (`int`, `float`, `double`) [Hor82,AB86] and it seems easier and more efficient to use a powerfull processor than to interface a floating-point coprocessor with 1-bit processor, as in the Connection Machine 2. In this last case, 1-bit PEs are no longer used for scientifical computations...

Unfortunately, no commercial 32-bit SIMD PE exists such as the GAPP [Gap84] for 1-bit SIMD machines or the Transputer (an MIMD PE [Inm89]).

We consider that PEs for non dedicated SIMD machines must have the same qualities as classical processors. An intermediate choice could be to design a rather small PE with all the necessary hardware required to micro-code efficiently the floating-point operations (like the MasPar machine [Bla90a]).

The consequences of this coarse-grain choice are important because this seems to suppose the development of a very complex PE. We need very broad competences to be able to design competitive 32-bit PE with floating-point and only large semiconductors companies can cope with such developments. The problem is not limited to chip design but also to the development of all the software environment. Developing the PE is not the good solution.

Let us summarize the requirements for the PE:

- a lot of MIPS: an efficient integer ALU,

- a lot of MFLOPS: an efficient floating-point ALU,
- indirect access to local memory: a data address generator,
- local flow control: a local enable mechanism,
- communications: an efficient network and a routing mechanism.

In fact it is very similar to a classical processor.

4.2 Why not use an off-the-shelf processor ?

Such an approach had already be done for the PASM computer [SSIK84]. The advantages of using a commercial processor are clear:

- we need not develop a PE, it is cheaper and less time-consuming,
- a C compiler is available for our PE,
- we can benefit from every improvement of the processor (this is very important since the speed of RISC processors is regularly doubled inside a common architecture).
- if we remain rather independent of the processor (particularly in the software domain), it is possible to change processors when it appears that a more suitable architecture has been introduced on the market.

We can drastically limit our developments and provide an easy evolution for our machine. The general concept is to choose the best processor at any given time.

Four sensitive points have to be coped with:

- We have to broadcast an instruction to every processor. This is easier if the chosen PE has a Harvard architecture[2].
- We have to keep every PE synchronous. Each instruction must take the same time independently of the data processed. This is mostly the case in RISC processors (as opposed to microcoded processors), provided that all accesses to the memory last the same time. This prohibits the use of PEs with caches.
- We have to independently freeze every PE to process the **where** statement. It is possible if an "instruction not ready" mechanism is implemented on the code bus. This is the case every RISC processor with off-chip cache.
- We have to provide each PE with an access to the network and some facilities to communicate with the scalar processor. It requires special hardware and which will be discussed in the section 5.

In 1991, there exists one processor presenting the required characteristics: the Motorola 88100 [Mot88]. We chose it as the PE of POMP.

---

[2]a special input bus for the instructions.

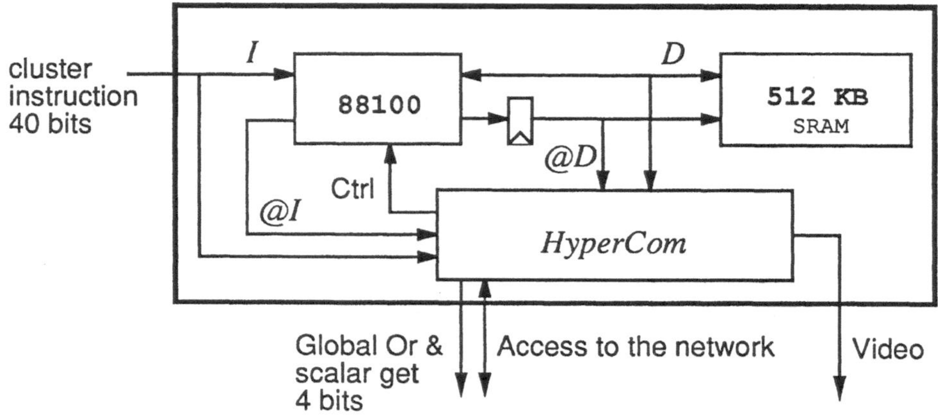

Figure 2: The basic cluster.

4.3 The Processing Element

We can now present the basic cluster for the PE (figure 2). It contains mainly:

- the 88100 at 20MHz (17 MIPS and 7 MFLOPS);
- 128K × 32-bit static RAM with a 35ns access time in 4 chips;
- *Hypercom*, a chip to customize the CPU to its SIMD environment.

The SIMD approach permits us to use only 9 integrated circuits per PE.

A 40-bit instruction bus broadcasts the instructions and the control of the *Hypercom* chip to the cluster. The *Hypercom* provides the following mechanisms:

- the activity management. Depending on the current activity loaded in the *Hypercom*, The control bus defines for each instruction if it is executed. The eventuality of n nested `where` seems to require an n-depth stack to save the current activity. In fact, it can be implemented with a counter [Ker89,Lev90] which is convenient to cope with the complicated activity handling required when using `break`, `continue`, `return`, `case` and `default`.
- the network access. It consists in limited routing capabilities and shift registers. The section 5 is dedicated to the network and will give some indications on the hardware required in each *Hypercom* chip for the access to the network.
- the hardware required for the communications between the PEs and the scalar processor: a 4-bit open-collector bus to get the *global or* of a distributed variable. This also allows to send a vector element to the scalar processor (useful for the

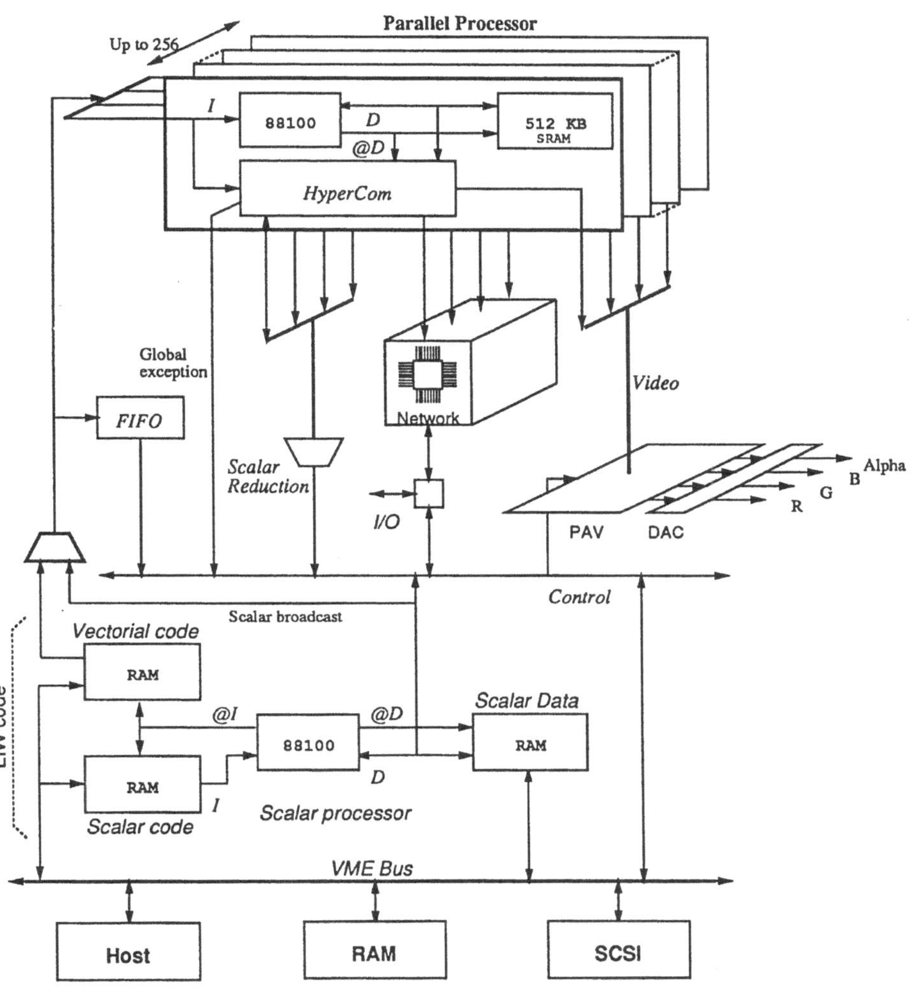

Figure 3: The global architecture of POMP.

scalar get and for the final stage of the associative scalar concentration) by nibbles of 4 bits as in [Bla90a], which is a good compromise.

- the hardware required to correctly recover from an exception or an interruption.

4.4 The controller and the scalar processor

Some SIMD machines use an independent sequencer to run scalar code or to expand microinstructions generated by an host computer [Tmc87]. The use of the host computer as scalar processor facilitates the software development but requires high input/output bandwidth for the broadcast of the code from the host to the PEs. It is possible only if an intermediate sequencer expands some high-level instructions into microcode (typically 32-bit instructions expanded into 1-bit microinstructions, when 1-bit PEs are used). We cannot use such a structure because we need a 20 MHz 40-bit instruction rate. The scalar processor must be directly located in the SIMD machine. This choice has been made in [Bla90a,AB86].

As we use a commercial processor for the PE, it is natural to use the same processor as scalar processor for code orthogonality and for easier synchronization between PEs and the scalar processor. It simultaneously fetches its own 32-bit instruction and the 40-bit instruction broadcasted to the clusters. The whole machine is driven by a 72-bit Long Instruction Word (*LIW*). The figure 3 presents the global architecture of POMP.

An history of the cluster instructions is saved in a FIFO: when an exception occurs, the PEs can correctly resume execution. The scalar processor can access a register to override some fields of vectorial instructions, enabling scalar broadcasts of values.

Since the most important argument claimed by the SIMD defenders is the removal of synchronization issues, we think that the implicit LIW is a way to go further in the synchronization of scalar code with the parallel code, allowing a more global code optimization.

5 The interconnection network

Choosing a network consists in choosing the best trade-off between performance and cost for a given class of applications.

5.1 Measuring the performance

Applications may require different communication types:

- random access,
- 1-neighbour access,
- all neighbours according to a multidimensional mesh,

and different object granularities for the network:

- size of the packets (1 bit to 1 Kbit),
- number of physical processors,
- number of virtual processors per physical processor (*vp-ratio*).

To measure the needs of a target application, we have to evaluate the occurrence of every combination above. A unit system is required for such measurements to compare the performance of the network with the performance of the PEs. We have decided to speak in terms of:

$$\frac{\text{time required for the communication of 32 bits for each virtual processor}}{\text{time required for the addition of 32 bits for each virtual processor}}$$

5.2 Measuring the cost

The global cost of the machine depends on the cost of the PEs and the cost of the network. The latter is not easy to evaluate because it is not a linear function of the performance. This cost grows by step when the implementation must move from one technology to another at different hierarchical levels [FWT82]:

- the number of transistors required for the network by each PE,
- the number of pins required by each PE,
- the density on each motherboard (the number of routing levels on motherboards),
- the number of connections between motherboards.

5.3 Choosing the Network

Many network designs have been described in the literature and can be classified according to some criteria such as operation mode, control strategy, switching method and network topology [Fen91,Gil86,Kot87].

In our case, the network is synchronous (SIMD machine) and the control is distributed (for scalability and simplicity). The choice of the switching method is not obvious:

- packet switching requires local storage and more complex hardware,
- circuit switching needs to establish a connection through several physical links.

The network topology is probably the major issue in parallel computer designing because it depends on the applications and on almost all the machine parameters.

5.4 Implementation of a hybrid interconnection network

We propose a network for applications which require mostly random accesses (required for image synthesis with distributed data-base) but also simultaneous accesses to all neighbours on a multidimensional mesh.

These aspects seem incompatible and would require respectively a dynamic (switched) and a static network. Existing machines demonstrate it:

static network CM-2 [Tmc87], ILLIAC IV [Hor82], MPP [Bat80],

dynamic network PASM [SSIK84], OPSILA [AB86],

static network and dynamic network MasPar [Bla90b].

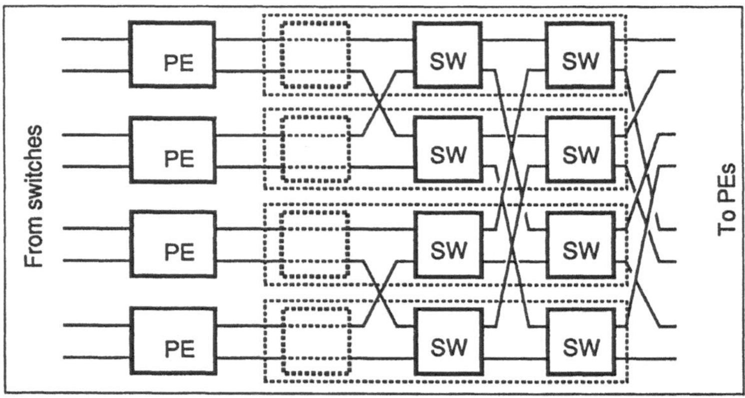

Figure 4: The network for $n = 4$ and $p = 2$.

A candidate for the static network is the hypercube network and a candidate for the dynamic network is the indirect binary cube MIN (multistage interconnection network) [Sch91].

But since a dynamic network is a spatial unfolding of a static network, it must be possible to use the physical wires between switches as a static network instead of using two separated networks like in MasPar.

Each stage of our hybrid MIN can be seen as a dimension of the hypercube. If such a MIN is built with $n \log_p n$ switches instead of $\frac{n}{p} \log_p n$ for a cube MIN with n PEs and $p \times p$-crossbars[3] (ie n lines of $\log_p n$ crossbars instead of $\frac{n}{p}$ lines), it is possible to partition the MIN into n similar subsets mapped on a hypercube, as seen on figure 4 for $n = 4$ and $p = 2$.

The classical design approach leads to the development of an ASIC (Application Specific Integrated Circuit) for the *Hypercom*. This is not convenient because we are quite obliged to redesign the ASIC if the number of processors or the network change.

In order to follow our minimalist philosophy, the *Hypercom* circuit can be implemented with some reprogrammable LCA (Logic Cell Arrays) such as the new 4000 family of Xilinx [Xil90], which offers the required performance, complexity and pin count. Each switch is reversible, offers broadcast capabilities and uses a destination tag algorithm to establish a connection.

For communications on a mesh, a control bit enable changing from the dynamic to the static network. Thus routing overheads are avoided.

5.5 Performance and cost

This study is illustrated for the case of 256 + 1 processors packaged as:

[3]We consider a generalized hypercube pattern with p PEs totally interconnected on each dimension. $p = 2$ for the standard hypercube.

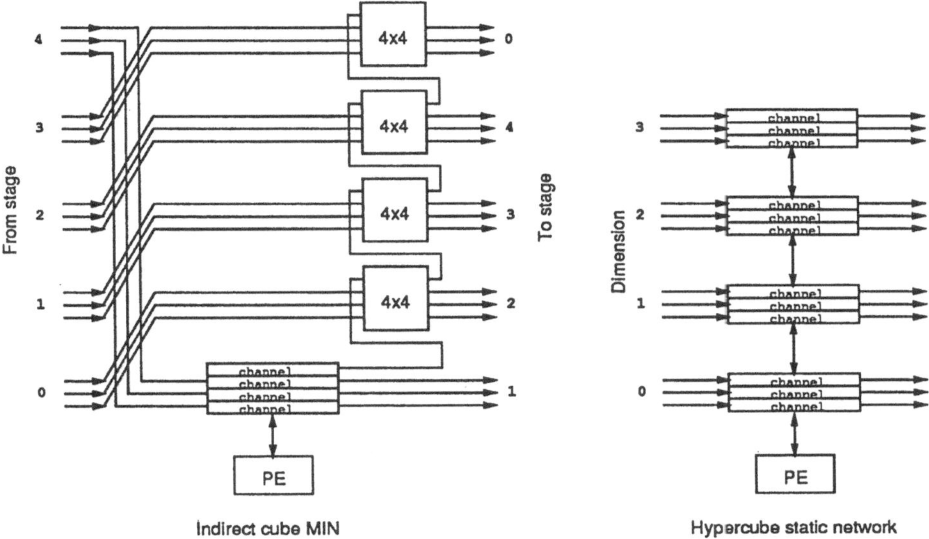

Figure 5: The two configurations of the network with $n = 256$ processors and $p = 4$.

Network type	*Average efficiency*	*Average #cycles per 32 bits*				*Peak throughput*[a]
		int	double	256 bits	∞	
8 stages, 2×2	0.30	66.8	63.4	60.8	60	1.3 GB/s
4 stages, 4×4	0.37	41.5	38.5	36.2	35.4	2.6 GB/s
2 stages, 16×16	0.48	21.4	15	10.2	8.6	10 GB/s
1 stages, 256×256	0.63	15.8	11.1	7.5	6.3	10 GB/s[b]

[a]For regular routing, like matrix multiplication.
[b]The throughput is limited by the PE data bus.

Table 2: Performance of some hybrid networks.

- 1 controller board,
- 16 motherboards of 16 PEs

in a 19" *Triple Europe* rack.

The performance evaluation of the network for random routing is complex, contrary to neighbourhood communications. We have simulated the random routing for a high *vp_ratio*, as shown in table 2, with 1-bit datapaths. Related costs are represented in table 3.

Network type	*#Links /PE*	#Communication *pins/PE*	#switches */PE*	*#Wires between motherboards*
8 stages, 2×2	2	18	32	158
4 stages, 4×4	4	30	64	282
2 stages, 16×16	16	90	512	960
1 stages, 256×256	256	960	65536	15360

Table 3: Costs of some hybrid networks.

8-stage and 4-stage (figure 5) networks present a correct trade-off between performance for random routing and cost. They are both small enough to be implemented in the *Hypercom* with a reprogrammable LCA, even with 4-bit datapath, for the 8-stage MIN, which is then very performant.

6 The code generation for POMP

Figure 6 illustrates the code generation process for POMP. The final instruction is 72-bit wide and consists in the following fields:

- a 32-bit instruction for the scalar processor;
- a 32-bit instruction broadcasted to the PEs;
- an 8-bit instruction to control the *Hypercoms*.

The global idea of our code generation consists in using commercial compiler and assembler, which is coherent with our philosophy to develop as little software as possible.

The most complex part for this generation is the splitting of the POMPC source file into two C files. This is the first phase of a compiler. It is necessary to develop a parser for POMPC. A typechecker identifies the collection of each expression and each statement. An instruction breaker cuts the different parts of expressions to separate instructions with scalar side effects (scalar assignment, scalar increments and decrements, function calls,...) from purely vectorial instructions which depend on an activity and are repeated as many times as necessary to handle the virtual processing management[4]. Consecutive instructions depending on the same collection are then associated to share the same virtual management loop. Vectorial local variables declared in block handled in a single virtual management loop are relegated to the *processor* collection: local arrays are transformed into single elements which are commonly compiled in registers (this confirms the usefulness of a RISC processor for the PEs). The final step of the program is the code generation. This part is simple because the generated code is C which is a symbolic language. Three kinds of generators are used:

[4]This is the virtual management loop.

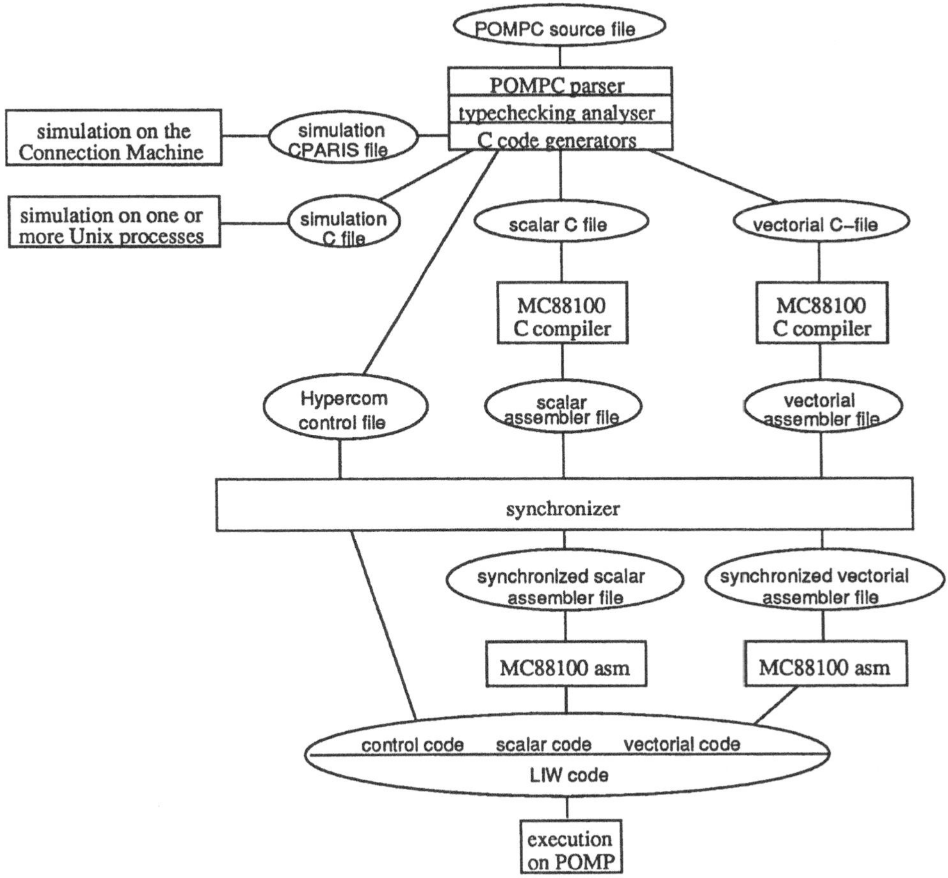

Figure 6: The code generation diagram

- the generation of the code for POMP. The different files are generated with synchronization points declared in the C codes by dummy function calls to pseudo-functions (`synchro_1()`, `synchro_2()`,...)

- the generation of a C file for one or more processes simulation: it is in fact the same generation of code but in a single file. Each PE is simulated by a Unix process which also runs the scalar code. A shared memory segment allows to synchronize the processes and to simulate the communication. Thus it is possible to develop in POMPC the communication routines for POMP and for the simulation. It allows to measure the performance of the network. The number of processes is defined by an environment variable. The monoprocess simulation is a multiprocess simulation with only one PE.

- the generation of a CPaRIS code (the C Parallel Instruction Set of the Connection Machine [Tmc87]) which allows to perform real-time simulations on the Connection Machine[5].
- other code generators can be thought of; for instance it could be interesting to write some for the Intel Hypercube or for the Sequent Machine and to study the interest of a SPMD language to program MIMD machines.

The second program to be developed is the synchronizer. It takes as inputs the three files for each field and resynchronizes them. This program must understand the assembly code of the target processor (here the MC88100) in order to identify the synchronization pseudo-function calls. Synchronization is achieved by inserting nops in the code to be delayed. This program must also take into account the pipeline structure of the controller and the internal pipeline of the MC88100. The internal scoreboarding of the chip must be managed at compile time to avoid the desynchronization of the whole machine at run time.

The third program is a simple loader with a parallel symbolic debugger.

Only the last two programs depend on the type of the choosen processor and must be rewritten if we choose another processor. This limits the complexity of the development for today and for tomorrow.

7 Conclusion

Choosing to limit the developments does not lead to poor performances.

As concerns the hardware, we have only to develop the controller board, which is easy thanks to the use of a commercial processor, the replicated module of the cluster (a very small board with 9 circuits) and the interconnections of the 16 mother boards.

Software developments are limited to the development of a POMPC preprocessor (20,000 lines of C) and the synchronizer (5,000 lines of C).

It is possible for programs requiring little networking (at most one global indirection every 50 instructions) to reach the full efficiency of the machine: 4000 MIPS and 1700 MFLOPS with a small machine ($\approx$ 1 kW).

8 Current Work

A 3-processor machine is now under development. It will demonstrate the feasibility of the controller and of the programming concepts. As soon as credits can be found (we need 1 MFF in commercial chips) a prototype with 257 processors will be built.

The POMPC compiler is written. Simulations on the Connection Machine 2 (located at the ETCA) and on Unix work. Small applications like a One-Step-Relaxation

[5]This language is used by people programming on the Connection Machine

electrical simulator have been developed in POMPC and run on the Connection Machine and on the Unix simulators. Some aspects of the semantic of POMPC have been studied by Luc Bougé and Jean Luc Levaire [Bou90,Lev90]. A ray-tracer is under development using the spatial coherence of the rays with beam tracing techniques developed in [Thi90].

References

[AB86] M. Auguin and F. Boeri. The OPSILA Computer. In INRIA, editor, *Parallel Algorithms & Architectures*, pages 143–153, North-Holland, 1986.

[AJ88] Kurt Akeley and Tom Jermoluk. High-Performance Polygon Rendering. In *Computer Graphics (SIGGRAPH '88)*, ACM, August 1988. Volume 22, Number 4, pp 239–246.

[Bat80] Kenneth E. Batcher. Architecture of a Massively Parallel Processor. In *SIGARCH 80*, pages 168–173, IEEE, 1980.

[BCJ89] Edward C. Bronson, Thoms L. Casavant, and Leah H. Jamieson. Experimental application-driven architecture analysis of an SIMD/MIMD parallel processing system. In *International Conference on Parallel Processing*, pages 59–67, IEEE, Academic Press, 1989.

[BDW85] John Beetem, Monty Denneau, and Don Weingarten. The GF11 Supercomputer. In *SIGARCH 85*, pages 108–115, IEEE, 1985.

[Bla90a] Tom Blank. The Design of the MasPar MP-1, A Cost-Effective Massively Parallel Computer. In IEEE, editor, *IEEE Compcon Spring 1990*, February 1990.

[Bla90b] Tom Blank. The MasPar MP-1 Architecture. In IEEE, editor, *IEEE Compcon Spring 1990*, February 1990.

[Bou90] Luc Bougé. *On the Semantics of Languages for Massively Parallel SIMD Architecture.* Technical Report LIENS-90-13, Laboratoire d'Informatique de l'Ecole Normale Supérieure, June 1990.

[Chr90] Peter Christy. Software to Support Massively Parallel Computing on the MasPar MP-1. In IEEE, editor, *IEEE Compcon Spring 1990*, February 1990.

[Fen91] Tse Yun Feng. A Survey of Interconnection Networks. *Computer*, 14(12):12–27, December 1991. IEEE.

[FP81] Henry Fuchs and John Poulton. Pixel–Plane: a VLSI–oriented design for a raster graphics engine. *VLSI Design*, 2(3), 1981.

[FPE*89] Henry Fuchs, John Poulton, John Eyle, Trey Greer, Jack Goldfeather, David Ellsworth, Steve Molnar, Greg Turk, Brice Tebbs, and Laura Israel. Pixel–Plane 5: A Heterogeneous Multiprocessor Graphics System Using Processor–Enhanced Memories. In *Computer Graphics (SIGGRAPH '89)*, ACM, July 1989. Volume 23, Number 4, pp 79–88.

[FWT82] Mark A. Franklin, Donald F. Wann, and William J. Thomas. Pin Limitation and Partitionning of VLSI Interconnection Networks. *IEEE Transactions on Computers*, C-31(11):1109–1116, November 1982.

[Gap84] *Geometric arithmetic parallel processor NCR45CG72.* NCR, 1984.

[Gil86] Wolfgang K. Giloi. Interconnection networks for massively parallel computer systems. In *Future Parallel Computers*, pages 321–348, Springer-Verlag, 1986.

[Hor82] R. Michael Hord. *The ILLIAC IV, The First Supercomputer.* Computer Science Press, 1982.

[Inm89] *The Transputer Databook.* INMOS, 1989.

[JHH80] James-H.Clark and Mark-R. Hannah. Distributed processing in a high performance smart image memory. *Lambda*, 1(4), 1980.

[Ker89] Ronan Keryell. *POMP2: D'un Petit Ordinateur Massivement Parallèle.* Rapport de Magistère, LIENS — Ecole Normale Supérieure, octobre 1989.

[Kot87] S. C. Kothari. *Multistage Interconnection Networks fo Multiprocessor Systems*, pages 155–199. Volume 26, Academic Press, 1987.

[KR78] Brian W. Kernighan and Dennis M. Ritchie. *The C programming language.* Prentice-Hall, 1978.

[KV90] David Kirk and Douglas Voorhies. The Rendering Architecture of the DN10000VS. In *Computer Graphics (SIGGRAPH '90)*, ACM, August 1990. Volume 24, Number 4, pp 299–307.

[Lev90] Jean-Luc Levaire. *Deux sémantiques opérationnelles pour POMPC.* Diplôme d'Etude Approfondie, LIENS, Paris, September 1990.

[Mlc90] *The multiC Programming Language: Extending C to Accomodate Data Parallel Processing.* Technical Report, Wavetracer Inc., 1990.

[Mot88] *MC88100 RISC processor user's manual.* MOTOROLA, 1988.

[Par90] Nicolas Paris. *Définition de POMPC (Version 1.5).* Technical Report, LIENS, février 1990.

[Sch91] Isaac D. Scherson. Orthogonal Graphs for the Construction of a Class of Interconnection Networks. *IEEE Transactions on Parallel and Distributed Systems*, 2(1):3–19, January 1991.

[SSIK84] Howard Jay Siegel, Thomas Schwederski, Nathaniel J. Davis IV, and James T. Kuehn. PASM: A Reconfigurable Parallel System For Image Processing. *ACM SIGARCH Newsletter*, 12(4):7–19, September 1984.

[Thi90] Jean-Philippe Thirion. *Interval Arithmetic for High Resolution Ray Tracing.* Technical Report LIENS-90-4, Laboratoire d'Informatique de l'Ecole Normale Supérieure, February 1990.

[Tmc87] *Connection Machine Model CM-2 Technical Summary.* Technical Report HA87-4, Thinking Machine Corporation, April 1987.

[Tuc90] Russ Tuck. *Porta-SIMD: An Optimaly Portable SIMD Programming Language.* PhD thesis, University of North Carolina at Chapel Hill, May 1990.

[Xil90] *XC 4000 Logic Cell™ Array Family.* XILINX, 1990. Technical Data.

The Function Processor: An Architecture for Efficient Execution of Recursive Functions

Jonas Vasell   Jesper Vasell
Department of Computer Engineering
Chalmers University of Technology
S–412 96 Göteborg
Sweden

Abstract

The Function Processor is a wavefront array architecture, i.e., a regular structure of locally interconnected processing elements called Function Cells, which operate according to the data flow execution principle. By means of a compilation method developed for this architecture, data flow graphs for functional programs can be created and mapped onto the processor array, so that each Function Cell is assigned the execution of one graph node. The main result presented in this paper is a Function Cell architecture which has been designed to support the functionality required by these data flow graphs. Some implementation results are also presented.

1 Introduction

Due to the rapid development of VLSI technology in recent years, array architectures have become an increasingly interesting alternative for very fast algorithm implementations. An array processor consists of a large number of simple processing elements. Each processing element is directly connected only to a few neighbour cells, and repeatedly performs a single operation on data arriving from these neighbours. Systolic arrays [Kun82, MMU87] is a well known example of these architectures. Another example is wavefront arrays [KAGER82, KLJH87, Kun88, KMPS88]. These architectures differ in the way the processing elements are synchronized. In wavefront arrays, a processing element works in a way similar to the data flow principle [Arv80, Den80, Vee86, McG89], i.e., it performs its operation as soon as all necessary operands have arrived from its neighbours.

Traditionally, array architectures have mostly been used for highly regular computations in signal processing and image analysis. A regular computation is always performed in the same way, independently of the actual input data. This limits the computations that may be performed, but it makes it possible to exploit data parallelism to a large extent. An example of a regular computation is multiplication of fixed-size matrices. Our

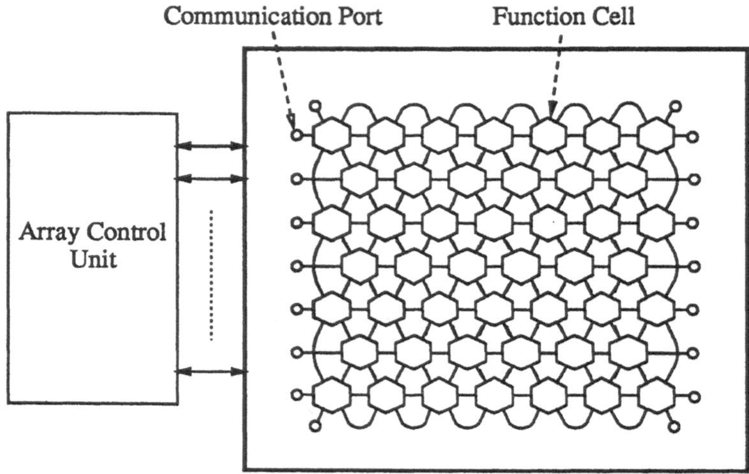

Figure 1: Function Processor Architecture

aim has instead been to try to use array architectures to support irregular computations. It is then more difficult to exploit data parallelism, but much is still to be gained from pipelining parallelism and a very low interpretation and communication overhead. We have also wanted to find support for fast execution of symbolic computations expressed in functional programming languages [Bac78, Hug89, Rea89]. Simulations have shown that typically a speed-up of between five and fifty times can be achieved with wavefront arrays compared to sequential execution on state-of-the-art workstations (see section 5).

We will here present an architecture for a wavefront array processing element suitable for these purposes. These processing elements are called *Function Cells.* The *Function Processor* (see figure 1) is an architecture consisting of an array of Function Cells and an Array Control Unit. Several different types of systems can be built around the Function Processor architecture. For instance, a system can consist of several Function Processors and a conventional processor which performs administrative tasks. Another possibility is to use the Function Processor as an accelerator for a host processor running a conventional implementation of a functional programming language. In this case the Function Processor should execute one or several critical functions in the program, while the rest of the program is executed on the host processor. The Function Processor could also be used in a specialized system, for instance a signal processing system, where it executes a limited number of different algorithms.

It is important that a new architecture is developed together with efficient programming methods. We have therefore developed a method to create data flow graphs, DFG:s, for functional programs, which can be mapped onto the Function Processor. By *mapping* a graph onto an array architecture, we mean assigning one processing element to each graph node in a way that allows intermediate results to be transferred between nodes via the processing element communication links.

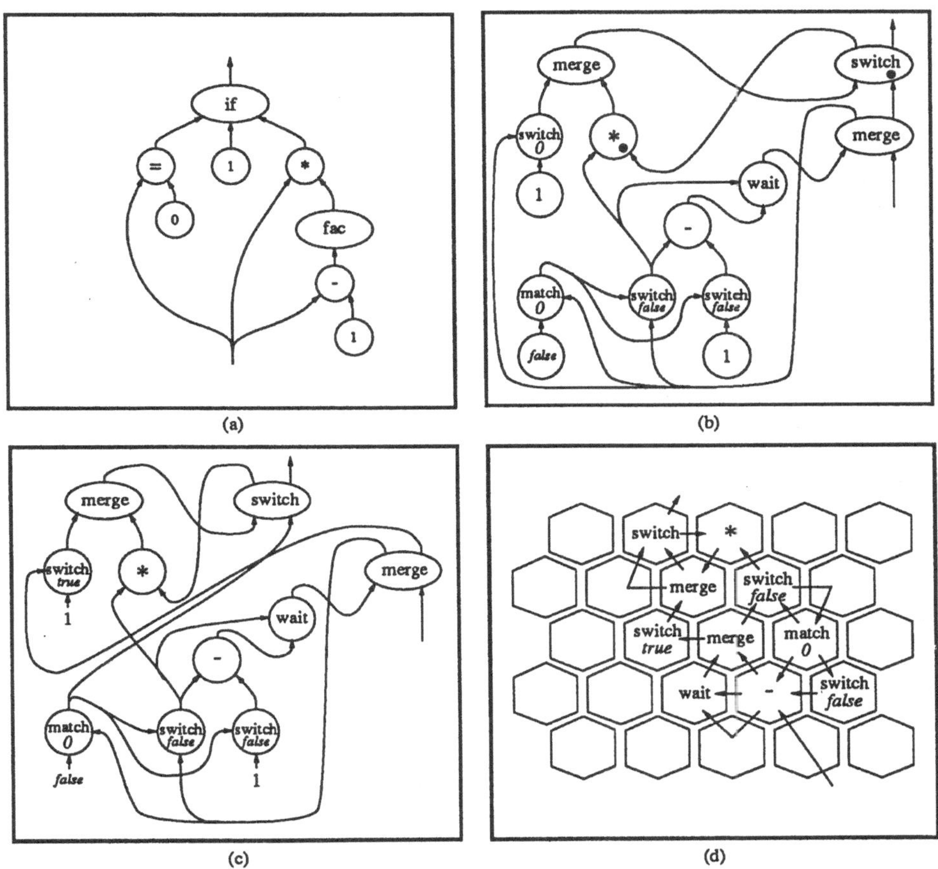

Figure 2: Programming Example

In figure 2, an overview of the whole programming process is shown for a simple example, the factorial function:

```
fac n = if n==0 then 1
        else n*fac (n-1)
```

The programming method takes a functional program, i.e., a set of (possibly mutually) recursive function definitions, as input. From these function definitions a DFG is created. The DFG for the factorial function is shown in figure 2(a).

In general, such a DFG can not be directly mapped onto the processor array. This can be for many reasons, but the most important is that the DFG has to be static, i.e., it must not contain any nodes which represent an expansion of the graph. The reason for this is naturally that the graph can not be allowed to expand during execution since the hardware configuration is fixed and finite. We have defined four requirements that a DFG have to fulfill before it can be mapped onto the Function Processor. We say that a DFG that meet these requirements is on *hardware implementable form.* Thus, the next step in the programming or compilation process is to transform the original DFG to this form (figure 2(b)).

Usually the execution order imposed by the hardware implementable DFG is not the most efficient, and a number of optimizations are therefore applied on the graph. For the factorial function, one important optimization takes advantage of the fact that multiplication is a commutative and associative operation. Therefore, the order of the successive multiplications can be reversed. This results in a graph which corresponds to a tail-recursive version of the function in which the calculation starts immediately, rather than after all recursive calls have been made and operands have been pushed on a stack (figure 2(c)).

Finally, the optimized DFG is mapped onto the array (figure 2(d)). This sometimes requires that some Function Cells are used only to route data between other cells.

The use of array architectures for irregular functional computations has also been proposed by *I. Koren, G. Silberman et al* (see [KMPS88]). They do, however, not especially address the problems in supporting functional languages and symbolic computations. Instead, they have concentrated on the problem of mapping DFG:s onto wavefront arrays.

M. Sheeran presents a method for synthesis of algorithm-specific array processors from functional specifications in [She85]. This method has a set of predefined higher order operators as its starting point.

2 Hardware Implementable Data Flow Graphs

Traditionally, array architectures have been used for highly regular computations, i.e., computations which proceed in exactly the same way independently of input data. The DFG:s of these computations have a very regular structure and lend themselves very well to execution on an array architecture since they use each processing element in exactly the same way all the time. But if we limit ourselves to such regular computations, we also limit the usefulness of array processors.

If we compile a functional program to a DFG we are, however, not guaranteed to get a DFG which can be used directly as a program for an array processor, since usually the computations are irregular.

We say that a DFG which can be used as a program for an array architecture is *hardware implementable.* The intuitive meaning of this is that it can be seen as a description of a machine performing a specific computation. We have defined the following criteria for a Hardware Implementable DFG (HIDFG):

- *Finiteness*: The number of nodes in the DFG must be finite and statically determined, and must not exceed the number of available processing elements.

- *Realizability*: Each node in the graph must be executable in the resources available in one processing element.

- *Repeatability*: The time from the start of one terminating computation in the DFG until the next computation can be started must be finite. This means that no unnecessary subcomputations must be started, unless they are known to be terminating, and do not interfere with any other computation.

- *Representability*: The DFG must only handle data objects which can be represented in a way that the processing elements can interpret.

As can be seen, these criteria are defined relative to properties of a target architecture.

The DFG of the factorial function shown in figure 2(a) contains a node representing a function application. This node causes a problem since we want to statically assign each node, or rather the operation of each node, to a processor. But the application represents a dynamic expansion of the graph, which in this case is further complicated by the fact that it is a recursive application, and there is no possibility to decide how many expansions of the graph are necessary unless we know the value of the argument. This means that the DFG is not finite, since the number of expansions, and thus the number of nodes, can not be statically determined. The only possible solution to this problem is to let all applications of a function be executed in the same instance of the function graph. We could, of course, try to execute the application node in one processing element. This would, however, not be realizable in most architectures, since it would require that a processing element is capable of changing the operations of other processing elements during execution.

Another problem with the DFG in figure 2(a) is the if-node which requires a boolean value, and then chooses to output either the value of the then-branch or the else-branch. Regardless of which branch the if-node eventually will choose, both branches are computed. This means that computations are performed which might not be needed. The problem is that if an unnecessary computation is performed it might give rise to an incorrect result, e.g., if it contains a recursive application, or it might turn out to be a non-terminating computation which blocks any further computations in the graph. It can be seen from this discussion that the presence of an if-operation may give rise to a DFG which is not repeatable, since it requires unnecessary subcomputations to be started, which may lead to non-termination or even an incorrect result. In fact, handling conditionals is a major issue in the design and use of array architectures, since they are the reason why a computation becomes irregular.

The DFG in figure 2(b) solves the problems described above, and is thus hardware implementable. The problem with the recursive function application is solved by allowing all applications of the function to execute in the same graph. In order to make this possible we introduce nodes which keep track of where the results of different applications should be sent. To make this as simple as possible, the order in which different recursive applications are executed is statically determined. Executing multiple function applications in the same graph also creates a need for saving the state of a function application. This is accomplished by introducing stacks in the nodes that contain part of the state when a function application is performed. In the figure, stacks are indicated by a dot at the input of a node.

The solution described above is used for recursive applications, while non-recursive applications are handled by substituting the graph of the applied function for the application node.

Another possible approach to solve the problem with recursive applications would have been a scheme where all data is tagged with instance numbers. This is a method which has been used in other data flow architectures, such as TTDA [Arv80, AN87] and ETS (Monsoon) [CP90]. However, though this is a very flexible solution which makes synchronization simpler, it requires much more complex hardware to be implemented in the processing elements. This would make it more difficult to build sufficiently large arrays. It would also increase the amount of data communicated between cells. Thus, we have chosen a different approach which we believe is better suited for array processors.

We have developed a method to compile functional programs into HIDFG:s. This method has been implemented in a compiler. Although it is intended to be as general as possible, the method currently places the following restrictions on the source code:

- The method assumes that the source code has been lambda-lifted [Joh85], so that it consists of a set of global recursive function definitions and a result expressions, which contain no local function definitions or lambda abstractions.
- It is also assumed that all pattern matching and case expressions, have been transformed into case-expressions using only simple patterns, i.e., patterns which are variables or constructors applied to variables. A method to do this transformation has been described by Augustsson [Aug85].
- Only a set of predefined types is available, and therefore can no new type declarations be made. The compilation method does not restrict the use of other data types, but the set is restricted by the data type support provided by the target architecture. For the Function Processor this means integers, booleans, pairs, character strings, and lists of these types.
- The methods to handle higher order functions, i.e., functions taking other functions as arguments or producing functions as results, are not yet fully developed. A set of frequently used higher order functions has been made available as predefined functions. These functions include the well-known *map*, *filter*, and *reduce* functions, and the set could easily be extended with many other functions.

Since there exist methods for lamda-lifting and compilation of pattern matching, it is only the restrictions on data types and higher order functions which can be considered to be real restrictions on the source code. The programming method has been described in [Vas90, VV91].

3 Architectural Requirements

The requirements on the Function Cell architecture are to a large extent determined by the types of nodes that may appear in a HIDFG since it must be able to perform the function of any such node.

In general, the nodes have one or two inputs and one or two outputs. They operate according to the data flow execution principle, i.e., as soon as they have received all necessary operands they perform their operation. The following are the node types that are used in mapped HIDFG:s.

Merge. This node has two inputs (left and right), and two outputs (data and selection).

When a data object is available at either input it is copied to the data output of the node. If the object came from the left input, a boolean "true" value is sent out at the selection output, and if it came from the right a boolean "false" is sent out. The selection output is not always used. In those cases, it will be omitted in the figures.

Switch. This node has two inputs (data and select), and two outputs (left and right). A data type constructor is also always specified. The node requires data to be available at both inputs. If the object available at the selection input is built with the specified constructor, the object at the data input is copied to the left output, otherwise it is copied to the right output. Sometimes only one output will be used in the figures. It is then always the left output, i.e., the output used when the selection object matches the specified constructor.

Match. This node has two inputs (match and select) and one output. A data type constructor is also specified. If the object at the select input matches the specified constructor, a boolean "true" is sent out as the result. Otherwise, the object at the match input is copied to the output.

Split. This node has one input, two outputs (first and second), and a specified constructor. All objects arriving at the input must match the specified constructor. The constructors should have two elements. The first element of the input object is copied to the first output, and the second element is copied to the second output.

Wait. This node has two inputs (data and control) and one output. When objects are available at both inputs, the object at the data input is copied to the output. In the graphs, it is always the input in line with the output that is the data input. This node type is special in that it can be specified to be in an initial state where one object already is available at the control input. These nodes are used to synchronize the flow of data at places where the order of computations is significant.

Route. This node can have a variable number of input-output pairs. A data object received at one input is copied to the corresponding output. These nodes appear in the physical mapping of a DFG when two nodes which are connected in the graph are assigned processing elements which are not neighbours in the array, and thus do not have any direct connection.

Constructors and operators. The rest of the nodes belong to this class. These nodes perform some specific operation or constructor function on its inputs, or simply produce a constant object.

Some of these node types may operate on different data types. Therefore the Function Cell must be able to recognize and distinguish the types of the data objects which are communicated in a DFG. More specifically it must be able to recognize the type of a data object received on an input as well as whether all parts of it has been received. As many data types as possible should be supported, but there are some which are especially well suited for support in an array architecture. These are data types which easily can be represented as sequences of bit-groups with a very small overhead for coding and decoding. We wish to support four such data types which will be described in detail later.

They are scalars, pairs, strings and lists. Scalars are single word objects such as integers and characters. A pair is constructed from two scalar values. Both strings and lists are sequences of other data objects.

Furthermore, some of the nodes in a DFG may have a LIFO buffer (a stack) at one of its inputs. Also, the data types which are represented by variable length sequences of data objects sometimes cause a need for FIFO buffering of some node operands.

When creating a physical mapping of a DFG, i.e., assigning a processor to each node and finding a communication path corresponding to each arc, it is necessary to introduce *route*-nodes in order to make it possible to connect nodes which are not neighbours in the array. This can be simplified if it is possible to have a Function Cell perform not only the operation of a specific node type, but at the same time route a data object between two physical ports. This can be achieved since each Function Cell has six physical ports but the node types use at most four of them. Thus, the remaining two ports could be used for routing.

Since each Function Cell within an array must be able to handle any of the node types and data types it must be reconfigurable, i.e., it must be possible to load and store a configuration within a cell.

To make it possible to adapt the array size to a given application, it is important to make the architecture truly scalable. It should be possible to enlarge an existing array simply by adding more Function Cells. The main problem that prevents scalability is that it is difficult to provide all cells with a common clock. This problem can be solved by making the communication between cells asynchronous. This implies that each Function Cell should have its own clock.

The main part of the Function Processor architecture is the array of Function Cells. The size of this array imposes a limit on the size of programs that may be executed on it, i.e., the number of nodes in a physical mapping of a HIDFG must not exceed the number of Function Cells in the array. In order to make this restriction less severe, it must also be possible to store multiple configurations within one array, and to change quickly between them.

In the following sections we will discuss the architecture of the Function Cell in more detail.

The second part of the Function Processor architecture is the Array Control Unit (ACU). The ACU makes it possible for a Function Processor to communicate with other parts of a system, e.g., a host processor, a memory or other Function Processors. The ACU is responsible for loading configuration data into the array, and for handling input and output of data during execution. It consists of a number of independently programmable communication channels, which are connected to the array.

4 Function Cell Architecture

The Function Cell architecture we are going to describe here can be divided into five separate parts, each of which is separately described below. An overview of the architecture is shown in figure 3.

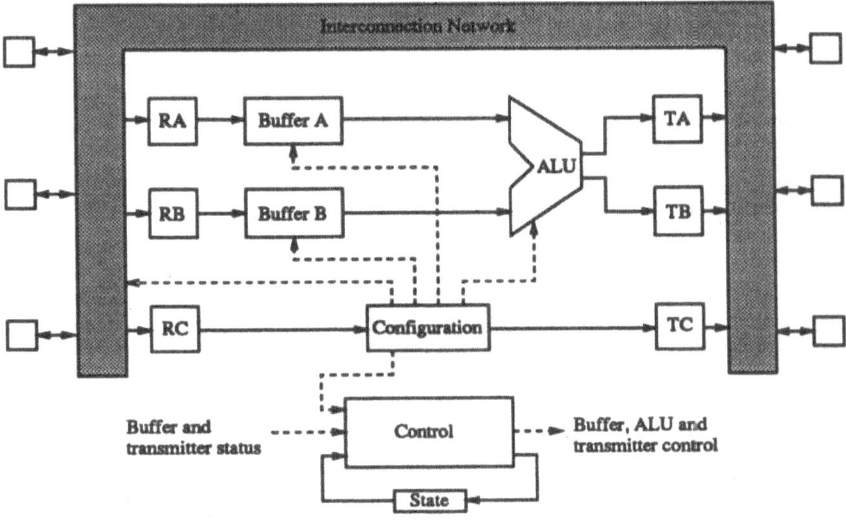

Figure 3: Function Cell Architecture

4.1 Ports

The Function Cell can be divided into two major parts; a functional unit and an interconnection network. The functional unit has three inputs called *logical input ports A, B and C*, and three outputs called *logical output ports A, B and C*. The logical input and output ports A and B are used to implement the inputs and outputs of the different DFG node types. The C ports are used for configuration and routing. At the logical ports are receivers and transmitters (RA, RB, RC, and TA, TB, TC, respectively) placed. These are responsible for the asynchronous communication between logical ports on different cells via *interconnection buses*.

The *interconnection network* is a flexible reconfigurable switch capable of connecting each of the logical ports to any one of six *physical ports*. Any physical port can act either as an input port or as an output port. In the Function Processor, every physical port on one cell is connected to exactly one physical port on another cell, or to one of the external communication ports (see figure 1). Despite this very rigid physical structure, the flexible interconnection network in the cells makes the task of mapping DFG:s onto the array much easier. The interconnection network configuration is programmed at the same time as the functionality of the cell is programmed.

Data are communicated between cells in units of *words*. A word is also the amount of data that the ALU can process in one operation. The number of bits in a word, the *wordwidth*, will not be fixed in this architecture description. It can vary between different implementations and applications, but a practical lower limit is probably 16 bits.

4.2 Configuration

The Function Cell is programmable, or rather, reconfigurable. Each configuration specifies the DFG node operation performed by the cell, as well as through which physical ports the cell communicates with its neighbours. The configuration is stored as a set of control words in a number of configuration registers. Several configurations can be stored in the cell, but only one configuration can be active during the execution of one function, i.e., the configurations should not be seen as instructions in a sequential program. By storing several configurations in the cells, the Function Processor can quickly switch between different tasks by means of a global signal telling the cells which configuration to use. The exact number of configurations and control words that the cell can store is implementation dependent, but the total number of bits required for one configuration can be estimated to 48 plus the width of one word. One of the words in the configuration is always a constant value used by the input buffers, either to recognize data objects, or as a constant input.

The Function Cell can be either in *execution mode* or in *configuration mode*. The mode is selected by means of an external signal. In configuration mode, the cell shifts configuration data from the logical C input into the currently selected configuration registers, whose earlier contents simultaneously are shifted out via the logical C output. This operation is independent of the current configuration, and the logical C ports are routed to predetermined physical ports. This means that, by putting all cells in an array in configuration mode, their C input and output ports form a chain through which configuration data can be shifted to all cells. After a new configuration has been stored or selected, the input buffers and the control unit are emptied and reset. The cell can then start its new operation in execution mode. In this mode, the C input is directly connected to the C output, bypassing the configuration registers. This direct link can be used to route data through the cell independently of its other functions. Thus, mapping of DFG:s onto a cell array is made easier.

4.3 Input Buffers

The input buffers are the parts that contribute the most to the special characteristics of the Function Cell. Their main purpose is to asynchronously receive operands to the node function implemented by the cell, and to inform the control unit when operands are available. In accordance with the architectural requirements, one of the buffers can be configured as a multi-word FIFO or LIFO (stack) buffer. Simulations of benchmark functions indicate that the buffer size should be at least 256 words. An overview of the input buffer can be seen in figure 4.

We have, however, also chosen to let the input buffers provide support for a number of different data types. The idea is to let the input buffers be responsible for recognizing objects of any specified data type, thereby making the control algorithms for different node types independent of the operand types. The *merge* node, for instance, is implemented by an algorithm that only has to specify that as soon as one data object is available at any input it should be copied to one of the outputs. It does not have to be concerned about whether the object is a list that consists of several words, or a simple scalar value represented by a single word. These functions are performed by a special part of the input buffer called *object detector*. The object types to be expected at the input buffers

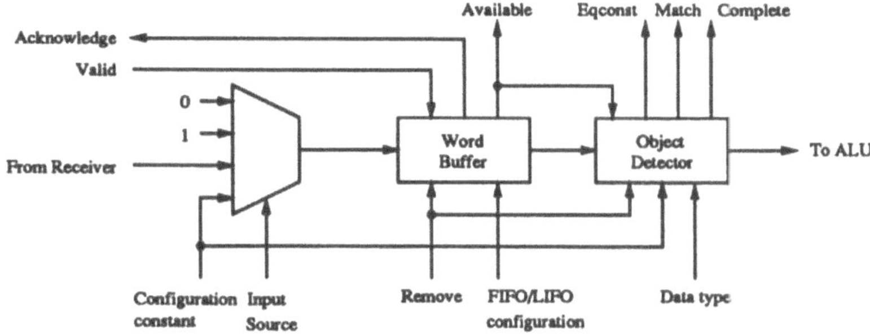

Figure 4: Input Buffer

are specified as a part of the cell configuration. This information is taken from an analysis of the DFG that has been mapped onto the Function Processor.

The object detectors can be configured to recognize one of four different data types. Objects of all types are represented by sequences of elements, sometimes terminated by a word with a special value. The elements can be either objects of another type less complex than the type of the object they are part of, or single words. The objects of a type can be built up in different ways, with different constructors. One of the constructors for each type is designated the *primary constructor* of that type. The object detector is capable of recognizing objects built with the primary constructor.

The least complex data type supported is the *scalar type*. A scalar object always consists of a single word. If this word is equal to the configuration constant, the object is recognized as the primary constructor. The next more complex data type is the *pair type*. A pair is built up either by two scalar values, in which case it is formed with the primary pair constructor, or a single word equal to the configuration constant.

The third data type is the *string type* which can have elements of scalar or pair type. A string object consists of zero or more elements followed by the configuration constant (note that no element must contain the configuration constant as this will be interpreted as the string terminator). Typically a string can be a null-terminated character string, i.e., a sequence of character codes followed by character code zero. A string containing at least one element is built up with the primary string constructor. In a way, the string type is many different types, each with a different termination word. Therefore, a string can also contain elements of any other string type as long as no string element contains the termination word of the string it is an element of. The most complex type is the *list type*. A list is very similar to a string, except that it can contain elements of any of the other types, and that the termination word for lists has a predefined value called *nil-token*, which is equal to the smallest two-complement value a word can contain, i.e., $-2^{wordwidth-1}$. A non-empty list is built with the primary list constructor (often called *cons*). The list has the advantage over other types that the object detector does not depend on the configuration constant which therefore can be used for other purposes.

The object detectors output four status signals used by the control unit to determine its actions. The first signal, *available*, indicates if any word (any part of an object) is available in the buffer. The second signal, *eqconst*, indicates if the word available at the buffer output (if any) is equal to the constant specified in the configuration. The third signal, *complete*, indicates when all the words in a complete data object of the specified type have been received. The fourth signal, *match*, indicates if a complete object was constructed with the primary constructor of the specified type.

The input buffers are so autonomous that the control algorithms only have to control a single operation on them. This operation is to remove the word currently available at the buffer output, called the topmost element, and replace it with the next word in the buffer if there is one. This is controlled by the *remove* signal. When, for instance, an addition node has added two scalar operands, these are removed from the input buffer, and the cell starts to wait for two new operands to become available.

Usually, an input buffer receives data asynchronously from a receiver with which it communicates via a handshaking protocol (*valid, acknowledge*). It can, however, also be configured to continuously receive a constant value; zero, one or the configuration constant. Zero and one are frequently used constants, and are therefore made available even in configurations where the configuration constant is used for other purposes.

4.4 ALU

The Arithmetic and Logic Unit (ALU) is responsible for all data processing in the Function Cell. It takes the two operands coming from input buffers A and B as input, and it produces two results on separate, independent outputs. Each result can be equal to either of the two operands, or be the result of an arithmetic or logic operation on the two operands. The results are selected by the control unit. The operation performed by the ALU is determined directly by the contents of the configuration registers.

In this way, the control algorithms become independent of the ALU operation and the number of different node types supported by the Function Cell is kept down. The set of operations that the ALU can perform can vary between different implementations, but traditional arithmetic operators (add, subtract, multiply), comparison operators (which produce boolean results), and the boolean constants true (-1) and false (0) should be supported. There should be no need to support unary operators since either operand can be set to a constant value in the input buffers.

The ALU outputs are connected to the transmitters at the logical outputs A and B. The transmitters are controlled directly by the control unit which informs them when valid data are available at the ALU outputs. The transmitters also produce status signals informing the control unit when the last valid data have been sent out. Data are not sent by a transmitter before all the receivers it is connected to have informed it via the interconnection buses that they have passed on the last transmitted word to their input buffers. This information is exchanged between the transmitters and the control unit by means of *valid* and *ready* signals.

4.5 Control Unit

The control unit is a finite state machine that implements control algorithms for all node types supported by the Function Cell. After a cell has been configured, it is reset. This means that the control unit enters a special start state in which it inspects the configuration registers to determine what node type has been selected. It then enters

The *switch* node reads an object from input A. If this object is a primary constructor object of the input A type, one data object is read from input B and sent to output A, otherwise it is sent to output B. State 1 is the initial state.

```
State 1. Read a complete object from input A:
         If any word of input A object is available
              If input A object is complete go to state 2
              Else remove one word of input A object and go to state 1
         Else go to state 1
State 2. Send a complete object from input B to the selected output
         If any word of input B object is available
              If input B object is complete
                   If input A object matches primary constructor
                        If transmitter A is ready
                             Remove one word from inputs A and B
                             Output the word to transmitter A
                             Go to state 1
                        Else go to state 2
                   Else
                        If transmitter B is ready
                             Remove one word from inputs A and B
                             Output the word to transmitter B
                             Go to state 1
                        Else go to state 2
              Else
                   If input A object matches primary constructor
                        If transmitter A is ready
                             Remove one word from input B
                             Output the word to transmitter A
                             Go to state 2
                        Else go to state 2
                   Else
                        If transmitter B is ready
                             Remove one word from input B
                             Output the word to transmitter B
                             Go to state 2
                        Else go to state 2
         Else go to state 2
```

Figure 5: Control Algortihm for the *switch* Node

the initial state in the control algorithm for this node type. A control algorithm usually consists of a few states which typically correspond to how much of the necessary operands that has arrived. In figure 5, the control algorithm for a *switch* node is given as an example.

The control unit chooses its actions according to the status signals it receives from the input buffers and the transmitters. These status signals are sampled by the control unit at regular intervals. As mentioned above, the status signals from each of the input buffers are *available, eqconst, complete and match*, and from each of the transmitters *ready*. In each state, the control unit also outputs a set of control signals which have been described above. These signals are the input buffer *remove* signals, the ALU output select signals, and the *valid* signals to the output transmitters.

5 Implementation Results

A single chip VLSI implementation of the Function Cell architecture presented here, has been made using a standard cell silicon compiler. The technology used is a double metal layer $1.5\mu m$ CMOS process. The implementation uses 16-bit words and contains a 512-word buffer. The largest parts in terms of chip area is the buffer and the interconnection network including the transmitters and receivers. The interconnection buses are only 6 bits wide (4 bits data, 2 bits handshaking) in order to reduce the size of the interconnection network and the chip pin-count. This means that a 16 bit word is communicated as four 4-bit groups. A whole 16-bit word can, however, be communicated in parallel with the completion of one operation cycle which takes approximately $100ns$.

The configuration data for this cell consist of 61 bits, so the whole configuration is divided into four words. Thus, it takes at least $400ns$ to configure each cell in an array. The number of cells in a practical implementation of the Function Processor should be at least 100–200, which would result in a total configuration time of at least $40 - 80\mu s$, assuming that cells are configured sequentially. This implementation allows only one configuration to be stored in a Function Cell. However, it can easily be extended to allow multiple configurations to be stored in each function cell. That would make it possible to quickly switch between different configurations by means of a global signal which tells the cells which configuration to use. The number of configurations stored in a cell could probably be in the range 8–16 without making the hardware significantly more complex.

The Function Processor has been simulated, executing automatically compiled and optimized HIDFG:s for a number of different functions. The simulated functions are:

fac: computes factorial number 20 (20!).

numbers: generates a list of integers from 1 to 200.

fib: computes the 15:th Fibonacci number.

sum: computes the sum of the list of integers from 1 to 20.

sort: builds the descending list of numbers from 20 to 1, and sorts that list in ascending order using the insertion sort algorithm.

Function	*Nodes*	*Cells*	*Execution Time (μs)*			*Speedup over*	
			Function Processor	*SPARC*	*RT/PC*	*SPARC*	*RT/PC*
fac	10	13	12.9	280	700	21.7	54.3
numbers	10	18	140.7	2320	6300	16.5	44.8
fib	15	17	1813.5	14085	37300	7.8	20.6
sum	21	32	64.2	437	1900	6.8	29.6
sort	36	57	873.6	3090	9300	3.5	10.6
primes	42	83	179.1	990	2400	5.5	13.4
queens	114	–	450.3	9770	24000	21.7	53.3
substitute	141	–	36.6	870	–	23.8	–

Table 1: Results from Simulated Performance Measurements.

primes: produces a list of all prime numbers smaller than 20, using the Sieve of Erathostenes method.

queens: generates a solution to the problem of placing 7 queens in safe positions on a 7 by 7 chess board.

substitute: finds and replaces the first occurrence of a search string in a text.

The main results of these simulations are summarized in table 1. The first column contains the number of nodes in the HIDFG for the function. The effect of mapping the HIDFG onto the Function Processor is shown in the second column, which contains the number of Function Cells that are needed for the mapping, i.e., the number of HIDFG nodes plus the number of cells used only for routing. These mappings, except for the mapping *sort*, have been made by hand. The mapping of *sort* has been made automatically using an experimental version of a program currently being developed especially for mapping HIDFG:s onto the Function Processor. We have not yet been able to produce any satisfactory mappings for the *queens* and *substitute* functions. More information about the problem of mapping irregular graphs onto this type of architecture can for example be found in [KMPS88, WSS89].

The next three columns show execution times for the functions on three different machines. The first execution time is for the Function Processor executing the mapped HIDFG (except for the last two functions which have been simulated without routing delays). This has been taken from a simulator and assumes a 100*ns* cycle time. To compare this with state-of-the-art single processor performance, the next two columns show execution times for two conventional workstations; the SUN Sparcstation 1 and the IBM RT/PC. On these machines, the functions were written in LML (Lazy ML) and compiled with the LML compiler [Joh87, Aug87] which produces graph reduction code. The last two columns show how much faster the Function Processor is compared to the other two machines.

For the mapped functions, the execution times for the unmapped graphs have also been measured. This shows that the routing delays caused by the mapping increases the execution time by 7–75%. There does, however, not appear to be any correlation between the execution time increase and the graph size for these examples.

6 Conclusions

The Function Cell architecture presented here fulfills the requirements stated in section 3. So far, the implementation results have shown that the architecture is realizable and in accordance with the assumptions made in earlier stages of the project.

The most important restriction in the architecture presented here has to do with the data types. There is no support for data structures which can not easily be represented as a sequence. One solution to this problem is to use the Array Control Unit to access parts of data structures to make it possible to handle them as pointers within the array. It is, however, not yet clear how this solution would affect the Function Processor performance.

The compilation method does currently not support general higher order functions. It does, however, support a set of predefined higher order functions, and it is also possible to extend this to cover some user-defined functions as well.

Since the Function Processor is not a general-purpose architecture, its usefulness is dependent on the type of system it is used in. We do, however, feel that the performance results obtained so far, are encouraging and motivate investigations of different types of systems using the Function Processor.

7 Acknowledgements

We would like to thank Tony Nordström for his contributions to this project. The project has been financially supported by the Swedish Board for Technical Development, STU.

References

[AN87] Arvind and Rishiyur S. Nikhil. Executing a program on the MIT tagged-token dataflow architecture. In *Proceedings of the PARLE Conference*, June 1987.

[Arv80] Arvind. A data flow architecture with tagged tokens. Technical report, MIT, Cambridge, Massachusetts, June 1980.

[Aug85] L. Augustsson. Compiling pattern matching. In *Proceedings 1985 Conference on Functional Programming Languages and Computer Architecture*, Nancy, France, 1985.

[Aug87] L. Augustsson. *Compiling Lazy Functional Languages, Part II.* PhD thesis, Dept. of Computer Science, Chalmers University of Technology, Göteborg, Sweden, November 1987.

[Bac78] J. Backus. Can programming be liberated from the von Neumann style? A functional style and its algebra of programs. *Communications of the ACM*, 21:280–294, August 1978.

[CP90] David E. Culler and Gregory M. Papadopoulos. The explicit token store. *Journal of Parallel and Distributed Computing*, 10(4):289–308, December 1990.

[Den80] Jack B. Dennis. Data flow supercomputers. *IEEE Computer*, pages 48–56, November 1980.

[Hug89] J. Hughes. Why functional programming matters. *The Computer Journal*, 32(2):98–107, 1989.

[Joh85] T. Johnsson. Lambda lifting: Transforming programs to recursive equations. In *Proceedings 1985 Conference on Functional Programming Languages and Computer Architecture*, Nancy, France, 1985.

[Joh87] T. Johnsson. *Compiling Lazy Functional Languages.* PhD thesis, Dept. of Computer Science, Chalmers University of Technology, Göteborg, Sweden, February 1987.

[KAGER82] S-Y. Kung, K.S. Arun, Ron J. Gal-Ezer, and D.V. Bhaskar Rao. Wavefront array processor: Language, architecture and applications. *IEEE Transactions on Computers*, C-31(11):1054–1066, November 1982.

[KLJH87] S-Y. Kung, S.C. Lo, S.N. Jean, and J.N. Hwang. Wavefront array processors - concept to implementation. *IEEE Computer*, pages 18–33, July 1987.

[KMPS88] Israel Koren, Bilha Mendelson, Irit Pedel, and Gabriel M. Silberman. A data-driven VLSI array for arbitrary algorithms. *IEEE Computer*, pages 30–43, October 1988.

[Kun82] H.T. Kung. Why systolic architectures? *IEEE Computer*, (1):37–46, January 1982.

[Kun88] S-Y. Kung. *VLSI Array Processors.* Information and System Sciences Series. Prentice Hall, Englewood Cliffs, New Jersey 07632, USA, 1988.

[McG89] J.R. McGraw. Data flow computing: System concepts and design strategies. In S.P. Kartashev and S.I. Kartashev, editors, *Designing and Programming Modern Computer Systems, Vol.III*, chapter 2, pages 73–189. Prentice Hall, 1989.

[MMU87] W. Moore, A. McCabe, and R. Urquhart, editors. *Systolic Arrays.* Adam Hilger, 1987.

[Rea89] Chris Reade. *Elements of Functional Programming.* Addison-Wesley, 1989.

[She85] Mary Sheeran. Designing regular array architectures using higher order functions. In *Proceedings 1985 Conference on Functional Programming Languages and Computer Architecture*, volume 201, pages 220–237. Springer Verlag, 1985.

[Vas90] Jesper Vasell. Implementing functional programming languages on wavefront arrays. Licentiate thesis 84L, Department of Computer Engineering, Chalmers University of Technology, 412 96 Göteborg, Sweden, April 1990.

[Vee86] Arthur H. Veen. Dataflow machine architecture. *ACM Computing Surveys*, 18(4), December 1986.

[VV91] Jonas Vasell and Jesper Vasell. A functional programming technique for programmable wavefront arrays. In E.F. Deprettere, editor, *Algorithms and Parallel VLSI Architectures*. Elsevier, Amsterdam, The Netherlands, (to appear in 1991).

[WSS89] Shlomit Weiss, Ilan Spillinger, and Gabriel M. Silberman. Architectural improvements for data-driven VLSI processing arrays. In *Proceedings of the 1989 Conference on Functional Programming Languages and Computer Architecture*, pages 243–259, London, 1989.

THE G-LINE
A DISTRIBUTED PROCESSOR FOR GRAPH REDUCTION

R.Milikowski and W.G.Vree

Computer Systems Department, University of Amsterdam,
Kruislaan 409, 1098 SJ Amsterdam, the Netherlands,
milikows@fwi.uva.nl.

Abstract

The G-line is a horizontally coded graph reducer. In lazy functional languages much time is used to manage the graph. A method has been developed to perform construction of a subgraph in a single parallel access to the graph memory. With a simulation, using infinite hardware, we have shown that the architecture performs well. We argue that a realistic implementation is possible.

1. Introduction

The distributed G-processor is specially designed to perform lazy graph reduction. It is a hardware implementation of the abstract Göteborg machine [1]. In the distributed G-processor the price of laziness is small as managing the graph scarcely causes cost.

Laziness is a property in which computational work is postponed until it is really necessary. Until that moment the representation of that work is stored in the graph. Often evaluation is never called for. In the second place laziness avoids doing the same work more than once. This is implemented in the graph by means of sharing. The core of a machine which implements lazy graph reduction generally consists of a heap containing the graph and a stack containing pointers into the graph. During the reduction process data is moved between the program code, the graph and the pointerstack and inside the pointerstack and the graph. Operations on the graph are mostly seen as a burden. Constructing it, keeping it up to date, consulting it and collecting garbage in it costs a lot of time [2]. Research is done to limit the work on the graph and gains have been made in different ways. Compared to the 'standard combinators' of Turner the 'program derived combinators' as used in the G-machine are more efficient allowing larger steps in the reduction process and accessing the memory substantially less [3]. Compiler directed approaches extract as much information as possible in advance from the expression to be evaluated. By means of strictness analysis building of graphs can be prevented, and by sharing analysis unnecessary updates can be avoided. However, in programs of a realistic size only a part of graph construction can be avoided.

In recently developed abstract machines more efficient representations of expressions in the heap are used. These are the 'TIM-machine' which uses closure like representations and the 'spineless tagless machine', and the most recent version of the G-machine - the <ν,G> machine [4,5,6,7]. Implementations of these reducers run on existing processors by further compilation. However, there are also well known examples of special processors designed for graph reduction. The first machines reduced SKI-combinators: SKIM and NORMA [8,9]. To assess the performance of supercombinators simulators have been built. Kieburtz designed a RISC architecture mainly based on the Göteborg machine. It contains a special designed instruction-fetch-unit and uses caching [10,11]. It makes efficient execution of sequential G-code possible. At the moment by far most the architectural work on reduction machines is directed towards building parallel reducers and concentrates on the communication between the processors. Little attention is paid in current research projects to the processor itself [12].

The distributed G-processor, designed by the first author, is based on parallel access of the graph memory and the pointerstack. Constructing a subgraph only takes a single access to the heap. This is realized by using horizontal microcode. This micro parallelism has to be distinguished from parallel redex reduction. The latter means reduction of different redexes of the graph on different machines. The micro parallelism in the G-line is parallelism inside the reduction of a single redex. First we have implemented a brute force version in which we do not bother about the amount of hardware and in which the parallelism is unbounded. The main goal is to study the validity of the architecture idea and to gather statistics about performance. The results are presented in this paper. In a next phase of our research we apply this idea to develop a realistic distributed design that could be realized in hardware.

2. Basic graph operations

The reduction process consists of various parts. Before a function is called, pointers to the arguments are collected on the top of the pointerstack. Sometimes pointers to the arguments are present elsewhere on the pointerstack. In that case these pointers are copied to the top of the pointerstack. In other cases the pointers must be fetched from the graph by an operation called 'unwinding'. Pointers to the arguments being on the top of the pointerstack, the function is called to do its work. The called function terminates its work by updating the root of the redex with the result. The result may be a value (present in a node created recently) or the root of a constructed subgraph. It may be necessary to evaluate the arguments of the function themselves before the function can use them. In that case the evaluation mechanism is executed recursively. Many variations of these operations are possible and indeed used but we will concentrate on the simplest versions. In fig 2.1 - 2.3 reduction operations in a graph with binary nodes are shown.

Fig 2.1 A subgraph representing the expression 'add (hd(cons b nil)) a' is constructed. The left part of the figure shows the pointerstack before this operation. Afterwards the top of the pointerstack points to the root of the subgraph just constructed.

The pointers a and b point to nodes somewhere in the graph. These nodes may themselves be the root of subgraphs. The graph construction is done by moving data from the program and from the pointerstack to the heap. After constructing a new subgraph a pointer to its root is at the top of the pointerstack.The most extensively used graph operations (graph construction, unwind and update) are pictured in fig 2.1. In the left of the figures the situation on the pointerstack before the operation is shown and in the right the situation afterwards.

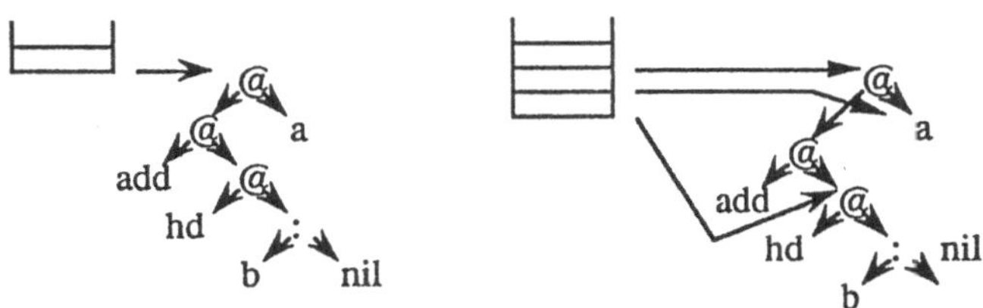

Fig 2.2 Building argument stack. On the left part of the figure the top of the pointerstack points to the root of a redex.. On the right part pointers to the arguments of the function add have been pushed.

Before unwinding starts the top of the pointerstack points to the root of a redex (fig 3.2). A redex is a function f applied to a number of arguments. These arguments must be copied to the pointerstack and will then be used by the code of the function f which is called after unwinding.

The arguments are found by inspecting the type field (tag field) of the left child node, starting at the root of the redex. If it is an apply node (@) then the right child is moved to the pointerstack and the left child is entered. This is repeated until the left child is a function node.

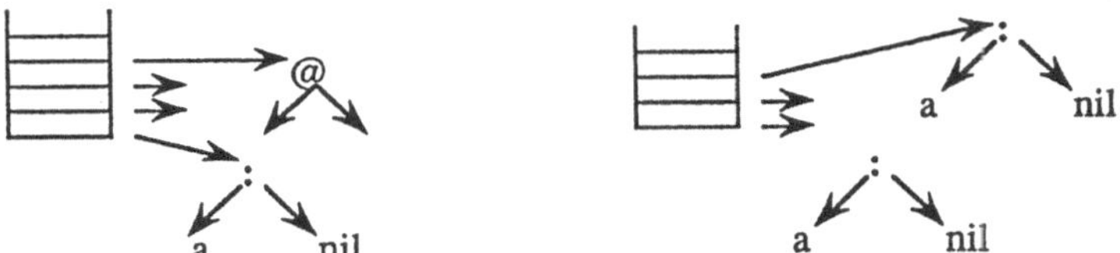

Fig 2.3 Updating; on the left is the result of some computation pointed to by the top of the pointerstack. on the right this result has been copied in the root node of the redex.

The update is done by reading both pointers of the pointerstack and then moving the contents from one node to another (fig 2.3). If not shared the source node will become garbage.

The G-code is a specially designed graph directed language to perform the graph operations [1]. Usually, this code is in a next step compiled to code for some existing machine.

2.1. Memory memory bottle-neck

Execution of the basic graph operations consumes much time of the reduction process. Different performance data have been published, using different ways to measure the performance giving an insight in the way work is spent during the reduction process.

Results have been published by Kieburtz on his S-machine - an optimized version of the G-machine concerning real execution times [13]. The ALU operations generally consume a very limited portion of time, in this benchmark from 12% to 2 % of the execution time. Nearly all the rest are moves of data from heap to heap, from stack to stack, from stack to heap from, the program to the heap and vice versa.Hartel has run a realistically sized benchmark on a variant of the G-machine [14]. He counted memory accesses, resulting in similar conclusions. We will call this the memory memory bottle-neck. One may expect parallel operations on the memory will speed up the execution.

3. Constructing a subgraph

As a running example we will take append, which is is lazy in cons.

append x y = if (null x) then y else (cons (hd x) (append (tl x) y))

Entering the else part of the expression causes construction of a suspension of append because cons is not strict in its arguments. The work stored in the else part is only done if the 'need to print' forces so. Until that moment a representation of this expression (a suspension) is stored in a graph, shown in fig 3.1. The suspensions of (hd x) and (tl x) are also constructed, though advanced compiler techniques could avoid this. Each node of the graph consists of three fields. The first field of a node - the tagfield - contains information about the node type and some additional information. Both other fields may be used to store values, e.g. floats. In function nodes the second field contains the address of the code. The second and third fields of an apply- and a cons node contain pointers. In the example the crossed fields are unused. More compact representations exist. However, we start our design from this conventional memory representation.

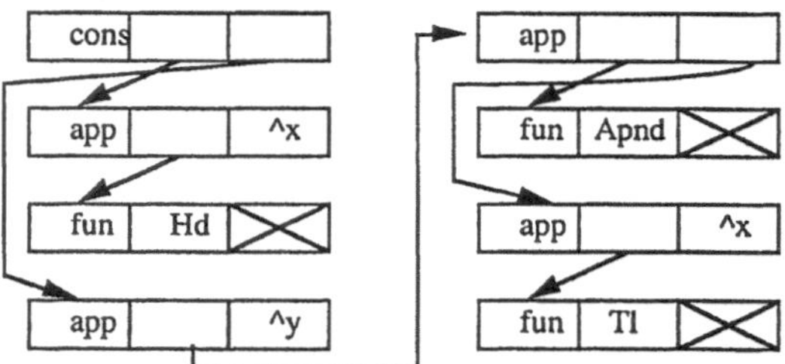

Fig 3.1. Representation of a suspension in the heap as written after execution of a G-code sequence.

The work to be done when writing this graph includes 21 accesses of the graph memory (all fields without a cross). The contents of the tag fields are known at compile time and are normally stored in the program code. The offsets inside the reduct can be computed by the compiler in advance. At the moment the address is known at which construction of this piece of graph starts all 'local pointers' (pointers inside the subgraph) can be computed. Only the pointers ^x and ^y are unknown at compile time and must be fetched from the pointerstack at runtime. Writing this subgraph in a conventional machine requires multiple accesses to the program code, the heap and the pointerstack.

4. Micro parallel architecture

A subgraph to be written will be considered as a vector of which the elements are the fields of the subgraph. The architecture of the G-line is organized in such a way the elements of this vector can be written in parallel. In contrast to vectors in numerical computing the fields of these vectors are not constructed uniformly. The contents of the various fields of a single subgraph may originate from the operand of the code, from the pointerstack or from an address computed at runtime. The mechanism to write a single field is rather complex, actually nearly as complex as a whole conventional G-machine.

In the G-line the graph memory is divided in memory banks. All memory banks are managed by their own G-machine. The global buses that connect these G-machines are at the same time the address and data bus of the heap. These G-machines are equal and execute synchronous nearly the same code when a graph access has to be done. A write on the memory can be handled locally by the appropriate machine. No communication with the other G-machines is required. In that way it is possible that a number of writes can be done in parallel which is the case in graph construction. A read on the heap in the conventional architecture means the data is moved from the heap via the bus to the G-machine. In the G-line data is moved from one of the memory banks via the global buses to all replicas of the G-machine. That G-machine that manages the specific memory bank to be read performs locally the addressing of the memory bank and takes control of the global buses. All local machines are *nearly* complete copies of the whole G-machine in hardware and in software. They differ by the fact they manage different memory banks. Furthermore those sections of the program that deal with graph writing are distributed over the different machines, as the graph writing itself is distributed over the different machines. Local G-machines contain replicas of the valuestack, the pointerstack, the ALU , the program counter (PC) and the control store. Care has to be taken that the replicas of the stacks and the PC remain identical in the local G-machines. This invariant will be called the *shadow condition* which has to hold always after execution of an instruction. There is no hierarchy between the machines and no scheduler. In the next paragraphs this architecture and the way a program runs on this distributed processor will be described.

4.1. G-line

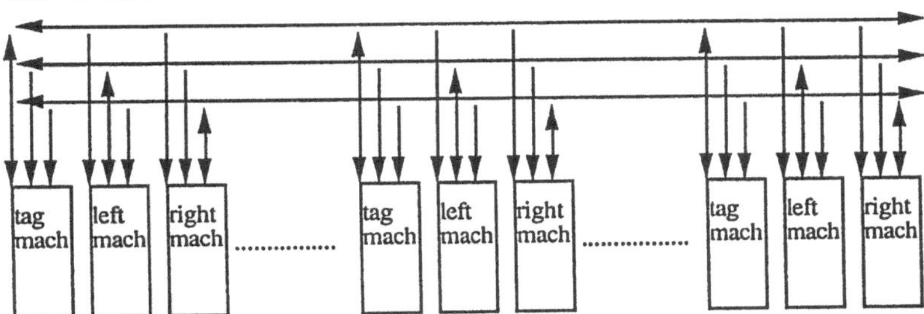

Fig 4.1 G-line. Each of the machines is a replica of the G-machine containing a memory bank of the heap.

The number of G-machines in the G-line architecture of fog 4.1 is equal to the maximum number of fields of a subgraph that will possibly be written in a specific program.

As each node consists of three fields, three G-machines together forming a cluster, are needed to write those fields in parallel.

First, we discuss the different units of the G-machines. The tag, left and right machines are not exactly equal. Fig 4.2 shows a left machine.

The heap: The heap in fig 4.2 contains the graph and consists of 3n memory banks (sometimes called local memories), in which n is the number of clusters. In an alternating order these are memories containing tag fields, left fields and right fields. A heap cell consists of a word from all three local memories in a cluster on a single global address. The tagfield normally contains the tag and the number of arguments of a function, and some runtime information. According to the shadow condition the heap pointers are equal in all local machines. To guarantee that the heap pointers in all machines are equal after garbage collection, a compacting garbage collector will be used. The heap has bi-directional communication with the global buses. The tag memory with the tag bus etc.

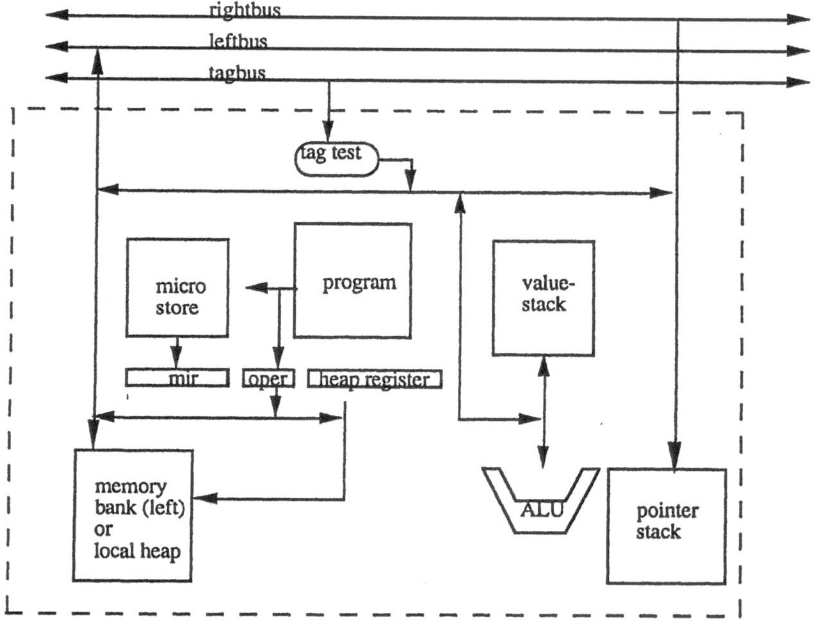

Fig 4.2 Left machine

Program memories are identical in each machine. The start address of a specific function are the same everywhere. The code is equal, except for the MKGRAPH instruction, which is introduced in the next section.

The microstore: Dividing the execution of instructions over a number of machines, makes the operation to be done in each G-machine very simple, especially after some minor optimizations, which we treat in the final paragraph. UNWIND remains the most complex instruction.

The heap register: The heap register contains the local heap pointer which is equal in all G-machines. During graph construction it may be necessary to write locally a remote pointer (to a node constructed in parallel in a different cluster). It is computed locally from the heap pointer and the operand of the MKGRAPH instruction, in this case containing the number of the remote cluster.

The pointerstack: Each machine contains a copy of the pointerstack. The pointerstacks are kept identical during execution of the program. The pointerstack has only input from the global right bus and only locally delivers output

ALU and valuestack: The ALU only operates on data present on the valuestack. The valuestack does not communicate with the outside world.

The buses: The distributed G-processor has three global buses. Each G-machine has access to the tag the left and the right bus. Each time only a single machine will do write operations to the buses. It is analogous to the way data in a memory consisting of different banks are put on the bus. An address on the address bus will cause only a single memory bank to become active. All three buses are data buses. Global addressing is only used in case of memory indirections. The address then originates from data, in casu a left field. So global addressing is only done by the left bus. Locally only the heap pointer addresses the heap.

In the G-line only the contents of heap words are transported over the global buses. The state of the machine uniquely determines from which global heap address this word is fetched and so which local machine will access the buses. No contention is possible and no arbitration is required. The code produced by the compiler and the synchronous execution of the processes guarantee this. The tag bus is often used for testing as a part of the program control.

The clock: Besides the buses the clock is the only global hardware. The clock synchronizes the G-machines.

5. Programming the G-line

To present the working of the machine we show the append program as it is compiled for the the conventional G-machine, using the compilation scheme of Johnsson [1].

Instructions 7 - 17 construct the subgraph of fig 3.1. As each G-machine in the G-line has to write only a single field of a specific subgraph, this sequence will be replaced in each G-machine by code to write only that field.

```
0   FAPND   PUSH    0   || copy pntr to x to top of pointerstack
1           EVAL        || evaluate x
2           NULL        || test if x = null
3           JFALSE  1
```

```
                PUSH        1         || copy pntr to y to top of pointerstack
                EVAL                  || evaluate it
                JMP         2

      LABEL 1   PUSHFUN     FHD       || else; construction of suspension
                PUSH        1         || copy pntr to y to top of pointerstack
                MKAP                  || make apply node
                PUSHFUN     FAPND     || push pointer to append node
                PUSHFUN     FTL       || push pointer to tail node
                PUSH        3         || copy pntr to y to top of pointerstack
                MKAP                  || make apply node
                MKAP                  || make apply node
                PUSH        3         || copy pntr to y to top of pointerstack
                MKAP                  || make apply node
                CONS                  || make cons node

      LABEL 2   UPDATE      3         || update root of redex with result
                RET         2         || pop arguments and return to caller
```

Fig 5.1 Sequential G-code of the function append x y = if (null x) then y else (cons (hd x) append (tl y)).

This varying code will be the instruction 'MKGRAPH_Action Field'. Four different MKGRAPH_ opcodes exist, according to the the kind of action needed to write a field.

MKGRAPH_heap n: construct global heap address by prefixing local heap pointer by and write this into new local heap field
MKGRAPH_stack n: copy n-th element relative to top of pointerstack in new heap field
MKGRAPH_copy n move n into new heap field
MKGRAPH_nop machine performs nop on heap

The nodes of a new subgraph are written in parallel as a vector on the same local addresses in the different clusters of the G-line. An offset between two nodes in a subgraph as in fig 3.1 is translated into an distance between clusters in the G-line. This distance, or the number of the remote processor, is stored in the operand of the MKGRAPH_addr instruction. As the local heappointers are equal a global pointer to the remote node can be constructed at runtime locally by concatenating the contents of the heap pointer register with the operand of the MKGRAPH_addr instruction. If a operand has to be stored in the graph, a 'MKGRAPH_copy n' instruction is used. This is the case for values in the left and the right fields. Tag fields are always constructed by a MKGRAPH-copy instruction. Four MKGRAPH_ instructions are defined.

The transformed program is shown in fig 5.2.

Fig 5.3 shows all local MKGRAPH instructions derived from the sequential program. The first column denotes the machine (number of the machine cluster and machine in the cluster). In the tag machines always the operand of the MKGRAPH instruction is written into the heap. One may consult fig 3.1 to see how each MKGRAPH instruction constructs a field of this subgraph. The only difference is that the arity of the function is added in the operand.

```
0   FAPND      PUSH              0
1              EVAL
2              NULL
3              JFALSE            1

4              PUSH              1
5              EVAL
6              JMP               2

7   LABEL 1    MKGRAPH_Action    Number    || construct field subgraph
                                              operand differs per machine
8   LABEL 2    UPDATE            3
9              RET               2
```

Fig 5.2 Code of append in G-machines of the G-line.

MACHINE	OPCODE	OPERAND		COMMENTS
0 tag machine	MKGRAPH_copy	cons		
1 tag machine	MKGRAPH_copy	app		
2 tag machine	MKGRAPH_copy	fun	1	
3 tag machine	MKGRAPH_copy	app		
4 tag machine	MKGRAPH_copy	app		
5 tag machine	MKGRAPH_copy	fun	2	
6 tag machine	MKGRAPH_copy	app		
7 tag machine	MKGRAPH_copy	fun	1	
0 left machine	MKGRAPH_heap	1		|| addr(1,hp)
1 left machine	MKGRAPH_heap	2		|| addr(2,hp)
2 left machine	MKGRAPH_copy	17		|| code addr FHD
3 left machine	MKGRAPH_heap	4		|| addr(4,hp)
4 left machine	MKGRAPH_heap	5		|| addr(5,hp)
5 left machine	MKGRAPH_copy	0		|| code addr FAPND
6 left machine	MKGRAPH_heap	7		|| addr(7,hp)
7 left machine	MKGRAPH_copy	10		|| code addr FTL
0 right machine	MKGRAPH_heap	3		|| address(3,hp)
1 right machine	MKGRAPH_stack	0		|| copy pntr to x from pstack
2 right machine	MKGRAPH_nop			|| unused field
3 right machine	MKGRAPH_stack	1		|| copy pntr to y from pstack
4 right machine	MKGRAPH_heap	6		|| address(6,hp)
5 right machine	MKGRAPH_nop			|| unused field
6 right machine	MKGRAPH_stack	0		|| copy pntr to x from pstack
7 right machine	MKGRAPH_nop			|| unused field

Fig 5.3 Instructions to construct the subgraph of FAPND in a distributed way.

In the comment is denoted which field of the subgraph of fig 3.1 is written to the local memory. By address(3,hp) is meant the address that is constructed by adding 3 as most significant bits to the local heap pointer. The common code address of a function is the address where the code of this function starts. In this example, though not shown, the code of FTL starts at 10 and of FHD at 17.

5.1. Running the program

All G-machines are started to run together. All execute, synchronized by the global clock, the same instruction. The way the G-instruction is executed in the different G-machines depends on the type of the machine (tag left or right) and on the kind of instruction it concerns. The G-instructions can be split into three groups, according to their implementation in the G-machines of the G-line.

no_heaps: ADD (and the other arithmetic and logical operations), JMP, JFALSE, LABEL, SLIDE n, PUSH n. These instructions do not access the heap, they only use pointerstack, valuestack and ALU and are executed identical in all G-machines.
old_heaps: PRINT, GET, UPDATE, EVAL, UNWIND, NULL, HEAD, TAIL. Instructions access *existing* graph in the heap that is globally addressed. Depending from the fact if the addressed heap cell is in their own local memory the execution is partly different over the G-machines.
new_heaps: MKGRAPH_ Field. It creates new graph in the heap and accesses only the local memory. Is executed differently in different G-machines

Only the old_heaps communicate with the other G-machines in the G-line. Some examples will show the way the execution of the code is implemented.

ADD: All machines arrive at the same time at the instruction ADD and then their local ALU adds the top two elements of their valuestacks, which were the same in all machines, and replace these by the result of the addition. After execution of ADD all valuestacks are identical again and the shadow condition is fulfilled.

GET. The instruction GET in the conventional G-machine reads the value (integer or boolean) from the heap node pointed at by the top of the pointerstack. This value is pushed on the valuestack. In the distributed machine each G-machine now looks if the address on top of the pointerstack points in its own local memory. If this is actually the case it moves the value from this node to the global left bus else it waits. Next, the rest of the GET instruction again is executed identically by all G-machines. Each one pushes the contents of the global left bus on its own local valuestack. Again: the shadow condition remains fulfilled.

EVAL AND UNWIND EVAL starts like GET, resulting in the root of the redex being on the buses. EVAL proceeds in all G-machines with a test on the tag bus. If the tag is a "cons" , a "number" etc, the contents of the node is available on left and right bus for processing by the next instruction. If it is an apply node each G-machine starts execution of UNWIND:

The contents of the root of the redex is still on the buses. Two actions are undertaken by each G-machine:
- The contents of the right bus is pushed on the pointerstack (it is the pointer to an argument)
- The contents of the left bus is inspected. It contains a pointer to a node one step deeper down in the spine. If it points in the local memory the contents of that memory location is moved to the appropriate bus. All G-machines repeat the UNWIND algorithm until a function-node appears on the buses. Then the contents of the left bus - the code address of this function - is moved to the PC in each G-machine.

All pointerstacks always remain identical (shadow condition).

MKGRAPH Field We have already described to which micro operations the variable part of the instruction is decoded. Furthermore, the address of the root, computed from the local heap pointer, is pushed on the pointerstack in all machines. The local heap pointers are incremented all over the G-line.

Graph construction instructions from the conventional G-machine (MKAP, MKCONS, PUSHINT i, PUSHNIL, PUSHFUN f) are superseded by MKGRAPH in the G-line.

Special attention has to be paid to the UNWIND instruction in case the depth of the spine (d) is larger than the arity k of the function F at the bottom of the spine. In the conventional G-machine unwinding is done by first pushing pointers to the spine nodes on the pointerstack. In the next step those spine pointers belonging to the current redex are replaced by pointers to arguments. This is called restructuring of the pointerstack. In case d > k the top part of the spine will not be restructured. Restructuring is a costly operation. Our strategy is inverse restructuring or rewinding. Always pointers to the arguments are pushed on the pointerstack. In case d > k the arguments have to be replaced by the spine node pointers. Starting at the top node an inverse restructuring is performed. The argument pointer is replaced by the spine node pointer. This operation is repeated (d - k) times. An inverse restructuring operation certainly is not more costly than restructuring in the G-line. Moreover restructuring always is required in the conventional G-machine. As in our benchmark in less than 3 % of all UNWIND operations d > k, inverse restructuring is required in only few cases.

6. Generating the code

The G-code for the G-line is derived from the sequential G-code by a backend. This backend simulates the execution on the stacks and the heap. Those parts of the G-code which construct a subgraph (instruction 7 - 17 in the example) will do this in the simulated heap resulting in a piece of memory like the append subgraph in fig 3.1. Next we assign each field of this graph to a G-machine and MKGRAPH instructions are generated to construct this field.

7. Efficiency

We have compared the number of memory accesses in the conventional architecture with those in the G-line. To count the number of memory accesses in the conventional reducer is, with some minor differences, similar to the way Hartel has counted [14]. To each G-instruction a number of accesses to the different stores (heap and stacks) is assigned. The representation of the a subgraph in fig 3.1 also contains function nodes (for Hd, Tl and Apnd). This is not necessary as these nodes are preloaded by the machine and the subgraph only contains a pointer to those function nodes. We have corrected our measurements to account for this optimization. The result of this way of counting in the sequential G-machine is in the first column of fig 7.1. We have used peep hole optimization concerning the accesses on the valuestack. If a boolean is stored on the valuestack (p.e. after a tag test) and in the next instruction is popped again (in executing JFALSE) a short cut is used. The second column contains the results of the G-line. To count the number of accesses in the G-line we proceeded as follows. If all G-machines perform an identical memory operation (e.g read on a stack) in parallel, it is counted as a single memory access. If a specific G-machine reads the memory and the other machines are idle counting is continued in an active machine. This is the case when executing the old-heaps. The maximum number of accesses by MKGRAPH in a single G-machine is one heap access, one access to the pointerstack if a data from this stack has to be moved to the heap and one access to the pointerstack to push a pointer to the root of the subgraph. To make sure the maximum work is measured, counting during execution of MKGRAPH is done in the the cluster that constructs the root of the subgraph.

We have used a set of benchmark programs that are compiled from SASL, a run time typed functional language, to Johnsson's intermediate language and then to G-code. 'Schedule' calculates an optimum schedule of 7 parallel jobs with a branch and bound algorithm [15]. 'Hamming' prints in ascending order the first 50 natural numbers whose primes are 2,3 and 5 only. 'Paraffine' enumerates in order of increasing size the first six paraffine molecules.

	sequential	*micro parallel*	*optimized*
		paraffine	
dump (context switch)	52 K	52 K	52 K
valuestack	77	77	77
pointerstack	429	209	143
heap	377	132	78
total	936	471	351
		hamming	
dump	98 K	98 K	98 K
valuestack	78	78	78
pointerstack	590	299	198
heap	519	192	109
total	1284	668	484
		schedule	
dump	58 K	58 K	58 K
valuestack	47	47	47
pointerstack	539	301	189
heap	498	188	97

Fig 7.1 Memory accesses (in thousands) in a conventional G-machine and in the G-line

In fig 7.1 we show the number of register saves, labeled "dump" due to a context switch (two register saves when a function is called and two more when returning from the function). The dump is not treated in the architecture because it is often implemented on the valuestack.

7.1. Optimizations

Some optimizations fit nicely in the architecture of the G-line because they are implemented very simply in the hardware or because they make further use of the micro parallelism. Two examples show this.

1) Care is taken that after returning from a function call the resulting value is on the buses. This is guarantied by the G-machine that does the update. As the result data is moved via a local bus to the local heap, the only thing to do is to enable the appropriate bus register. This means that after returning from EVAL no instruction needs to access the heap. This concerns instructions like GET, HEAD, TAIL which now become more simple.

2) In most reducers of supercombinators updating is optimized. E.g the function f x y = x + y is compiled to the G-code sequence PUSH 0, EVAL, GET, PUSH 2, EVAL, GET, ADD, MKINT, UPDATE 3, RET 2. MKINT reads the result of the addition from the valuestack, creates a new node with this value in the heap. This node is immediately used to update the root of the redex and made to garbage. The instructions MKINT, UPDATE 3 will be combined into something like UPDINT 3 [2]. This instruction directly does an update with the value at the top of the valuestack. In the G-line this can easily be implemented. The (left)machine containing the node to be updated fetches the value from its local value stack and moves it to its local heap. The tag-machine belonging to the same cluster moves the tag of type 'int' to the tag heap.

The same can be done with MKGRAPH Field, UPDATE n. Then creating a new node for the root of the subgraph can be avoided. As the value with which to update is available in each G-machine it can immediately be written in the root of the redex which now has been reduced. The cost of the combined execution of these instructions varies between 2 heap access + 2 stack access and 1 heap access + 1 stack access. The results of these optimizations are recorded in the third column of fig 7.1. These optimizations do not affect the compiler that generates the sequential G-code. In the experiment described here we have not at all done any static compiler analysis. It would be fair to compare the optimized parallel simulation results with simulation results from an optimized sequential simulation. This depends however from the machine architecture one would choose.

7.2. Arguments

We have counted the number of arguments that are accessed in the heap when running the program. These include arguments that first have to be evaluated before they can be used for some ALU or list constructing operation.

	heap arguments used	*heap accesses (optimized)*
paraffine	33954	78.10^3
hamming	51418	109.10^3
schedule	41326	97.10^3

Fig 7.2 Heap arguments used in relation to heap accesses.

8. Conclusions about the G-line and future work

The efficiency gain seems to be rather independent of the program. The number of accesses to the pointerstack have been reduced to about one third and to the heap to about one fifth. In average the cost of using an argument (including those that consist of 2 fields as floats and cons nodes) in the heap is 2 - 3 accesses. Only slightly more than once storing a value in the heap and once loading it. The cost of laziness thus has disappeared. We conclude that our architecture performs well.

The use of the graph memory has become an advantage because it offers the possibility of a horizontally coded reducer.

We used infinite hardware in this simulation of the G-line. When hardware is reduced to a feasible size some of the micro parallelism is lost. The following restrictions on hardware can be applied to obtain a realistic machine.

- The number of G-machines is limited. If the size of the subgraph is larger than the number of G-machines, the subgraph is wrapped around the G-line. Some G-machines have to write more than one memory field. If the number of G-machines is N and the size of a subgraph S it now requires (S DIV N) memory cycles to construct a subgraph. We have measured the size of the subgraphs constructed during execution of the programs.

size	*schedule*	*paraffine*	*hamming*
0 - 4	16760	2384	12272
5 - 8	2281	4508	4980
9 - 16	1488	741	0
> 16	0	3	0

Fig 8.1 Size of constructed subgraphs.

Looking at these programs a size of 8 clusters will not cause much loss of parallelism.

- A cluster of tag- left- and right machine can share a number of resources. Only a single valuestack and ALU and a single connection with each bus is needed per cluster. No speed will be lost in this way. Sharing of the pointerstack might cause some loss in parallelism. To construct a cons node or an apply node two different elements of the pointerstack need to be accessed. It costs one more memory cycle in the construction of some of the subgraphs.

8.1. No more replicas

A more radical change of this architecture is realized by splitting the G-line in a global machine and a distributed machine. Making of replicas is then avoided. ALU, valuestack and pointerstack only are present in the global machine. The shared part of the program code resides in the global machine and the MKGRAPH instruction is split in a distributed and a global part. The no-heap instructions only are executed in the global machine, UNWIND and UPDATE in cooperation between global and distributed machine. MKGRAPH is executed in the distributed machine with some support of the global machine. To execute UNWIND, UPDATE and MKGRAPH the local machines contain micro code.

This architecture is subject of current research in which trade offs are made between the amount of micro parallelism and the amount of hardware used. We expect, however, that some loss in parallelism will be outweighed by typical hardware optimizations which generally are not yet included in the design of the G-line.

8.2. To realistic software

We used code generated from an untyped language. The compilation to G-code was straight forward using Johnsson's compilation rules [1]. No compiler optimizations were applied. In our current low level simulations, we use G-code derived from LML, a typed functional language [16]. The G-code generated from typed languages is more efficient than the G-code we used in the experiments described above and better suited to derive parallel microcode.

9. Acknowledgements

We wish to thank Pieter Hartel, Henk Muller and Rutger Hofman for the comments on the draft(s) of this paper.

10. References

[1] T.Johnsson; Efficient compilation of lazy evaluation; Sigplan Notices 19(6):58,69, June 1984.

[2] Simon L.Peyton Jones; Implementation of graph reduction; Prentice Hall; London, 1987.

[3] D.A. Turner, "A new implementation technique for applicative languages"; Software Practice and Experience 9(1) pp 31-49 (A jan. 1979).

[4] J.Fairbairn, S.Wray; TIM: A simple lazy abstract machine to execute supercombinators; Third conference on functional programming and computer architecture; Portland, Oregon, USA; Proceedings 34-35, September 1987.

[5] Guy Argo; Improving the three instruction machine; Proceedings of the FPLCA, pp 100-115, London 1989.

[6] Simon L.Peyton Jones, Jon Salkild; The spineless tagless G-machine; Proceedings of the FPLCA, pp 184-201, London 1989.

[7] L.Augustsson, T.Johnsson; Parallel graph reduction with the <nu,G>-machine; Proceedings of the FPLCA,pp 202-213, London 1989.

[8] W.R.Stoye, T.J.W.Clarke, A.C.Norman; Some practical methods for rapid combinator reduction; Proc. of 1984 ACM Conf on Lisp and Functional Prog, pp 159-166, aug 1984.

[9] M.Scheevel, NORMA, a normal-order combinator reduction machine, colloquium presented at Oregon Graduate Center, july 1984.

[10] R.B.Kieburtz; The G-machine, a fast graph-reduction evaluator; Second Conference on functional Languages and Computer Architecture, LNCS 201, Nancy, 1985; Springer Verlag.

[11] R.B.Kieburtz; A Risc architecture for symbolic computation; Sigplan Notices 22(10):146-155, October 1987.

[12] Willem Vree, Design considerations for a parallel reduction machine; Ph. D. thesis, pp 27-50, Amsterdam, 1989

[13] R.B.Kieburtz and B. Agapiev; Optimizing the evaluation of suspensions, pp 267 - 282 in Proceedings of the Workshop of functional languages, ed. T.Johnsson,

S.L.Peyton Jones, K.Karlsson, Dept of Computer Science, Chalmers Univ. of Technology, Göteborg, Aspensas, 1988.

[14] P.H.Hartel; Performance of lazy combinator graph reduction; To be published in "Software practice and experience".

[15] W.G Vree, P.H.Hartel; Parallel graph reduction for divide-and conquer applications; Part II -program performance; Internal Report D-20, PRM-Project, December 1988.

[16] T.Johnsson; Compiling Lazy Functional Languages, Ph.D. thesis, Göteborg, 1987.

Appendix

SPECIFICATION OF THE G-LINE

M: local machine, o: output, c: codestack, s: pointerstack, v: valuestack,
G: graph node, E: environment, D: dump, H: local heappointer.

<M_m,o,EVAL.c,n.s,v,G_a[n=int,i,-],Bus,E,D,H> ->
<M_a,o,EVAL.c,n.s,v,G,Bus[int,i,-],E,D,H> ->
<M_m,o,c,s,v,G,Bus[int,i,-],E,D,H>
(similarly for booleans)

<M_m,o,EVAL.c,n.s,v,G_a[n=app,n_l,n_r],E,D,Bus,> ->
<M_a,o,EVAL.c,n.s,v,G,Bus[app,n_l,n_r],E,D,H> ->
<M_m,o,UNWIND.(),n_r.n.(),v,G,Bus[app,n_l,n_r],E,(c,s).D,H>

<M_m,o,UNWIND.c,n_j.n_{j-1}...n_0.(),v,G,Bus[app,n_l,n_r],E,(c,s).D,H> ->
<M_m,o,UNWIND.(),n_r.n_j.n_{j-1}...n_0.(),v,G,Bus[app,n_l,n_r],E,(c,s).D,H> ->
<M_a,o,UNWIND.(),n_r.n_j.n_{j-1}...n_0.(),v,G_a[n_l=tag,left,right],E,D,Bus[tag,left,right],H>

<M_m,o,UNWIND.(),n_j.n_{j-1}...n_0.(),v,G,Bus[fun k, c'],E,(c,s).D,H> ->
<M_m,o,c',s',v,G,Bus[fun k, c'],E,D,H> *k=j+1*

<M_m,o,UNWIND.(),n_j.n_{j-1}...n_0.s,v,G,Bus[fun k, c'],E,(c,s).D,H> ->
<M_m,o,EVAL.c,s,v,G,Bus[fun,c',k],E,D,H> *k>j+1*

<M_m,o,UNWIND.(),n_j.n_{j-1}...n_0.(),v,G,Bus[fun k, c'],E[fun c',k],(c,s).D,H> ->
<M_m,o,REWIND.c',n_j.n_{j-1}..n_0.(),v,G_a[n_0=app,p_l,p_r],Bus[app,p_l,p_r],E[fun c',k],(c,s).D,H>
k<j+1

<M_m,o,REWIND.c,n_j...n_{j-k}..n_1.n_0.s,v,G,Bus[app,p_{le},p_{ri}],E[fun c',k],D> ->
<M_m,o,REWIND.c',n_j..n_{j-k}..n_{le}.n_0.s,v,G,Bus,E[fun c',k],D>
<M_a,o,REWIND.c',n_j.n_{j-1}...n_{le}.n_0.s,v,G_a[n_{le}=app,p_l,p_r],Bus[app,p_l,p_r],E[fun c',k],D,H> ->
k>j+1

<M_m,o,REWIND.c',n_j...n_{j-k}..n_0.s,v,G,Bus[app,p_l,p_r],E[fun c',k],D> ->
<M_m,o,c',n_j.n_{j-1}...n_0.s,v,G,Bus[app,p_l,p_r],E[fun c',k],D,H>
k=j+1

<M_m,o,PRINT.c,n.s,v,G_a[n=int,i,-],Bus,E,D,H> ->
<M_a,o.i,c.PRINT,n.s,v,G,Bus[int,i,-],E,D,H> ->
<M_m,o.i,c,s,v,G,Bus[int,i,-],E,D,H>
(similarly for booleans)

<M_m,o,PRINT.c,n.s,v,G_a[n=cons,p_l,p_r],Bus,E,D,H> ->
<M_a,o.i,c.PRINT,n.s,v,G,Bus[cons,p_l,p_r],E,D,H> ->
<M_m,o,PRINT.EVAL.PRINT.EVAL.c,p_l.p_r.s,v,G,Bus[cons,p_l,p_r],E,D,H>

<M_m,o,UPDATE k.c, n_0...n_k.s,G_a[n_0=tag,left,right],Bus,E,D,H> ->
<M_a,o,UPDATE k.c, n_0...n_k.s,G,Bus[tag,left,right],E,D,H> ->
<M_m,o,c, n_1...n_k.s,G,Bus[tag,left,right],E,D,H> ->
<$M_{a'}$,o,c, n_1...n_k.s,$G_{a'}$[n_k=tag,left,right],Bus[tag,left,right],E,D,H>

<M_m,o,RET n.c,n_1...n_n.n_0.s,G,Bus[tag,left,right],E,D,H> ->
<M_m,o,RET n.c,n_0.s,G_a[n_0=tag,left,right],Bus[tag,left,right],E,D,H> ->
<M_m,o,c,n_0.s,G,Bus[tag,left,right],E,D,H> $tag \neq app,fun$

<M_m,o,RET n.c,n_1...n_n.n_0.s,G,Bus[tag,left,right],E,D,H> ->
<M_k,o,RET n.c,n_0.s,G_0[n_0=tag,left,right],Bus[tag,left,right],E,D,H> ->
<M_m,o,EVAL.c,n_0.s,G,Bus,E,D,H> tag = app,fun

<M_m,o,PUSH k.c,n_0....n_k.s,v,G,Bus,E,D,H> ->
<M_m,o,c,n_k.n_0..n_k.d.s,v,G,Bus,E,D,H>

<M_m,o,MKINT.c,s,i.v,G,E,D,Bus,H> ->
<M_0,o,c,n.s,i.v,G_0[n=int,i,-],Bus[int,i,-],E,D,H[lhp]> ->
<M_m,o,c,lhp.s,v,G,Bus,E,D,H[lhp+1]>
(similarly for booleans)

<M_a,o,MKGRAPH_copy n.c,s,v,G,Bus,E,D,H[lhp]> ->
<M_a,o,c,lhp.s,v,G[..,n,..],Bus,E,D,H[lhp+1]>

<M_a,o,MKGRAPH_heap n.c,s,v,G,Bus,E,D,H[lhp]> ->
<M_a,o,c,lhp.s,v,G[..,n.lhp,..],Bus,E,D,H[lhp+1]>

<M_a,o,MKGRAPH_stack n.c,p_1..,p_n.s,v,G,Bus,E,D,H[lhp]> ->
<M_a,o,c,lhp.p_1..,p_n.s,v,G[..,p_n,..],Bus,E,D,H[lhp+1]>

<M_a,o,MKGRAPH_nop .c,s,v,G,Bus,E,D,H[lhp]> ->
<M_a,o,c,lhp.s,v,G,Bus,E,D,[lhp+1]>

<M_m,o,ALLOC.c,s,v,G,Bus,,E,D,H[lhp]> ->
<M_0,o,c,lhp.s,G_0[hole,_,_],Bus,E,D,H[lhp+1]>
<M_m,o,c,n_k..n_1.s,G,Bus,E,D,H> $m \neq 0$

<M_m,o,PUSH k.c,n_0....n_k.s,v,G,Bus,E,D,H> ->
<M_m,o,c,n_k.n_0..n_k.d.s,v,G,Bus,E,D,H>

<M_m,o,c.GET.c,n.s,v,G_a[n=int,i,-],E,D,Bus,E,D,H> ->
<M_a,o,GET.c,n_a.s,v,G,Bus[int,i,-],E,D,H> ->
<M_m,o,c,s,i.v,G,Bus[int,i,-],E,D,H>
(similarly for booleans)

<M_m,o,ADD.c,s,i_2.i_1.v,G,Bus,E,D,H> ->
<M_m,o,c,s,i_2+i_1.v,G,Bus,E,D,H>
(similarly for all arithmetic and logical operations)

<M_m,o,JFALSE .c,s,true.v,G,Bus,E,D,H> ->
<M_m,o,JMP.c,s,v,G,Bus,>

<M_m,o,JMP l,....LABEL l.c,s,v,G,Bus,E,D,H> ->
<M_m,o,LABEL.c,s,v,G,Bus,E,D,H>

<M_m,o,LABEL l.c,s,v,G,Bus,E,D,H> ->
<M_m,o,c,s,v,G,Bus,E,D,H>

<M_m,o,HD.c,n.s,v,G_a[n=cons,p_1,p_2],E,D,Bus,E,D,H> ->
<M_a,o,HD.c,n,s,v,G,Bus[cons,p_1,p_2],E,D,H> ->
<M_m,o,c,p_1.s,v,G,Bus[cons,p_1,p_2],E,D,H>

<M_m,o,TAIL.c,n.s,v,G_a[n=cons,p_1,p_2],E,D,Bus,E,D,H> ->
<M_a,o,TAIL.c,n.s,v,G,Bus[cons,p_1,p_2],E,D,H> ->
<M_m,o,c,p_2.s,v,G,Bus[cons,p_1,p_2],E,D,H>

<M_m,o,c.NULL,n.s,v,G_a[n=cons,p_1,p_2],Bus,E,D,H> ->
<M_a,o,NULL.c,n.s,v,G,Bus[cons,p_1,p_2],E,D,H> ->
<M_m,o,c,s,false.v,G,Bus[cons,p_1,p_2],E,D,H>

<M_m,o,c.NULL,n.s,v,G_a[n=nil,-,-],Bus,E,D,H> ->
<M_a,o,NULL.c,n_a.s,v,G,Bus[nil,-,-],E,D,H> ->
<M_m,o,c,s,true.v,G,Bus[nil,-,-],E,D,H>

Remarks

1)'Bus' denotes the contents of all three buses. Mind that the contents of the buses on entrance or on exit of the rule sometimes is denoted. E.g on exit of EVAL this is the case, so after executing this instruction the next one can be sure the result of EVAL can be read from the buses. UNWIND makes use of this.

2) In G_a[n=_,_,_] the subscript of G_a specifies the machine (and local graph) in which n is pointing.

3) The transition rules contain no optimizations like contraction of MKINT and UPDATE

4) We have separately defined MKINT etc and ALLOC. They could be special cases of the MKGRAPH instruction.

5) An special instruction REWIND is introduced in this diagram to clarify the action in case of unwinding a spine with number of arguments smaller than the depth of the spine.

The Derivation of Distributed Termination Detection Algorithms from Garbage Collection Schemes

(Extended Abstract)

*Gerard Tel**
Department of Computer Science, Utrecht University,
P.O. Box 80.089, 3508 TB Utrecht, The Netherlands.
(Email: gerard@cs.ruu.nl)

Friedemann Mattern
Department of Computer Science, Kaiserslautern University,
P.O. Box 3049, D 6750 Kaiserslautern, Fed. Rep. Germany.
(Email: mattern@informatik.uni-kl.de)

Abstract. It is shown that the termination detection problem for distributed computations can be modeled as an instance of the garbage collection problem. Consequently, algorithms for the termination detection problem are obtained by applying transformations to garbage collection algorithms. The transformation can be applied to collectors of the "mark-and-sweep" type as well as to reference counting garbage collectors. As an example, the scheme is used to transform the weighted reference counting protocol.

1 Introduction

A substantial amount of the research efforts in distributed algorithms design has been devoted to the problem of detecting when a distributed computation has terminated. There are several reasons for the impressive number of publications on this subject. First, as the problem has shown up under varying model assumptions and there are several solutions for each model, a really large number of different algorithms has emerged. All these algorithms were published separately, because unifying approaches, treating a number of algorithms as a class, have been rare. Second, the problem of termination detection, being sufficiently easy to define and yet non-trivial to solve, has been seen as a good candidate

*The work of this author was supported by the ESPRIT II Basic Research Actions Program of the EC under contract no. 3075 (project ALCOM).

to illustrate the merits of design or proof methods for distributed algorithms. Third, it has been observed that the fundamental difficulties of the termination detection problem are the same as those of other problems in distributed computing. Termination detection algorithms are related to algorithms for computing distributed snapshots [CL85], and detecting deadlocks [CMH83]. Thus the problem is seen to be important both from a practical, algorithmical, and from a theoretical, methodological point of view.

From both points of view we consider it useful to recognize general design paradigms for distributed termination detection algorithms. One such paradigm was described in [Te90]. A new paradigm is presented in this paper: it is shown that termination detection algorithms are obtained as suitable instantiations of garbage collection algorithms. A connection between the two problems was pointed out before. Tel, Tan, and Van Leeuwen [TTL88] have shown that garbage collection algorithms (of the so-called *mark-and-sweep* type, see section 1.2) can be derived from termination detection algorithms. Using a different transformation, garbage collection algorithms of the *reference counting* type can also be derived from termination detection algorithms, see section 4.1. The results in this paper further strengthen this connection by presenting a transformation in the reverse direction.

Subsections 1.1 and 1.2 introduce the termination detection problem and the distributed garbage collection problem. Section 2 describes how the termination detection problem can be formulated as garbage collecting one hypothetical object and derives the algorithmical transformation. Section 3 provides an example of the transformation. Section 4 contains some additional remarks and comments.

1.1 The Termination Detection Problem

The problem of termination detection is described formally as follows. A collection P of *processes* is considered, communicating by message passing. For the sake of simplicity it is assumed that P is a fixed collection, but the results in this paper are easily generalized to handle process creation and deletion as well. A process is either *passive* or *active*. *Active* processes can send messages, but *passive* processes cannot. An *active* process can spontaneously become *passive*, but a *passive* process can become *active* only on receipt of a message. Formally, the allowed actions of the processes are described as follows. (In all programs to follow, actions are *atomic* and braces ("{" and "}") enclose a guard for an action.)

$\mathbf{S}_p$: $\{\ state_p = active\ \}$
send a message $\langle M \rangle$

$\mathbf{R}_p$: { A message has arrived }
receive message $\langle M \rangle$; $state_p := active$

$\mathbf{I}_p$: { $state_p = active$ }
$state_p := passive$

Define the *termination condition* as:

No process is *active* and no messages are in transit.

When processes behave as described, this condition is stable: once true, it remains so. The problem of termination detection now is to superimpose on the described *basic* computation a *control* computation which enables one or more of the processes to detect when the termination condition holds. To this end a new special state *terminated* is introduced for each process. The following two criteria specify the correctness of the control algorithm.

T1 *Safety.* If some process is in the *terminated* state then the termination condition holds.

T2 *Liveness.* If the termination condition holds, then eventually some process will be in the *terminated* state.

A *passive* process may take part in this control computation, and receiving control messages does not make a *passive* process *active.*

Solutions to the termination detection problem are non-trivial, mainly due to the possibility that a process becomes *active* after being observed as *passive* by the control algorithm. Several classes of solutions are known. The most important ones are those based on *probes* and those based on *acknowledgements.* The best known example of the former class is [DFG83], and a general treatment is given in [Te90]. Examples of the latter class are [DS80, SF86]. Solutions based on counting sent and received messages are proposed in [Ma87].

1.2 The Distributed Garbage Collection Problem

As our approach for deriving termination detection algorithms is based on solutions to the garbage collection problem, we shall now describe this problem in a model which is close to the model of Lermen and Maurer [LM86]. The advantage of this model is that it abstracts from aspects which are not relevant to our purposes, such as processors, memory cells, and the difference between "local" and "remote" references.

An (object–oriented) distributed system consists of a collection O of cooperating *objects.* A subset of O is designated as *root objects.* Objects are able to hold *references* to other objects. These references can be transmitted in messages, see below. A reference to an object r will be called an r–reference. An object r is a *descendant* of q if q holds an r–reference or a message containing an r–reference is in transit to q. An object is *reachable* if it is a root object or a descendant of a reachable object. An object p holding an r–reference may *delete* it, after which p no longer holds this reference. Also, a reachable object p holding an r–reference may *copy* the reference to another object q, by sending an r–reference in a message to q. Object q will hold an r–reference after receipt of this message. An object can have multiple references to the same target object.

An object is called *garbage* if it is not reachable. As only reachable objects copy references, only references to reachable objects are copied, and thus a garbage object remains garbage forever. For reasons of memory management it is required that garbage objects are identified and collected. This task is taken care of by a garbage collecting algorithm. The following two criteria define the correctness of a garbage collecting algorithm.

G1 *Safety.* If an object is collected, it is garbage.

G2 *Liveness.* If an object is garbage, it will eventually be collected.

Many solutions have been proposed to the distributed garbage collection problem, most of which fall into one of two categories: collectors of the *reference counting* type and collectors of the *mark–and–sweep* type. Both types of solutions have been known for over 30 years for classical, non–distributed systems [Co60, McC60].

Collectors of the first type [LM86, WW87, Be89] maintain for each non–root object a count of the number of references in existence to that object. References in other objects as well as references in messages are taken into account. The reference count is incremented when a corresponding reference is copied, and decremented when such a reference is deleted. When the count for an object drops to zero, it can be concluded that the object is garbage and consequently the object can be collected. Reference counting garbage collectors are unable to collect *cyclic garbage* (a collection of garbage objects pointing to each other). As will be seen at the end of section 2, this does not render our transformation invalid for reference counting garbage collectors.

Collectors of the second type [Dij78] mark all reachable objects as such, starting from the roots and recursively marking all descendants of marked objects. In this way all reachable objects become marked eventually. The design of the marking algorithm is complicated by the possibility that references are copied and deleted during its operation. The objects must cooperate with the marking algorithm, e.g., by also marking objects when references are copied, cf. [Dij78]. When the marking phase is terminated a sweep through all objects is made, in which all unmarked objects are collected. These two phases are repeated as long as necessary.

2 Termination Detection Using Garbage Collection

In this section we describe how the termination detection problem in general can be modeled as an instance of the garbage collection problem. As a result, solutions to the termination detection problem can be derived from garbage collection algorithms, of which an example will be shown in section 3. First the collection O of objects used for this purpose is described, as well as the behavior of these objects. Next it is shown that the termination condition is equivalent to one particular object becoming garbage, so termination can be detected by a garbage collection algorithm.

Recall that P is the set of processes whose termination is to be detected. The collection O of objects consists of one root object A_p for every process p in P, and a single *indicator object* Z. Object A_p mimics the behavior of process p as far as the basic computation is concerned (it sends and receives p's basic messages, and has all the variables p has). Messages may contain a reference, in which case the message is a copy message for that reference. Object A_p is called *passive* (*active*) when the mimicked process p is *passive* (*active*). As A_p is a root object, it is always reachable.

The indicator object Z is not a root object. Its only purpose is to indicate the termination condition with its reachability status by the following equivalence, which will be maintained during execution.

(IND) Z is garbage $\Leftrightarrow$ the termination condition holds.

Theorem 2.1 *IND holds when the following two rules are observed:*

R1 *An object holds a Z-reference if and only if it is active.*

R2 *Each message of the basic computation contains a Z-reference.*

Proof. Z is garbage is equivalent to: Z is not a descendant of any of the A_p. By definition, this means that no A_p holds a Z-reference, and to no A_p a message is in transit containing a Z-reference. By R1 and R2 this is equivalent to: no A_p is *active* and to no A_p a message (of the basic computation) is in transit. This is the definition of the termination condition. □

It remains to be shown how R1 and R2 can be maintained. It is possible to ensure through proper initialization that R1 and R2 hold initially. To this end, assume that *active* objects are initialized with the necessary Z–reference, and *passive* objects without it, and that messages in transit initially contain the reference also. To maintain R1 and R2 during the distributed computation, each transmission of a message copies the Z–reference, and processes delete their Z–reference when they become *passive*. More explicitly, the actions to be carried out by A_p are modified as follows:

S_p: { $state_p = active$ }
send a message $\langle M, Z \rangle$

R_p: { A basic message has arrived }
receive message $\langle M, Z \rangle$; $state_p := active$;
insert Z in the references of A_p

I_p: { $state_p = active$ }
$state_p := passive$;
delete Z from the references of A_p

With these modifications R1 and R2 are maintained indeed. R1 is maintained because Z–references are deleted in action $\mathbf{I_p}$, and inserted in action $\mathbf{R_p}$. The latter is possible because the message contains a Z–reference by R2. R2 is maintained because in action $\mathbf{S_p}$ a Z-reference is included in every message. This is possible because only *active* objects send messages, and these objects contain a Z–reference by R1. Thus R1 and R2 are maintained during computation, and by theorem 2.1 IND holds. To arrive at a termination detection algorithm, superimpose upon the objects as described a garbage collection algorithm to detect that Z is garbage. The garbage collection algorithm is then modified so as to inform the objects A_p when Z is identified as garbage. (On receiving this notice, the root objects enter the *terminated* state. We omit this (trivial) operation from the description of the algorithms that will follow.)

Theorem 2.2 *The algorithm as constructed satisfies conditions T1 and T2.*

Proof. Assume any process enters the *terminated* state. This happens upon notice that Z is collected. By the correctness of the garbage collection algorithm (condition G1) this implies that Z is garbage. By IND the termination condition holds.

Assume the termination condition holds. By IND, Z is garbage, hence, by the liveness of the garbage collector (condition G2) Z will eventually be collected. Notice of this will be sent to the processes, and these will enter the *terminated* state in finite time. □

Garbage collectors of the reference counting type are not able to collect cyclic structures of garbage, which may possibly harm the liveness of the termination detection algorithm. It is, however, easily seen that Z is not part of such a cyclic structure, and in fact the following, stronger equivalence holds.

There are no references to $Z \Leftrightarrow$ the termination condition holds.

Summary of the transformation. The construction of a termination detection algorithm is summarized in the following four steps.

1. Form the set $\mathbf{O}$ of objects, consisting of the root objects A_p and one indicator object Z.
2. Superimpose upon the actions of the basic computation the handling of the Z–reference.
3. Superimpose upon this combined algorithm a garbage collection algorithm.
4. Replace the collection of Z (or its identification as garbage) by a notification of termination.

3 An Example of the Transformation

The transformation described in section 2 can in principle be applied to any garbage collection scheme, of the reference counting as well as the mark–and–sweep type, or working according to other principles. In this section we consider the transformation of a garbage collection algorithm based on weighted reference counting. The resulting termination detection algorithm turns out to be an already known algorithm: it was proposed in [Ma89]. More derivations, yielding new and non–trivial termination detection algorithms, are found in the full paper [TM90].

In a weighted reference counting scheme, each reference has an associated positive weight. Each object o maintains a reference count, which equals (barring certain update delays) the total weight of existing o–references. (The term "reference weight accumulator" might be more appropriate for this variable, but in accordance with the existing literature we shall continue to use the word "count".) When a reference is copied, its weight is *split* among the existing and the new reference. Thus, although the *number* of references increases, the *weight* remains the same, and the reference count need not be incremented and no message need be sent to the referenced object. When an object deletes an r–reference, a decrement message is sent to r, reporting the weight of the deleted reference. Upon receipt of this message, r subtracts this weight from its reference count (and is collected if the count drops to zero).

3.1 Description of the Scheme

Distributed weighted reference counting schemes have been given by Watson and Watson [WW87], Bevan [Be89], and others. In the description below the mechanism to create new objects is omitted, because in the transformation no new objects are ever created. An o–reference is a tuple (o, w), where w denotes the weight of the reference. Initially for each non–root object o, the reference count RC_o equals the sum of the weights of all existing o–references. The following (atomic) actions can take place. ($\mathbf{CR}_p$ represents the sending of a copy message, $\mathbf{RR}_p$ the receipt of such a message, $\mathbf{DR}_p$ the deletion of a reference and the associated sending of a decrement–weight message, and $\mathbf{RD}_o$ the receipt of such a message.)

$\mathbf{CR}_p$: { p holds reference (o, w) }
send $\mathbf{cop}(o, w/2)$ to q ; $w := w/2$

$\mathbf{RR}_p$: { A message $\mathbf{cop}(o, w)$ has arrived at p }
receive $\mathbf{cop}(o, w)$;
if p has an o–reference
  then add w to its weight
  else insert the o–reference with weight w

$\mathbf{DR}_p$: { p holds reference (o, w) }
send $\mathbf{dec}(o, w)$ to o ; delete the o–reference

$\mathbf{RD}_o$: { A $\mathbf{dec}(o, w)$ message has arrived at o }
receive $\mathbf{dec}(o, w)$; $RC_o := RC_o - w$;
if $RC_o = 0$ **then** collect o

Action $\mathbf{RR}_p$ guarantees that in this scheme an object has at most one reference to each other object. A correctness proof and analysis of the scheme is found in [WW87] or [Be89] and is based on invariance of the following two assertions:

1. Each reference has a positive weight; each delete message contains a positive weight.
2. $RC_o = \sum_{R=(o,w)} w + \sum_{D=\mathbf{dec}(o,w)} w$, where R ranges over all o–references in existence (including **cop** messages) and D ranges over all delete messages in transit.

3.2 Transformation into a Termination Detection Algorithm

To transform the garbage collection scheme into a termination detection algorithm we apply the four-step construction of section 2.

1. The set **O** of objects consists of the objects A_p and the indicator object Z.

2. Superimpose upon the actions of the basic computation the handling of the Z-reference. This yields the following program text.

   S_p: { $state_p =$ *active* }
   send a message $\langle M, Z\rangle$

   R_p: { A basic message has arrived }
   receive message $\langle M, Z\rangle$; $state_p :=$ *active* ;
   insert Z in the references of A_p

   I_p: { $state_p =$ *active* }
   $state_p :=$ *passive* ;
   delete Z from the references of A_p

3. Superimpose the reference counting scheme upon these actions. To this end, action **CR_p** is included in action **S_p**, action **RR_p** is included in action **R_p**, and action **DR_p** is included in action **I_p**. For o the object Z is substituted. This results in the following program text.

   S_p: { $state_p =$ *active* and p holds reference (Z, w) }
   send a message $\langle M, \mathbf{cop}(Z, w/2)\rangle$; $w := w/2$

   R_p: { A basic message has arrived }
   receive message $\langle M, \mathbf{cop}(Z, w)\rangle$; $state_p :=$ *active*;
   if p has a Z-reference
   then add w to its weight
   else insert the Z-reference with weight w

   I_p: { $state_p =$ *active* }
   $state_p :=$ *passive* ;
   send $\mathbf{dec}(Z, w)$ to Z ; delete the Z-reference

RD$_Z$: { A **dec**(Z, w) message has arrived at Z }
receive **dec**(Z, w) ; $RC_Z := RC_Z - w$;
if $RC_Z = 0$ **then** collect Z

4. Replace the collection of Z by a notification of termination. Some more simplifications can be made in addition: the actual handling of the Z reference can be removed; instead we equip every process p with a variable W_p, representing the weight of p's (virtual) Z–reference (0 if p has no such reference). The subscript Z is dropped. This finally results in the following algorithm.

S$_p$: { $state_p = active$ }
send a message $\langle M, W_p/2 \rangle$; $W_p := W_p/2$

R$_p$: { A basic message has arrived }
receive message $\langle M, W \rangle$; $state_p := active$;
$W_p := W_p + W$

I$_p$: { $state_p = active$ }
$state_p := passive$;
send **dec**(W_p) to Z ; $W_p := 0$

RD: { A **dec**(W) message has arrived at Z }
receive **dec**(W) ; $RC := RC - W$;
if $RC = 0$ **then** send **term** to all A_p

The initial conditions for this algorithm are: $W_p = 0$ if p is *passive*; $W_p > 0$ if p is *active*; $RC = \sum_p W_p$; and no messages are in transit. (Or, if there are messages, RC correctly reflects their weight.)

The termination detection algorithm that has just been derived is known as the *Credit Recovery* algorithm [Ma89]. The algorithms discussed in this section face the problem of so–called *weight underflow*. When weights are represented in a finite number of bits, there exists a smallest positive value a weight can take, and it is not possible to split this weight in two positive parts. Furthermore, the accumulation of small fragments may cause problems. Solutions to these problems and variants of the scheme may be found in [Be89, Ma89].

4 Conclusions

In this paper we have presented a transformation of garbage collection schemes into termination detection algorithms. Applying the transformation to the weighted reference counting scheme, we have derived the Credit Recovery algorithm for termination detection. Virtually all garbage collection schemes can be transformed into sensible termination detection algorithms. The full paper [TM90] contains derivations of more termination detection algorithms, including three new ones: the Activity Counting algorithm, the Generational termination detection algorithm, and a "dual-tour" token algorithm for a ring of processes. It also contains a discussion of several related aspects, of which we only sketch two here.

4.1 Reverse Transformation

It is also possible to transform a termination detection algorithm into a reference counting garbage collection scheme. The aim of a reference counting algorithm is to collect an object o when all o–references (in objects) have been deleted and no more o–references are in transit (in copy messages).

An object is defined to be *o–active* if it holds an o–reference and *o–passive* otherwise, and a message is called an o–activation message if it carries an o–reference. Under these definitions, an *o–passive* object becomes *o–active* only upon receipt of an o–activation message, and only *o–active* objects send o–activation messages. Now the *o–termination condition*, defined as:

No process is *o–active* and no o–activation messages are in transit

is stable and can be detected by a termination detection algorithm. Furthermore

(RT) There are no o–references $\Leftrightarrow$ the o–termination condition holds.

To arrive at a reference counting garbage collection algorithm, a termination detection algorithm is superimposed on the o–reference handling. When o–termination is detected, o is collected. For each object a separate instance of the termination detection algorithm is executed concurrently.

Although this transformation could be applied to any termination detection algorithm, the resulting reference counting garbage collection scheme would not be feasible in all cases. A complete algorithm along these lines, based on the algorithm in [DS80], was proposed by Rudalics [Ru90].

4.2 Deadlock Detection

The termination detection problem is an instance of a class of detection problems in distributed systems. Communication deadlock detection is a generalization where also a part of the network can be terminated. In this problem, for each *passive* process a subset of the processes is determined at the moment it becomes *passive*. The process can become *active* only by receiving a message from a process in this subset. The termination detection problem is obtained, when each process always chooses the full set of processes. We are currently investigating how the approach in this paper can be generalized to derive (mark–and–sweep) deadlock detection algorithms from garbage collection algorithms.

Acknowledgements: We want to thank Martin Rudalics and Jörg Richter for their discussions of the paper and numerous suggestions and comments. We want to thank the Eindhoven Tuesday Afternoon Club and Reinhard Schwarz for their careful revision of the text.

References

[Be89] Bevan, D.I., *An Efficient Reference Counting Solution to the Distributed Garbage Collection Problem,* Parallel Computing 9 (1989) 179–192.

[CL85] Chandy, K.M., L. Lamport, *Distributed Snapshots: Determining Global States of Distributed Systems*, ACM Trans. on Computer Systems 3 (1985) 45–56.

[CMH83] Chandy, K.M., J. Misra, L.M. Haas, *Distributed Deadlock Detection*, ACM Trans. on Computer Systems 1 (1983) 144–156.

[Co60] Collins, G.E., *A Method for Overlapping and Erasure of Lists*, Comm. ACM 3 (1960) 655–657.

[DFG83] Dijkstra, E.W., W.H.J. Feijen, A.J.M. van Gasteren, *Derivation of a Termination Detection Algorithm for Distributed Computations,* Inf. Proc. Lett. 16 (1983) 217–219.

[Dij78] Dijkstra, E.W., L. Lamport, A.J. Martin, C.S. Scholten, E.F.M. Steffens, *On-the-fly Garbage Collection: An Exercise in Cooperation*, Comm. ACM 21 (1978) 966–975.

[DS80] Dijkstra, E.W., C.S. Scholten, *Termination Detection for Diffusing Computations*, Inf. Proc. Lett. 11 (1980) 1–4.

[LM86] Lermen, C.–W., D. Maurer, *A Protocol for Distributed Reference Counting*, ACM Conference on Lisp and Functional Programming, Cambridge, 1986, pp. 343–354.

[Ma87] Mattern, F., *Algorithms for Distributed Termination Detection*, Distributed Computing 2 (1987) 161–175.

[Ma89] Mattern, F., *Global Quiescence Detection Based on Credit Distribution and Recovery,* Inf. Proc. Lett. 30 (1989) 195–200.

[McC60] McCarthy, J., *Recursive Functions of Symbolic Expressions and Their Computation by Machine*, Comm. ACM 3 (1960) 184–195.

[Ru90] Rudalics, M., *Implementation of Distributed Reference Counts*, Technical Report (forthcoming), Research Institute for Symbolic Computation, J. Kepler University, Linz, 1990.

[SF86] Shavit, N., N. Francez, *A New Approach to Detection of Locally Indicative Stability,* in: L. Kott (ed.), *Proceedings ICALP 1986*, Lecture Notes in Computer Science 226, Springer–Verlag, 1986, pp. 344–358.

[Te90] Tel, G., *Total Algorithms*, Technical Report RUU–CS–88–16, Dept. of Computer Science, Utrecht University, 1988. Also in: Algorithms Review 1 (1990) 13–42.

[TM90] Tel, G., F. Mattern, *The Derivation of Distributed Termination Detection Algorithms from Garbage Collection Schemes*, Technical Report RUU–CS–90–24, Dept. of Computer Science, Utrecht University, 1990.

[TTL88] Tel, G., R.B. Tan, J. van Leeuwen, *The Derivation of Graph Marking Algorithms from Distributed Termination Detection Protocols*, Science of Computer Programming 10 (1988) 107–137.

[WW87] Watson, P., I. Watson, *An Efficient Garbage Collection Scheme for Parallel Computer Architectures*, in: J.W. de Bakker, A.J. Nijman, P.C. Treleaven (eds.), Proceedings Parallel Architectures and Languages Europe, vol. II, Lecture Notes in Computer Science 259, Springer–Verlag, 1987, pp. 432–443.

Indirect Reference Counting: A Distributed Garbage Collection Algorithm *

José M. Piquer †
INRIA - École Polytechnique
Domaine de Voluceau, B. P. 105
78153 LE CHESNAY CEDEX (France).
email: piquer@inria.inria.fr

Abstract

This paper exposes a Garbage Collection (GC) algorithm for loosely-coupled multi-processors. The algorithm is based on reference counting and is designed to reclaim distant-pointed objects. It behaves like *weighted reference counting*, using only *decrement* messages. The main advantages are that it never creates indirect cells (so accesses to distant pointed objects are always done in constant time) and it does not need synchronization with the proprietary site of the object. Object migration is also supported with only one *decrement* message, involving only the source and destination sites. On the other hand, two extra fields are needed in every remote pointer and *decrement* messages can be generated in cascades of arbitrary size, making it less predictable.

A first version adapted to a distributed Lisp running on a group of Transputer[1] processors will be presented along with some measures. Each processor has its own local Garbage Collector (of the Mark-Scan type) and the indirect reference counting for remote pointers has been implemented without having to change the local GC.

In this paper the algorithm is presented, it is compared with the *weighted reference counting* garbage collector and some experimental results are shown along with some implementation issues.

Keywords: Garbage Collection, Reference Counting, Distributed Systems.

1 Introduction

Many garbage collecting algorithms have been proposed for distributed systems. They can be classified as Mark-Scan or as Reference Count (excepting some work existing on distributed copying [Ruda 86]). The first ones are rather complex, as they must

*This work was partially supported by the University of Chile, INRIA and the french government.

†Author's current address: Departamento de Computación, Universidad de Chile, Casilla 2777, Santiago, Chile.

[1]Transputer is a trademark of INMOS Ltd.

be "on-the-fly" [Dijk 78] and distributed [Huda 82, Augu 87, Derb 90], adding the termination problem [Dijk 80] and the detection of global states [Chan 85] to the parallelism problem. The reference counting algorithms have the problem of being unable to reclaim cyclic structures (even though they can be extended to do it), but are much simpler than the Mark-Scan ones and do not require synchronization between the collectors and the mutators. We have chosen the reference count scheme mainly for its simplicity and because it could be implemented without modifying the local garbage collectors. In this way we have two completely independent Garbage Collection algorithms for local objects (a Mark-Scan) and for remote pointers (Reference Counting), an idea already advocated in [Fost 89]. Their only interaction is at the end of each local garbage collection, to find which remote pointers have been deleted.

The environment in which the algorithm is designed to run is a loosely-coupled multi-processor system with independent memories, and a reliable point to point message passing system. The message passing is very expensive and remote pointers already have many fields of information concerning the remote site, address, etc. In this kind of machine, we can accept to lose memory if extra messages can be avoided when requiring remote access, or when doing garbage collection. The proposed algorithm requires two extra fields per remote pointer, and one reference count per remote-pointed object, but it guarantees that the remote accesses are always done without indirections and it never requires synchronization with another site. This is important in a language with mutable data such as Lisp, because the maximum number of indirect cells created by algorithms like *weighted reference counting* cannot generally be estimated, as is done for pure functional languages in [Beva 87]. The algorithm is easy to extend to support object migration, which is used by the distributed Lisp to mutate distant objects.

2 The Model

The basic model is composed of the set of processors P and the set of all objects O. Given an object $o \in O$, it resides at a processor $p \in P$: p is called the *owner* of o. The set of all the remote pointers to o is denoted $\mathrm{RP}(o)$. It is assumed that each object has one and only one *owner* at a given time (although it may change).

A remote pointer to an object o is called an o-reference[2]. For every object $o \in O$, $\mathrm{RP}(o)$ includes all of the existing o-references, including the remote pointers to o contained in messages already sent but not yet received. Only remote pointers are considered in this model, so the local pointers to o are not included in $\mathrm{RP}(o)$ and they are not called o-references.

The model requires also that, given an object o residing at site p, every other site can hold at most one o-reference. Obviously, there are no o-references at p. Remote pointers are usually handled this way: if there are multiple pointers at a site to the same remote object o, they pass by a local indirection. This indirection represents for us the only o-reference at the site.

Any garbage collection algorithm must detect the objects which are not remote-pointed from any other site. For this, every remote pointer operation must be con-

[2] The o-reference terminology and the basis of this model were proposed in [Lerm 86].

sidered, and the model includes only four operations:

1. Creation of an *o*-reference

   A site *p*, where an object *o* resides, transmits an *o*-reference to another site *q*. This operation does not imply the creation of the object, and happens each time that the ***owner*** of an object *o* sends a message with an *o*-reference to another site.

2. Duplication of an *o*-reference

   A site *q*, which already has an *o*-reference to an object on another site *p*, transmits the *o*-reference to a third site *r*. This operation differs from creation because the ***owner*** of the object (*p*) is not involved. So, it does not know that a new *o*-reference was created.

3. Deletion of an *o*-reference

   A site *q*, holding an *o*-reference to an object located on a distant site *p*, discards it.

4. Migration of an object *o*

   A site *q*, holding an *o*-reference to an object located at a distant site *p*, wants to become the new *owner* of the object (with the permission of *p*). The object *o* is going to change its *owner*, and every *o*-reference is going to change along with it, while the ***old owner*** transforms its local pointer to *o* into an *o*-reference to the ***new owner***. This is done with a distributed protocol using, for example, multicast or broadcast messages. This migration protocol is independent of the GC algorithm. However, the GC protocol must support asynchronous owner changes, with messages in transit, without losing consistency.

   Existing distributed garbage collection algorithms do not consider this operation. It will be shown how these algorithms can be extended to handle it.

The model abstracts from the remote pointers implementation and the way to decide when to delete an *o*-reference. On the other hand it distinguishes the *o*-reference creation from its duplication which pose completely different problems to the reference counting algorithms as is explained later.

The distributed garbage collection algorithm should specify how to decide when an object *o* can be deleted locally (i.e. how to detect that RP(*o*) is empty) and what actions are to be performed in the four cases above.

As the model only considers remote pointers, it is presumed that there are local garbage collectors keeping track of the local references and objects. Thus, when an object *o* can be deleted locally, it is the responsibility of the local garbage collector to do so. The only restrictions on interactions between the distributed and local garbage collectors are that the local collector must never reclaim remote-pointed objects (it is enough to keep them in a local list), and that at the end of each local garbage collection, every *o*-reference locally reclaimed could be found (to detect the deletion operation).

3 The Existing Reference Counting Algorithms

Existing reference counting algorithms (initially proposed for Lisp systems in [Coll 60]) are based on maintaining, for each object o, the total number of existing pointers to it, called the reference count of o. For this, every object o has an extra field, Ref_cnt(o), which, in the classic algorithm, keeps the reference count of o. In our model, it is the number of elements of RP(o), denoted card(RP(o)).

In the shared memory version, the mutator adjusts every object's Ref_cnt(o) each time a pointer is created, duplicated or deleted. When the reference count reaches zero, the object can be reclaimed, which means that all of the pointers contained in it can be deleted. This scheme does not detect self-referential structures which are not accessible from the outside, but extensions to handle cycles have been designed for purely functional languages[Bobr 80], combinator graph reduction machines[Brow 85] and distributed systems[Beck 86]. However, in this paper only the basic algorithm is considered.

When trying to map the simple reference counting algorithm onto our distributed model, there are new problems and new algorithms. In this section we present two known distributed reference counting algorithms and their problems.

3.1 The Naive Algorithm

The first algorithm obtained from direct extension of the shared memory version is to keep the reference count of each object o equal to the total number of o-references in the system. Note that only remote pointers are considered.

Thus, for every object o we have:

$$\text{card}(\text{RP}(o)) = \text{Ref\_cnt}(o).$$

The actions associated with each operation are:

1. Creation of an o-reference

   When p sends an o-reference to q, it increments locally the Ref_cnt(o).

2. Duplication of an o-reference

   When q duplicates an o-reference, it sends an *increment* message to the owner site p to make it increment Ref_cnt(o).

3. Deletion of an o-reference

   When an o-reference is discarded at site q, it sends a *decrement* message to the owner site p to make it decrement Ref_cnt(o).

4. Migration of an object o

   Migrations are not handled in the original algorithm. However, it is important to see what kind of problems may be encountered if objects can change their *owner*. A simple solution is to perform one distant pointer creation (from the *old owner* to the *new owner*) and every time an o-reference is changed (at the other sites), send a *decrement* message to the *old owner* and an *increment* message to the *new owner*. If there are n o-references in the system, this algorithm uses $2(n-1)$

messages for one migration. This is extremely expensive in messages, but the algorithm can be improved, noting that the reference count of the object remains the same (since no o-reference is created or deleted). Therefore, the object can migrate along with its reference count, without spending any messages.

To consider messages in transit on the network, the invariant must be extended. Denote by DM(o) the set of all the *decrement* messages concerning o not yet received, and by IM(o) the set of the *increment* messages. The extended invariant becomes:

For every object o we have:

$$\text{card}(\text{RP}(o)) = \text{Ref\_cnt}(o) - \text{card}(\text{DM}(o)) + \text{card}(\text{IM}(o)).$$

The garbage collection rule is: "an object o can be deleted when Ref_cnt(o) reaches zero".

The problem of this algorithm is that it does not work correctly if the messages in the network are not guaranteed to be totally time-ordered (which is expensive). When there exists asynchronous communication between sites, messages are only partially ordered as was noted by the well-known paper [Lamp 78].

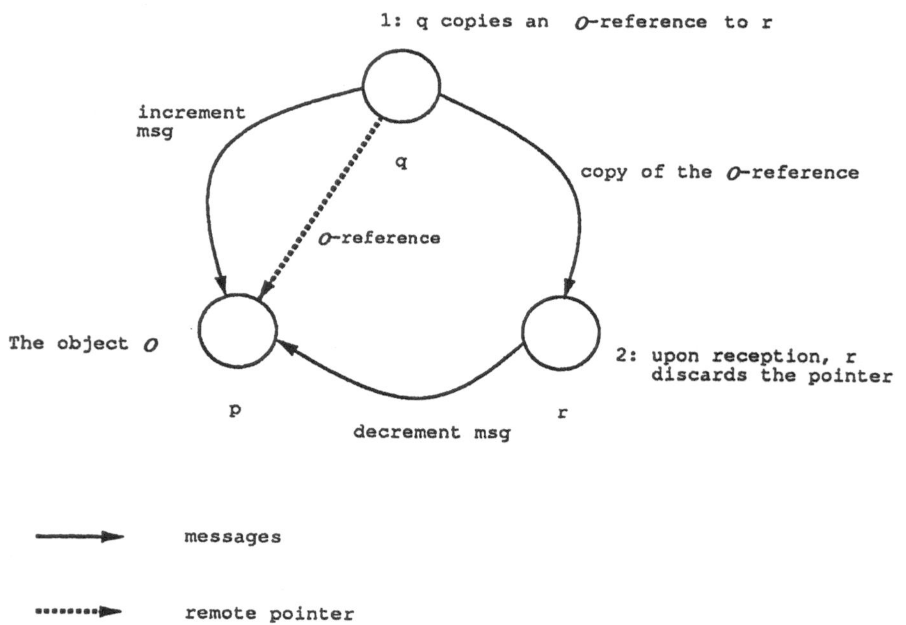

Figure 1: Two contradictory messages on their way to p

The co-habitation of two kinds of messages (*increment* and *decrement*) introduces a problem which invalidates the algorithm. Since it cannot be guaranteed that two messages sent at different times from different sites will arrive to a third site time-ordered, a *decrement* message can arrive first even if an *increment* message is on its way to the site. This situation can be seen in Figure 1.

The problem exists even if the messages can be guaranteed to be ordered between two given sites, because q can send a copy of an o-reference (for an object residing at p) to r, with r discarding it immediately after reception. In this case there are two messages in transit to p from two different sites and it cannot be known which will arrive first. Obviously, the worst case is the *decrement* message arriving first, which can cause a premature zero of the object's reference count at p. Some other solutions have been proposed ([Lerm 86, Gold 89]) but they use more messages or consume more memory.

Furthermore, the improved migration protocol (migrating the object along with its reference count) is incorrect in an asynchronous environment. Messages affecting the reference count of an object can be in transit while the object is migrating. If using the improved migration algorithm, a site can receive *increment* or *decrement* messages for an object already migrated to another site. This can be handled keeping track of migrated objects and forwarding messages to the *new owner* site. However, this indirection must be maintained until every message has been received, which is a global condition and thus, difficult to detect.

3.2 Weighted Reference Count

A better solution to the distributed reference counting problem is the *weighted reference count* [Beva 87, Wats 87]. This algorithm avoids the problem of the premature zero by simply eliminating the *increment* messages. The algorithm is elegant and simple: each o-reference rp has a *weight* associated with it (noted Weight(rp)) and each object has, as always, a reference count. The difference is that now the Ref_cnt(o) does not keep the number of pointers to o but the sum of their weights. So, the invariant is:

For every object o:

$$\sum_{rp \in RP(o)} \text{Weight}(rp) = \text{Ref\_cnt}(o).$$

To preserve this invariant, the actions associated with the remote pointer operations are:

1. Creation

   The new o-reference takes an initial weight, say w (with $w > 0$), and the reference count of o is incremented by w. (It is supposed that, at object creation, its reference count is zero.)

2. Duplication

   If o resides at p, and q is sending a copy of the o-reference to r, the original o-reference at q takes half of its old weight and the newly created copy takes the other half (this is why usually the initial value w is a power of two) [3]. In this

[3] In fact, this is an optimization. Theoretically, if w is the old weight of the pointer, it is enough to find two new weights u and v such that $u + v = w$.

way, the invariant is preserved without having to send a message to the *owner*, p.

3. Deletion

   When an o-reference is deleted, a *decrement* message is sent to the *owner* along with the weight of the deleted pointer. Upon reception of a *decrement* message, the reference count is decremented by this amount.

4. Migration

   This operation was not considered in the original papers. When an object o migrates from p to q, we can migrate the object along with its reference count and the new distant pointer (from p to q) can take the weight associated with the old pointer (from q to p). If every pointer is changed at the same time, the invariant is preserved.

In order to consider *decrement* messages in transit, the original invariant must be extended to consider the weights of the *decrement* messages not yet received. Denoting as DM(o) the set of existing *decrement* messages related to remote pointers to o, the extended invariant is:

$$\sum_{rp \in \mathrm{RP}(o)} \mathrm{Weight}(rp) = \mathrm{Ref\_cnt}(o) - \sum_{m \in \mathrm{DM}(o)} \mathrm{Weight}(m).$$

The big problem with this algorithm is the copying of an o-reference with weight 1. One solution is to send an *increment* message to the proprietary site [Beck 86] and wait for an acknowledge, which means a synchronization with this site before proceeding with the local computation. Another solution is to create an *indirection cell* [Wats 87, Beva 87] with a new reference count of w, but the accesses are now done to this cell and must be forwarded to the original object. In a distributed environment, this can create remote indirections which are very expensive, and also affects accesses to the data. Even worse, references can be copied many times from one processor to another, having to create indirection cells many times.

The migration protocol also poses a problem if considering messages in transit: *decrement* messages can arrive at the wrong site (the *old owner*). The *old owner* site can keep track of the migration to forward these messages, but, again, the global condition "every o-reference has been changed" is not easy to detect.

4 Indirect Reference Counting

In this section the proposed algorithm is presented, which is based on avoiding the *increment* messages by maintaining a *distributed reference count*[Piqu 90a].

The *weighted reference* algorithm eliminates the *increment* messages using a new definition of the field Ref_cnt(o) which only decreases. The *indirect reference* algorithm eliminates the *increment* messages by always maintaining enough information at each node to do the increments locally, without any communication. This is true even for duplications.

The basis of the algorithm is to maintain a tree structure representing the diffusion tree of the pointer throughout the system. In fact, this structure is equivalent to the tree used by [Dijk 80] to detect termination, and we are using it to detect the end of the pointer's diffusion. This structure contains every o-reference and the object itself (which is always the root of the tree). When an o-reference is created or duplicated, a node is added as a child of the creator. When an o-reference is deleted, the corresponding node is deleted from the tree only if it was a leaf. If not, the descriptor is kept in the structure until it becomes a leaf (to avoid a distributed deletion, the leaf case being trivial). Obviously, when the root is the only node in the tree, there are no more o-references in the system.

The implementation is very simple: an inverted tree is used, where each node keeps one pointer to its parent and a counter with the number of children[4]. The structure can be seen in Figure 2.

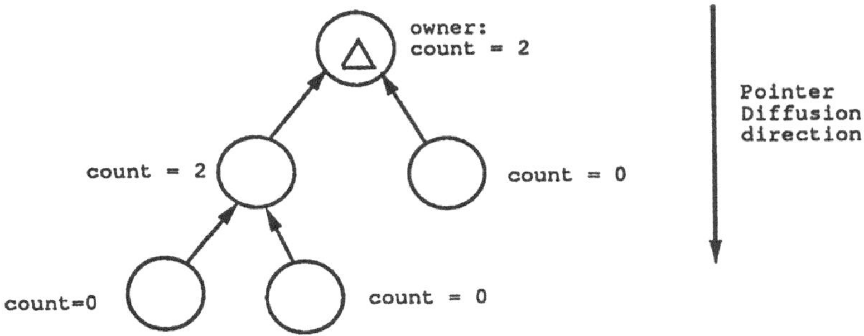

Figure 2: The inverted tree

The nodes of the tree are the o-references themselves, which are extended with two new fields: **copy_cnt** (the number of children) and **parent** (the pointer to the parent in the tree). We suppose that an o-reference always contains the **owner** field (which in our case is a pointer to the root of the tree).
Each remote pointer is a record:

```
remote_pointer = record
       integer owner;          /* owner, root of the tree */
                               /* necessary information for accesses */
       .
       .
       integer copy_cnt;       /* number of duplications */
       integer parent;         /* from which we received it */
end;
```

The object o (the root of the tree) has a similar descriptor, where **owner** shows that the local site is the owner, **parent** is **NIL**, and the **copy_cnt** keeps the number of creations done locally. Upon record creation, the **parent** initial value is **NONE**, to distinguish it from the root descriptor (with **parent** value **NIL**).

The invariant of the algorithm is to preserve the tree structure consistently, with only one root, and with the correct number of children. It is easy to see that if the

[4] This is the Indirect Reference Count because, when it reaches zero (the node becomes a leaf), the node can be deleted from the structure if the local o-reference is deleted.

tree is correct, when the copy_cnt of the root is zero, the object can be deleted locally. For every o-reference or object rp at site p, we denote Children(rp, p) the set of all o-references with parent p. At each node p, the invariant is:

$$\text{card}(\text{Children}(rp,p)) = \texttt{copy\_cnt}(rp).$$

The total number of o-references on the system is equal to the total number of nodes on the tree:

$$\text{card}(\text{RP}(o)) = \sum_{rp \in \text{RP}(o) \bigcup \{o\}} \texttt{copy\_cnt}(rp).$$

To preserve this structure, we must avoid cycles, so every received o-reference already known is refused, replying with a *decrement* message. The actions performed upon operations on remote pointers are:

1. Creation

   - at p, when a message containing an o-reference is sent:

     ```
     o.copy_cnt = o.copy_cnt + 1;
     ```

   - at q, upon reception of a message with an o-reference rp:

     ```
     if ( rp.parent == NONE )
     {
        rp.parent = p;
        rp.copy_cnt = 0;
     }
     else
        /* the reference already belonged to the tree */
        ''send decrement message to p'';
     ```

2. Duplication (it is as a Creation)

   - at q, when a message with a copy of an o-reference rp is sent:

     ```
     rp.copy_cnt = rp.copy_cnt + 1;
     ```

   - at r, upon reception of a message with a copy of an o-reference rp:

     ```
     if ( rp.parent == NONE )
     {
        rp.parent = q;
        rp.copy_cnt = 0;
     }
     else
        /* the reference already belonged to the tree */
        ''send decrement message to q'';
     ```

3. Deletion

   When an *o*-reference *rp* is deleted at *p*, and its copy_cnt is zero, it can be deleted from the tree. Leaf deletion is trivial: a *decrement* message is sent to the parent. If the count is not zero, the descriptor must be maintained waiting for other *decrement* messages, but it is marked as *deleted*:

```
if( rp.copy_cnt == 0)
  {
    ''send decrement message to rp.parent'';
    Delete(rp);
  }
else
    Mark_deleted(rp);
```

   Upon reception of the *decrement* message for an object *o*, the number of children must be decremented. If the count reaches zero, and if the local *o*-reference was deleted, a *decrement* message is sent to the parent:

```
/* reception of a decrement message for a remote pointer rp */

    rp.copy_cnt = rp.copy_cnt - 1;
    if (rp.copy_cnt == 0 && Test_deleted(rp))
      {
        ''send decrement message to rp.parent'';
        Delete(rp);
      }
```

   An *o*-reference marked as *deleted* can be unmarked if the same *o*-reference is received again, before its copy_cnt reaches zero. However, a *decrement* message must be sent to the *o*-reference's sender, because the descriptor already has a valid parent. Thus, an Unmark_deleted directive must be added to the code handling the reception of any *o*-reference.

4. Migration

   The migration of an object *o* from *p* to *q* means a change of the root in the diffusion tree. This operation is trivial on an inverted tree if the old root is known: the new root (at *q*) is extracted from the tree (along with its sub-tree) and the old root (at *p*) is added as a son of the new root. The extraction costs one *decrement* message, and the addition is done locally at the respective nodes (the new and old roots):

   - at *p*, when it changes the owner of the object o:

     o.parent = *q*;

- at q, upon reception of a message authorizing it to be the new owner of an object pointed by `rp`:

```
''send decrement message to rp.parent'';
rp.parent = NIL;
rp.copy_cnt = rp.copy_cnt + 1;
```

The condition to locally delete an object o is simply that the `copy_cnt` of o equals zero.

This system sends only *decrement* messages, as the increments are always done locally. The sites are informed of the deletion of locally sent o-references, only when the reference and all of its remote duplicated copies have been deleted. Also, accesses to distant objects are always done directly [5].

Migrations are handled neatly, with a simple modification of the tree structure. The migration operation complicates the other GC algorithms because it asynchronously modifies the `owner` pointers which are used by the *decrement* messages. If the objects were to migrate with their reference counts, messages could arrive at invalid destinations from which they would have to be forwarded. If not, many messages should be used to assure the invariant at any time.

The *indirect reference counting* does not use the `owner` pointers, but rather the `parent` pointers, to send its *decrement* messages. However, `parent` pointers are also modified during migration (the other operations are only allowed to initialize the `parent` field, not to change it) but the reference counts never move: objects migrate alone and they get the local reference counts. Therefore, any *decrement* message in transit will always arrive at the correct destination (which is the site holding the reference count to be decremented).

A shortcoming of the GC algorithm is that the deletion of an o-reference could generate many messages in cascades of arbitrary size, if the pointer was forwarded from one site to another many times. However, this only happens if all of the `copy_cnt`'s are reaching zero at once.

5 Implementation Issues

The *indirect reference counting* algorithm has been implemented on a distributed Lisp system called TransPive[Piqu 90b], based on Le-Lisp version 15.2 [Chai 84], extended to support remote pointers. In this system, remote pointers can be accessed for writing (for functions such as `rplaca`) which causes data migration. The current implementation uses the following organization:

- remote pointers

  Remote pointers are implemented adding an extra field on every Lisp object. This field, called `descriptor`, points to an object descriptor which contains all the information about the object: `obj_id` is a unique object identifier, `owner` is the object's owner, and `copy_cnt` and `parent` are the extra fields for the garbage collector. This descriptor also exists if the object is local.

[5] The `parent` is a new field and it is not used upon accesses to the values. The remote pointers always point directly to the site where o resides (as there are no indirection cells) using the `owner` field.

- object identifier

  Each site has a local hash table with the descriptor of every locally known object. The keys are the unique object identifiers. We have a function `(search_desc obj_id)` which returns a pointer to the descriptor.

- local garbage collectors

  The local GC always invokes a function **gcalarm** [Chai 84] after each execution. This permits us to search in the hash table all the deleted objects. We have a function called **garbagep** to detect descriptors of deleted objects. This descriptors are kept in the hash table until their `copy_cnt` field is zero.

  Also a `GC_list` is constructed to retain every object with a `copy_cnt` greater than zero, to prevent them from being garbage collected (the hash table does not keep a pointer to the object itself, only to the descriptor). Every time an object gets a `copy_cnt` equal to zero, it is deleted from this list.

The Lisp codification of the algorithm is (ignoring semaphores and some details):

```
; every Lisp object has a new field:
; (descriptor obj)

; (descriptor obj) returns a structure with fields:
;       (obj_id desc)       ;     unique object identifier
;       (owner  desc)       ;     owner of this object
;       (copy_cnt desc)     ;     indirect reference count
;       (parent desc)       ;     from which it came to us
;

; Each time a pointer is sent in a message the copy_ptr routine is called.
; This is done both for creation and duplication of a remote pointer

; The actions performed are:
;    add the object to the GC_list (the first time)
;        to prevent the local GC from collecting it
;    increment its copy_cnt

(de copy_ptr (obj)
    (let ((desc (descriptor obj)))
        (when (= (copy_cnt desc) 0)
            (setq GC_list (cons obj GC_list)) )

        (copy_cnt desc (+ (copy_cnt desc) 1)) ))

; After a local GC
(de gcalarm ()
    ; update the GC_list
    (setq GC_list (del_ptrs GC_list))
```

```
    ; update the hash_tab
    (foreach (desc hash_tab)
        (when (and (garbagep desc) (= (copy_cnt desc) 0))
            ; this is a leaf, we can just delete it
            (send_decr_msg (obj_id desc) (parent desc))
            (delete_from_htab hash_tab desc) )))

; delete from the list every object with ref_cnt = 0
(de del_ptrs (list)
    (if (null list)
        ()
        (if (= (copy_cnt (descriptor (car list))) 0)
            (del_ptrs (cdr list))
            (cons (car list) (del_ptrs (cdr list))) )))

; The GC process:

; upon reception of a decrement message
(de decr_msg (obj_id)
    (let ((desc (search_desc obj_id)))
        (copy_cnt desc (- (copy_cnt desc) 1)) ))


; Migrations:

; We are the new owner of the object
(de new_owner (desc)
    (send_decr_msg (obj_id desc) (parent desc))
    (parent desc ())
    (copy_cnt desc (+ (copy_cnt desc) 1)) )

; We are the old owner
(de migrate_to (desc new_owner)
    (parent desc new_owner) )
```

We can see in this implementation that, upon the reception of a *decrement* message, we just perform the decrement. The test to see if the object was locally deleted and its copy_cnt is already zero is done only after local garbage collection (in gcalarm).

6 Measures and Results

We have done some measures on three distributed applications: a merge-sort (using physical modification routines, thus causing migrations and being very inefficient), a search on a graph and a *tic-tac-toe* playing program. The applications were run on four Transputers, excepting the *tic-tac-toe* which was run on five Transputers.

Applic.	Execution Time (s)		
	with GC	no GC	overhead (%)
sort(1000)	4.1/3.1	3.8/2.8	7/9
sort(2000)	7.1/6.0	6.6/5.2	7/14
search	18.9/10.0	18.5/9.7	2/3
tic-tac-toe	11.3	11.2	1

Table 1
The GC overhead

We have measured the total execution time of the same application with or without the GC. The results are shown in Table 1, and we can see that the total time overhead of the GC is, on average, 10%. There are two numbers for each application, the first time and the second time it runs, because the second time the TransPive cache system has kept the data copied locally and so it runs faster. The *tic-tac-toe* application runs at the same speed both times.

The memory available for each Lisp was 1 Megabyte, and each execution generated, on average, one local GC at each node.

7 Conclusions

A new reference counting algorithm has been designed and implemented for distributed systems, based on a distributed tree structure of the pointers to each object, keeping *indirect reference counts* at each node. It spends two extra fields in each pointer (`parent` and `copy_cnt`) and can generate more than one message when a reference is deleted, but in total there is always at most one message per deleted reference. It never creates indirect cells nor needs to synchronize with any other site, and migrations are supported neatly with only one overhead message for the GC.

The algorithm is very simple and has a time overhead of approximatively 10% in the cases studied. Obviously, it is not suited to do garbage collection of local references, as two fields per pointer is unrealistic. However, when only remote pointers are concerned it becomes feasible and very simple to add to any local garbage collection system.

Generating messages in cascades of arbitrary size can be a serious problem if used on real-time systems (which was not our case).

Sharing a typical disadvantage of reference counting algorithms, the algorithm, as presented, is unable to collect distributed cyclic structures.

8 Acknowledgements

I am greatly indebted to Dr. Christian Queinnec who always suggested interesting ideas. This work was made possible thanks to the constant support of INRIA, the University of Chile and the French-Chilean cooperation program.

References

[Augu 87] Lex Augusteijn, *Garbage Collection in a Distributed Environment*, LNCS 259, PARLE Proceedings Vol. I, Eindhoven, Springer Verlag, June 1987.

[Beck 86] M. J. Beckerle, K. Ekanadham, *Distributed Garbage Collection with no Global Synchronization*, IBM Research Report, RC 11667 (#52377), January 1986.

[Beva 87] D. I. Bevan, *Distributed Garbage Collection Using Reference Counting*, LNCS 259, PARLE Proceedings Vol. II, Eindhoven, Springer Verlag, June 1987.

[Bobr 80] D. G. Bobrow, *Managing Reentrant Structures Using Reference Counts*, ACM Trans. on Programming Languages and Systems, Vol. 2, No. 3, pp. 269-273, July 1980.

[Brow 85] D. R. Brownbridge, *Cyclic Reference Counting for Combinator Machines*, Functional Programming Languages and Computer Architecture, LNCS 201, pp. 273-288, Springer Verlag, September 1985.

[Chai 84] J. Chailloux, M. Devin, J. M. Hullot, *Le-Lisp: A Portable and Efficient Lisp System*, Proc. 1984 ACM Symposium on Lisp and Functional Programming, August 1984.

[Chan 85] K. M. Chandy, L. Lamport, *Distributed snapshots: Determining global states of distributed systems*, ACM Trans. on Computer Systems, Vol. 3, No. 1, February 1985.

[Coll 60] G. E. Collins, *A Method for Overlapping and Erasure of Lists*, Comm. of the ACM, Vol. 3, No. 12, pp. 655-657, December 1960.

[Derb 90] M. H. Derbyshire, *Mark Scan Garbage Collection On A Distributed Architecture*, Lisp and Symbolic Computation, Vol. 3, No. 2, pp. 135-170, April 1990.

[Dijk 78] E. W. Dijkstra, L. Lamport, A. J. Martin, C. S. Scholten, E. F. M. Steffens, *On-the-fly Garbage Collection: an exercise in cooperation*, Comm. of the ACM Vol. 21, No. 11, November 1978.

[Dijk 80] E. W. Dijkstra, C. S. Scholten, *Termination Detection for Diffusing Computations*, Information Processing Letters, Vol. 11, No. 1, August 1980.

[Fost 89] I. Foster, *A Multicomputer Garbage Collector for a Single-Assignment Language*, Int. Journal of Parallel Programming, Vol. 18, No. 3, 1989.

[Gold 89] B. Goldberg, *Generational Reference Counting: A Reduced-Communication Distributed Storage Reclamation Scheme*, SIGPLAN Conference on Programming Languages Design and Implementation, Portland, Oregon, June 1989.

[Huda 82] P. Hudak, R. M. Keller, *Garbage Collection and Task Deletion in a Distributed Applicative Processing System*, 1982 ACM Symposium on Lisp and Functional Programming, 1982.

[Lamp 78] L. Lamport, *Time, Clocks, and the Ordering of Events in a Distributed System*, Comm. ACM, Vol. 21, No. 7, pp. 558-565, July 1978.

[Lerm 86] C. W. Lermen, D. Maurer, *A Protocol for Distributed Reference Counting*, Proc. 1986 ACM Conference on Lisp and Functional Programming, Cambridge, Massachussets, August 1986.

[Piqu 90a] J. M. Piquer, *Un GC parallèle pour un Lisp distribué*, Journées francophones des langages applicatifs, La Rochelle, January 1990. BIGRE 69, July 1990 (french).

[Piqu 90b] J. M. Piquer, *Sharing Data Structures in a Distributed Lisp*, Proc. High Performance and Parallel Computing in Lisp Workshop, Twickenham, London, UK, November 1990.

[Ruda 86] M. Rudalics, *Distributed Copying Garbage Collection*, Proceedings of the ACM Conference on LISP and functional prog., Cambridge, Massachussets, August 1986.

[Wats 87] P. Watson, I. Watson, *An Efficient Garbage Collection Scheme for Parallel Computer Architectures*, LNCS 259, PARLE Proceedings Vol. II, Eindhoven, Springer Verlag, June 1987.

Periodic Multiprocessor Scheduling

Jan Korst[1], Emile Aarts[1,2], Jan Karel Lenstra[2,3] and Jaap Wessels[2]

1. Philips Research Laboratories, P.O. Box 80.000, 5600 JA Eindhoven, the Netherlands
2. Eindhoven University of Technology, P.O. Box 513, 5600 MB Eindhoven, the Netherlands
3. CWI, P.O. Box 4079, 1009 AB Amsterdam, the Netherlands

Abstract

A number of scheduling and assignment problems are presented involving the execution of periodic operations in a multiprocessor environment. We consider the computational complexity of these problems and propose approximation algorithms for operations with identical periods as well as for operations with arbitrary integer periods.

Keywords: periodic scheduling, periodic assignment, cyclic scheduling, nonpreemptive scheduling

1 Introduction

This paper deals with the problem of scheduling periodic operations, i.e., operations that have to be repeated at a constant rate over an infinite time horizon. Periodic scheduling problems naturally arise in such diverse areas as real-time processing, process control, vehicle scheduling, personnel scheduling and preventive maintenance scheduling; see Section 2 for references. Our interests in periodic scheduling originate from the field of real-time video signal processing, where the samples of a video signal have to be processed at a constant high frequency (10 - 100 MHz) on a network of processors. Due to the high frequencies, the processing of successive samples necessarily overlaps in time. The intrinsic periodic nature of video signal processing gives rise to a periodic scheduling formulation. This application area poses some specific constraints, resulting in a class of optimization problems that so far have received little attention in the literature. In this paper we discuss this class of problems by examining their computational complexity, introducing approximation algorithms, and indicating relevant results presented in the literature.

We aim to keep the discussion as general as possible by proposing solution strategies that are also applicable in other application areas. Many papers on periodic scheduling are concerned with specific applications, proposing solution strategies that are often strongly tailored to the application at hand, a notable exception being the paper by Serafini & Ukovich [1989], which presents a general mathematical model for periodic scheduling problems. However, their emphasis is on periodic scheduling subject to precedence constraints. In our paper, the emphasis is on periodic scheduling subject to resource constraints. In that respect, our work is complementary to theirs.

The organization of the paper is as follows. Section 2 briefly surveys the literature on periodic scheduling. Section 3 gives a mathematical model of periodic scheduling, from which a number of interrelated optimization problems are derived. The computational complexity of these problems is examined in Section 4. Section 5 gives approximation algorithms and bounds on their worst-case performance, if available. Section 6 contains some concluding remarks.

2 Survey of the Literature

In the literature, the notion 'scheduling' refers to planning in time as well as planning in time and space. In this paper, we take the latter interpretation. We divide the literature on scheduling periodic operations into two main areas of interest, namely

(i) *Periodic Scheduling*: assigning start times and processors to periodic operations so as to minimize the number of processors, possibly subject to precedence constraints, and

(ii) *Periodic Assignment*: assigning processors to periodic operations so as to minimize the number of processors for periodic operations with fixed start times.

Clearly, periodic assignment is a subproblem in periodic scheduling. Next, we briefly describe some results obtained in both areas. We do not aim to give a complete overview.

2.1 Periodic Scheduling

Most of the literature on scheduling periodic operations in time is restricted to preemptive scheduling. Preemptive scheduling allows interruption of an execution on a given processor at some time and its resumption at the same time on a different processor or at a later time on any processor.

2.1.1 Preemptive Periodic Scheduling

Preemptive periodic scheduling problems are usually modelled as follows. Given a set of operations $O = \{o_1, \ldots, o_n\}$, any operation $o_i \in O$ is periodically requested to be executed with a given period $p(o_i)$ between two successive requests of operation o_i. Once requested at time t an execution of o_i is required to be completed at time $t + d(o_i)$, called its deadline. The objective is then to find a feasible schedule that requires a minimal number of processors, where a schedule is called feasible if all deadlines are met. Leung & Merrill [1980] prove that the problem of deciding whether a feasible schedule exists on m processors is NP-complete, even for $m = 1$. However, this problem can be solved in polynomial time if the deadline of each execution coincides with the next request for the operation. For $m = 1$, Liu & Layland [1973] and Labetoulle [1974] prove that, if a feasible schedule exists, then it is obtained by the so-called *deadline driven algorithm*, which is a dynamic-priority algorithm that schedules executions with earliest deadlines as soon as possible. Liu & Layland also give a fixed-priority scheduling algorithm for $m = 1$, known as *rate-monotonic priority assignment*, which is optimal in the sense that the algorithm finds a feasible schedule whenever a feasible fixed-priority schedule exists. Dhall & Liu [1978] present two fixed-priority scheduling algorithms for $m \geq 1$, and discuss their worst-case performance. Leung & Whitehead [1982] study the complexity of preemptive fixed-priority scheduling. Lawler & Martel [1981] show that a feasible preemptive schedule exists if and only if a feasible periodic schedule exists with a period equal to the least common multiple of the periods of the individual operations. Bertossi & Bonuccelli [1983, 1985] consider preemptive scheduling on multiprocessor systems consisting of 'processors of different speeds'. Scheduling periodic operations together with 'sporadic time-critical operations' is examined by Chetto & Chetto [1989].

2.1.2 Nonpreemptive Periodic Scheduling

So far, nonpreemptive periodic scheduling has received little attention in the literature. To schedule periodic operations nonpreemptively, it is usually assumed that the operations have to be executed with a fixed time between successive executions of the same operation. Gonzalez & Soh propose an optimization algorithm for nonpreemptively scheduling periodic operations for the rather special case that the period of the ith operation is half the period of the $(i+1)$th one. Serafini & Ukovich [1989] discuss nonpreemptive periodic scheduling subject to precedence constraints and show that this problem is NP-complete. Park & Yun [1985] give an ILP formulation of a nonpreemptive scheduling problem. They consider a set of independent periodic operations, where each execution requires a given number of resources during one unit of time, and aim to minimize the maximum required amount of resources. They show how this problem can be partitioned into a set of independent subproblems, which can be optimized independently. The partitioning divides the operations into subsets such that the periods of operations in different subsets are relatively prime. A problem related to nonpreemptive periodic scheduling is the problem of inscribing regular polygons in a circle so as to maximize the minimum distance between two vertices on the circle. Burkard [1986] solves this problem for a set of regular polygons that includes only two different types of polygons. Vince [1989] presents a more general approach to this problem.

2.2 Periodic Assignment

Periodic assignment deals with the problem of assigning the executions of periodic operations to a minimal number of processors, assuming that the executions are fixed in time. As we show in the next sections, this problem is closely related to that of colouring circular arcs. Circular-arc colouring has been studied by several authors. Garey, Johnson, Miller & Papadimitriou [1980] prove that circular-arc graph colouring is NP-hard. Tucker [1975] gives upper bounds on the number of colours needed to colour various types of circular-arc graphs. Orlin, Bonuccelli & Bovet [1981] and Shih & Hsu [1989] give efficient algorithms for the polynomially solvable subproblem of colouring proper circular-arc graphs.

Bartholdi, Orlin & Ratliff [1980] consider the periodic assignment problem under the assumption that the availability of resources is also periodic. This problem naturally arises in the area of personnel scheduling, where periodic jobs have to be assigned to persons having periodic working hours. Bartholdi [1981] proposes a linear programming round-off algorithm and gives its worst-case deviation from optimum. Orlin [1982] discusses the periodic assignment problem under the assumption that processors require a setup time s_{ij} to switch from execution i to execution j. This problem naturally arises in the area of vehicle scheduling, where a vehicle has to be transported from the end point of route i to the starting point of route j before it can start traversing route j.

3 Problem Description

In this section we give a formal description of a number of interrelated periodic scheduling and assignment problems. We restrict ourselves to nonpreemptive scheduling and do not consider precedence constraints.

Let $O = \{o_1, \ldots, o_n\}$ be a set of n periodic operations. For each $o \in O$ an execution time $e(o) \in \mathbb{N}$ and a period $p(o) \in \mathbb{N}$ are given. We assume that $p(o) \geq 1$ and $e(o) \leq p(o)$ for each $o \in O$. Once an execution of an operation o is started at a time unit $t \in \mathbb{Z}$, it is completed without interruption on the same processor. Note that in this paper time is measured in time units, i.e., time periods of equal length. If an operation o with execution time $e(o)$ is said to start at time unit t, it starts at the beginning of time unit t and completes at the end of time unit $t + e(o) - 1$. Similarly, a time interval $[t_1, t_2]$ denotes a set of consecutive time units, given by $\{t_1, t_1 + 1, \ldots, t_2\}$. The kth execution of operation o is denoted by $o[k]$. If execution $o[k]$ is started at time unit t, then execution $o[k+1]$ is started at time unit $t + p(o)$. The set of all executions is given by

$$E = \{o[k] \mid o \in O,\ k \in \mathbb{Z}\}.$$

So, each operation $o \in O$ is started exactly every $p(o)$ time units. Consequently, if for an operation o the start time of an arbitrary execution is fixed, then all executions of o are fixed in time. Without loss of generality, the executions of operation o are uniquely specified by a start time $s(o)$, with $0 \leq s(o) < p(o)$. Hence, a schedule $\mathcal{S}$ of the operations in O is uniquely determined by an n-tuple $(s(o_1), s(o_2), \ldots, s(o_n))$, with $0 \leq s(o_i) < p(o_i)$ for all $o_i \in O$. Furthermore, the operations are considered independent, i.e., there are no precedence constraints between executions of different operations.

Scheduling periodic operations naturally leads to periodic schedules. A schedule $\mathcal{S}$ is called periodic with period P if for each time unit $t \in \mathbb{Z}$ and each $o \in O$ the following holds:

operation o is executed at time unit t if and only if it is executed at time unit t + P.

Clearly, in order for a schedule to be periodic with period P, it is required that $p(o) \mid P$, for each $o \in O$. Consequently, the minimal period P of a schedule is given by $\text{lcm}(p(o_1), \ldots, p(o_n))$, i.e., the least common multiple of the periods of the individual operations.

Let M denote the set of processors. The processors are supposed to be identical, i.e., each operation $o \in O$ can be executed on any processor $m \in M$ and the time to execute operation o does not depend on the processor. Furthermore, a processor can only execute one operation at a time. We aim to minimize the number of processors necessary for the execution of the operations in O. Given a schedule $\mathcal{S}$, we can define the *thickness function* $T_\mathcal{S} : \mathbb{Z} \rightarrow \mathbb{N}$ which assigns to each $t \in \mathbb{Z}$ the number of operations that are being executed at that time unit. Since a processor can only execute one operation at a time, $\max_t T_\mathcal{S}(t)$ gives, for a given schedule $\mathcal{S}$, a lower bound on the number of processors that is required to carry out schedule $\mathcal{S}$. If schedule $\mathcal{S}$ is periodic with period P, then the thickness function $T_\mathcal{S}$ is also periodic with a period P', for which $P' \mid P$. Hence, to determine $\max_t T_\mathcal{S}(t)$, it suffices to consider time units $t \in \{1, \ldots, P\}$.

With respect to the assignment of executions to processors we consider two different cases, namely

(i) the *constrained* case, where all executions of an operation o have to be assigned to the same processor, for all $o \in O$, i.e., an assignment from O to M is required, and

(ii) the *unconstrained* case, where each execution $o[k]$ can be assigned to a different processor, i.e., an assignment from E to M is required.

An assignment of each execution in E to a processor in M may be difficult to specify, since E is a (countably) infinite set. We therefore restrict ourselves to periodic assignments. An assignment is called periodic with period $P \in \mathbb{N}$ if for each time unit $t \in \mathbb{Z}$, each $o \in O$, and for each $m \in M$ the following holds:

m executes o at time unit t if and only if m executes o at time unit t + P.

If for a periodic schedule S with period P the corresponding assignment is periodic with period P', then necessarily $P \mid P'$. In the constrained case, i.e., if all executions of an operation are assigned to the same processor, the assignment is necessarily periodic with period $P' = P$. For the unconstrained case, restricting oneself to periodic assignments does not lead to the use of extra processors as long as the length of period P' is not restricted. This is shown in the following theorem.

Theorem 1 *For each periodic schedule S a periodic assignment exists requiring only* $\max_t \mathrm{T}_S(t)$ *processors.*

Proof We have seen that $\max_t T_S(t)$ gives a lower bound on the required number of processors. Now a finite set of executions can be optimally assigned to $\max_t T_S(t)$ processors, using an $O(n \log n)$ algorithm [Hashimoto & Stevens, 1971; Gupta, Lee & Leung, 1979], where n denotes the number of executions. The algorithm assigns the executions in order of increasing start times to the first available processor, i.e., to the available processor with the smallest index number. Let us consider the assignment of a finite set of executions, namely the set of all executions in the time interval $[0, mP-1]$, with $m \in \mathbb{N}$ and $P = \mathrm{lcm}(p(o_1), \ldots, p(o_n))$. We show that, if m is chosen sufficiently large, the assignment necessarily becomes periodic with some period $m'P$, $m' < m$. Let us examine the assignment in intervals $[lP, (l+1)P-1]$, with $0 \leq l < m$. The assignment can attain only a finite set of different solutions in such an interval $[lP, (l+1)P-1]$, since a finite set of executions can be assigned to a finite set of processors. Consequently, if m is chosen sufficiently large, then in two intervals $[lP, (l+1)P-1]$ and $[l'P, (l'+1)P-1]$, with $0 \leq l < l' < m$, the assignment must necessarily be identical. Hence, the assignment necessarily becomes periodic with period $(l'-l)P$, using only $\max_t T_S(t)$ processors, which completes the proof of the theorem. ∎

The minimum period for which a periodic assignment uses $\max_t T_S(t)$ processors may generally be very large. For reasons of simplicity, we restrict ourselves in this paper to periodic assignments with periods of minimal length, i.e., with a period $P = \mathrm{lcm}(p(o_1), \ldots, p(o_n))$. In this way, an operation o is executed on at most $P/p(o)$ different processors. For the unconstrained case, an assignment is thus completely specified if the processor is given for $P/p(o)$ successive executions of each operation $o \in O$, denoted by $o[1], o[2], \ldots, o[P/p(o)]$, where $o[1]$ is defined to be the first execution starting at a time unit $t \geq 0$.

Given the definitions and assumptions described above, we can define the following periodic assignment problems. We formulate these problems as decision problems.

Unconstrained Periodic Assignment (UPA)
Given a schedule S for a set O of periodic operations with an execution time $e(o) \in \mathbb{N}$ and a period $p(o) \in \mathbb{N}$ for each $o \in O$, and an integer k, does an unconstrained periodic assignment with period $P = \mathrm{lcm}(p(o_1), \ldots, p(o_n))$ exist that uses at most k processors?

Constrained Periodic Assignment (CPA)
Given a schedule S for a set O of periodic operations with an execution time $e(o) \in \mathbb{N}$ and a period $p(o) \in \mathbb{N}$ for each $o \in O$, and an integer k, does a constrained periodic assignment exist that uses at most k processors?

Likewise, we define the following periodic scheduling problems.

Unconstrained Periodic Scheduling (UPS)
Given a set O of periodic operations with an execution time $e(o) \in \mathbb{N}$ and a period $p(o) \in \mathbb{N}$ for each $o \in O$, and an integer k, does a schedule exist for which an unconstrained periodic assignment with period $P = \text{lcm}(p(o_1), \ldots, p(o_n))$ uses at most k processors?

Constrained Periodic Scheduling (CPS)
Given a set O of periodic operations with an execution time $e(o) \in \mathbb{N}$ and a period $p(o) \in \mathbb{N}$ for each $o \in O$, and an integer k, does a schedule exist for which a constrained periodic assignment uses at most k processors?

With respect to CPS the following theorem gives a necessary and sufficient condition for scheduling the executions of two operations on the same processor.

Theorem 2 *The executions of two periodic operations o_i and o_j can be scheduled on the same processor if and only if*

$$\gcd(p(o_i), p(o_j)) \geq e(o_i) + e(o_j). \tag{1}$$

Proof Let $g = \gcd(p(o_i), p(o_j))$. We first prove that (1) is a sufficient condition. This is shown as follows. Choosing the start times $s(o_i) = 0$ and $s(o_j) = e(o_i)$, operation o_i is executed in a subset of the set I_i of intervals, defined by $[lg, lg + e(o_i) - 1]$, $l \in \mathbb{Z}$, and operation o_j is executed in a subset of the set I_j of intervals, defined by $[lg + e(o_i), lg + e(o_i) + e(o_j) - 1]$, $l \in \mathbb{Z}$. Hence, if $g \geq e(o_i) + e(o_j)$, then no intervals of I_i and I_j overlap, which proves the sufficiency of (1).

We prove the necessity of (1) by showing that, if $g < e(o_i) + e(o_j)$, operation o_i and o_j cannot be scheduled on the same processor. So, assume that $g < e(o_i) + e(o_j)$. Without loss of generality we may assume that $s(o_i) = 0$. We now have to prove that integers x, y exist for which

$$[xp(o_i), xp(o_i) + e(o_i) - 1] \cap [s(o_j) + yp(o_j), s(o_j) + yp(o_j) + e(o_j) - 1] \neq \emptyset$$

or, equivalently,

$$[xp(o_i) - yp(o_j), xp(o_i) - yp(o_j) + e(o_i) - 1] \cap [s(o_j), s(o_j) + e(o_j) - 1] \neq \emptyset.$$

From elementary number theory it is known that integers w, z exist for which $wp(o_i) + zp(o_j) = g$. If we choose $x = lw$ and $y = -lz$, with $l \in \mathbb{Z}$, it suffices to show that for some integer l

$$[lg, lg + e(o_i) - 1] \cap [s(o_j), s(o_j) + e(o_j) - 1] \neq \emptyset.$$

Clearly, this must be the case since the free intervals between the intervals $[lg, lg + e(o_i) - 1]$, $l = 0, 1, \ldots$, are of length $g - e(o_i)$, while the intervals $[s(o_j), s(o_j) + e(o_j) - 1]$ are of length $e(o_j)$. Hence, the assumption that $g < e(o_i) + e(o_j)$ implies that some integer l necessarily exists for which $[lg, lg + e(o_i) - 1]$ and $[s(o_j), s(o_j) + e(o_j) - 1]$ overlap. This completes the proof of the theorem. ■

A similar condition can be derived for CPA, as is shown in the following theorem.

Theorem 3 *For CPA, two periodic operations o_i and o_j with given start times $s(o_i)$ and $s(o_j)$, can be executed on the same processor if and only if*

$$e(o_i) \leq (s(o_j) - s(o_i)) \bmod g \leq g - e(o_j), \tag{2}$$

where $g = \gcd(p(o_i), p(o_j))$.

Proof Without loss of generality we may assume that $s(o_i) = 0$. This is true since, if $s(o_i) \neq 0$, then the start times of o_i and o_j can be shifted such that $s(o_i)$ becomes zero, without affecting possible overlap. The sufficiency of (2) is shown as follows. Let us consider time intervals $[0 + kg, g - 1 + kg]$, with $k \in \mathbb{Z}$. The first $e(o_i)$ time units of each of these intervals can be allocated for executions of o_i, and the remaining $g - e(o_i)$ time units for executions of o_j. Now, if (2) holds, then the allocated time units surely suffices to execute o_i and o_j. The first $e(o_i)$ time units of the intervals are only used to execute o_i once every $p(o_i)/g$ intervals. The remaining $g - e(o_i)$ time units are only (partly) used to execute o_j once every $p(o_j)/g$ intervals.

The necessity of (2) is shown as follows. Let us again consider the time intervals $[0+kg, g-1+kg]$, with $k \in \mathbb{Z}$. If (2) does not hold then the execution of o_j overlaps the first $e(o_i)$ time units once every $p(o_j)/g$ time intervals. We have already seen that the first $e(o_i)$ time units of the intervals are used for the execution of o_i once every $p(o_i)/g$ time units. Now, by definition, $\gcd(p(o_i)/g, p(o_j)/g) = 1$. Hence, if (2) does not hold, then operatiôns o_i and o_j cannot be executed on the same processor. This completes the proof of the theorem. ∎

Note that Theorem 2 can be considered a corollary of Theorem 3, since (1) directly follows from (2). In the next section we examine the computational complexity of the problems defined above.

4 Computational Complexity

To examine the complexity of the periodic assignment problems CPA and UPA, we focus our attention on the subset of problem instances for which $p(o) = p$ for all $o \in O$. Note that under this restriction CPA and UPA are identical. If we prove that this subset of instances is NP-complete, then both CPA and UPA have been proved to be NP-complete.

Theorem 4 *CPA and UPA are NP-complete.*

Proof It is ėasily verified that CPA and UPA are in $\mathcal{NP}$. Now the NP-completeness is proved by a reduction from circular-arc colouring, which has been shown to be NP-complete by Garey, Johnson, Miller & Papadimitriou [1980]. We first define circular-arc colouring. Let a set of circular arcs $A = \{a_1, \ldots, a_n\}$ be given, where each arc a_i, specified by an ordered pair (l_i, r_i), with $l_i, r_i \in \{0, 1, \ldots, 2n-1\}$, is an arc on a circle with circumference $2n$ that stretches clockwise from point l_i to point r_i, containing both endpoints, and let an integer k be given. The problem is now: is A k-colourable, i.e., does a function $f : A \to \{1, \ldots, k\}$ exist such that $f(a_i) \neq f(a_j)$ whenever a_i and a_j overlap? Any instance of circular-arc colouring can be transformed to a periodic assignment instance as follows. For each arc a_i we define a periodic operation with period $p(o_i) = 2n$, start time $s(o_i) = l_i$, and execution time $e(o_i) = r_i - l_i + 1$ if $r_i \geq l_i$ and $e(o_i) = r_i - l_i + 2n + 1$ if $r_i < l_i$. Now two periodic operations can be assigned to the same processor if and only if the corresponding circular arcs can be coloured with the same colour. Consequently, the circular arcs can be coloured using k colours if and only if the periodic operations can be assigned to k processors. Evidently, this is a polynomial-time transformation, which completes the proof of the theorem. ■

Note that the transformation from circular-arc colouring defines an equivalence between circular-arc colouring and the problem of assigning operations with identical periods, which we will use in Section 5.1.

To consider the complexity of CPS and UPS we again focus our attention on the subset of problem instances for which $p(o) = p$ for all $o \in O$. Again notice that this subset is in the intersection of the CPS and UPS problem instances.

Theorem 5 *CPS and UPS are NP-complete in the strong sense.*

Proof It is easily verified that CPS and UPS belong to $\mathcal{NP}$. We now prove the NP-completeness by a reduction from bin packing, which is NP-complete in the strong sense [Garey & Johnson, 1979]. An instance of bin packing is specified as follows. Let a finite set $A = \{a_1, \ldots, a_n\}$ of items be given, with for each item $a_i \in A$ a positive integer size $s(a_i)$, a positive bin capacity B and a positive integer k. Can A be partitioned into k disjoint subsets $A_1, \ldots, A_k$, such that the sum of the sizes in each subset A_i does not exceed the bin capacity B? Any instance of bin packing can be directly transformed into an instance of CPS or UPS as follows. For each item a_i we define a periodic operation o_i with execution time $e(o_i) = s(a_i)$ and period $p(o_i) = B$. Clearly, a number of periodic operations can be executed on the same processor if the corresponding items can be packed in one bin, and vice versa. Hence, the items $a_1, \ldots, a_n$ can be packed into k bins if and only if the operations $o_1, \ldots, o_n$ can be scheduled on k processors. Since the above transformation is polynomial, CPS and UPS are both NP-complete in the strong sense. ■

An alternative reduction from 3-partition can be constructed, showing that the problems remain NP-complete in the strong sense for the case that only one processor is available. Hence, this gives a stronger result. We have chosen, however, to give the reduction from bin packing since this reduction defines an equivalence between bin packing and the problem of scheduling periodic operations with identical periods, which we will use in Section 5.1.

5 Approximation Algorithms

All problems presented in Section 3 are NP-complete. This means that, unless $\mathcal{P} = \mathcal{NP}$, efficient optimization algorithms do not exist for these problems. We therefore focus our attention on approximation algorithms, i.e., algorithms which do not guarantee to find an optimal solution for every instance but attempt to find near-optimal solutions. In the remainder of this paper we present approximation algorithms for the periodic scheduling and assignment problems presented in Section 3 and, to some extent, analyse their performance. An interesting subclass of problems arises if we assume that the operations all have identical periods. We first consider approximation algorithms for this subclass of problems.

5.1 Periodic Operations with Identical Periods

In Section 4 we already indicated the equivalence between bin packing and the problem of scheduling periodic operations with identical periods. Hence, approximation algorithms for bin packing can be directly applied to this problem. A large number of approximation algorithms exist for bin packing, ranging from simple approximation algorithms called *first fit* and *first fit decreasing*, which have asymptotic performance ratios of $\frac{17}{10}$ and $\frac{11}{9}$, respectively, to approximation schemes. An extensive survey of the literature on approximation algorithms for bin packing is given by Coffmann, Garey & Johnson [1984]. A bin packing algorithm gives a partitioning of the operations into subsets such that the operations in the same subset can be assigned to the same processor. A feasible schedule can then easily be obtained by scheduling the operations in each subset one after the other, in some arbitrary order. The wealth of approximation algorithms for bin packing provided by the literature surely suffices to effectively handle this subclass of periodic scheduling problems.

To present approximation algorithms for the assignment of periodic operations with identical periods we refer to its equivalence with the problem of colouring circular arcs, as indicated in Section 4. To the best of our knowledge, Tucker [1975] is the only author who considers the subject of approximation algorithms for colouring circular arcs, in order to give an upper bound on the number of colours necessary for colouring circular arcs. Elaborating on this result, we present the following 2-step approximation algorithm for colouring circular arcs, called *sort&match*.

1. Partition the set of arcs into two subsets A and B, where A contains all arcs that cover one specific point $t \in \{0, 1, \ldots, 2n-1\}$ for which the thickness function attains a minimum value, and B contains all remaining arcs. Consequently, $|A| = \min_t T_S(t)$. Now the arcs in B can be optimally assigned using the assignment algorithms of Hashimoto & Stevens [1971] or Gupta, Lee & Leung [1979] using $\max_t T_S(t)$ colours: the arcs a_i in B are sorted in order of their starting point l_i and they are assigned in this order to the first available colour, i.e., the available colour with the smallest index number.

2. Determine a maximum subset A' of arcs in A which can be coloured with a colour that is already used in step 1 to colour arcs in B. This problem can be formulated as a maximum-cardinality matching problem in a bipartite graph, which can be solved efficiently using an augmenting path algorithm [Edmonds, 1965; Hopcroft & Karp, 1973]. Finally, each remaining arc in $A - A'$ is given a different free colour.

Tucker [1975] only considers the first step of the algorithm presented above. Clearly the algorithm requires at most $\max_t T_S(t) + \min_t T_S(t)$ colours. Since $\max_t T_S(t)$ is a lower bound on the number of required colours, *sort&match* has a worst-case performance ratio of 2. This worst-case performance ratio already holds for the first step of the algorithm (assuming that all arcs in A are given a different free colour), which Tucker already showed. The worst-case performance bound can be shown to be tight [Korst, Aarts, Lenstra & Wessels, 1991]. The average-case performance of *sort&match* is much better. Experimental results indicate that the algorithm almost always finds solutions that are within 10% of the optimum for randomly generated instances [Korst, Aarts, Lenstra & Wessels, 1991].

5.2 Periodic Operations with Arbitrary Periods

In this subsection we discuss possible approximation algorithms for the UPA, CPA, UPS and CPS problems, for the case that operations have arbitrary integer periods.

Approximation Algorithm for UPA

Sort&match, presented in Section 5.1, can also be used as an approximation algorithm for UPA by associating an arc with each execution that is contained in a time window of length $P = \text{lcm}(p(o_1), \ldots, p(o_n))$. Note, however, that here the number of arcs is not polynomially bounded by the number of operations. The performance bound of *sort&match* clearly remains unaffected. Circular arcs can be efficiently coloured if they are proper, i.e., if no arc is completely contained in another arc [Orlin, Bonuccelli & Bovet, 1981;Shih & Hsu, 1989]. Hence, if periodic operations all have identical execution times, they can be optimally assigned to processors in a time that is polynomial in the number of executions.

Approximation Algorithms for CPA

Using Theorem 3 we can easily determine for each pair of periodic operations whether they can be assigned to the same processor. Consequently, we can define a graph $\mathcal{G} = (\mathcal{V}, \mathcal{E})$, where each $v_i \in \mathcal{V}$ is associated with a periodic operation o_i. Two vertices v_i and v_j are adjacent if the associated operations o_i and o_j cannot be assigned to the same processor. The resulting graph $\mathcal{G}$ is called a *periodic-interval graph*. Now it is easy to see that solving a CPA instance is identical to colouring the vertices of the corresponding periodic-interval graph with a minimum number of colours. A periodic-interval graph can be considered to be a generalization of a circular-arc graph in the case that all periods are identical. To the best of our knowledge no graph colouring algorithms are presented in the literature that are tailored to colouring periodic-interval graphs. However, approximation algorithms for colouring arbitrary graphs might give satisfactory results in practice.

Approximation Algorithms for UPS

Experimental results indicate that *sort&match* is able to find solutions for UPA that are often close to $\max_t T_S(t)$. It therefore seems tempting to handle UPS using the following two-step approach:

1. first determine start times for the operations such that $\max_t T_S(t)$ is minimized, and

2. next use *sort&match* to find a feasible assignment.

Now the problem of finding a schedule such that $\max_t T_S(t)$ is minimized can be shown to be NP-complete. This immediately follows from the fact that UPS remains NP-complete for the single processor case. Consequently, we can restrict ourselves to constructing an approximation algorithm for the problem of finding start times that minimize $\max_t T_S(t)$. Note that for a set O' of periodic operations with $\gcd(p(o_i), p(o_j)) = 1$ for all $o_i, o_j \in O'$, we have $\max_t T_S(t) = |O'|$ for any possible choice of start times. This is a corollary of Theorem 2; see also [Park & Yun, 1985]. Consequently, the set of periodic operations O can be partitioned into a number of disjoint subsets $O_1, O_2, \ldots, O_l$ such that $\gcd(p(o_i), p(o_j)) = 1$ for each pair of operations o_i, o_j that have been assigned to different subsets, and $\max_t T_{S(O_i)}(t)$ can be minimized independently for each subset O_i. The total thickness $\max_t T_S(t)$ is then given by $\sum_{O_i} \max_t T_{S(O_i)}$. This partitioning approach will reduce the size of the problem.

We now restrict ourselves to minimizing $\max_t T_S(t)$ for a given subset O_i. This can be done as follows. First select a subset $O_{i'}$ of O_i, for which $\gcd(p(o_i), p(o_j)) = 1$ for all $o_i, o_j \in O_{i'}$, such that $O_{i'}$ is as large as possible. This is done by using some independent set heuristic. The operations in $O_{i'}$ are given arbitrary start times. Next, the remaining operations must be given start times subject to the start times of the operations in $O_{i'}$. If the number of operations in $O_i - O_{i'}$ is small, an enumeration is most appropriate. Otherwise, some constructive or local search approach can be used.

Approximation Algorithm for CPS

In the case of CPS we observe the following. If one or more periodic operations are assigned to a processor, then the time that the processor remains idle can be expressed as one or more periodic intervals, each with a period and a duration. For example, if a periodic operation o_i with period $p(o_i)$ and execution time $e(o_i)$ is assigned to an idle processor, then the remaining idle time can be expressed as a periodic interval with period $p(o_i)$ and a duration $p(o_i) - e(o_i)$. We can thus consider the problem of assigning periodic operations to processors as the problem of assigning periodic operations to periodic intervals. For reasons of simplicity we denote a periodic operation o_i with period $p(o_i)$ and execution time $e(o_i)$ by the ordered pair (p_i, e_i) and a periodic interval with period p_j and duration d_j by the ordered pair $[\![p_j, d_j]\!]$. From Theorem 2 we derive that a periodic operation (p_i, e_i) can be assigned to a periodic interval $[\![p_j, d_j]\!]$ if and only if $\gcd(p_i, p_j) \geq e_i + (p_j - d_j)$. Let $g = \gcd(p_i, p_j)$ and $e_j = p_j - d_j$; then by assigning periodic operation (p_i, e_i) to periodic interval $[\![p_j, d_j]\!]$, the remaining idle time can be expressed as a set of periodic intervals in a number of alternative ways. We assume that a periodic operation is always started at the begin of the periodic interval to which it is assigned. Consequently, the remaining idle time can be expressed as one of the following three alternatives.

1. $p_i/g - 1$ periodic intervals $[\![p_i, g - e_j]\!]$,
   $p_j/g - 1$ periodic intervals $[\![p_j, e_j]\!]$, and
   1 periodic interval $[\![p_i, g - e_i - e_j]\!]$
2. $p_i/g - 1$ periodic intervals $[\![p_i, e_i]\!]$,
   $p_j/g - 1$ periodic intervals $[\![p_j, e_j]\!]$, and
   1 periodic interval $[\![g, g - e_i - e_j]\!]$
3. $p_i/g - 1$ periodic intervals $[\![p_i, e_i]\!]$,
   $p_j/g - 1$ periodic intervals $[\![p_j, g - e_i]\!]$, and
   1 periodic interval $[\![p_j, g - e_i - e_j]\!]$

In all three cases the number of periodic intervals is given by

$$\frac{p_i + p_j}{\gcd(p_i, p_j)} - 1.$$

Note that, if $p_i = p_j$, the three alternatives are identical, leading to only one periodic interval. Otherwise, if $p_i | p_j$ or $p_j | p_i$, then the three alternatives reduce to two essentially different ones.

Based on this observation, we propose the following iterative approximation algorithm. In each iteration all possible assignments of periodic operations to periodic intervals are considered and the one that is considered best is selected to be scheduled. The 'goodness' of an operation-to-interval assignment is defined by the amount of idle time that remains after assigning the periodic operation to the periodic interval. In each iteration the assignment of (p_i, e_i) to $[\![p_j, d_j]\!]$ is selected for which $d_j/p_j - e_i/p_i$ is minimal, provided that the assignment is feasible. Clearly, the amount of idle time that remains after assigning an operation (p_i, e_i) to an idle processor is given by $1 - e_i/p_i$. Consequently, the algorithm will not assign a periodic operation to an idle processor as long as the periodic operation can be assigned to a periodic interval of a processor that is already in use. After each iteration, the remaining idle time is expressed as one or more periodic intervals using one of the three alternatives mentioned above. Which alternative is selected is determined by considering how well the unassigned operations fit in the periodic intervals. This can be considered as a maximum-weight matching problem on a bipartite graph, which can be handled efficiently.

A detailed analysis of the algorithm is beyond the scope of the paper. We mention that, in the case of periodic operations with identical periods, solutions are found that are identical to the ones obtained by *first fit decreasing* for bin packing.

6 Conclusions

A number of closely interrelated optimization problems have been discussed from the field of nonpreemptive periodic scheduling. The complexity of these problems has been examined. We have derived Necessary and sufficient conditions for executing two periodic operations on a single processor. Finally, approximation algorithms have been proposed for periodic scheduling and periodic assignment problems, for the constrained case as well as the unconstrained case.
The material presented in this paper leaves the following open problems:

- Which constraints do we have to impose on the problems discussed in this paper to allow for efficient optimization algorithms?
- Do approximation algorithms exist for colouring periodic-interval graphs that have a constant worst-case performance ratio?
- Do approximation algorithms exist for colouring circular-arc graphs with a worst-case performance ratio smaller than two?
- Is it possible to give a constant worst-case performance ratio for the approximation algorithms for CPS and UPS?

Bibliography

Bartholdi, J.J. [1981], A guaranteed-accuracy round-off algorithm for cyclic scheduling and set covering, *Operations Research* **29**, 501-510.

Bartholdi, J.J., J.B. Orlin, and H.D. Ratliff [1980], Cyclic scheduling via integer programs with circular ones, *Operations Research* **28**, 1074-1085.

Bertossi, A.A. and M.A. Bonuccelli [1983], Preemptive scheduling of periodic jobs in uniform multiprocessor systems, *Information Processing Letters* **16**, 3-6.

Bertossi, A.A. and M.A. Bonuccelli [1985], A polynomial feasibility test for preemptive periodic scheduling of unrelated processors, *Discrete Applied Mathematics* **12**, 195-201.

Burkard, R.E. [1986], Optimal schedules for periodically recurring events, *Discrete Applied Mathematics* **15**, 167-180.

Chetto, H. and M. Chetto [1989], Scheduling periodic and sporadic tasks in a real-time system, *Information Processing Letters* **30**, 177-184.

Coffmann, E.G., Jr., M.R. Garey, and D.S. Johnson [1984], Approximation algorithms for bin packing - an updated survey, in: G. Ausiello, M. Lucertini, and P. Serafini (Eds.), *Algorithms Design and Computer System Design*, CISM Courses and Lectures 284, Springer, Vienna, 49-106.

Dhall, S.K. and C.L. Liu [1978], On a real-time scheduling problem, *Operations Research* **26**, 127-140.

Edmonds, J. [1965], Paths, trees and flowers, *Canadian Journal of Mathematics* **17**, 449-467.

Garey, M.R. and D.S. Johnson [1979], *Computers and Intractability: A Guide to the Theory of NP-Completeness*, W.H. Freeman and Co., San Francisco.

Garey, M.R., D.S. Johnson, G.L. Miller, and C.H. Papadimitriou [1980], The complexity of coloring circular arcs and chords, *SIAM Journal on Algebraic and Discrete Methods* **1**, 216-227.

Gonzalez, M.J. and J.W. Soh [1975], Periodic job scheduling in a distributed processor system, *IEEE Transactions on Aerospace and Electronic Systems* **12**, 530-536.

Gupta, U.I., D.T. Lee, and J.Y.-T. Leung [1979], An optimal solution for the channel-assignment problem, *IEEE Transaction on Computers* **28**, 807-810.

Hashimoto, A. and J. Stevens [1971], Wire routing by optimizing channel assignment with large apertures, *Proceedings of the 8th Design Automation Conference*, 155-169.

Hopcroft, J.E. and R.M. Karp [1973], An $n^{5/2}$ algorithm for maximum matchings in bipartite graphs, *SIAM Journal on Computing* **2**, 225-231.

Korst, J.H.M., E.H.L. Aarts, J.K. Lenstra, and J. Wessels [1991], Periodic Assignment and Graph Colouring, *Philips Research Manuscript.*

Labetoulle, J. [1974], Some theorems on real time scheduling, in: E. Gelenbe and R. Mahl (Eds.), *Computer Architecture and Networks*, North-Holland, Amsterdam, 285-293.

Lawler, E.L. and C.U. Martel [1981], Scheduling periodically occurring tasks on multiple processors, *Information Processing Letters* **12**, 9-12.

Leung, J.Y.-T. and M.L. Merrill [1980], A note on preemptive scheduling of periodic, real-time tasks, *Information Processing Letters* **11**, 115-118.

Leung, J.Y.-T. and J. Whitehead [1982], On the complexity of fixed-priority scheduling of periodic, real-time tasks, *Performance Evaluation* **2**, 237-250.

Liu, C.L. and J.W. Layland [1973], Scheduling algorithms for multiprogramming in a hard real-time environment, *Journal of the Association for Computing Machinery* **20**, 46-61.

Orlin, J.B. [1982], Minimizing the number of vehicles to meet a fixed periodic schedule: an application of periodic posets, *Operations Research* **30**, 760-776.

Orlin, J.B., M.A. Bonuccelli, and D.P. Bovet [1981], An $O(n^2)$ algorithm for coloring proper circular arc graphs, *SIAM Journal on Algebraic and Discrete Methods* **2**, 88-93.

Park, K.S. and D.K. Yun [1985], Optimal scheduling of periodic activities, *Operations Research* **33**, 690-695.

Serafini, P. and W. Ukovich [1989], A mathematical model for periodic scheduling problems, *SIAM Journal on Discrete Mathematics* **2**, 550-581.

Shih, W.-K. and W.-L. Hsu [1989], An $O(n^{1.5})$ algorithm to color proper circular arcs, *Discrete Applied Mathematics* **25**, 321-323.

Tucker, A [1975], Coloring a family of circular arcs, *SIAM Journal on Applied Mathematics* **29**, 493-552.

Vince, J. [1989], Scheduling periodic events, *Discrete Applied Mathematics* **25**, 299-310.

Embeddings of shuffle-like graphs in hypercubes *

M. Baumslag§, M.C. Heydemann†, J. Opatrny‡, D. Sotteau†
§Comp. and Inf. Science Dept, Univ. of Massachusetts, USA
†LRI, UA 410 CNRS, bât 490, Univ. Paris-Sud, 91405 Orsay France
‡Dept of Computer Sciences, Concordia Univ., Montréal, Canada

Abstract

Let G and H be two simple undirected graphs. An *embedding* of the graph G in the graph H is an injective mapping f from the vertices of G into the vertices of H together with a mapping P_f of edges of G into paths in H. The *dilation* of the embedding is the maximum taken over all the lengths of the paths $P_f(x,y)$ associated with the edges (x,y) of G.
One challenge pointed out in [9] is to find embeddings of the de Bruijn graph in the hypercube of the same order which have a low dilation. For a de Bruijn graph of diameter D we give an embedding in a hypercube of the same diameter of dilation $2\lceil D/5\rceil$, and determine the edge-congestion and vertex-congestion of this embedding. Similar results are given for the shuffle-exchange graphs.

1 Introduction

A parallel algorithm can be represented by a graph, say G, in which the nodes represent the processes, and the edges represent the communications among the processes. Similarly, a parallel computer can be represented by a graph, say H, in which the nodes represent the processors and the edges represent the communication links among the processors. An important problem in parallel computations is how to map G into H such that H can carry out efficiently all communications specified by G. This problem has two components. First, how to map the processes of G into the processors of H, and second, how to assign a physical communication path in H to each edge of G so that the parallel algorithm can be efficiently executed. This problem is known in graph theoric terms as graph embedding problem. We can define it more precisely as follows.

Let G and H be two simple undirected graphs. An *embedding* of the graph G in the graph H is an injective mapping f from the vertices of G into the vertices of H together with a mapping P_f of edges of G into paths in H. For any edge (x,y) of G, $P_f(x,y)$ denotes the path between the vertices $f(x)$ and $f(y)$ in H assigned by P_f.

*The work was supported partially by NSERC of Canada and by PRC C3 of France and was partially done while the third author was visiting the University of Paris-Sud.

From among the parameters that have been used in measuring the efficiency of embeddings (see for example [10]), we will restrict our attention to the following ones.

The *dilation* of a given embedding f, denoted by $dil(f)$, is the maximum of the lengths of the paths $P_f(x,y)$ in H associated with all edges (x,y) of G. The minimum dilation of an embedding of G in H, denoted $dil(G,H)$, is the minimum of $dil(f)$ taken over all the embeddings f of G in H. Clearly the minimum will be reached in particular if we take for $P_f(x,y)$ shortest paths between x and y in H. In this paper we will always do so.

The *expansion* of the embedding f is the ratio of the number of vertices in H to the number of vertices in G. Here we only consider embeddings with expansion equal to 1.

The *edge-congestion* of f, denoted by $econg(f)$ is the maximum, over all edges e of H, of the number of edges of G mapped to a path of H which includes e.

The *vertex-congestion* of f, denoted by $vcong(f)$ is the maximum, over all vertices v of H, of the number of edges of G mapped to a path of H containing v as internal vertex.

Determining the computational power of hypercubes is a central problem in the theory and practice of parallel networks. In particular, we would like to know what communication patterns the hypercube can simulate efficiently. The following embeddings in hypercubes are known (for other results see the survey [12]):

- Any binary tree can be embedded in the smallest hypercube big enough to contain it with constant dilation and edge-congestion [2].
- Any d-dimensional mesh can be embedded in the smallest hypercube big enough to contain it with dilation $O(d)$ [4].
- Any butterfly-like graph (*i.e.*, the cube-connected cycles, the butterfly or the FFT graphs) can be embedded in the smallest hypercube big enough to contain it with dilation 2 and unit edge-congestion [9].

Also, Winkler [14] has characterized the graphs that can be embedded as an *isometric subgraph* of the hypercube, and Greenberg and Bhatt [8] have recently studied the problem of embedding *multiple copies* of the above classes of graphs into hypercubes.

One challenge pointed out by D. Greenberg, L. Heath and A. Rosenberg [9] is to find low dilation embeddings with expansion equal to 1 of shuffle-like networks, such as the de Bruijn graphs and the shuffle-exchange graphs, in hypercubes. De Bruijn networks have been studied as a possible choice for designing large communication networks , and very efficient, general sorting algorithms have been developed for them [1]. A de Bruijn network is being constructed for the NASA's Galileo space mission to be used as a signal decoder [5].

A different approach to the problem of a simulation of one network by another is through the notion of **work-preserving emulations** (see, for example, [11]). A guest network G is said to have a work-preserving emulation on a host network H if any T computation steps of G can be emulated in $O(T|G|/|H|)$ steps on H. In this model, computations of G can be *replicated* at several nodes of H (this may be visualized as a one-to-many embedding). Schwabe [13] recently proved that (in this model) any T steps of an N-node de Bruijn network can be emulated in $O(T)$ steps on an N-node hypercube, provided that $T \geq \log N$. However, this result does not give an embedding of a de Bruijn network in a hypercube with expansion 1.

The D-dimensional hypercube, denoted by $H(D)$, has for vertex set the set of all binary words of length D. There is an edge between any two words that differ in exactly one position. The distance between two vertices x and y of $H(D)$ is denoted by $d_D(x,y)$.

A de Bruijn digraph $B(d,D)$ of order d^D, out and indegree d has been defined in [6] as follows. Its vertex set is the set of all words of length D on an alphabet A of size d. There is an arc from any vertex $x_1x_2\cdots x_D$ to the d vertices $x_2x_3\cdots x_D\lambda$, where λ is any letter of A. The undirected de Bruijn graph $UB(d,D)$ is obtained from the de Bruijn digraph by taking the underlying graph and deleting self loops and multiple edges. In this paper we will be concerned with the undirected de Bruijn graph on the alphabet $A = \{0,1\}$. We will denote $UB(2,D)$ by $B(D)$ for short.

The binary shuffle-exchange graph of diameter D, denoted by $S(D)$, is the graph whose vertices are all binary words of length D and whose edges are of two types. A *shuffle edge* connects any vertex $x_1x_2\cdots x_D$ to the vertex $x_2x_3\cdots x_Dx_1$. An *exchange edge* connects any vertex $x_1x_2\cdots x_D$ to the vertex $x_1x_2\cdots\overline{x_D}$ where $\overline{x_D} = 1 - x_D$.

The cartesian product of two graphs G and G', denoted by $G\Box G'$, is the graph whose vertices are all the pairs (u,v) where u is a vertex of G and v is a vertex of G'. We will denote here by uv the pair (u,v). The notation (u,v) will be reserved for the edge between the vertices u and v if it exists. Two vertices uv and $u'v'$ are connected in $G\Box G'$ if and only if $u = u'$ and (v,v') is an edge of G' or $v = v'$ and (u,u') is an edge of G.

Some heuristics have been given in [3] for embeddings of de Bruijn graphs in hypercubes. As far as the present authors can ascertain, there were no other known non-trivial embeddings of the deBruijn network into the hypercube until now.

In this paper we give constructions of embeddings of de Bruijn and shuffle-exchange graphs in hypercubes and deduce upper bounds for the parameters defined above. Notice that the dilation of an embedding of these shuffle-like graphs which contain odd cycles is at least 2 since the hypercube is bipartite (this is the case for $B(D)$, $D \geq 2$, and $S(D)$, $D \geq 3$). But we don't know any better lower bound than 2 for the dilation or 1 for the congestions of these embeddings.

In section 2 we obtain embeddings of the de Bruijn graph $B(D)$ in the hypercube $H(D)$ with dilation less than or equal to $2\lceil D/5\rceil$ and edge-congestion 2. From these embeddings we deduce in section 3 a bound for the dilation of embeddings of shuffle-exchange graphs.

In the appendix we give embeddings of shuffle-like graphs for graphs of diameters 2 to 6.

2 Upper bounds for the parameters of embeddings of de Bruijn graphs

Proposition 2.1 *Let f and f' be embeddings of $B(D)$ and $B(D')$ respectively in $H(D)$ and $H(D')$. Then there exists an embedding g of $B(D+D')$ in $H(D+D')$ with*

$$dil(g) \leq dil(f) + dil(f')$$

$$econg(g) \leq max\ (dil(f),\ dil(f'),\ 2)$$

$$vcon(g) \leq vcong(f) + vcong(f') + 2$$

Thus, for any strictly positive D and D', we have

$$dil(B(D+D'), H(D+D')) \leq dil(B(D), H(D)) + dil(B(D'), H(D')),$$

$$econg(B(D+D'), H(D+D')) \leq max\ (econg(B(D), H(D)), econg(B(D'), H(D')), 2)$$

$$vcong(B(D+D'), H(D+D')) \leq vcong(B(D), H(D)) + vcong(B(D'), H(D')) + 2.$$

Proof: The hypercube $H(D+D')$ is isomorphic to the cartesian product of the hypercubes $H(D)$ and $H(D')$. Any vertex $u = x_1x_2\cdots x_{D+D'}$ of $H(D+D')$ can be written as u_1u_2 where $u_1 = x_1x_2\cdots x_D$ and $u_2 = x_{D+1}\cdots x_{D+D'}$. When u_1, u_2 is fixed, the vertices u_1u, uu_2 of $H(D+D')$ span an induced subgraph isomorphic to $H(D')$, $H(D)$ respectively. If $s = s_1s_2$ and $t = t_1t_2$ are two vertices of $H(D+D')$ then

$$d_{D+D'}(s,t) = d_D(s_1,t_1) + d_{D'}(s_2,t_2). \tag{1}$$

Let f and f' be embeddings of $B(D)$ and $B(D')$ in $H(D)$ and $H(D')$, respectively. We define an embedding g of $B(D+D')$ in $H(D+D')$ as follows. For any vertex $x = x_1x_2\cdots x_{D+D'}$,

$$g(x_1x_2\cdots x_{D+D'}) = f(x_1x_2\cdots x_D)f'(x_{D+1}x_{D+2}\cdots x_{D+D'})$$

The paths of $H(D+D')$ associated with the edges of $B(D+D')$ will be specified later. This embedding could be considered as the composition of two embeddings. First, an embedding of $B(D+D')$ in the cartesian product of $B(D)$ and $B(D')$, where any vertex $u = x_1x_2\cdots x_{D+D'}$ of $B(D+D')$ is mapped on the vertex $x_1x_2\cdots x_Dx_{D+1}\cdots x_{D+D'}$ of $B(D)\square B(D')$ where $u_1 = x_1x_2\cdots x_D$ is a vertex of $B(D)$ and $u_2 = x_{D+1}\cdots x_{D+D'}$ is a vertex of $B(D')$. Second, an embedding of $B(D)\square B(D')$ in $H(D)\square H(D')$ defined by the embeddings f of $B(D)$ in $H(D)$ and f' of $B(D)$ in $H(D')$. Embeddings of cartesian products have been studied in [10], but here we will not take this approach since it is much simpler to study g directly.

Consider an edge (u,v) of $B(D+D')$ with $u = u_1u_2 = x_1x_2\cdots x_Dx_{D+1}\cdots x_{D+D'}$ and $v = v_1v_2 = x_2x_3\cdots x_{D+1}x_{D+2}\cdots x_{D+D'}y$. Note that (u,v) is an edge of $B(D+D')$ if and only if (u_1,v_1) is an edge of $B(D)$ and (u_2,v_2) is an edge of $B(D')$.

By the definition of the embedding g, $g(u) = f(u_1)f'(u_2)$ and $g(v) = f(v_1)f'(v_2)$ in $H(D+D')$. Therefore, by (1) we get

$$d_{D+D'}(g(u), g(v)) = d_D(f(u_1), f(v_1)) + d_{D'}(f'(u_2), f'(v_2)).$$

Thus, for $D \geq D' \geq 1$,

$$dil(g) \leq dil(f) + dil(f').$$

We will now study the congestion of g.
To complete the definition of the embedding g, let us define the path $P_g(u,v)$ in $H(D+D')$ between $g(u)$ and $g(v)$, for every edge (u,v) of $B(D+D')$. Without loss of generality, we can assume that $u = u_1u_2$ with $u_1 = \gamma u'_1$, $u_2 = \alpha u'_2$, and $v = v_1v_2$ with $v_1 = u'_1\alpha$, $v_2 = u'_2\beta$ where $\alpha, \beta, \gamma \in \{0,1\}$.
Let $w = u_1v_2$. The path $P_g(u,v)$ is obtained as the concatenation of two paths.
The first one is a path from $g(u) = f(u_1)f'(u_2)$ to $g(w) = f(u_1)f'(v_2)$ in the subhypercube isomorphic to $H(D')$ obtained by making the first D coordinates equal to $f(u_1)$, a path isomorphic to $P_{f'}(u_2,v_2)$.
The second one is a path from $g(w) = f(u_1)f'(v_2)$ to $g(v) = f(v_1)f'(v_2)$ in the subhypercube isomorphic to $H(D)$ obtained by making the last D' coordinates equal to $f'(v_2)$, a path isomorphic to $P_f(u_1,v_1)$.

Remark: If we know two vertices from $\{g(u), g(v), g(w)\}$, we can determine the third one uniquely except if $D = 1$ or $D' = 1$. Also, given any vertex $g(w)$ (or equivalently w since g is bijective) we can determine $g(u)$ and $g(v)$ up to the parameter α (where $u_2 = \alpha u'_2$ and $v_1 = u'_1\alpha$). Thus, there are only two paths $P_g(u,v)$ going through $g(w)$.

Let us consider the load induced on any edge of $H(D+D')$ by all the paths $P_g(u,v)$. The edges of type (a_1a_2, a_1b_2), in a subhypercube isomorphic to $H(D')$, are only loaded by paths $P_{f'}(u_2,v_2)$ for all pairs of adjacent vertices u_2 and v_2 of $B(D')$, except if $D' = 1$. In this case the load of the unique edge (a_10, a_11) can be two. Similarly the edges of type (a_1a_2, b_1a_2), in a subhypercube isomorphic to $H(D)$, are only loaded by paths $P_f(u_1,v_1)$ for all pairs of adjacent vertices u_1 and v_1 of $B(D)$ except if $D = 1$. Indeed, from the above remark, a given path $P_{f'}(u_2,v_2)$ is connected to a unique path $P_f(u_1,v_1)$ (and conversely) in a unique path $P_g(u,v)$ except if D or D' equals 1. Therefore,

$$econg(g) \leq max(econg(f), econg(f'), 2).$$

Let us now consider any vertex of $H(D+D')$. Since f and f' are bijective, we can write such a vertex as $f(a_1)f'(a_2)$ (where a_1, a_2, is a vertex of $B(D)$, $B(D')$ respectively). Vertex $f(a_1)f'(a_2)$ is loaded by paths $P_{f'}(u_2,v_2)$ for all pairs of adjacent vertices u_2 and v_2 of $B(D')$ in the subhypercube isomorphic to $H(D')$ having the first D coordinates equal to $f(a_1)$. It is also loaded by paths $P_f(u_1,v_1)$ for all pairs of adjacent vertices u_1 and v_1 of $B(D)$ in the subhypercube isomorphic to $H(D)$ having the last D' coordinates equal to $f'(a_2)$. Finally, $f(a_1)f'(a_2)$ is also loaded as the connecting vertex $g(w)$ of two paths $P_g(u,v)$ for $w = a_1a_2$, by the above remark. Therefore

$$vcong(g) \leq vcong(f) + vcong(f') + 2. \quad \square$$

Lemma 2.2 *For $1 \leq D \leq 5$ there exist embeddings of the de Bruijn graph $B(D)$ in the hypercube $H(D)$ with the following parameters.*

D	1	2	3	4	5
dilation	1	2	2	2	2
edge congestion	1	2	2	2	2
vertex congestion	0	1	1	2	2

Proof: Embeddings giving these values are described in the appendix.

Notice that the dilation of embeddings of graphs in the above lemma matches the lower bound, but we don't know, for $D \geq 4$, if the congestions are optimal.

Theorem 2.3 *For $D \geq 2$, there exists an embedding of the de Bruijn graph $B(D)$ in the hypercube $H(D)$ with parameters bounded as follows.*

D	$\equiv 1 \pmod 5$	$\equiv 2$ *or* $3 \pmod 5$	$\equiv 4$ *or* $0 \pmod 5$
dilation	$\leq 2\lceil D/5 \rceil - 1$	$\leq 2\lceil D/5 \rceil$	
edge congestion		$= 2$	
vertex congestion	$\leq 4\lceil D/5 \rceil - 4$	$\leq 4\lceil D/5 \rceil - 3$	$\leq 4\lceil D/5 \rceil - 2$

Proof: For $D \leq 5$ we use the result of lemma 2.2. For $D \geq 6$ we use inductively the result of proposition 2.1. If $D = 5(k-1)+r$, $1 \leq r \leq 5$, we embed $B(D)$ in the cartesian product of $(k-1)$ copies of $B(5)$ and a copy of $B(r)$. The hypercube $H(D)$ is the cartesian product of $(k-1)$ copies of $H(5)$ and a copy of $H(r)$. We embed each $B(5)$ in a distinct copy of $H(5)$ and $B(r)$ in $H(r)$. From proposition 2.1 we get immediately

$$\begin{aligned} dil(B(D),H(D)) &\leq (k-1)dil(B(5),H(5)) + dil(B(r),H(r)) \\ &\leq 2(k-1)+2 \quad \text{if } r \neq 1 \\ &\leq 2(k-1)+1 \quad \text{if } r = 1 \end{aligned}$$

$$\begin{aligned} econg(B(D),H(D)) &\leq max_{1 \leq r \leq 5}\ econg(B(r),H(r)) \\ &\leq 2 \end{aligned}$$

$$\begin{aligned} vcong(B(D),H(D)) &\leq (k-1)vcong(B(5),H(5)) + vcong(B(r),H(r)) + 2(k-1) \\ &\leq 4(k-1) + vcong(B(r),H(r)) \\ &\leq 4\left\lceil \tfrac{D}{5} \right\rceil - 4 \quad \text{if } r = 1 \\ &\leq 4\left\lceil \tfrac{D}{5} \right\rceil - 3 \quad \text{if } r = 2,3 \\ &\leq 4\left\lceil \tfrac{D}{5} \right\rceil - 2 \quad \text{if } r = 4,5. \quad \square \end{aligned}$$

3 Dilation of embeddings of shuffle-exchange graphs

The following result was proved recently by Feldmann and Unger [7].

Theorem 3.1 *For any D the shuffle-exchange graph $S(D)$ is isomorphic to a subgraph of the binary de Bruijn graph $B(D)$.*

So we immediately get

Corollary 3.2

$$dil(S(D), H(D)) \leq dil(B(D), H(D))$$

Therefore, using theorem 2.3 we have the following.

Corollary 3.3 *For $D \geq 2$, $dil(S(D), H(D)) \leq 2\lceil D/5 \rceil$. If, furthermore, $D \equiv 1$ (mod 5), then $dil(S(D), H(D)) \leq 2\lceil D/5 \rceil - 1$.*

Obviously the bounds of theorem 2.3 on the edge and vertex congestion are also valid for the embedding of the shuffle-exchange graph.

However for small values of D we get better values for the parameters as showed in the following array.

Lemma 3.4 *For small values of D there exist embeddings of the shuffle-exchange graph $S(D)$ in the hypercube $H(D)$ with the following parameters.*

D	*1*	*2*	*3*	*4*	*5*	*6*
dilation	*1*	*1*	*2*	*2*	*2*	*2*
edge congestion	*1*	*1*	*1*	*1*	*1*	*2*
vertex congestion	*0*	*0*	*1*	*1*	*1*	*2*

Proof: The embeddings are given in the appendix.
Notice that for these small values of D, the parameters of the embeddings are optimal except possibly the edge-congestion for $D = 6$.

Remark 3.5 : Using the results of the previous section we can give a direct proof of corollary 3.3. Indeed, let $D \geq 7$, $D = 5(k-1) + r$, $k \geq 2$, $1 \leq r \leq 5$. Consider the embedding g of $B(D)$ in $H(D)$ described in the proof of theorem 2.3,

$$g(x_1x_2\cdots x_D) = f_5(x_1x_2\cdots x_5)f_5(x_6x_7\cdots x_{10})\cdots f_5(x_{5(k-2)+1}x_{5(k-2)+2}\cdots x_{5(k-1)})f_r(x_{5(k-1)+1}\cdots x_D)$$

where, for $1 \leq i \leq 5$, f_i is an embedding of $B(i)$ in $H(i)$ from lemma 2.2.

Let g' be the embedding of $S(D)$ in $H(D)$ such that $g' = g$. We will prove that $dil(g') \leq max\ (4, dil(g))$, which will give us an upper bound on $dil(S(D), H(D))$ identical to the upper bound on $dil(g)$ from theorem 2.3 for $D \geq 7$.

Let us consider the two types of edges (u,v) of S(D). If (u,v) is a shuffle edge of $S(D)$ then (u,v) is also an edge of $B(D)$, and by the proof of theorem 2.3, we get

$$d_D(g'(u),g'(v)) \le d_D(g(u),g(v)) \le dil(g).$$

If (u,v) is an exchange edge, that is $u = x_1x_2\cdots x_D$ and $v = x_1x_2\cdots x_{D-1}\overline{x_D}$ with $\overline{x_D} = 1 - x_D$, then u and v are mapped into the same subhypercube $H(r)$ of $H(D)$, and we have

$$d_D(g'(u),g'(v)) \le d_r(f_r(x_{5(k-1)+1}\cdots x_D), f_r(x_{5(k-1)+1}\cdots \overline{x_D})).$$

For $1 \le r \le 4$, since we have $d_r(u',v') \le 4$ for any vertices u', v' of $H(r)$, we obtain

$$d_D(g'(u),g'(v)) \le 4.$$

For $r = 5$, by calculating the distances $d_5(f_5(u'), f_5(v'))$ for any exchange edge (u',v') of $S(5)$, we get

$$d_5(f_5(u'), f_5(v')) \le 4.$$

Therefore, $d_D(g'(u),g'(v)) \le \max(4, dil(g))$. □

Remark 3.6 We believe that a more direct proof could give better bounds for the parameters of the embedding of a shuffle-exchange graph in the hypercube than for those of the embedding of a de Bruijn graph.

4 Conclusion

The dilation of embeddings of binary shuffle-like graphs in hypercubes presented in this paper is better for $D \ge 5$ than those obtained by heuristics in [3]. However, the dilation of our embeddings is increasing linearly with the diameter. It is still an open problem whether the shuffle-like graphs can be embedded in the hypercubes with expansion 1, dilation $O(1)$ and edge-congestion $O(1)$.

As mentioned in [12], a de Bruijn graph can be embedded with dilation 2 into the shuffle-exchange graph of the same diameter, and from theorem 3.1 the shuffle-exchange graph is a subgraph of the de Bruijn graph. This gives

$$dil(S(D),H(D)) \le dil(B(D),H(D)), \text{ and } dil(B(D),H(D)) \le 2\, dil(S(D),H(D)).$$

Therefore it would be sufficient to find embeddings of one of these two types of graphs with dilation $O(1)$ in hypercubes, in order to find good embeddings for the other type.

Let us notice that if we could find an embedding of $S(D)$ in $H(D)$ with dilation $dil(S(D),H(D))$ where the exchange edges are embedded on edges of $H(D)$, then we would have

$$dil(B(D),H(D)) \le dil(S(D),H(D)) + 1.$$

On the other hand we think that it might be possible to construct an embedding f'_{D+1} of $S(D+1)$ in $H(D+1)$ by splitting each vertex $x_1x_2\cdots x_D$ of $B(D)$ in two vertices $x_1x_2\cdots x_D0$ and $x_1x_2\cdots x_D1$. If f_D is an embedding of $B(D)$ in $H(D)$ of minimum dilation, the set of any two vertices $x_1x_2\cdots x_D0$ and $x_1x_2\cdots x_D1$ is embedded by f'_{D+1} on the set of the two vertices $f_D(x_1x_2\cdots x_D)0$ and $f_D(x_1x_2\cdots x_D)1$. If that is the case (as obtained for $D \leq 5$ in the appendix 5.2), we would get

$$dil(S(D+1), H(D+1)) \leq dil(B(D), H(D)).$$

Clearly, whenever we find an embedding of a de Bruijn graph G of fixed diameter $D \geq 6$ in the hypercube $H(D)$, whose dilation is smaller than the one from theorem 2.3, we can use G in the proof of the theorem, and improve the results of the theorem 2.3 and corollary 3.3 appropriately. However, the bound on the dilation would remain linearly proportional to the diameter of the graph. Thus, any attempt to show that the dilation of the embeddings of shuffle-like graphs is $O(1)$ must use a method different from ours.

5 Appendix: embeddings for small values of the diameter

We give here embeddings for $B(D)$ and $S(D)$ for $D \leq 6$.
Any vertex of each of the graphs considered is specified by the integer whose binary representation is the corresponding word of $B(D)$, $S(D)$, or $H(D)$. In the right column we give a vertex u of $B(D)$, $S(D)$, respectively, and in the left column we give the image of u in $H(D)$.

5.1 Embeddings of B(D)

For $D = 1$, $B(1)$ and $H(1)$ are isomorphic and we take the identity function as the embedding.

We give in the following table embeddings f_D of $B(D)$ in $H(D)$.

H(2)	B(2)
0	0
1	1
2	2
3	3

H(3)	B(3)
0	0
1	1
2	4
3	2
4	7
5	3
6	6
7	5

H(4)	B(4)
0	3
1	1
2	9
3	8
4	2
5	0
6	4
7	10
8	6
9	7
10	12
11	14
12	11
13	5
14	13
15	15

H(5)	B(5)
0	17
1	3
2	12
3	6
4	2
5	1
6	5
7	0
8	24
9	20
10	18
11	9
12	8
13	16
14	4
15	10
16	14
17	7
18	25
19	19
20	23
21	15
22	11
23	31
24	28
25	26
26	22
27	13
28	29
29	30
30	27
31	21

Table 1: Embeddings f_D of $B(D)$ in $H(D)$ with dilation 2.

Although it does not show in the tables, the given embeddings have "some" symetry, but we have not found a general rule for it.
We have explored three directions to find these embeddings, hoping to find a general pattern. The first one is by computer, but it gives embedding without any regularity. The second one is by trying to find symetries based on the binary representation of the vertices; this is the case for f_D, $D \leq 4$. In the third one, we try to first optimize the dilation of the embeddings of the cycles of $B(D)$ of the form $(x_1x_2\cdots x_D,\ x_2x_3\cdots x_Dx_1,\ \cdots,x_Dx_1\cdots x_{D-1})$ in $H(D)$. We obtained f_5 in that way (the first embedding of $B(5)$ in $H(5)$ with dilation 2 was found by computer).

5.2 Embeddings of S(D)

For $D = 1$, $S(1)$ and $H(1)$ are isomorphic and we take the identity function as the embedding.

We give in the following table embeddings f'_D of $S(D)$ in $H(D)$ for $2 \leq D \leq 6$.

H(2)	S(2)
0	0
1	3
2	1
3	2

H(3)	S(3)
0	0
1	1
2	5
3	4
4	3
5	2
6	6
7	7

H(4)	S(4)
0	1
1	2
2	0
3	3
4	14
5	7
6	15
7	6
8	8
9	4
10	9
11	5
12	13
13	11
14	12
15	10

H(5)	S(5)
0	6
1	7
2	18
3	19
4	3
5	2
6	17
7	16
8	12
9	13
10	24
11	25
12	15
13	14
14	29
15	28
16	5
17	4
18	9
19	8
20	0
21	1
22	20
23	21
24	23
25	22
26	27
27	26
28	10
29	11
30	30
31	31

H(6)	S(6)	H(6)	S(6)
0	34	32	48
1	6	33	40
2	35	34	49
3	7	35	41
4	5	36	17
5	3	37	33
6	4	38	16
7	2	39	32
8	24	40	36
9	12	41	18
10	25	42	37
11	13	43	19
12	10	44	9
13	0	45	20
14	11	46	8
15	1	47	21
16	29	48	57
17	15	49	53
18	28	50	56
19	14	51	52
20	46	52	58
21	30	53	60
22	47	54	59
23	31	55	61
24	51	56	45
25	39	57	27
26	50	58	44
27	38	59	26
28	23	60	54
29	63	61	43
30	22	62	55
31	62	63	42

Table 2: Embeddings f'_D of $S(D)$ in $H(D)$ with dilation 2 for $3 \leq D \leq 6$.

The embeddings f'_{D+1}, for $1 \leq D \leq 5$, are obtained from the embeddings f_D of $B(D)$ in $H(D)$ by splitting every vertex $x_1x_2 \cdots x_D$ of $B(D)$ in two vertices $x_1x_2 \cdots x_D0$ and $x_1x_2 \cdots x_D1$ of $S(D+1)$ which are mapped on adjacent vertices of $H(D+1)$.

Furthermore for each $D \leq 5$ we have $dil(f'_{D+1}) \leq dil(f_D)$. It would be interesting to know if this construction is always possible, for it would give $dil(S(D+1), H(D+1)) \leq dil(B(D), H(D))$.

References

[1] J.-C. Bermond and C. Peyrat. de Bruijn and Kautz networks: a competitor for the hypercube? *Proceedings of the 1st European Workshop on Hypercubes and Distributed Computers, Rennes, North Holland, F.Andre and J.P.Verjus ed.*, pages 279–293, 1989.

[2] S. Bhatt, F. Chung, T. Leighton, and A. Rosenberg. Optimal simulations of tree machines. *IEEE*, pages 274–282, 1986.

[3] M. Bouabdallah and J.-C. Konig. Embedding de Bruijn networks in the hypercube. *Preprint*, 1990.

[4] M. Chan. Embedding of *d*-dimensional grids into optimal hypercubes. *1st ACM Symposium on Parallel Algorithms and Architectures*, pages 52–57, 1989.

[5] O. Collins, S. Dolinar, R. McEliece, and F. Pollara. A VLSI decomposition of the de Bruijn graph. *Preprint*, 1989.

[6] N. de Bruijn. A combinatorical problem. *Koninklijke Nederlandsche Akademie van Wetenschappen Proc.*, A 49:758–764, 1946.

[7] R. Feldmann and W. Unger. The cube connected cycle network is a subgraph of the butterfly network. Technical report, University of Paderborn, 1991.

[8] D. Greenberg and S. Bhatt. Routing multiple paths in hypercubes. *Proceedings of SPAA*, pages 45–54, 1990.

[9] D. S. Greenberg, L. S. Heath, and A. Rosenberg. Optimal embeddings of butterfly-like graphs in the hypercube. *Mathematical Systems Theory*, 23:61–77, 1990.

[10] C.-T. Ho and S. L. Johnson. Embedding meshes in boolean cubes by graph decomposition. *Journal of Parallel and Distributed Computing*, 8:325–339, 1990.

[11] R. Koch, T. Leighton, B. Maggs, S. Rao, and A. Rosenberg. Work-preserving emulations of fixed-connection networks. *Preprint.*

[12] B. Monien and H. Sudborough. Comparing interconnection networks. *Preprint*, 1988.

[13] E. J. Schwabe. On the computational equivalence of hypercube-derived networks. *2nd Symposium on Parallel Algorithms and Architectures*, pages 388–397, 1990.

[14] P. M. Winkler. The metric structure of graphs. *Surveys in Combinatorics, (C. Whitehead, ed.), London Math. Soc. Lecture Notes Series*, 123:197–221, 1987.

MAPPING UNIFORM RECURRENCES ONTO SMALL SIZE ARRAYS

Vincent Van Dongen
Philips Research Laboratory
4 Av. Albert Einstein, B-1348 Louvain-la-Neuve, Belgium
e-mail: vvd@prlb.philips.be

Abstract

Given a regular application described by a system of uniform recurrence equations, systolic arrays are commonly derived by means of an affine transformation; an affine schedule determines when the computations are performed and an affine processor allocation where they are performed. Circuit transformations are then applied on the resulting circuit when the application needs to be mapped onto a smaller size array. This method is in two steps and thus can hardly be optimized globally.

We hereafter present a different method for designing small size arrays. We derive them in one step by means of an affine schedule and a near-affine processor allocation. By doing so, we can generalize the optimization technique for affine mapping to be applicable here. The method is illustrated on the band-matrix multiplication and on the convolution algorithms.

1 Introduction

Systolic arrays are particular circuits made of identical processing elements connected in a local and regular manner; a high throughput is achieved by making extensive use of parallelism and pipelining. The regularity and locality of their connections make them ideally suited for a VLSI implementation. Such arrays are applicable to problems in signal processing, numerical computing, graph theory and other areas [KL78].

Given an application described by a system of uniform recurrence equations, systolic arrays are derived when using affine space-time transformations; an affine schedule determines when the computations are performed and an affine processor allocation where they are performed [Qui84,Rao85]. This space-time mapping technique is generalized in this paper to derive small size arrays, i.e. arrays containing fewer cells than the ones obtained when using affine mappings.

The problem of realizing automatically small size arrays is important; its applications can be classified as follows:

1. *Software compilation for general-purpose arrays.*
   The problem is encountered in the mapping of applications, here systems of URE's, on fixed size general-purpose arrays, e.g. a number of Transputers [The89] or the Warp [AAG*87].

2. *Hardware compilation.*
   Small size circuits may be required due to area constraints.

The common approach for deriving small systolic arrays consists in two steps [Bu90,Cla90,ND88,DI88,GN89]. An array is first derived by means of an affine space-time mapping. This array usually contains too many processors. The schedule is then slowed down so that less parallelism is achieved, and the processor allocation is modified correspondingly. This second step is a particular circuit transformation.

Our approach is different. we derive small size arrays directly from the behavioral description by means of a unique space-time mapping , as shown in figure 1. The advantage of this direct approach is that it can be optimized. Its drawback is that it only works with a particular partitioning strategy known as the *Locally Sequential Globally Parallel* (LSGP) partitioning scheme [Bu90,ND88,Kun87,GN89]. But this scheme is ideal when using a general-purpose processor array as an array of Transputers [The89,Bra90].

In the next section, systems of uniform recurrences are introduced. Then, in section 3, near-affine mappings are defined. In particular, it is shown that these mappings can be viewed as a set of affine transformations defined on lattices of the index space. In section 4, constraints for dealing with these mappings are derived. In section 5, we recall how an affine schedule can be optimized. In section 6, we generalize the optimization method to deal with near-affine mappings. We summarize the complete methodology in section 7, and we conclude in section 8.

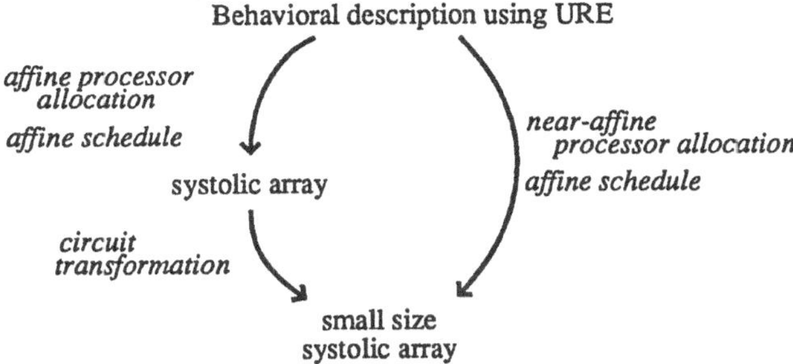

Figure 1: The common approach for deriving small systolic arrays consists in two steps. In this paper, small size arrays are derived directly from the behavioral description by means of a unique space-time mapping.

2 Uniform recurrence equations

We follow the suggestion given by Karp, et. al. [KMW67] of describing "regular" algorithms (for which systolic arrays are suited) using uniform recurrence equations (URE). Quinton was first in recognizing that such a description can be used in the synthesis of systolic arrays [Qui84].

Definition 2.1 A system of *uniform recurrences* is a set of m recurrences of the form

$$\begin{array}{ll} \mathbf{z} \in \mathcal{D}_s \rightarrow & O_1(\mathbf{z}) = f_1(O_{1_1}(\mathbf{z}-\vartheta_{1,1}), O_{1_2}(\mathbf{z}-\vartheta_{2,1}), \dots, O_{1_{l_1}}(\mathbf{z}-\vartheta_{l_1,1})) \\ \mathbf{z} \in \mathcal{D}_s \rightarrow & O_2(\mathbf{z}) = f_2(O_{2_1}(\mathbf{z}-\vartheta_{1,2}), O_{2_2}(\mathbf{z}-\vartheta_{2,2}), \dots, O_{2_{l_2}}(\mathbf{z}-\vartheta_{l_2,2})) \\ \dots & \\ \mathbf{z} \in \mathcal{D}_s \rightarrow & O_m(\mathbf{z}) = f_n(O_{m_1}(\mathbf{z}-\vartheta_{1,m}), \dots, O_{m_{l_m}}(\mathbf{z}-\vartheta_{l_m,m})) \end{array} \tag{1}$$

where

$$\forall v \in [1,m], \forall u \in [1,l_v] \rightarrow \vartheta_{u,v} \in \mathbf{Z}^e, \tag{2}$$

and $\mathcal{D}_s$ is a convex polyhedron of $\mathbf{Z}^e$ parameterized with s (the number of operands of f_v is noted l_v). ∘

The vectors $\vartheta_{u,v}$ will be called the *dependence vectors* of (1), and $O_v(\mathbf{z})$ will be called an *instance* of the variable O_v.

Example 2.1 The convolution algorithm (for FIR filtering) can be implemented with the following system of URE:

$$
\begin{array}{lll}
i \geq 1,\ 1 \leq k \leq N & \rightarrow & y(i,k) = y(i,k-1) + p(i,k) \\
i \geq 1,\ 1 \leq k \leq N & \rightarrow & p(i,k) = w(i,k) \times x(i,k) \\
i \geq 1,\ 1 \leq k \leq N & \rightarrow & w(i,k) = w(i-1,k) \\
i \geq 1,\ 1 \leq k \leq N & \rightarrow & x(i,k) = x(i-1,k-1)
\end{array} \tag{3}
$$

The values of $x(i,0)$ are initialized with the input signal x_i, and the outputs are $y_i = y(i,N)$.

□

Example 2.2 The band-matrix multiplication can be computed with the following system of URE:

$$
\begin{array}{l}
C(i,j,k) = C(i,j,k-1) + A(i,j,k) \times B(i,j,k) \\
A(i,j,k) = A(i-1,j,k) \\
B(i,j,k) = B(i,j-1,k)
\end{array} \tag{4}
$$

The index domain is defined by the following constraints:

$$
0 \leq i \leq N-1, 0 \leq j \leq N-1, 0 \leq k \leq N-1, -w_1 \leq i-k \leq w_1, -w_2 \leq j-k \leq w_2. \tag{5}
$$

The domain is a polyhedron having 16 vertices. The size parameters w_1 and w_2 represent half band widths; A is of band width $2.w_1 + 1$ and B is of band width $2.w_2 + 1$.

□

3 Space-time mapping

We suggest to use space-time mappings of the form

$$
\mathbf{Z}^e \rightarrow \mathbf{Z}^e : (\mathbf{z}, v) \rightarrow \left(\begin{array}{ll}
t(\mathbf{z},v) & = \mathbf{T}.\mathbf{z} + \alpha_v \\
p_1(\mathbf{z},v) & = (\mathbf{P}_1.\mathbf{z} + \beta_{v,1}) \text{ div } d_1 \\
p_2(\mathbf{z},v) & = (\mathbf{P}_2.\mathbf{z} + \beta_{v,2}) \text{ div } d_2 \\
\dots & \\
p_{e-1}(\mathbf{z},v) & = (\mathbf{P}_{e-1}.\mathbf{z} + \beta_{v,e-1}) \text{ div } d_{e-1} \\
v' & = v
\end{array} \right) \tag{6}
$$

where " div " denotes the integer division, and $d_l \in \mathbf{N}_0$, $\forall l \in [1, e-1]$. In matricial form, the mapping is written (we omit $v' = v$):

$$
\begin{pmatrix} t(\mathbf{z},v) \\ \mathbf{p}(\mathbf{z},v) \end{pmatrix} = \begin{pmatrix} \mathbf{T}.\mathbf{z} + \alpha_v \\ (\mathbf{P}.\mathbf{z} + \beta_v) \textbf{ div } \mathbf{d}_p, \end{pmatrix} \tag{7}
$$

where $\mathbf{P} \in \mathbf{Z}^{e-1} \times \mathbf{Z}^e$, $\beta_v \in \mathbf{Z}^{e-1}$, $\mathbf{d}_p = (d_1, d_2, ..., d_e - 1) \in \mathbf{N}_0^{e-1}$ and "div" represents the component wise integer division. By definition, the processor allocation $\mathbf{p}(\mathbf{z}, v)$ of

(7) is *near-affine*, a sub-class of quasi-affine mappings [Van91]. . The schedule $t(\mathbf{z}, v)$ is affine. Note that when $d_1 = d_2 = ... = d_{e-1} = 1$, the mapping (6) is affine.

With affine mappings, the computations that are processed by the same processor are on a line perpendicular to $\mathbf{P}_1$, $\mathbf{P}_2$, etc. The hyperplane perpendicular to $\mathbf{P}_l$ contains all the computations performed by the processors with the same l^{th} component. With a *near-affine* allocation, the computations performed by the processors with the same l^{th} component are on d_l consecutive hyperplanes perpendicular to $\mathbf{P}_l$. Thus, "bands" of indexed points are mapped onto the same processor. Within each band, the computations are performed in sequence by making use of the local memory of the processor.

Example 2.1 (cont'd) With the mapping

$$\begin{pmatrix} t(i,k,v) \\ p(i,k,v) \end{pmatrix} = \begin{pmatrix} 2.i + k \\ k \text{ div } 2 \end{pmatrix}, \tag{8}$$

one maps the uniform recurrences (3) with $N = 6$ onto the circuit shown in figure 2. It works as follows. Each input to the combinational logic is a two-input switch. Every clock cycle, the switches change of input. This is achieved with a control signal whose value alternates from one to zero. The control signal can either be broadcasted or it can be pipelined through the array as shown in figure 2. The signal values can be generated on line by a simple circuit that initializes its values when the *reset* is on. This part of the circuit is called the *control signal generator* in figure 2. The rate of the circuit is one-half; every two clock cycles, a new input signal enters the array, and an output value is being produced.

This can be easily generalized as follows. With the affine schedule

$$t(i,k,v) = R.i + k,$$

and the processor allocation

$$p(i,k,v) = k \text{ div } R.$$

where $R \geq 2$, one maps the uniform recurrences (3) onto an array of $M = \lceil N/R \rceil$ processors. Every R clock cycles, a new input value enters the array, and a new output is produced; the I/O rate of the array is $1/R$. Each processor contains one latch for y, and $R + 1$ latches for x. The number of switches is one for x, and one for y. Note that in the complete array, the total number of latches is $N + M$ for x; it is proportional to N for any value of M. When $R \geq N$, the architecture is sequential.

The processors communicate with their neighbours once every R clock cycles only; they are loosely coupled which is ideal when using a general-purpose fixed size array with expensive inter-processor communication.

□

We now show that the near-affine mapping (7) transforms the system of URE (1) into a finite set of systems of URE defined on lattices of $(t, \mathbf{p}) \in \mathbf{Z}^e$. Let the *rational approximant* of (7) be:

$$\begin{pmatrix} t(\mathbf{z},v) \\ \tilde{\mathbf{p}}(\mathbf{z},v) \end{pmatrix} = \begin{pmatrix} \mathbf{T}.\mathbf{z} + \alpha_v \\ (\mathbf{P}.\mathbf{z} + \beta_v)/\mathbf{d}_p, \end{pmatrix} \tag{9}$$

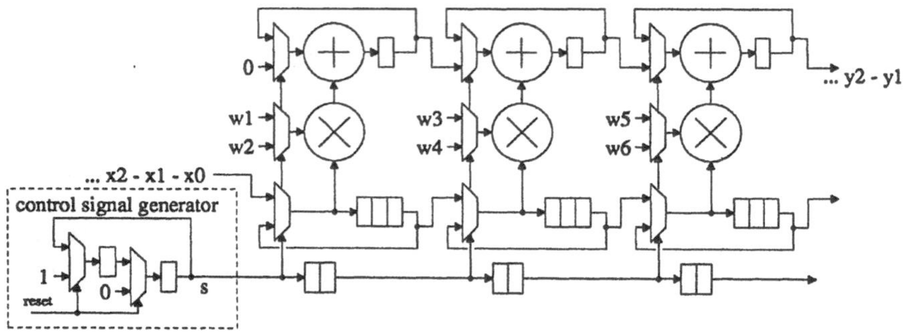

Figure 2: LSGP partitioning: the circuit is a systolic array that uses some internal memory and a simple control mechanism.

The mapping (7) can be written as:

$$\begin{pmatrix} t(\mathbf{z},v) \\ \mathbf{p}(\mathbf{z},v) \end{pmatrix} = \begin{pmatrix} \mathbf{T}.\mathbf{z} + \alpha_v \\ (\mathbf{P}.\mathbf{z} + \beta_v - \mathbf{r}_p)/\mathbf{d}_p \end{pmatrix} = \begin{pmatrix} t \\ \tilde{\mathbf{p}} \end{pmatrix} - \begin{pmatrix} 0 \\ \mathbf{r}_p/\mathbf{d}_p \end{pmatrix}, \tag{10}$$

with

$$\mathbf{r}_p = (\mathbf{P}.\mathbf{z} + \beta_v) \bmod \mathbf{d_p}. \tag{11}$$

The components of $\mathbf{r}_p$ are $(r_1, r_2, ..., r_{e-1})$, and for all $l \in [1, e-1]$, $r_l \in [0, d_l - 1]$. So, $\mathbf{r}_p$ has $d1 \times d_2 \times ... \times d_{e-1}$ distinct values. For each value of $\mathbf{r}_p$, the mapping (10) is affine over the lattice of $\mathbf{z}$ defined by (11). Thus, a near-affine mapping is a finite set of affine mappings defined over lattices of $\mathbf{z}$. More precisely, it is a rational affine mapping followed by a translation that depends on a lattice of $\mathbf{z}$; the rational approximant is followed by a *perturbation function.*

The inverse of a near-affine mapping is a finite set of affine mappings defined over lattices of $(t, \mathbf{p})$. The latter lattices can be computed as follows. The inverse of (10) is:

$$\begin{aligned} \mathbf{z} &= \tfrac{1}{a} \cdot adj \begin{pmatrix} \mathbf{T} \\ \mathbf{P} \end{pmatrix} \cdot \begin{pmatrix} t - \alpha_v \\ \mathbf{d}_p \times \mathbf{p} - \beta_v + \mathbf{r}_p \end{pmatrix} \\ &= \tfrac{1}{a} \cdot adj \begin{pmatrix} \mathbf{T} \\ \mathbf{P} \end{pmatrix} \cdot \begin{pmatrix} t - \alpha_v \\ \mathbf{d}_p \times \mathbf{p} - \beta_v \end{pmatrix} + \tfrac{1}{a} \cdot adj \begin{pmatrix} \mathbf{T} \\ \mathbf{P} \end{pmatrix} \cdot \begin{pmatrix} 0 \\ \mathbf{r}_p \end{pmatrix}, \end{aligned} \tag{12}$$

where a is the determinant of $\begin{pmatrix} \mathbf{T} \\ \mathbf{P} \end{pmatrix}$ and $adj\begin{pmatrix} \mathbf{T} \\ \mathbf{P} \end{pmatrix}$ denotes the adjoint matrix of $\begin{pmatrix} \mathbf{T} \\ \mathbf{P} \end{pmatrix}$. The index $\mathbf{z}$ is in $\mathbf{Z}^e$ if and only if

$$adj \begin{pmatrix} \mathbf{T} \\ \mathbf{P} \end{pmatrix} \cdot \begin{pmatrix} t - \alpha_v \\ \mathbf{d}_p \times \mathbf{p} - \beta_v + \mathbf{r}_p \end{pmatrix} \bmod \begin{pmatrix} a \\ \mathbf{a} \end{pmatrix} = \begin{pmatrix} 0 \\ 0 \end{pmatrix}. \tag{13}$$

For a fixed value of $\mathbf{r}_p$, (12) is an affine mapping and (13) defines a lattice of $(t,\mathbf{p})$. More precisely, it is a rational affine transformation followed by a translation that depends on the lattice of $(t,\mathbf{p})$.

Example 2.1 (cont'd) The mapping (8) can be rewritten as:

$$\begin{pmatrix} t(i,k,v) \\ p(i,k,v) \end{pmatrix} = \begin{pmatrix} 2.i+k \\ (k-r)/2 \end{pmatrix},$$

where $r = k \bmod 2$. This mapping consists of two affine mappings defined on lattices of (i,k). Figures 3(a0) and 3(b0) show the first lattice and the result of the associated transformation. Figures 3(a1) and 3(b1) show the second lattice and the result of the associated transformation.

The inverse of (8) is:

$$\begin{pmatrix} i \\ k \end{pmatrix} = \frac{1}{2} \cdot \begin{pmatrix} 1 & -1 \\ 0 & 2 \end{pmatrix} \cdot \begin{pmatrix} t \\ 2.p+r \end{pmatrix}. \tag{14}$$

The two lattices in $(t,\mathbf{p})$ are defined by $(t-2.p-r) \bmod 2 = 0$.

□

Because of this, each uniform recurrence of (1) is transformed into a set of uniform recurrences, one for each lattice of $(t,\mathbf{p})$. Furthermore, the domain $\mathcal{D}_s$ is a convex polyhedron of $\mathbf{Z}^e$, and the image of $\mathcal{D}_s$ becomes the union of polyhedrons defined on lattices of $(t,\mathbf{p})$.

Finally, let us consider the control that must be added to the circuit to take into account the changes in the connections upon the lattices of $(t,\mathbf{p})$. It can be verified that any hyperplane $z_i = c$, where z_i is the ith component of $\mathbf{z}$ and c is some constant, becomes

$$\mathbf{L}_i \cdot \begin{pmatrix} t-\alpha_v \\ \mathbf{p}-\beta_v+\mathbf{r}_p \end{pmatrix} = c.a,$$

where $\mathbf{L}_i$ is the ith line of $adj \begin{pmatrix} \mathbf{T} \\ \mathbf{P} \end{pmatrix}$. This hyperplane of $(t,\mathbf{p})$ is exactly the one used to define the lattice in (13). Thus, a control signal can run along it to specify this lattice. At most e control signals will be required in the final array, one per axis.

Example 2.1 (cont'd) Consider the recurrence $x(i,k) = x((i,k)-(1,1))$ of (3). When $k \bmod 2 = 0$, (i,k) is in the first lattice and $((i,k)-(1,1))$ is in the second one. The application of the corresponding affine transformations gives:

$x(2.i+k, k/2) = x(2.(i-1)+(k-1), ((k-1)-1)/2)$ which is equivalent to:

$x(t,p) = x(t-3, p-1)$. On the other hand, when $k \bmod 2 = 1$, we obtain:

$x(2.i+k, (k-1)/2) = x(2.(i-1)+(k-1), (k-1)/2)$ which is equivalent to:

$x(t,p) = x(t-3,p)$. Figure 4 shows these two cases.

Furthermore, the image of $\mathcal{D}_s$ is the union of polyhedrons defined on lattices of $(t,\mathbf{p})$. Consider the inequality $i \geq 0$ defining a boundary of the domain of (3). For the first

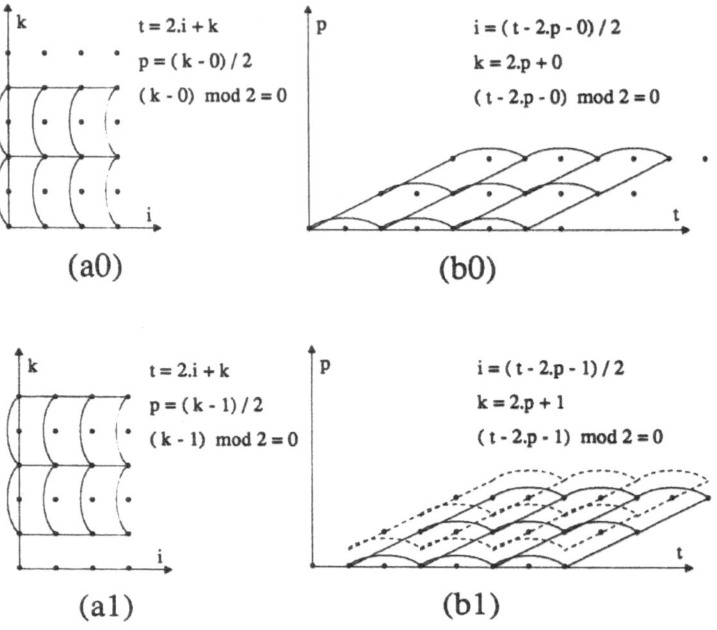

Figure 3: A near-affine mapping is a finite set of affine mappings: on the lattice shown in (a0), the first affine transformation gives the lattice shown in (b0), while on the lattice shown in (a1), the second transformation gives the lattice shown in (b1). It can also be obtained with a rational affine transformation followed by a translation that depends on a lattice of z; the result of the rational transformation is shown in dash in (b1).

lattice of (t, p), $i = (t-2.p-0)/2$, and the condition becomes: $(t-2.p-0)/2 \geq 0$. For the second lattice of (t, p), $i = (t-2.p-1)/2$, and the condition becomes: $(t-2.p-1)/2 \geq 0$. Figure 4 shows the two resulting boundaries in (t, p).

For the control, the hyperplane $i = 0$ becomes $t - 2.p - r = 0$. A control signal can run along it, i.e. along $(t,p) = (2,1)$. (See figure 2.) The hyperplane $k = 0$ becomes $2.p + r = 0$. No control signal is needed along that plane since it does not define the lattices. (See figure 3.)

□

4 Constraints

The LSGP partitioning scheme is achieved with an affine schedule. The basic constraints on the schedule are [Qui84]:

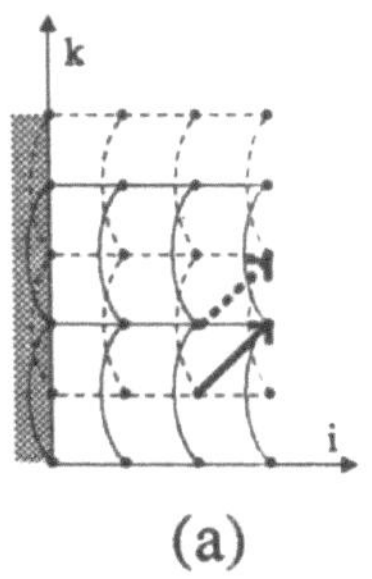

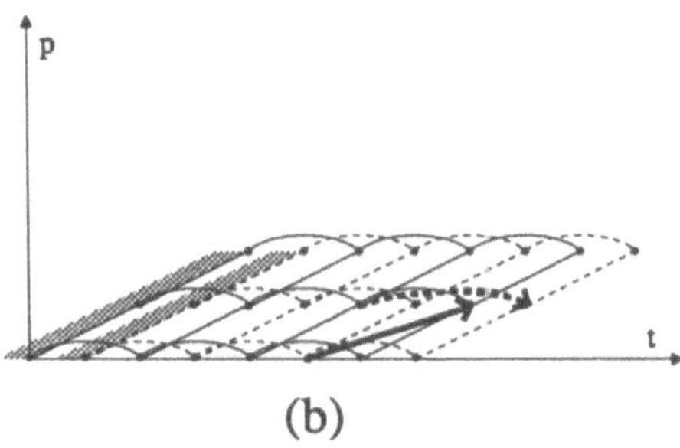

Figure 4: A near-affine mapping transforms a uniform recurrence into a finite set of uniform recurrences defined on lattices of $(t, \mathbf{p})$. It transforms a linear constraint of $\mathcal{D}_s$ into a finite set of linear constraints in $(t, \mathbf{p})$.

$$\begin{aligned} \forall v \in [1, m], \forall u \in [1, l_v] &\rightarrow \mathbf{T} \cdot \vartheta_{u,v} + \alpha_v - \alpha_{v_u} \geq 1, & (15) \\ \forall v \in [1, m] &\rightarrow \alpha_v \geq -min_{\mathbf{V}_k \in \mathcal{D}_s}(\mathbf{T}.\mathbf{V}_k), & (16) \end{aligned}$$

where the $\mathbf{V}_k$'s are the vertices of $\mathcal{D}_s$. Furthermore, if the domain is semi-infinite along a ray $\mathbf{R}$, the schedule is subject to the additional constraint:

$$\mathbf{T} \cdot \mathbf{R} \geq 1. \quad (17)$$

On the other hand, the constraint on the processor allocation, when the domain is semi-infinite along $\mathbf{R}$, is

$$\forall l \in [1, e-1] \rightarrow \mathbf{P}_l \cdot \mathbf{R} = 0. \quad (18)$$

This must be satisfied with both affine and near-affine processor allocations.

Any two different instances of O_v that are computed in parallel must be performed on distinct processors. In other words, the mapping (6) must be *one-to-one*, We now consider this problem.

Let "w" denote the number of d_l's that are equal to one; $w \in [0, e-1]$. When $w = 0$, all d_l's are different from one, while when $w = e - 1$, the mapping is affine. In the following, let us assume that $d_1 = d_2 = ... = d_w = 1$ and that $d_{w+1}, d_{w+2}, ..., d_{e-1}$ are different from one. (One can always order the $e - 1$ axes of the processor allocation so that it is verified.) Furthermore, let g_l denote the greatest common divisor of the components of $\mathbf{P}_l$ (i.e. $g_l = gcd(\mathbf{P}_l)$), and $\mathbf{P}'_l = \mathbf{P}_l / g_l$. Also, we define g_t as $gcd(\mathbf{T})$ and $\mathbf{T}'$ as $\mathbf{T}' = \mathbf{T}/g_t$. Thus, $\mathbf{T}'$ and $\mathbf{P}'_l$ are integral vectors whose components are relatively prime. Let us summarize the results given in [Van91].

- The values g_t, g_1, ..., g_w have no effect on the condition for having a one-to-one mapping.

- For the mapping (6) to be one-to-one, the absolute value of the determinant of $(\mathbf{T}', \mathbf{P}'_1, ..., \mathbf{P}'_w, \mathbf{P}_{w+1}, ..., \mathbf{P}_{e-1})$ must be greater than $\prod_{l=1}^{e-1} d_l$.
- When $e = 2$, the mapping (7) is one-to-one if and only if

$$\left\| \begin{matrix} \mathbf{T}' \\ \mathbf{P}_1 \end{matrix} \right\| \geq d_1. \tag{19}$$

- When $e \geq 3$ and $w = 1$, a systematic method polynomial with e can be used to check that the mapping is one-to-one. check that the mapping is one-to-one when only one value of d_l is different from one.
- In the general case, the mapping (7) is one-to-one if and only if, $\forall v \in [1, m]$, there exists no pair of points in some reduced domains $\mathcal{R}_v$ that have the same image.

Because a near-affine mapping is a set of affine transformations on lattices of $\mathbf{z}$, another method for checking that it is one-to-one consists in verifying that the images of these lattices do not intersect. First, one can compute the non-empty lattices defined by (13), by solving systems of diophantine equations. Second, one can check that any pair of non-empty mappings do not intersect. This is simple to check since two lattices defined by (13) and characterized by $\mathbf{r}_p = \mathbf{r}$ and $\mathbf{r}_p = \mathbf{r}'$ do not intersect if and only if the following system of diophantine equations of vector variable $\mathbf{k}$ is empty:

$$\mathbf{A}.(\mathbf{r} - \mathbf{r}') = \mathbf{k}.\mathbf{a},$$

where $\mathbf{A}$ is the $e \times (e-1)$ matrix formed with the $e-1$ last columns of $adj \begin{pmatrix} \mathbf{T} \\ \mathbf{P} \end{pmatrix}$, and $\mathbf{a}$ is a vector whose e entries are equal to a.

Example 2.1 (cont'd) Consider the near-affine mapping (8) and its inverse (14). Here, $\mathbf{A} = \begin{pmatrix} -1 \\ 2 \end{pmatrix}$. Its two lattices do not intersect if and only if the system

$$\begin{pmatrix} -1 \\ 2 \end{pmatrix}.(0 - 1) = \begin{pmatrix} k_1 \\ k_2 \end{pmatrix}.(2)$$

has no solution, which is the case since $1 = 2.k_1$ has no solution.

□

Unidirectional arrays have specific properties. In particular, the so-called Locally Parallel Globally Sequential (LPGS) partitioning scheme is directly applicable on such arrays. Its advantage is to use the external memory instead of the local memory as with the LSGP scheme [Kun87,Mol83]. Unidirectional arrays can be automatically realized, by means of additional constraints. For example, the lth component of $\mathbf{c}_{u,v}$ is positive if and only if

$$\mathbf{P}_l.\vartheta_{u,v} + \beta_{v,l} - \beta_{u,l} \geq 0.$$

5 Optimization of the schedule

In a system of URE, the index domain $\mathcal{D}_s$ is parameterized with s. Let $\mathcal{D}$ denote an instance of $\mathcal{D}_s$ when s has a given value. Given a finite dependence graph and an affine schedule, the *latency*, which will be noted Δt, is

$$\begin{aligned}\Delta t &= max_{\mathbf{z}\in\mathcal{D}}(t(\mathbf{z})) - min_{\mathbf{z}\in\mathcal{D}}(t(\mathbf{z})) + 1\\ &= max_{j\in[1,h]}(\mathbf{T}.\mathbf{V}_j) - min_{i\in[1,h]}(\mathbf{T}.\mathbf{V}_i) + max_{v\in[1,m]}(\alpha_v) - min_{u\in[1,m]}(\alpha_u) + 1\\ &= \Delta T + max_{v\in[1,m]}(\alpha_v) - min_{u\in[1,m]}(\alpha_u)\end{aligned}$$

with

$$\Delta T = max_{j\in[1,h]}(\mathbf{T}.\mathbf{V}_j) - min_{i\in[1,h]}(\mathbf{T}.\mathbf{V}_i) + 1$$

For large computation domains $\mathcal{D}$, the minimization of ΔT leads to a minimal value of Δt.

The problem of minimizing ΔT can be solved by means of integer programming. Let $D_{i,j}$ be the set of all non-null vectors $\mathbf{X}$ such that

$$\forall \mathbf{z} \in \mathcal{D} \rightarrow \mathbf{X}.\mathbf{V}_i \leq \mathbf{X}.\mathbf{z} \leq \mathbf{X}.\mathbf{V}_j.$$

The later condition is equivalent to the following set of linear constraints:

$$\begin{cases} \forall a \in [1, h_i] \rightarrow (\mathbf{V}_{i_a} - \mathbf{V}_i).\mathbf{X} \geq 0 \\ \forall b \in [1, h_j] \rightarrow (\mathbf{V}_j - \mathbf{V}_{j_b}).\mathbf{X} \geq 0 \end{cases},$$

where $\mathbf{V}_{i_a}$ and $\mathbf{V}_{j_b}$ denote the vertices of $\mathcal{D}$ connected respectively to $\mathbf{V}_i$ and to $\mathbf{V}_j$. By definition of $D_{i,j}$ it comes:

$$\forall \mathbf{T} \in D_{i,j} \rightarrow \Delta T = \mathbf{T}.(\mathbf{V}_j - \mathbf{V}_i) + 1. \tag{20}$$

Thus, the optimization can be achieved for any possible pair of vertices (V_i, V_j) of $\mathcal{D}$, and the global minimum will simply be the minimum of the local solutions.

Example 2.2 (cont'd) Assume that all $\Delta_{u,v}$'s are one. It can be shown that the only pair of vertices whose $D_{i,j}$ gives a solution is $(\mathbf{V}_i, \mathbf{V}_j) = ((0,0,0),(N-1,N-1,N-1))$. The optimal schedule is found by minimizing

$$\mathbf{T}.(1,1,1)$$

under the constraints $\mathbf{T}.(1,0,0) \geq 1$, $\mathbf{T}.(0,1,0) \geq 1$, and $\mathbf{T}.(0,0,1) \geq 1$. The solution is given by:

$$t(i,j,k) = i + j + k + \alpha_v, \tag{21}$$

with $\alpha_A = \alpha_B = 0$ and $\alpha_C = 1$. The associated number of time steps is $3.(N-1)+1$. □

Instead of looking for $\mathbf{T}$ that minimizes ΔT, one may want to find $\mathbf{T}$ that minimizes some other cost function. This can easily be achieved when the solution space of $\mathbf{T}$ is bounded. In that case, one can compute all possible solutions, and find the one that minimizes the cost.

The solution space of $\mathbf{T}$ can be bounded by imposing ΔT to be in a given range. Let $[\Delta T^u, \Delta T^v]$ denote this range. Because of (20), any slope $\mathbf{T}$ of $D_{i,j}$ has its ΔT in that range if and only if:

$$\Delta T^u - 1 \leq \mathbf{T}.(\mathbf{V}_j - \mathbf{V}_i) \leq \Delta T^v - 1. \tag{22}$$

Example 2.2 (cont'd) Let us for example compute all the slopes $\mathbf{T}$ whose $\Delta T \in [30, 40]$, when $N = 8$. The set of solutions is:

$$\{(1,1,3),(1,2,2),(1,3,1),(2,1,2),(2,2,1),(3,1,1)\}. \tag{23}$$

□

The same technique is also applicable to infinite computation domains. Yet, in that case, the quantity

$$max_{\mathbf{z}\in\mathcal{D}}(t(\mathbf{z})) - min_{\mathbf{z}\in\mathcal{D}}(t(\mathbf{z})) + 1$$

is always infinite. Hence, it cannot be used for minimizing the schedule. Given an infinite domain $\mathcal{D}$ of vertices $\mathbf{V}_1$, $\mathbf{V}_2$, ..., $\mathbf{V}_h$, and of ray $\mathbf{R}$, we define $\mathcal{D}'$ as the finite convex domain of vertices $\mathbf{V}_k$ and $\mathbf{V}_{h+k} = \mathbf{V}_k + \mathbf{R}$, $k \in [1, h]$. The edges of $\mathcal{D}'$ are the edges of $\mathcal{D}$, plus the edges between $\mathbf{V}_k$ and $\mathbf{V}_{h+k}$, plus the edges between $\mathbf{V}_{h+i}$ and $\mathbf{V}_{h+j}$ whenever there is an edge between $\mathbf{V}_i$ and $\mathbf{V}_j$.

The value of Δt can be defined on this restricted domain as

$$\Delta t = max_{\mathbf{z}\in\mathcal{D}'}(t(\mathbf{z})) - min_{\mathbf{z}\in\mathcal{D}'}(t(\mathbf{z})) + 1.$$

According to this definition, Δt is now a finite quantity which can be used in the comparison of different schedules of infinite dependence graphs.

Example 2.1 (cont'd) The restricted domain $\mathcal{D}'$ is here a rectangle of vertices $(1,1)$, $(1,N)$, $(2,N)$ and $(2,1)$. Let us assume that all $\Delta_{u,v}$'s are one. The only pair of vertices which lead to a valid schedule is $(\mathbf{V}_i, \mathbf{V}_j) = ((1,1),(2,N))$. The schedule that minimizes ΔT is found by minimizing $\mathbf{T}.(1, N-1)$ under the constraints $\mathbf{T}.(1,0) \geq 1$, $\mathbf{T}.(0,1) \geq 1$, and $\mathbf{T}.(1,1) \geq 1$. The solution is given by:

$$t(i,k) = i + k + \alpha_v,$$

with $\alpha_x = \alpha_w = -2$, $\alpha_p = -1$ and $\alpha_y = 0$.

□

6 Processor allocation optimization

Let us define ΔP_l as follows:

$$\Delta P_l = \left\lceil \frac{max_{\mathbf{z}\in\mathcal{D}}(\mathbf{P}_l.\mathbf{z}) - min_{\mathbf{z}\in\mathcal{D}}(\mathbf{P}_l.\mathbf{z}) + 1}{d_l} \right\rceil . \tag{24}$$

When the components of $\mathbf{P}_l$ are relatively prime, ΔP_l represents the number of cells along l, for each value of v. Else, it is a good approximation of the number of cells. The value of $\mathbf{P}_l$ that minimizes

$$max_{\mathbf{z}\in\mathcal{D}}(\mathbf{P}_l.\mathbf{z}) - min_{\mathbf{z}\in\mathcal{D}}(\mathbf{P}_l.\mathbf{z}) + 1 \tag{25}$$

clearly minimizes (24). Thus, when d_l is fixed, the optimization of ΔP_l becomes similar to the one of ΔT. For a given $D_{i,j}$, we have:

$$\forall \mathbf{P}_l \in D_{i,j} \rightarrow max_{\mathbf{z}\in\mathcal{D}}(\mathbf{P}_l.\mathbf{z}) - min_{\mathbf{z}\in\mathcal{D}}(\mathbf{P}_l.\mathbf{z}) + 1 = \mathbf{P}_l.(\mathbf{V}_j - \mathbf{V}_i) + 1.$$

This later quantity can be minimized by means of integer programming.

Example 2.2 (cont'd) Assume again that $\mathbf{s} = (N, w_1, w_2) = (8, 2, 3)$. In that case, $\mathcal{D}$ has 16 vertices as shown in figure 4. Consider the pair $(\mathbf{V}_i, \mathbf{V}_j) = ((0, w_1 + w_2, w_1), (w_1 + w_2, 0, w_2))$. The associated integer programming problem is to minimize

$$(w_1 + w_2, -w_1 - w_2, w_2 - w_1).\mathbf{P}_l = (5, -5, 1).\mathbf{P}_l,$$

under the constraints

$$(5,-5,1).\mathbf{P}_l \geq 1 \text{ (to avoid } \mathbf{P}_l = \mathbf{0}),\ (1,1,1).\mathbf{P}_l \geq 0,\ (0,-1,0).\mathbf{P}_l \geq 0,$$
$$(0,-1,-1).\mathbf{P}_l \geq 0,\ (-1,-1,-1).\mathbf{P}_l \geq 0, (1,0,0).\mathbf{P}_l \geq 0,\ (1,0,1).\mathbf{P}_l \geq 0.$$

The optimal slope is $\mathbf{P}_l = (1, 0, -1)$. The associated ΔP_l is $2.w_1 + 1 = 5$. The optimal rational solution is therefore $\mathbf{P}_l/d_l = (1, 0, -1)/d_l$. It can be shown that this solution is a global optimum; no other pair of vertices can yield to a better solution. (There is only one equivalent solution which is $(-1, 0, 1)/d_l$.)

□

It is also possible to find all the slopes $\mathbf{P}_l$ such that ΔP_l is in a given range. Let $[\Delta P_l^u, \Delta P_l^v]$ denote this range. It can be verified that any slope $\mathbf{P}_l \in D_{i,j}$ has its associated ΔP_l in that range if and only if:

$$d_l.\Delta P_l^u - d_l \leq \mathbf{P}_l.(\mathbf{V}_j - \mathbf{V}_i) \leq d_l.\Delta P_l^v - 1. \tag{26}$$

This is direct generalization of what was done on the schedule, and an application of:

$$\begin{array}{l} \lceil a/b \rceil \geq c \iff a \geq b.(c-1)+1 \\ \lceil a/b \rceil \leq c \iff a \leq b.c \end{array} .$$

The constraint (26) bounds the solution space of $\mathbf{P}_l$.

Example 2.2 (cont'd) Consider the problem of finding all the slopes $\mathbf{P}_l$ such that ΔP_l is in $[5,10]$ and $d_l = 1$. When fixing $(\mathbf{V}_i, \mathbf{V}_j)$ to $((0, w_1 + w_2, w_1), (w_1 + w_2, 0, w_2))$, the set of solutions is:

$$\{(1,0,-1),(0,-1,1),(2,0,-2)\}.$$

When considering all pair of vertices, the complete set of solutions for $\mathbf{P}_l$ is:

$$\{\pm(1,0,-1),\pm(0,-1,1),\pm(2,0,-2),\pm(1,0,0),\pm(0,1,0),\pm(0,0,1)\}. \tag{27}$$

□

7 Design methodology

The problem is to find a mapping of the form (6) that verifies a particular number of design constraints. As explained in the previous sections, a set of slopes $\mathbf{T}$ that verifies a number of constraints and that minimizes ΔT can be found by means of integer programming. Also, the set of slopes whose ΔT is in a given range can be found with a similar technique. The same applies on each axis of the processor allocation separately (for each $\mathbf{P}_l$).

Once a set of solutions has been found for $\mathbf{T}$ and for each $\mathbf{P}_l$, the mappings of the form (6) can be found by simply combining the different solutions. For each combination, one can verify that the mapping is one-to-one, as explained in section 4; only the one-to-one mappings should be kept. If no compatible solution exists, one can either modify the value of one d_l or modify one range of values, and restart the process. The search for one-to-one mappings is clearly an iterative process.

Finally, a cost function can be evaluated on every one-to-one mapping to select the optimal solutions. The methodology is summarized in figure 5. It has been implemented with success in the design tool Presage [VP90].

Example 2.2 (cont'd) The minimal value of ΔT is 22. It is achieved with $\mathbf{T} = (1,1,1)$. When $d_1 = d_2 = 1$, the minimal value of ΔP_l is 5; $\mathbf{P}_l = \pm(1,0,-1)$. Yet, the mapping such that $\mathbf{T} = (1,1,1)$, $\mathbf{P}_1 = (1,0,-1)$ and $\mathbf{P}_2 = (-1,0,1)$ is not one-to-one. (The corresponding value of a' is zero.)

A first solution consists in relaxing the admissible range for ΔP_2. For example, if this range is $[5,10]$, the set of solutions for $\mathbf{P}_2$ is given by (27). Out of all combinations, four mappings only are one-to-one; they are all equivalent and given by

$$\begin{pmatrix} \mathbf{T} \\ \mathbf{P}_1 \\ \mathbf{P}_2 \end{pmatrix} = \begin{pmatrix} i+j+k+\alpha_v \\ \pm(1,0,-1).(i,j,k) \\ \pm(0,1,-1).(i,j,k) \end{pmatrix},$$

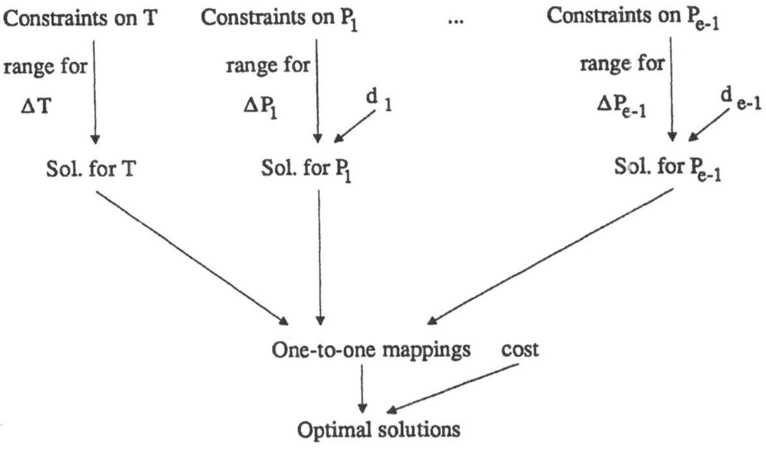

Figure 5: The main structure of the adopted methodology.

with $\alpha_A = \alpha_B = 0$ and $\alpha_C = 1$. All solutions lead to the same well-known systolic array presented by Kung and Leiserson in [MC80].

□

Example 2.2 (cont'd) The value of a' corresponding to the previous solution is 3, while $d_1.d_2 = 1$; each processor works once every three steps only. An implementation that requires a third of the cells can be achieved with $d_1 = 3$, e.g. with the mapping:

$$\begin{pmatrix} \mathbf{T} \\ \mathbf{P}_1 \\ \mathbf{P}_2 \end{pmatrix} = \begin{pmatrix} i+j+k+\alpha_v \\ (1,0,-1).(i,j,k) \text{ div } 3 \\ (0,1,-1).(i,j,k) \end{pmatrix}$$

It can be verified that this mapping is indeed one-to-one. When $w_1 = 2$ and $w_2 = 3$, the number of cells is $((2.2+1) \text{ div } 3) \times (2.3+1)$.

□

Example 2.2 (cont'd) The value of a' corresponding to the previous solution is 3, while $d_1.d_2 = 1$; each processor works once every three steps only. An implementation that requires a third of the cells can be achieved with $d_1 = 3$, e.g. with the mapping:

$$\begin{pmatrix} \mathbf{T} \\ \mathbf{P}_1 \\ \mathbf{P}_2 \end{pmatrix} = \begin{pmatrix} i+j+k+\alpha_v \\ (1,0,-1).(i,j,k) \text{ div } 3 \\ (0,1,-1).(i,j,k) \end{pmatrix}$$

It can be verified that this mapping is indeed one-to-one. When $w_1 = 2$ and $w_2 = 3$, the number of cells is $((2.2+1) \text{ div } 3) \times (2.3+1)$.

□

Example 2.2 (cont'd) In the previous example, an array of 2×7 cells is used. Assume that the problem is to be mapped on an array of size 8. One can either use an array of size 2×4 or a linear (i.e. one-D) array. In the first case, we can keep $d_1 = 3$ and $p_1(i,j,k) = (1,0,-1).(i,j,k)$ which give $\Delta P_1 = 2$. We can use $d_2 = 2$ and $p_2(i,j,k) = (0,1,-1).(i,j,k)$ div 2 to obtain $\Delta P_2 = 4$. The schedule that minimizes the latency can then be found as follows.

First, one can try with the minimal schedule (21), but the corresponding mapping is not one-to-one. One can then use a range for ΔT. The minimal value being $8.3-2 = 22$, one can first try with the range $[23, 30]$. It can be shown that the associated set of solutions is:

$$\{(1,1,2),(1,2,1),(2,1,1)\}.$$

But still, none of these solutions yield to a one-to-one mapping. One can then try with the range $[31, 40]$; the set of solutions for $\mathbf{T}$ is given by (23). Still, no schedule gives a one-to-one mapping. One can then try with the range $[41, 45]$ whose set of solutions is:

$$\{(1,1,4),(1,2,3),(2,1,3),(1,3,2),(2,2,2),(3,1,2),(1,4,1),(2,3,1),(3,2,1),(4,1,1)\}.$$

It can be verified that the any slope of $\mathbf{T}$ in the set:

$$\{(4,1,1),(1,3,2),(2,1,3),(2,3,1).\}$$

gives a one-to-one mapping.

For a linear array, one can use $d_1 = 5$ and $d_2 = 1$. In that case,

$$\begin{aligned} p_1 &= (1,0,-1).(i,j,k) \text{ div } 5 \\ p_2 &= (i,j,k).(0,1,-1) \end{aligned} \tag{28}$$

The same iterative process can be used to find a compatible schedule. It can be verified that any slope $\mathbf{T}$ in (23) yields to a one-to-one mapping.

□

8 Conclusion

Given a regular application described by a system of uniform recurrence equations, systolic arrays are commonly derived by means of an affine space-time mapping. In this paper, we generalized the associated methodology to design small size arrays, by using an affine schedule and a near-affine processor allocation, a sub-class of quasi-affine mappings [Van91]. We showed that a near-affine processor allocation that uses a given number of processors can be automatically derived; the optimization method is a direct extension to the one using affine mappings.

In the proposed approach, the schedule and each component of the processor allocation are found independently. Ranges for the number of time steps and for the number of processors are given by the user to delimit each solution space. Compatible solutions, i.e. one-to-one mappings, are then found by exhaustive search. A cost function finally selects the optimal solutions.

The associated partitioning method is the well-known *Locally Sequential Globally Parallel* scheme [ND88,GN89,Kun87]. This method requires large local memories for the processing elements. Other partitioning techniques are known, buth these cannot be found directly with our space-time mapping technique. Yet the advantage of our approach compared to the common one where circuit transformations are applied on systolic arrays derived with affine mappings is that optimal solutions can be found automatically.

As a particular case, we can optimally map e-D (i.e. e-dimensional) recurrences onto E-D arrays, where E is any value between 0 and $e-1$. One simply needs to fix the number of coordinates to 1 along $e-E-1$ axes of the processor allocation, and find a valid near-affine mapping. This method is to be compared with the *multiprojection* technique introduced in [WD85], which consists in applying an affine mapping k times to reduce the dimension of the array to $e-k$. Again, an optimal global solution can hardly be found with the latter technique, while it can when doing the space-time mapping in one pass.

Most of the theory presented in this paper has been implemented with success in the tool named "Presage" [VP90]. In fact, the illustrative mappings were derived with its use.

References

[AAG*87] M. Annaratone, E. Arnould, T. Gross, H.T. Kung, M. Lam, O. Menzilcioglu, and J.A. Webb. The warp computer : architecture, implementation, and performance. *IEEE Trans. on Computers*, C-36(12):1523–1538, December 1987.

[Bra90] H. Brams. *Adaptation du logiciel Presage à la génération de réseaux systoliques implémentables sur un réseau de Transputers.* Technical Report RR 90-23, Université Catholique de Louvain, Novembre 1990.

[Bu90] J. Bu. *Systematic Design of Regular VLSI Processor Arrays.* PhD thesis, Delft University of Technology, May 1990.

[Cla90] P. Clauss. *Synthèse d'Algorithmes Systoliques et Implantation Optimale en Place sur Réseaux de Processeurs Synchrones.* PhD thesis, Université de Franche-Comté, 1990.

[DI88] J.-M. Delosme and I.C.F. Ipsen. Sage and condense : a two-phase approach for the implementation of recurrence equations on multiprocessors architectures. In L.W. Hoevel, editor, *21st Annual Hawai Int. Conf. on System Sciences*, pages 126–130, 1988.

[GN89] M. Garcia and J. Navarro. Systematic hardware adaptation of systolic algorithms. In IEEE, editor, *The 16th Annual Int. Symp. on Comp. Architecture*, pages 96–104, Computer Society Press, 1989.

[KL78] H.T. Kung and C.E. Leiserson. Systolic arrays (for VLSI). In *Sparse Matrix Proc. 1978*, pages 256–282, Society for Industrial and Applied Mathematics, 1978.

[KMW67] R.M. Karp, R.E. Miller, and S. Winograd. The organization of computations for uniform recurrence equations. *Journal of the ACM*, 1967.

[Kun87] S.Y. Kung. *VLSI array processors*. Signal and Image Processing Institute, 1987.

[MC80] C. Mead and L. Conway. *Introduction to VLSI Systems*, chapter 8, Highly Concurrent Systems, pages 263–332. Addison-Wesley Series in Computer Science, 1980.

[Mol83] D.I. Moldovan. On the design of algorithms for VLSI systolic arrays. *IEEE Proceedings*, 1983.

[ND88] H. Nelis and E. Deprettere. Automatic design and partitioning of systolic/wavefront arrays for vlsi. *Circuits Systems and Signal Processing*, Vol. 7(2):235–252, 1988.

[Qui84] P. Quinton. Automatic synthesis of systolic arrays from uniform recurrent equations. In *Proc. IEEE 11-th Int. Sym. on Computer Architecture*, 1984.

[Rao85] S.K. Rao. *Regular iterative algorithms and their implementations on processor arrays*. PhD thesis, Information Systems lab., Standford University, 1985.

[The89] The transputer databook. 2nd ed. Inmos, Bristol, 1989.

[Van91] V. Van Dongen. *From Systolic to Periodic Array Design*. PhD thesis, Université Catholique de Louvain, January 1991.

[VP90] V. Van Dongen and M. Petit. Presage: a tool for the parallelization of nested loop programs. In L. Claesen (ed.), editor, *Formal VLSI Specification and Synthesis (VLSI Design Methods-I)*, pages 341–359, North-Holland, 1990.

[WD85] Y. Wong and J.M. Delosme. Optimal systolic implementations of n-dimensional recurrences. In *IEEE Int. Conf. on Computer Design: VLSI in Computers*, pages 618–621, Oct. 7-10 1985.

AREA COMPLEXITY OF MULTILECTIVE MERGING

PAVEL FERIANC
ONDREJ SÝKORA

Computing Centre, Slovak Academy of Sciences, Dúbravská 9
842 35 Bratislava, Czecho-Slovakia

Abstract. Lower bounds on the area A(n,m,k,r) required for merging of two sorted sequences of k-bit numbers with length n and m respectively, when the inputs can be replicated up to r times ($r \leq n$), are given:

$$A(n,m,k,r) = \begin{cases} \Omega\left(\frac{n}{r}\right) \text{ for } 2^k \geq \frac{n}{r} \text{ and } n \geq m \geq \frac{n}{r} \\ \Omega\left(m\left(\left(\log \frac{n}{rm}\right)+1\right)\right) \text{ for } 2^{\frac{3}{8}k} \geq \frac{n}{r} \text{ and } \frac{n}{r} \geq m \\ \Omega\left(m\left(\left(\log \frac{2^k}{m}\right)+1\right)\right) \text{ for } \frac{n}{r} \geq m \text{ and } \frac{n}{r} \geq 2^{\frac{3}{8}k} \text{ and } 2^{\left(\frac{3.(8^K)-1}{8^{K+1}-1}k\right)} \geq m \\ \qquad \text{where } K \geq 0 \text{ is the constant} \end{cases}$$

INTRODUCTION

The paper analyzes the following problem: let $m, n \in N$ and $m \leq n$, let $D_n = \{(x_1, ..., x_n) | x_1 \leq ... \leq x_n; x_i (i = 1, ..., n)$ consist of k bits $\}$ and let $D_m = \{(z_1, ..., z_m) | z_1 \leq ... \leq z_m; z_i (i = 1, ..., m)$ consist of k bits $\}$. The merging problem can be characterized by the function $f : D_n \times D_m \rightarrow D_{n+m}$, such that $f(x_1, ..., x_n, z_1, ..., z_m) = (y_1, ..., y_{n+m})$, where $X = \{x_1, ..., x_n\} \in D_n$, $Z = \{z_1, ..., z_m\} \in D_m$, $Y = \{y_1, ..., y_{n+m}\} \in D_{n+m}$ ($D_{n+m} = \{(y_1, \dots, y_{n+m}) | y_1 \leq \dots \leq y_{n+m}; y_i (i = 1, \dots, n+m)$ consist of k bits $\}$) and each $y_i (1 \leq i \leq n+m)$ satisfies either $y_i \in X$ or $y_i \in Z$. Without loss of generality let us assume that m, n, r, k be powers of 2.

The memory of our circuit consists of square units of area (each one with area λ^2 ($\lambda > 0$)) and at most one bit can be stored per unit of area. The i/o schedule is assumed to be when- and where-determinate. Determinate schedules, which require prespecified times and locations for the input and output of each bit, are discussed in [U]. Tight lower bounds on the area required for merging with semelective inputs (each data is read once) were shown in [PSV]. Our area bounds are proved for r-multilective input; i.e. data can be read more than once but at most r-times. All computations and temporary storage of data, however, must be done within the merging device. According to [Si] we can assume w.l.o.g. that in a time unit t at most one input bit is supplied or at most one output bit is delivered and only one input or output event occurs.

The motivation for the study of the multilective circuits is that it leads either towards more general techniques for searching area bounds or it brings more general results. It is evident that multilectivity enables substantially diminished area of circuit. For example, in the work [G], a language with the following property is described. The area required for its recognition is $\Omega(\sqrt{n})$, but if allow each input to be read twice, then the area for its recognition is O(1).

According to our knowledge until now there were proven only two results about nontrivial area lower bounds for concrete multilective problems. The area $A(n,k,r) = \Omega(\log n + 2^k \log \frac{n}{r2^{k-1}})$ for $2^k \leq \frac{n}{r}$; $A(n,k,r) = \Omega(\log n + \frac{n}{r} \log \frac{r2^{k+1}}{n})$ for $\frac{n}{r} \leq 2^k \leq n^{O(1)}$ for sorting n k-bit numbers is showed in [Si] and the area $A = \Omega(\frac{n^2}{r^2})$ for the matrix product of $n \times n$ matrices is shown in [Sa].

Now we give a brief introduction to the technique of our proofs(a similar technique was used in [Si]). A time interval τ of the entire time of computation is chosen. Call the inputs (outputs) which are read (delivered to output) during the interval τ τ-inputs (τ-outputs) and the inputs (outputs) for which no copy is read during the interval τ as non-τ-inputs (non-τ-outputs). Set the τ-inputs in a proper way so that the τ-outputs are dependent on the non-τ-inputs. The number of all various vectors of τ-outputs is equal to the number of all various vectors of corresponding non-τ-inputs, which is also the number of possible circuit states. The lower bound on the required area is the logarithm of the number of the circuit states.

LOWER BOUNDS

THEOREM 1. *If* $2^k \geq \frac{n}{r}$ *and* $n \geq m \geq \frac{n}{r}$, *then* $A(n,m,k,r) = \Omega(\frac{n}{r})$.

To prove this theorem we need the next lemma.

LEMMA. *Let* $b_1 ... b_t$, *where* t *is even, be a string of zeros and ones such that the number of zeros is equal to the number of ones. Then arbitrary* $s \leq t$ *satisfies: there exists* j $(0 \leq j \leq t-s)$ *such that substring* $b_{j+1} ... b_{j+s}$ *contains at least* $\lfloor \frac{1}{3}s \rfloor$ *ones.*

PROOF OF LEMMA: Three cases are possible:

(1) $\lceil \frac{3}{4}t \rceil \geq s > \frac{1}{2}t$

If there are at least $\lfloor \frac{1}{3}s \rfloor$ ones in the substring $b_{s+1} ... b_t$ then $j = t-s$. Otherwise the substring $b_1 ... b_s$ must contain at least $\frac{t}{2} - (\lfloor \frac{1}{3}s \rfloor - 1) > \lfloor \frac{1}{3}s \rfloor$ ones.

(2) $t \geq s > \lceil \frac{3}{4}t \rceil$

Let $t = s+i$. Hence $i < \lfloor \frac{1}{3}s \rfloor$. Let us assume the substring $b_{s+1} ... b_t$ contains only ones (it means i ones). Then the string $b_1 ... b_s$ must contain at least $\frac{s+i}{2} - i > \lfloor \frac{1}{3}s \rfloor$ ones.

(3) $s \leq \frac{1}{2}t$

Let us divide the string $b_1 ... b_t$ to three substrings so that two substrings of them have the length s and the last one has the length i. Let in each of these substrings be at most $(\lfloor \frac{1}{3}s \rfloor - 1)$ ones i.e. $3(\lfloor \frac{1}{3}s \rfloor - 1)$ ones together. However, in the whole string $b_1 ... b_t$ there must be $\frac{2s+i}{2}$ ones. Since $3(\lfloor \frac{1}{3}s \rfloor - 1) < \frac{2s+i}{2}$ there exists one substring, which contains at least $\lfloor \frac{1}{3}s \rfloor$ ones. ∎

PROOF OF THEOREM 1: W.l.o.g. assume that $n \geq 12r$. Let us divide the time interval $< 0, T >$, where T is the time of the computation, into 2r intervals so that in each of them exactly $\frac{n}{4r}$ least significant bits of variables $y_{v+1}, y_{v+2}, ..., y_{v+\frac{n}{2}}$ are delivered to output, where $v \geq \frac{m}{2}$ and $v + \frac{n}{2} \leq n$. There exists an interval τ such that during τ at most $\frac{m}{4}$ copies of the least significant bits of the variables $z_1, ..., z_{\frac{m}{2}}$ are read. This implies that at least $\frac{m}{4}$ least significant bits of variables $z_1, ..., z_{\frac{m}{2}}$ are non-τ-inputs. Let $z_{j_1,k}; ...; z_{j_{\frac{m}{4}},k}$, where $j_{\frac{m}{4}} \leq \frac{m}{2}$ and $j_1 < ... < j_{\frac{m}{4}}$, be these bits. Let us indicate the least significant bits of variables $y_{v+1}, y_{v+2}, ..., y_{v+\frac{n}{2}}$, delivered to output during the interval τ (τ-outputs) as $y_{i_1,k}; ...; y_{i_{\frac{n}{4r}},k}$, where $i_1 < ... < i_{\frac{n}{4r}}$.

Now assign

$$x_{1,k} = ... = x_{n,k} = 0$$

$$z_{l,k} = 0 \text{ for } l \in \{1, ..., m\} - \{j_1, ..., j_{\frac{m}{4}}\}.$$

We are going to set so the bits $z_{i,1}; ...; z_{i,k-1}$ for $i = 1, ..., m$ and $x_{j,1}; ...; x_{j,k-1}$ for $j = 1, ..., n$ that as many input bits $z_{j_1,k}; ...; z_{j_{\frac{m}{4}},k}$ as possible are delivered to the output as $y_{i_1,k}; ...; y_{i_{\frac{n}{4r}},k}$. So, if $z_{t,k}$ for $t \in \{j_1, j_2, ..., j_{\frac{m}{4}}\}$ is delivered to output as $y_{s,k}$ for $s \in \{i_1, i_2, ..., i_{\frac{n}{4r}}\}$ then:

$$\text{if } z_{t,k} = 0 \text{ then } y_{s,k} = 0$$

and

$$\text{if } z_{t,k} = 1 \text{ then } y_{s,k} = 1.$$

By lemma 1 there exists j $(0 \leq j \leq \frac{m}{2} - \frac{n}{4r})$ such that the string $z_{j+1,k}; ...; z_{j+\frac{n}{4r},k}$ contains $\lfloor \frac{1}{3} \frac{n}{4r} \rfloor = \lfloor \frac{n}{12r} \rfloor$ bits from the set $\{z_{j_1,k}; ...; z_{j_{\frac{m}{4}},k}\}$. Let us indicate these bits by $z_{q_1,k}; ...; z_{q_{\lfloor \frac{n}{12r} \rfloor},k}$, where $q_1 < ... < q_{\lfloor \frac{n}{12r} \rfloor}$. Now we must find a setting for bits $z_{i,1}; ...; z_{i,k-1}$ where $i = 1, ..., m$ and for the bits $x_{j,1}; ...; x_{j,k-1}$ where $j = 1, ..., n$ such that

$$z_{j+1,k} \text{ is delivered to the output as } y_{i_1,k}$$

$$z_{j+2,k} \text{ is delivered to the output as } y_{i_2,k}$$

...

$$z_{j+\frac{n}{4r},k} \text{ is delivered to the output as } y_{i_{\frac{n}{4r}},k}$$

and the bits $\{z_{q_1,k}; ...; z_{q_{\lfloor \frac{n}{12r} \rfloor},k}\} \subset \{z_{j+1}, ..., z_{j+\frac{n}{4r}}\}$ are non-τ-inputs, whereby the bit

$$z_{q_1,k} \text{ is delivered to the output as } y_{i_{d_1},k}$$

$$z_{q_2,k} \text{ is delivered to the output as } y_{i_{d_2},k}$$

...

$$z_{q_{\lfloor \frac{n}{12r} \rfloor},k} \text{ is delivered to the output as } y_{i_{d_{\lfloor \frac{n}{12r} \rfloor}},k}$$

for $d_1 < d_2 < \cdots < d_{\lfloor \frac{n}{12r} \rfloor}$ and $\{d_1, d_2, \ldots, d_{\lfloor \frac{n}{12r} \rfloor}\} \subset \{1, \ldots, \frac{n}{4r}\}$. Hence the bits $y_{i_{d_1},k}, \ldots,$

$y_{i_{d_{\lfloor \frac{n}{12r} \rfloor}},k}$ depend on the values of the bits $z_{q_1,k};\ldots;z_{q_{\lfloor \frac{n}{12r} \rfloor},k}$ which means that if the value $z_{q_j,k}$ is changed then the value of $y_{i_{d_j},k}$ is changed too. Let $w_{i,1}; w_{i,2};\ldots;w_{i,k-1}$ be the bits of the k bit variable w_i where $w_{i,1}$ is the most significant bit. We indicate the expression $w_{i,1}2^{k-2} + w_{i,2}2^{k-3} + \cdots + w_{i,k-1}2^0$ as H_{w_i}. Now we assign to the bits $z_{i,1};\ldots;z_{i,k-1}$ for $i = 1,\ldots,m$ and to the bits $x_{j,1};\ldots;x_{j,k-1}$ for $j = 1,\ldots,n$ such values that the following is fulfiled:

$$H_{z_i} = 0 \text{ for } i = 1,\ldots,q_1$$
$$H_{x_j} = 0 \text{ for } j = 1,\ldots,i_{d_1} - q_1$$
$$H_{z_i} = 1 \text{ for } i = q_1+1,\ldots,q_2$$
$$H_{x_j} = 1 \text{ for } j = i_{d_1} - q_1 + 1,\ldots,i_{d_2} - q_2$$
$$H_{z_i} = 2 \text{ for } i = q_2+1,\ldots,q_3$$
$$H_{x_j} = 2 \text{ for } j = i_{d_2} - q_2 + 1,\ldots,i_{d_3} - q_3$$
$$\ldots$$
$$H_{z_i} = (\left\lfloor \frac{n}{12r} \right\rfloor - 1) \text{ for } i = q_{\lfloor \frac{n}{12r} \rfloor - 1} + 1,\ldots,q_{\lfloor \frac{n}{12r} \rfloor}$$
$$H_{x_j} = (\left\lfloor \frac{n}{12r} \right\rfloor - 1) \text{ for } j = i_{d_{\lfloor \frac{n}{12r} \rfloor - 1}} - q_{\lfloor \frac{n}{12r} \rfloor - 1} + 1,\ldots,i_{d_{\lfloor \frac{n}{12r} \rfloor}} - q_{\lfloor \frac{n}{12r} \rfloor}$$
$$H_{z_i} = \left\lfloor \frac{n}{12r} \right\rfloor \text{ for } i = q_{\lfloor \frac{n}{12r} \rfloor} + 1,\ldots,m$$
$$H_{x_j} = \left\lfloor \frac{n}{12r} \right\rfloor \text{ for } j = i_{d_{\lfloor \frac{n}{12r} \rfloor}} - q_{\lfloor \frac{n}{12r} \rfloor} + 1,\ldots,n$$

and we set $z_{l,k} = 0$ for $l \in \{j_1,\ldots,j_{\frac{m}{4}}\} - \{q_1,\ldots,q_{\lfloor \frac{n}{12r} \rfloor}\}$. The vector $(z_{q_1,k};\ldots;z_{q_{\lfloor \frac{n}{12r} \rfloor},k})$ can attain $2^{\lfloor \frac{n}{12r} \rfloor}$ different values. So, various non-τ-inputs cause various τ-outputs and each of τ-inputs is set. This implies at least $2^{\lfloor \frac{n}{12r} \rfloor}$ various states for the circuit and therefore the area is at least $\log 2^{\lfloor \frac{n}{12r} \rfloor}$. Hence: $A(n,m,k,r) = \Omega(\frac{n}{r})$. ∎

REMARK 1. *The theorem 1 holds for* $n \geq m \geq \frac{n}{2r}$ *too. This means that for* $\frac{n}{r} \geq m \geq \frac{n}{2r}$ *it is* $A(n,m,k,r) = \Omega(m)$. *The constant does not play a role at all. Evidently for* $\frac{n}{r} \geq m \geq \frac{n}{Kr}$ *where* $K > 1$ *is an arbitrary constant* $A(n,m,k,r) = \Omega(m) = \Omega(\frac{n}{r})$.

THEOREM 2. *If* $2^{\frac{3}{8}k} \geq \frac{n}{r} \geq m \geq 8$, *then* $A(n,m,k,r) = \Omega(m((\log\frac{n}{rm}) + 1))$.

PROOF OF THEOREM 2: W.l.o.g. assume that $\frac{n}{8r} \geq m$. Divide the time of computation into $8r$ time intervals so that in each of them either $\lfloor \frac{n+m}{8r} \rfloor$ or $\lceil \frac{n+m}{8r} \rceil$ least significant bits of variables $y_1,\ldots,y_{n+m}$ are delivered to the output. Among these $8r$ intervals there exists an interval τ such that during τ at most $\frac{(k-1)mr}{8r} = \frac{(k-1)m}{8}$ of the $k-1$ most significant bits of variables $z_1,\ldots,z_m$ are input. This means at least $\frac{7}{8}m(k-1)$ of these bits are non-τ-inputs. There are at least $\frac{m}{2}$ variables $z_{l_1},\ldots,z_{l_{\frac{m}{2}}}$ from $z_1,\ldots,z_m$ such that at least $\lceil \frac{3}{4}(k-1) \rceil = \frac{3}{4}k$ bits of their $k-1$ most significant bits are non-τ-inputs. If there were $\frac{m}{2}+1$ variables whose $\frac{1}{4}k$ of the $k-1$ most significant bits were τ-inputs, then during interval τ there were read $(\frac{m}{2}+1)\frac{1}{4}k > \frac{(k-1)m}{8}$ bits, which contradicts the definition of the interval τ. Let us indicate the least significant bits which are τ-outputs as: $y_{i_1,k};\ldots;y_{i_{\lfloor\frac{n+m}{8r}\rfloor},k}$ where $i_1 < \cdots < i_{\lfloor\frac{n+m}{8r}\rfloor}$. We set $\frac{3}{8}k$ most significant bits of variables $z_1,\ldots,z_m$ in the following way(see Fig. 1.). We assign:

$$H_i = z_{i,1}2^{\frac{3}{8}k-1} + z_{i,2}2^{\frac{3}{8}k-2} + \cdots + z_{i,\frac{3}{8}k}2^0 = i-1$$

for $i = 1,\ldots,m$. This assignment ensures that: $z_1 < \cdots < z_m$. At least $\lceil\frac{3}{4}(k-1)\rceil - \frac{3}{8}k = \frac{3}{8}k$ bits among the bits:$z_{i,\frac{3}{8}k+1};\ldots;z_{i,k-1}$ for $i \in \{l_1,\ldots,l_{\frac{m}{2}}\}$ are such that they are non-τ-inputs and whatever choice their values maintains relation: $z_{l_1} < \cdots < z_{l_{\frac{m}{2}}}$. Further we set the bits $z_{i,j}$ for $i = 1,\ldots,m$ and $j = \frac{3}{8}k+1,\ldots,k$ in the following way

$$z_{1,k} = \cdots = z_{m,k} = 1$$

$$z_{i,j} = 0$$

for $i \in \{1,\ldots,m\} - \{l_1,\ldots,l_{\frac{m}{2}}\}$ and for $j = \frac{3}{8}k+1,\ldots,k-1$. Let us indicate by $z_{i,f_{i_1}},\ldots,z_{i,f_{i_{\frac{3}{8}k}}}$ for $i \in \{l_1,\ldots,l_{\frac{m}{2}}\}$ those bits from the bits $z_{i,\frac{3}{8}k+1};\ldots;z_{i,k-1}$ which are non-τ-inputs. Then we set $z_{i,j} = 0$ for $i \in \{l_1,\ldots,l_{\frac{m}{2}}\}$ and for $j \in \{\frac{3}{8}k+1,\ldots,k-1\} - \{f_{i_1},\ldots,f_{i_{\frac{3}{8}k}}\}$. Let us indicate by H_i' the following numbers:

$$H_i' = z_{i,f_{i_1}}2^{\frac{3}{8}k-1} + z_{i,f_{i_2}}2^{\frac{3}{8}k-2} + \cdots + z_{i,f_{i_{\frac{3}{8}k}}}2^0.$$

There exist $2^{\frac{3}{8}k}$ possible values for H_i' depending on the values of the variables $z_{i,f_{i_1}};\ldots;z_{i,f_{i_{\frac{3}{8}k}}}$. Now consider the output Y: $y_1 \le y_2 \le \cdots \le y_{m+n}$. Let us divide it to m (not necessary equally long) parts: $C_1 = \{y_1,\ldots,y_{i_{c_1}}\}$, $C_2 = \{y_{i_{c_1}+1},\ldots,y_{i_{c_2}}\}$, $\ldots$, $C_m = \{y_{i_{c_{m-1}}+1},\ldots,y_{i_{c_m}}\}$ where $1 < \cdots < i_{c_1} < \cdots < i_{c_2} < \cdots < i_{c_m}$, $i_{c_m} = n+m$ and $\bigcup_{i=1}^m C_i = Y$ so, that each part of them contains at least $\lfloor\frac{n+m}{8rm}\rfloor$ variables from $y_{i_1},\ldots,y_{i_{\lfloor\frac{n+m}{8r}\rfloor}}$ where $i_1 < \cdots < i_{\lfloor\frac{n+m}{8r}\rfloor}$ (see Fig. 1.) and so that the variables $y_{i_1},\ldots,y_{i_{\lfloor\frac{n+m}{8rm}\rfloor}}$ belong to the C_1, the variables $y_{i_{\lfloor\frac{n+m}{8rm}\rfloor+1}},\ldots,y_{i_{2\lfloor\frac{n+m}{8rm}\rfloor}}$ belong to the C_2 etc. and the variables $y_{i_{(m-1)(\lfloor\frac{n+m}{8rm}\rfloor)+1}},\ldots,y_{i_{\lfloor m(\frac{n+m}{8rm})\rfloor}}$ belong to the C_m. We proceed by the assignment of values to variables $x_1,\ldots,x_n$ as follows(see Fig. 1.): Let us also divide $x_1,\ldots,x_n$ into m parts so that the h-th part (for $h = 1,\ldots,m$) consists of $x_{(\sum_{q=1}^{h-1}(|C_q|-1))+1},\ldots,x_{(\sum_{q=1}^{h}(|C_q|-1))}$ where $|C_q|$ is the number of elements of the set C_q. So the h-th part corresponds to the set C_h in the output Y. This latter set contains at least $\lfloor\frac{n+m}{8rm}\rfloor$ output in τ variables:

$$y_{i_{(h-1)(\lfloor\frac{n+m}{8rm}\rfloor)+1}},\ldots,y_{i_{h\lfloor\frac{n+m}{8rm}\rfloor}}.$$

The assignment to the bits of the h-th part (for $h = 1,\ldots,m$) of the variables $x_1,\ldots,x_n$ is the following. The $\frac{3}{8}k$ most significant bits of all variables in the h-th part are assigned the values so that $x_{i,1};\ldots;x_{i,\frac{3}{8}k}$ fulfill the equality:

$$x_{i,1}2^{\frac{3}{8}k-1} + x_{i,2}2^{\frac{3}{8}k-2} + \cdots + x_{i,\frac{3}{8}k}2^0 = h-1.$$

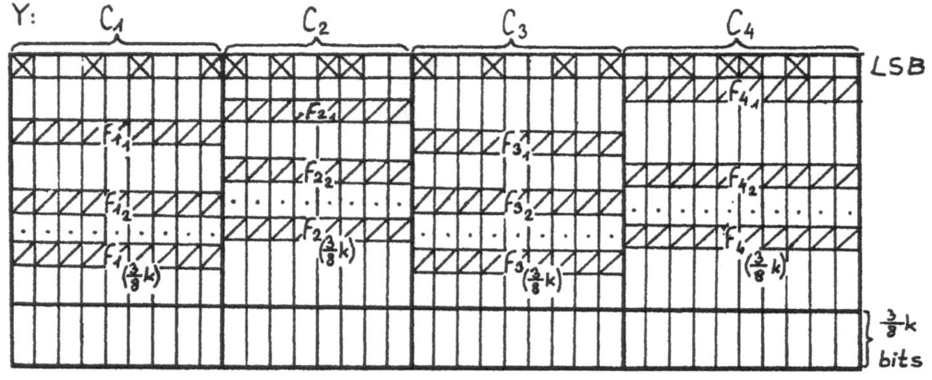

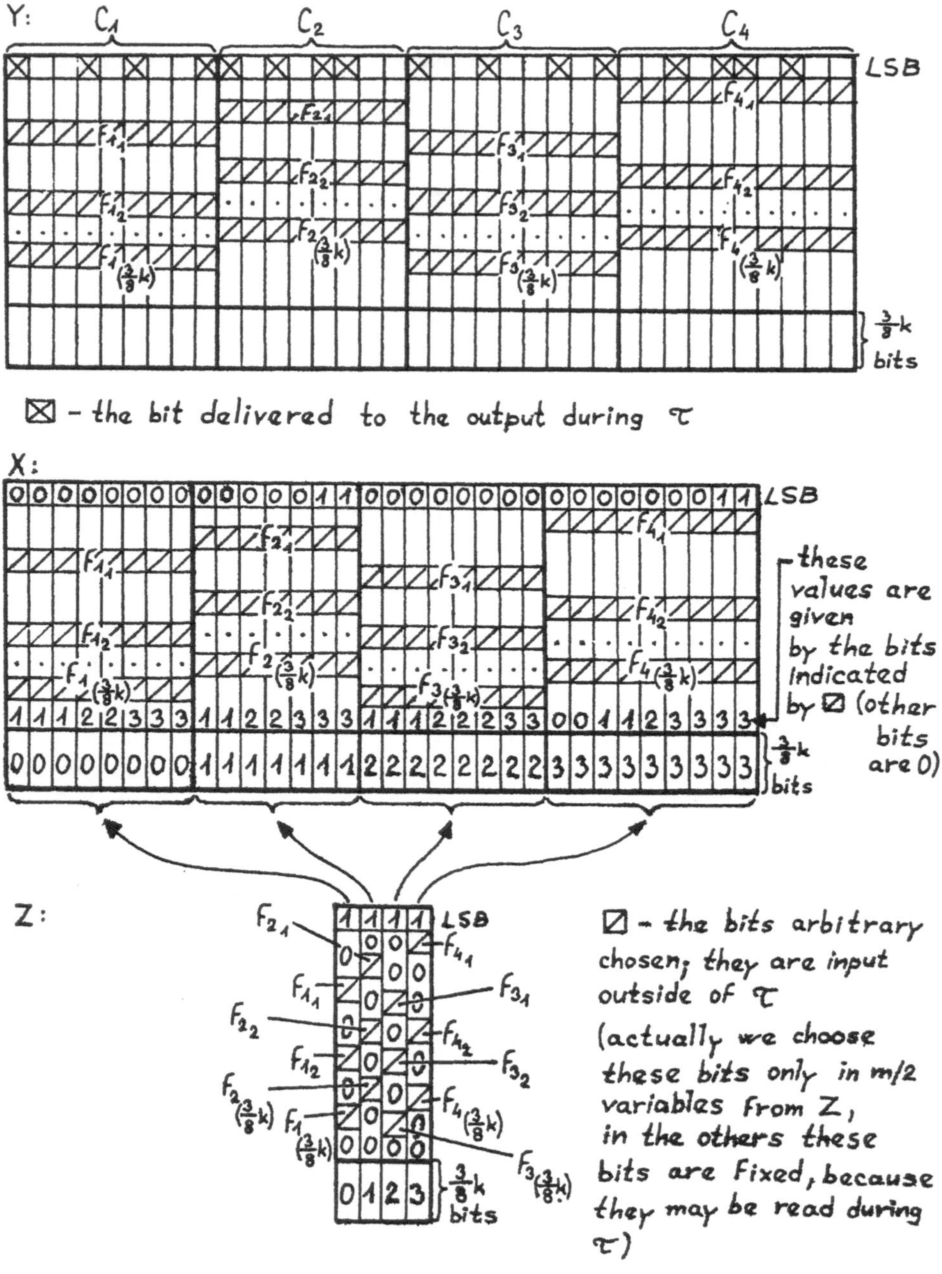

⊠ – the bit delivered to the output during τ

X:

LSB

these values are given by the bits indicated by ⧄ (other bits are 0)

$\frac{3}{8}k$ bits

Z:

LSB

$\frac{3}{8}k$ bits

⧄ – the bits arbitrary chosen; they are input outside of τ (actually we choose these bits only in $m/2$ variables from Z, in the others these bits are fixed, because they may be read during τ)

Fig. 1. The assignment to the inputs from X and Z for $m = 4$, $\frac{n}{r} = 16$, $n = 32$.

Let us define:

$$dif_h^j = i_{(h-1)\frac{n+m}{8rm}+j} - \sum_{q=1}^{h-1}(|C_q|), \; dif_h^0 = 1.$$

Let the positions $f_{h_1}, \ldots, f_{h_{\frac{3}{8}k}}$ be corresponding to the positions $f_{i_1}, \ldots, f_{i_{\frac{3}{8}k}}$ for $i = h$ in the variable $z_h (h = 1, \ldots, m)$. Let us assign the values to the $\frac{3}{8}k$ bits on these positions so that:

$$\text{for } \; j = 1, \ldots, \left\lfloor \frac{n+m}{8rm} \right\rfloor \; \text{ it holds}$$

$$x_{i,f_{h_1}} 2^{\frac{3}{8}k-1} + x_{i,f_{h_2}} 2^{\frac{3}{8}k-2} + \cdots + x_{i,f_{h_{\frac{3}{8}k}}} 2^0 = j - 1$$

$$\text{where } i = (\sum_{q=1}^{h-1}(|C_q| - 1)) + dif_h^{j-1}, \ldots, (\sum_{q=1}^{h-1}(|C_q| - 1)) + dif_h^j - 1$$

$$\text{and it holds } \; x_{i,f_{h_1}} 2^{\frac{3}{8}k-1} + x_{i,f_{h_2}} 2^{\frac{3}{8}k-2} + \cdots + x_{i,f_{h_{\frac{3}{8}k}}} 2^0 = \left\lfloor \frac{n+m}{8rm} \right\rfloor - 1$$

$$\text{where } i = (\sum_{q=1}^{h-1}(|C_q| - 1)) + dif_h^{\frac{n+m}{8rm}}, \ldots, \sum_{q=1}^{h}(|C_q| - 1).$$

The assignment to the least significant bits of the h-th part is the following

$$x_{j,k} = 0$$

$$\text{for } j = (\sum_{q=1}^{h-1}(|C_q| - 1)) + 1, \ldots, (\sum_{q=1}^{h-1}(|C_q| - 1)) + dif_h^{\lfloor \frac{n+m}{8rm} \rfloor} - 1$$

and

$$x_{j,k} = 1$$

$$\text{for } j = (\sum_{q=1}^{h-1}(|C_q| - 1)) + dif_h^{\lfloor \frac{n+m}{8rm} \rfloor}, \ldots, \sum_{q=1}^{h}(|C_q| - 1).$$

The other bits are set to 0. For $j = 1, \ldots, \frac{m}{2}$ we have: To variable z_{l_j} we assign a value $H'_{l_j} \in \{0, 1, \ldots, \lfloor \frac{n+m}{8rm} \rfloor - 1\}$. Then we get the following outputs. Let $L_j = (l_j - 1)(\lfloor \frac{n+m}{8rm} \rfloor)$ and $H'_{l_j} = u$. Then it holds:

$$y_{i_{L_j+1},k} = 0, \ldots, y_{i_{L_j+u},k} = 0, \; y_{i_{L_j+u+1},k} = 1, \; y_{i_{L_j+u+2},k} = 0, \ldots, y_{i_{L_j+\lfloor \frac{n+m}{8rm} \rfloor},k} = 0$$

There are $\lfloor \frac{n+m}{8rm} \rfloor$ different assignments of a value H'_{l_j} for $j \in \{1, \ldots, \frac{m}{2}\}$. It means there are $\lfloor \frac{n+m}{8rm} \rfloor^{\frac{m}{2}}$ various assignments for $H'_{l_1}, H'_{l_2}, \ldots, H'_{l_{\frac{m}{2}}}$ in summary and this implies the same number of different outputs. So, the circuit should differentiate among at least $\lfloor \frac{n+m}{8rm} \rfloor^{\frac{m}{2}}$ various states. Therefore $A(n, m, k, r) \geq \frac{m}{2} \log \lfloor \frac{n+m}{8rm} \rfloor$.

According to Remark 1 we can write

$$A(n, m, k, r) = \Omega \left(m \left(\left(\log \frac{n}{rm} \right) + 1 \right) \right). \; \blacksquare$$

REMARK 2. *One can prove in a similar way that* $A(n,m,k,r) = \Omega\left(\log \frac{n}{r}\right)$ *also for* $m < 8$.

THEOREM 3. *If* $\frac{n}{r} \geq 2^{\frac{3}{8}k} \geq m$, *then* $A(n,m,k,r) = \Omega(m((log\frac{2^{\frac{3}{8}k}}{m}) + 1))$.

PROOF OF THEOREM 3: The proof is analogous to the proof of Theorem 2. The difference is in consideration of only $2^{\frac{3}{8}k}$ (from $\lfloor \frac{n+m}{8r} \rfloor$ bits) least significant bits which are τ-outputs.

The groups C_i for $i = 1, ..., m$ (see the proof of Theorem 2) of variables $y_1, ..., y_{n+m}$ are such that each of them contains $\frac{2^{\frac{3}{8}k}}{m}$ variables (from the considered ones) that their least significant bits are τ-outputs. Therefore:

$$A \geq \log\left(\frac{2^{\frac{3}{8}k}}{m}\right)^{\frac{m}{2}} \Longrightarrow A = \Omega\left(m \log \frac{2^{\frac{3}{8}k}}{m}\right).$$

Similarly as in Theorem 2, we have:

$$A = \Omega\left(m\left(\left(\log \frac{2^{\frac{3}{8}k}}{m}\right) + 1\right)\right). \quad \blacksquare$$

COROLLARY 1. *If* $\frac{n}{r} \geq 2^{\frac{3}{8}k}$ *and* $\frac{3(8^K)-1}{8^{K+1}-1}k \geq \log m$ *where* $K \geq 0$ *then :*

$$A = \Omega\left(m\left(\left(\log \frac{2^k}{m}\right) + 1\right)\right).$$

PROOF OF COROLLARY 1: If $\frac{3(8^K)-1}{8^{K+1}-1}k \geq \log m$ then there exists $c > 0$ such that $cm \log \frac{2^k}{m} \leq m \log \frac{2^{\frac{3}{8}k}}{m}$. Let us take $c = \frac{1}{8^{K+1}}$. The inequality $\frac{3(8^K)-1}{8^{K+1}-1}k \geq \log m$ holds if and only if

$$\frac{1}{8^{K+1}} m \log \frac{2^k}{m} \leq m \log \frac{2^{\frac{3}{8}k}}{m}. \quad \blacksquare$$

CONCLUSION

Figure 2. contains the survey of the results of the paper and shows the open questions about the multilective merging.

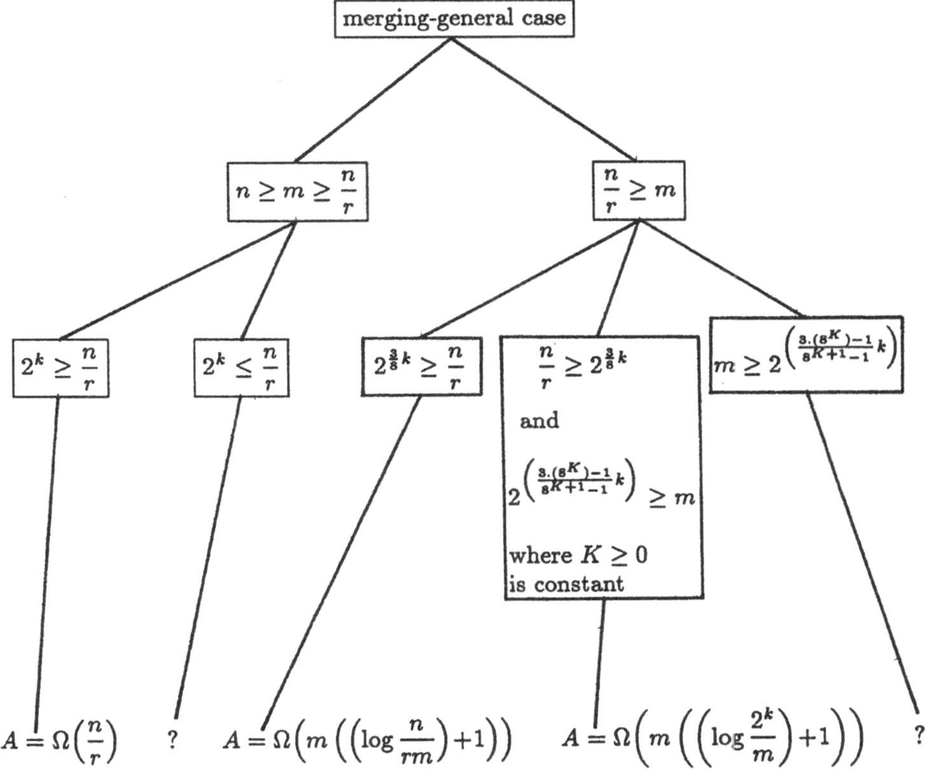

Fig. 2.

Our conjecture is that if $n \geq m \geq \frac{n}{r} \geq 2^k$ then $A = \Omega(2^k((\log \frac{\frac{n}{r}}{2^k}) + 1))$ and if $\frac{n}{r} \geq m \geq 2^k$ then $A = \Omega(2^k((\log \frac{m}{2^k}) + 1))$.

We hope we could prove all our lower bounds tight by using similar techniques as they are used for semelective merging **[PSV]**.

Acknowledgment

The second author thanks Professor Kurt Mehlhorn, Max–Planck–Institut für Informatik, Saarbrücken and Alexander von Humboldt Foundation, Bonn, Germany who partly supported this research.

REFERENCES

[U] Ullman,J.D., *Computational Aspects of VLSI*, Computer Science Press, Rockville, Md. 1983.

[Sa] Savage,J.E., *The Performance of Multilective VLSI Algorithms*, in "Journal of Computer and System Sciences Vol. 29, No. 2, October 1984," Academic Press, New York and London.

[Si] Siegel,A., *Tight Area Bounds and Provably Good AT^2 Bounds for Sorting Circuits.TR*, CS Dept.,New York University, New York 1984.

[G] Gubáš,X., *Close properties of the communication and the area complexity of VLSI circuits (in Slovak), Master thesis*, Comenius University, Bratislava 1988.

[PSV] Palko,V., Sýkora,O., Vrťo,I., *Area complexity of merging*, In: MFCS' 89, Springer Verlag 1989, 390–396.

Deriving Fully Efficient Systolic Arrays by Quasi-Linear Allocation Functions*

Xiaoxiong Zhong and Sanjay Rajopadhye
[xzhong, sanjay]@cs.uoregon.edu
Computer Science Department
University of Oregon
Eugene, Oregon 97403-1202

Abstract

We address the problem of deriving systolic arrays in which the processor utilization is 100%. We first address this problem in the context of *synthesis* from Uniform Recurrence Equations (UREs), and then generalize our result to deal with *arbitrary* systolic arrays (outside the context of synthesis). We show that in a systolic array, it is always possible to merge a parallelepiped of neighboring processors which are active at different clock cycles. The new array is fully efficient and its processors have almost the same cost as the original one. Such merging corresponds exactly to the transformation by a quasi-linear function. When the original array is derived by integral linear projections of systems of UREs, we give a method to mechanically determine the quasi-linear allocation function which yields the efficient array. The technique can also be extended to any (piece-wise) systolic array to derive a fully efficient array by "post-processing" it with a (piece-wise) quasi-linear function.

1 Introduction

In the past few years, a relatively mature synthesis technique for systolic arrays has emerged [Rao85, Qui87, Mol83]. The specification is a system of Uniform Recurrence Equations (UREs) [KMW67] (or Affine Recurrence Equations, AREs*). The standard technique is to transform these UREs into a systolic algorithm which has a direct correspondence to a systolic array implementation. Typically, one uses an integral linear transformation [CS84, Mol83, Qui87], and can be described as follows: **specify** the algorithm as a system of UREs; determine a **linear schedule** (represented by its *norm*

*Supported by NSF grant MIP-8802454

*It has been shown [RF90] that even if one starts from AREs, it is essential to localize the data dependencies and convert the ARE into a URE. Hence we assume that the specification has been processed in this manner.

λ); determine an **allocation function** (represented by an integral projection vector u) which does not conflict with the timing function (i.e. satisfies the constraint $\lambda u \neq 0$).

To evaluate the derived arrays, there are three common criteria—the computation time, the number of processors and the efficiency. The efficiency of the array derived from the above technique is $\eta = 1/\delta$ where $\delta = |\lambda^t u|$. The problem of finding optimal timing functions, under some standard assumptions, can be reduced to a linear programming problem [LW85, Qui87, Mol83]. Also, the problem of finding linear allocation functions which can yield minimum number of processors has been studied [WD89, ZWR90]. The third criterion, namely the **efficiency** of a systolic array, however, has not received much attention.

In this paper, we address the problem of deriving efficient systolic arrays. We first show that for any systolic array which is derived by the conventional linear transformations, it is always possible to **merge** every δ neighboring processors which are active at different time units to derive a new array. These δ processors form a parallelepiped which can be constructed by a new basis of the processor space. The **efficiency** of the new array is 100%. It has the same computation time as the original one and has only $1/\delta$ number of processors. Furthermore, the cost of the processor in the new array remains the same, except for a few additional registers and wires (no additional functional units are required). This method is equivalent to using allocation functions that are not integral but rational matrices, and then obtaining integral processor labels by using the floor function. Such functions are called *quasi-affine* functions by Quinton [Qui87].

Based on the standard view of the systolic array, we also show that the above mathematical technique can be applied to **any** systolic array to derive a fully efficient systolic array. This can also be easily generalized to transform piece-wise systolic arrays [Thi89] to fully efficient ones, by applying a *piece-wise* quasi-affine function.

The rest of this paper is arranged as follows. After a preliminary discussion of the standard synthesis methods and its constraints we prove a standard result about the efficiency of synthesized systolic arrays (Sec. 3). In Sec. 4, we will show that one can always "postprocess" the array derived by the conventional linear transformation to yield a new array which has less processors and is fully efficient. We will also show this corresponds exactly to an array derived by a quasi-linear allocation function applied to the original UREs. We extend these results to arbitrary systolic arrays in Sec 5, and conclude with a discussion and comparison.

2 Preliminaries

A systolic array consists of identical processors which are connected locally. Processors process data from input channels and send out the output through output channels to other processors at every clock cycle. To synchronize the data flow, it is possible that a processor has to be idle during some clock cycles. This leads to the concept of **extrinsic**

iteration interval [Rao85] which is defined as follows:

Definition 1 The **extrinsic iteration interval** of a systolic array is defined to be δ if every processor is active at exactly one out of every δ consecutive clock cycles. The **efficiency** η of the array defined as $\eta = 1/\delta$.

For example, in the Kung-Leiserson systolic array for banded matrix multiplication [KL80] (see Figure 1), the processors are active once every 3 clock cycles. Therefore, $\delta = 3$ and $\eta = 1/3$.

Definition 2 A $m \times n$ matrix U $(m \leq n)$ is said to be **e-unimodular** (for extended unimodular) if the gcd of the determinants of all its $m \times m$ submatrices is 1.

It is well known [Sch88] (pp. 47, Cor 4.1c), that a system of diophantine equations $Ux = I$ has an integral solution for *any* integral vector I, iff U is e-unimodular. One of the useful properties of e-unimodular matrices is that the column Hermite form of an e-unimodular $m \times n$ matrix, A, is* $[E_m\ 0]$. We say that a vector is a normalized vector if it is e-unimodular (i.e. the gcd of all its components is 1).

Throughout this paper, we will assume that the UREs are defined on an n-dimension integral domain $\mathcal{D}$. The standard linear transformation technique to synthesize systolic arrays from UREs involves in the two linear transformations, namely, the timing function (represented by its norm λ) and the allocation function (represented by a normalized projection vector u or an integral $(n-1) \times n$ matrix A (satisfying $Au = 0$, and u is the basis of the right null space of A). The following constraints must be satisfied.

- **Causality of the timing function:** If, evaluating $f(I+d)$ needs value $f(I)$, then $\lambda^t d > 0$.
- **Non-conflict:** The allocation function does not conflict with the timing function, i.e. no two points are mapped to the same processor at the same time. This is equivalent to $\lambda^t u \neq 0$.
- **Dense array** The derived array must be dense i.e. every integral point in the processor space must be the image of an integral point in the index space of the problem. This is equivalent to the constraint that A should be e-unimodular.

Besides the above constraints, we can make the following assumptions without any loss of generality.

- λ is a normalized vector. This is because if λ is not a normalized vector, it can be written as $s\lambda'$ for some positive integer, s and normalized vector, λ'. It is easy to verify that λ' still represents a valid timing function (see [RF90]) and it yields a faster schedule than λ.

---

*E_i denotes the $i \times i$ identity matrix.

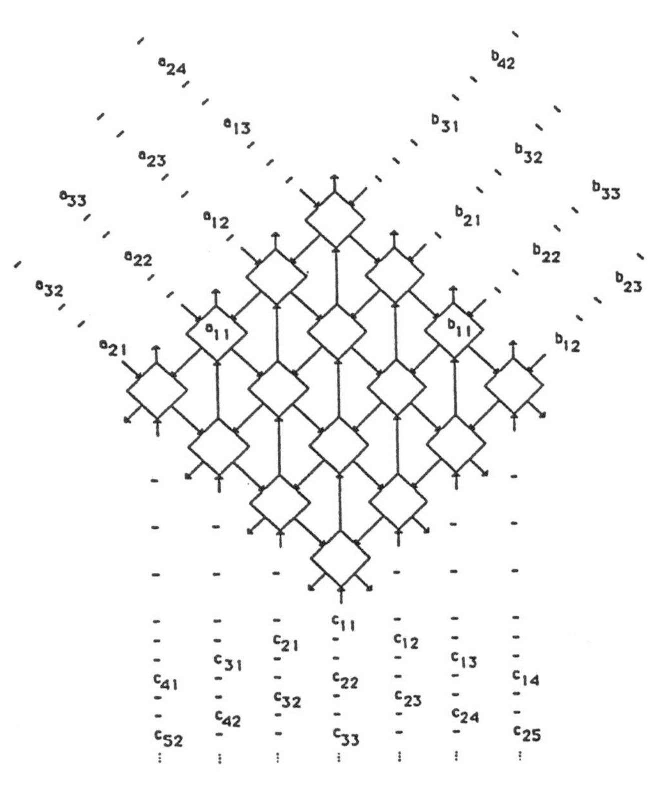

Figure 1: Kung-Leiserson Band Matrix Multiplication

- The dependency graph of the computation is connected. (Otherwise, the computation consists of more than one totally independent computation, and we can rewrite them as separate UREs). This constraint is satisfied iff the dependency matrix $(D_1 \ldots D_k)$ where $D_1, \ldots, D_k$ are all dependency vectors is e-unimodular (see [ZWR90]).

Geometrically, the timing and allocation functions can be unified as a single transformation from the computation domain $\mathcal{D}$ to the processor-time domain $\mathcal{T}$. This can be described by an $n \times n$ matrix $T = \begin{pmatrix} A \\ \lambda^t \end{pmatrix}$. We denote the processor space (i.e. the first $n-1$ dimensions of $\mathcal{T}$) as $\mathcal{P}$. Furthermore, we say a processor-time point $(P\,t)^t \in \mathcal{T}$ is *active* if it is the image of a computation point, i.e. there exists an $I \in \mathcal{D}$ such that $TI = (P\,t)^t$. Otherwise, $(P\,t)^t$ is said *inactive* (or "hole"). Intuitively, a processor-time point $(P\,t)^t$ is active iff the processor in the location P is active at time t.

3 Efficiency of Synthesized Arrays

We will now prove that the extrinsic iteration interval of the systolic array is $|\lambda^t u|$. Although this is well known, it is necessary to prove this in our formal setting. Furthermore, our proof establishes certain lemmata that are useful later.

The active points in a derived array can be characterized by certain properties of T. Let $C = (C_1 C_2 \ldots C_n)$ be an $n \times n$ unimodular matrix such that $AC = (E_{n-1}\,0)$ (since A is e-unimodular, its column Hermite form is $(E_{n-1}\,0)$). Let $v^t = \lambda^t(C_1 \ldots C_{n-1})$ and $k = \lambda^t C_n$ (thus, $\lambda^t C = (v^t\,k)$). Note that since $AC = (E_{n-1}\,0)$, we have $AC_n = 0$, i.e. C_n is a right null vector of A. Moreover, since C is unimodular, C_n must be normalized. Therefore, $C_n = \pm u$ and $k = \lambda^t C_n = \pm \lambda^t u$. The following lemma gives an important characterization of active points.

Lemma 1 A processor-time point $(P\,t)^t \in \mathcal{T}$ is active iff the following equation has an integral solution j.

$$v^t P + jk = t \tag{1}$$

Proof: Consider transformation matrix T. We have

$$T = \begin{pmatrix} A \\ \lambda^t \end{pmatrix} = \begin{pmatrix} (E_{n-1}0)C^{-1} \\ \lambda^t \end{pmatrix} = \begin{pmatrix} E_{n-1}\,0 \\ \lambda^t C \end{pmatrix} C^{-1}$$

So

$$T = \begin{pmatrix} E_{n-1} & 0 \\ v^t & k \end{pmatrix} C^{-1} \tag{2}$$

Now, a processor-time point $(P\,t)^t \in \mathcal{T}$ is active iff there exists $I \in \mathcal{Z}^n$ such that

$$\begin{pmatrix} P \\ t \end{pmatrix} = TI = \begin{pmatrix} E_{n-1} & 0 \\ v^t & k \end{pmatrix} C^{-1} I$$

Let $J = C^{-1}I$. This defines a bijection $\mathcal{Z}^n \to \mathcal{Z}^n$ because C (and hence, C^{-1}) is unimodular. Hence the above system of equations has an integral solution I iff the following system has an integral solution J.

$$\begin{pmatrix} E_{n-1} & 0 \\ v^t & k \end{pmatrix} J = \begin{pmatrix} P \\ t \end{pmatrix}$$

Letting $J = (j_1 \ldots j_{n-1}\, j)^t$, we see that the solution for $(j_1 \ldots j_{n-1})^t$ is simply P. Substituting this into the last equation and simplifying, yields Eqn. 1. ∎

Theorem 1 The extrinsic iteration interval δ, of the systolic array derived from a URE is $|\lambda^t u|$.

Proof: We first prove that if an processor-time point $P_{t_0} = (P\,t_0)^t$ is active, then for any scalar integer α, the processor-time point $(P\; t_0 + \alpha k)^t$ is also active. Since $(P\,t_0)^t$ is active, there exists an integer j_0 such that

$$v^t P + j_0 k = t_0$$

Adding αk to both sides,

$$v^t P + (j_0 + \alpha)k = t_0 + \alpha k$$

Hence, processor P is active at $t_0 + \alpha k$, and the extrinsic iteration interval, δ, must be a factor of k, i.e. $\delta | k$. To show that $|k|$ is exactly δ, we now prove that if processor P is active at t_1 and t_2, then $(t_1 - t_2)$ must be a multiple of k. Indeed, based on Lemma 1, there exist integers j_1 and j_2 such that

$$v^t P + j_1 k = t_1$$

and

$$v^t P + j_2 k = t_2$$

Hence,

$$(j_1 - j_2)k = t_1 - t_2$$

and therefore, the extrinsic iteration interval, δ, is $|k| = |\lambda^t u|$. ∎

Corollary 1 From Eqn 2, we further have $\det T = \pm\delta$.

The "holes" in the processor-time space occur when T is not bijective. This is true only when T is not a unimodular matrix. The above corollary conforms to this. Theoretically, it is always possible to select an integral vector u such that $\lambda^t u = \pm 1$. This is due to the following lemma.

Lemma 2 For any normalized n-dimensional vector w^t, it is always possible to choose an $(n-1) \times n$ integral matrix M such that the matrix $(w\ M^t)^t$ is unimodular.

Proof: Since $w^t = (w_1 \dots w_n)$ is a normalized vector, there always exists an integral vector u such that $w^t u = \gcd(w_1 \dots w_n) = 1$ and correspondingly we can choose an e-unimodular matrix M such that $Mu = 0$. It is easy to see that Eqn. 2 holds for M too. Therefore, $|M| = \pm w^t u = \pm$, and hence M is unimodular. ∎

In practice, however, there may be other factors which prohibit the choice of such allocation functions. The first factor is when the computation domain $\mathcal{D}$ is infinite. In this situation, there is only one projection vector (the extremal ray of $\mathcal{D}$) which yields a finite array, as shown by Quinton [Qui87] and this projection vector may not yield fully efficient array (i.e. $\delta = 1$). Consider banded matrix multiplication (for possibly infinitely large matrices), the computation domain is infinite and the only valid array derived by the above transformation is Kung-Leiserson array whose efficiency is only 1/3 (Figure 1).

The second factor is related to the minimization of the number of the processors. The projection vectors which yield more efficient arrays may not yield the arrays with fewer processors. For example, if we consider banded (but bounded) matrix multiplication, one of the projection vector yields a fully efficient array with n^2 processors while the Kung-Leiserson array uses only $w_1 w_2$ (w_1, w_2 are the bandwidths of the matrices) processors. Even though this array is not fully efficient, it may still be preferable to the other array.

Therefore, it is inherently restrictive to derive a fully efficient array by merely using integral linear transformations. An obvious extension to this is to permit the allocation function to be *rational*. But to guarantee that the transformation yields a regular array, we have to change the locations (labels) of the processors denoted by rationals to integers. One of the possibilities is to use the floor function. Such functions are called *quasi-linear* functions by Quinton [Qui87] and are used as timing functions in the same paper.

4 Synthesizing Fully Efficient Systolic Arrays

Our main idea in "postprocessing" the systolic array obtained by the integral linear transformation is to try to merge as many neighboring processors as possible. In order not to increase the functional units in the resultant processor, only those neighboring processors which are active at different clock cycles will be merged. Furthermore, while the shape of the clusters to be merged may be arbitrary, we are particularly interested in merging processors that form parallelepipeds in the processor space.

We therefore need $(n-1)$ directions $\nu_1, \ldots, \nu_{n-1}$ along which the parallelepipeds are formed. It is clear that we are only interested in those parallelepipeds of volume δ (i.e. parallelepipeds with δ integral points in them). To be more precise, we factorize δ such that $\delta = \delta_1 \ldots \delta_{n-1}$ and these factors together with the basis vectors define a family of parallelepipeds.

Definition 3 A *snapshot* at time t_0 is the space $\{(P\, t_0)^t \mid P \in \mathcal{P}\}$.

Definition 4 A factorization $\delta = \delta_1 \delta_2 \ldots \delta_{n-1}$ of δ, a matrix N of $(n-1)$ basis vectors $(N = (\nu_1, \ldots, \nu_{n-1})$, for the processor space $\mathcal{P}$, and a processor-time point $(P\, t)^t$ define a *parallelepiped*, denoted by $\Pr(N, P, t)$ of volume δ as follows.

$$\Pr(N, P, t) = \{(P + N(l_1 \ldots l_{n-1})^t \;\, t)^t \mid 0 \le l_i < \delta_i, i = 1, .., n-1\}$$

It should be noted that the requirement that $\nu_1, \nu_2, \ldots, \nu_{n-1}$ is a basis for the processor space $\mathcal{P}$ is necessary. Otherwise, there may be some integral points in the parallelepiped $\Pr(N, P, t)$ but they can not be written as $l_1\nu_1 + \ldots + l_{n-1}\nu_{n-1}$ for some $0 \le l_i < \delta_i$ (i.e. the number of integral points in the parallelepiped may be greater than δ). Our main result is that such clustering into parallelepipeds is always possible, provided that the factors of δ are mutually co-prime.

Let $k_i = \frac{\delta}{\delta_i}$. We will require our basis vectors to satisfy an additional constraint that* $v^t \nu_i = g c_i k_i$ where $c_1, \ldots, c_{n-1}$ are integers satisfying $\gcd(c_i, \delta) = 1$ for $i = 1, \ldots, (n-1)$ and $\gcd(c_1, \ldots, c_{n-1}) = 1$. This constraint will be used later. Thus each ν_i must be a solution of the i-th equation in the following system of diophantine equations.

$$\begin{aligned} v^t x_1 &= g c_1 k_1 \\ v^t x_2 &= g c_2 k_2 \\ &\vdots \\ v^t x_{n-1} &= g c_{n-1} k_{n-1} \end{aligned}$$

As shown by Banerjee ([Ban88], Th. 5.4.2, p.81), the general solution to the i-th equation is given by

$$x_i = U \begin{pmatrix} c_i k_i \\ t_1^i \\ \vdots \\ t_{n-2}^i \end{pmatrix}$$

where t_j^i for $j = 1, \ldots, n-2$ are arbitrary integers and U is a unimodular matrix satisfying $Uv = (g, 0, \ldots, 0)^t$. Such a U can always be found, and thus, our ν_i's satisfy the desired

*Recall that $\lambda^t C = (v^t\; k)$; we let g be the gcd of all components of v.

constraint iff the matrix

$$V = \begin{pmatrix} c_1k_1 & c_2k_2 & \dots & c_{n-1}k_{n-1} \\ t_1^1 & t_1^2 & \dots & t_1^{n-1} \\ \vdots & \vdots & \vdots & \vdots \\ t_{n-2}^1 & t_{n-2}^2 & \dots & t_{n-2}^{n-1} \end{pmatrix}$$

is unimodular because $(\nu_1 \dots \nu_{n-1}) = UV$. The following lemma gives us necessary and sufficient conditions for this.

Lemma 3 V is unimodular iff $\gcd(\delta_i, \delta_j) = 1$ for all $i \neq j$.

Proof:

If part. Let, $\gcd(\delta_i, \delta_j) = 1$, for all $i \neq j$. First, we prove that $(k_1 \dots k_{n-1})$ is a normalized vector. Suppose p is a common prime divisor for $k_1 \dots k_{n-1}$. Obviously, $p|k_1$. There exists $i \neq 1$ such that $p|\delta_i$, since $k_i = \delta_1 \dots \delta_{i-1}\delta_{i+1} \dots \delta_{n-1}$. Similarly, since $p|k_i$, there exists $j \neq i$ such that $p|\delta_j$. Because $\gcd(\delta_i, \delta_j) = 1$, p must be 1. Therefore, $(k_1, \dots, k_{n-1})$ is normalized. We now prove that $(c_1k_1 \dots c_{n-1}k_{k-1})$ is also normalized. Again, suppose p is a common prime divisor for all c_ik_i, then p either divides c_i or k_i. Since $\gcd(c_1, \dots, c_{n-1}) = 1$ and $\gcd(k_1, \dots, k_{n-1}) = 1$, there exist i, j $(i \neq j)$ such that $p|c_i, p|k_j$. But from $\gcd(c_i, \delta) = 1$, we know $\gcd(c_i, k_j) = 1$. Hence, $p = 1$. By Lemma 2, we know it is always possible to choose t_i^j's such that V is unimodular.

Only If Part. Without loss of generality, suppose $\gcd(\delta_1, \delta_2) = d \neq 1$. $d|k_1$ because $\delta_2|k_1$. Also, $d|k_i$ for $i \neq 1$ because $\delta_1|k_i$ for all $i \neq 1$. Therefore, we know $d|c_ik_i$ for all $i = 1, \dots, n-1$. Hence, V is not unimodular. ∎

We now prove that any snapshot of any parallelepiped formed by the basis vectors as chosen above contains exactly one active point.

Lemma 4 For any time instant t, there is exactly one active point in any parallelepiped $\Pr(N, P, t)$.

Proof: First, we prove that there is at most one active point in $\Pr(N, P, t)$. Suppose there are two active points, say, $(P_1\ t)^t$ and $(P_2\ t)^t$ in this parallelepiped. It is easy to see that the point $(P_1\ t)^t - (P_2\ t)^t$ is also active. This means that there are integers $l_1, l_2, \dots, l_{n-1}$ where $0 \leq l_i < \delta_i$ for $i = 1, \dots, n-1$ such that $\begin{pmatrix} l_1\nu_1 + \dots + l_{n-1}\nu_{n-1} \\ 0 \end{pmatrix}$ is active. From Lemma 1, we know that there is an integral solution to the Eqn. 1, i.e. there is an integer J_n such that

$$l_1 v^t \nu_1 + \dots + l_{n-1} v^t \nu_{n-1} + J_n k = 0$$

This can be further simplified to

$$g(l_1c_1k_1 + \dots + l_{n-1}c_{n-1}k_{n-1}) + J_n k = 0$$

Notice that for any $i \neq j$, $\delta_i | k_j$. Therefore, dividing both sides of the above equation by δ_i, we have $\delta_i | g l_i c_i k_i$. Because $\gcd(g, \delta) = 1$, hence $\gcd(g, \delta_i) = 1$. Also, $\gcd(c_i, \delta) = 1$, and hence $\gcd(c_i, \delta_i) = 1$. Furthermore, $\gcd(\delta_i, \delta_j) = 1$ (for $i \neq j$), and hence $\gcd(\delta_i, k_i) = 1$. So $\delta_i | l_i$. Because $0 \leq l_i < \delta_i$, l_i must be 0, i.e. P_t^1 is the same as P_t^2.

To prove that there is at least one active point in the parallelepiped, again, consider Eqn. 1 which can be simplified to

$$g(l_1 c_1 k_1 + \ldots + l_{n-1} c_{n-1} k_{n-1}) + J_n k = t - v^t P \tag{3}$$

We want to prove that there are integers $l_1, \ldots, l_{n-1}$ for $0 \leq l_i < \delta_i$ and some integer J_n as the solution to the above equation. To prove this, we first prove $\gcd(g c_1 k_1, \ldots, g c_{n-1} k_{n-1}, k) = 1$. Consider any prime common divisor p of these integers, we prove $p = 1$. Because $\gcd(g, k) = 1$, if $p | g$, then we already prove that $p = 1$ (because $p | k$ too). If p is not a factor of g, we have $p | c_i k_i$ for every $i = 1, \ldots, n-1$. Because $\gcd(c_1 k_1, \ldots, c_{n-1} k_{n-1}) = 1$, p must be 1.

Therefore, there exist integers $l'_1, \ldots, l'_{n-1}, J'_n$ such that

$$l'_1 g c_1 k_1 + \ldots + l'_{n-1} g c_{n-1} k_{n-1} + J'_n k = 1$$

Denoting $t - v^t P$ as t' and multiplying both sides of the equation by t', we have

$$t' l'_1 g c_1 k_1 + \ldots + t' l'_{n-1} g c_{n-1} k_{n-1} + J'_n t' k = t'$$

For $t' l'_i$, we can always find two integers q_i and l_i where $0 \leq l_i, \delta_i$ such that $t' l'_i = q_i \delta_i + l_i$ (i.e. l_i is the remainder of dividing $t' l_i$ by δ_i). Notice that $\delta_i k_i = \delta = \pm k$. The left hand side of the above equation becomes

$$(\sum_{i=1}^{n-1} l_i g c_i k_i) + J'_n t' k + q_1 g c_1 \delta_1 k_1 + \ldots + q_{n-1} g c_{n-1} \delta_{n-1} k_{n-1}$$

$$= (\sum_{i=1}^{n-1} l_i g c_i k_i) + (J'_n t' \pm (q_1 g c_1 + \ldots + q_{n-1} g c_{n-1}))k$$

Let $J_n = (J'_n t' \pm (q_1 g c_1 + \ldots + q_{n-1} g c_{n-1}))$, l_i and J_n together satisfy Eqn. 3. ∎

Therefore, We can merge all the δ processors in such parallelepipeds into one processor. The new processor does not need to have extra processing function units (but may need some additional links and registers) and it will be active in the whole computation. This leads to the following theorem.

Theorem 2 It is always possible to merge δ neighboring processors which form a parallelepiped in the processor space to derive a fully efficient array. The new array has the same computation time as the original one and has the same cost except for some additional links and registers.

It should be clear that there is much freedom to choose the basis vectors $\nu_1, \ldots, \nu_{n-1}$ to form the parallelepiped. In practice, the selection of the basis vectors should try to minimize the number of the extra links, and the number of specialized boundary processors required.

4.1 Quasi-Linear Allocation Functions

To place the resultant array onto the integral grid, we need an explicit allocation function which assigns computations to processors denoted by integral points. To find such a function for the resultant array, we observe that the procedure we described so far can be viewed as the composition of three parts: the original linear allocation function (represented by an $(n-1) \times n$ e-unimodular matrix A); a basis transformation function which transforms the derived processor space $\mathcal{P}$ into another space with $\nu_1, \ldots, \nu_{n-1}$ as the bases; and clustering of processors into rectangular parallelepipeds. This clustering is obtained by dividing the processor coordinate along the basis ν_i by δ_i and taking the floor function $\lfloor\ \rfloor$. Note that we have assumed that processor 0 is always in $\mathcal{P}$. If this were not so, we can simply add a constant to the allocation function derived here and we get a quasi-affine rather than a quasi-linear function.

Because the matrix $V = (\nu_1\ \ldots\ \nu_{n-1})$ is the basis transformation matrix from the new basis vectors to the old basis vectors, V^{-1} is the basis transformation from the old basis vectors to the new basis vectors. Therefore, we define the allocation function as follows:

For any $I \in \mathcal{D}$, $Q(I) = \lfloor DV^{-1}A(I) \rfloor$ where* D is an $(n-1) \times (n-1)$ diagonal matrix which is defined as $D = diag(1/\delta_1, \ldots, 1/\delta_{n-1})$. The following theorem states that the allocation function Q yields a dense array.

Theorem 3 The allocation function $Q = \lfloor DV^{-1}A \rfloor$ yields a dense array.

Proof: We need to show that Q is a surjective function from $\mathcal{Z}^n$ to $\mathcal{Z}^{n-1}$, i.e. for any $P = (p_1\, p_2\ \ldots\ p_{n-1})^t \in \mathcal{Z}^{n-1}$, there is at least an $I \in \mathcal{Z}^n$ such that $P = QI$. Because $DP_1 = P$ where $P_1 = (\delta_1 p_1\ \ \delta_2 p_2\ \ \ldots\ \ \delta_{n-1} p_{n-1})^t$ (actually, there are δ possible P_1s which are mapped to P under D, but for the purpose of our proof, we just choose one of them). Furthermore, there is $P_2 = VP_1 \in \mathcal{Z}^{n-1}$ (because V^{-1} is unimodular) such that $DV^{-1}P_2 = P$. Now, since A is e-unimodular and therefore, there always exists integral solution for the equation $P_2 = AI$ (pp. 47, Cor 4.1c, [Sch88]), therefore, Q is surjective. ∎

So far, we have shown that a particular class of quasi-linear functions (those that can be decomposed into the three factors above) can be used as allocation functions to

*$\lfloor M \rfloor$ for a matrix M is defined as taking floor for every entry in M

obtain efficient arrays. We hypothesize that one could use general quasi-linear functions as allocation functions. For this to be valid, we need to show that arrays obtained by such functions are still regular. An informal argument supporting this is as follows. First, we note that any $(n-1) \times n$ rational matrix Q can be written as $\frac{1}{d}P$ where d is an integer and P is an $(n-1)\times n$ integral matrix. Hence, if I depends on $I+D$ in the computation, then processor $\lfloor QI \rfloor$ should have a link from processor $\lfloor QI+QD \rfloor = \lfloor \frac{1}{d}(PI)+\frac{1}{d}(PD) \rfloor$. Let $QI = \lfloor QI \rfloor + f(I)$ where $f(I)$ is the fractional part of QI. We conjecture that $f(I)$ is periodic and bounded by d, and therefore, $\lfloor QI+QD \rfloor = \lfloor QI \rfloor + \lfloor f(I)+QD \rfloor$. Hence, processor QI has at most d different links (corresponding to dependency D) and this is true for any I. Therefore, the resultant array is regular. Furthermore, we also expect that one can always factorize any valid quasi-linear allocation function into the three parts as above.

5 Optimal Clustering of Arbitrary Systolic Arrays

So far, we have addressed the problem of deriving efficient systolic arrays in the context of synthesis. There are, however, many systolic arrays which are not derived from UREs by the conventional linear transformation. To transform such arrays into efficient ones, we study how to apply our theory to an arbitrary systolic array. First, let us recall the standard points of view of a systolic array.

By Rao and Kailath [RK86], a systolic array implements an RIA . This RIA is defined in a processor-time space. More precisely, suppose the processor space is defined in $\mathcal{P} \subset \mathcal{Z}^{n-1}$, then the RIA is defined in $\mathcal{Z}^n$ as follows:

If there is a link with a delay D_t ($D_t \geq 1$) from processor p_1 to processor $p_1 + D$ in the processor space, then for any time t, processor-time point $(p_1 + D\,t + D^t)^t$ depends on $(p_1\;t)^t$. Hence, there is a uniform dependency $\begin{pmatrix} D \\ D_t \end{pmatrix}$.

Generally, if $D_1, \ldots, D_k$ are all the link vectors in the processor space (we assume $k \geq n$, otherwise, the RIA can be transformed into a lower dimension space) and $D_1^t, \ldots, D_k^t$ are the time delayed along the i-th link respectively, then the RIA implemented by the array is defined by the matrix D formed by the dependency vectors as $\begin{pmatrix} D_1 & \ldots & D_k \\ D_1^t & \ldots & D_k^t \end{pmatrix}$.

It might seem straightforward to use the above technique because the array is derived by a projection of the RIA along the time axis. But if the array is not 100% efficient, then the computation dag of this RIA consists of k **disconnected** components in $\mathcal{Z}^n$, which violates the assumption we made in Section 2. Hence our previous results cannot be applied directly.

Let us examine processor-time points in $\mathcal{Z}^n$. Some of these points correspond to useful computations (i.e. the processors are active) and others do not. Moreover, if the extrinsic iteration interval is δ, then along the time axis, there is an active processor-time point every δ points.

Basically, if we assume that the origin point (i.e. $0 \in \mathcal{Z}^n$) is active, then any active point will be connected to the origin in the dag of the RIA. Thus, a processor-time point $(Pt)^t$ is active iff it can be represented by a linear combination of the dependency vectors of the dag. This leads to the following proposition.

Remark 1 A processor-time point $(Pt)^t$ is active iff there exists a k-dimensional integral vector J such that $DJ = (P\,t)^t$.

Let $D = \begin{pmatrix} D_p \\ D_t \end{pmatrix}$ where $D_p = (D_1 \ldots D_k)$ and $D_t = (D_1^t \ldots D_k^t)$. D_p is the $n-1 \times k$ connection matrix for the processor space. To guarantee that the array is dense (i.e. every integral point in $\mathcal{P}$ is a valid processor), D_p must be e-unimodular (Lemma 4, in [ZWR90]). Moreover, it is reasonable to assume that D_t is a normalized vector because, otherwise, we can get a faster array by just simply replacing D_t with D_t' where $D_t = cD_t'$ and D_t' is normalized. The matrix D is thus analogous to the transformation matrix T that we have studied so far, except that it is not square.

Because D_p is e-unimodular, there exists a $k \times k$ unimodular matrix $U = (U_1 \ldots U_k)$ such that $D_pU = (E_{n-1}\, 0)$. Define $w^t = D_t(U_1 \ldots U_{n-1})$ and $l_i = D_tU_i$ for $i = n, \ldots, k$, thus $D_tU = (w^t\, l_n \ldots l_k)$. We therefore have the following analogue of Lemma 1 (the proof is also analogous, and omitted for brevity).

Lemma 5 A processor-time point $(Pt)^t \in \mathcal{T}$ is an active point iff the following equation has an integral solution $J_n, \ldots, J_k$.

$$w^tP + J_nl_n + \ldots + J_kl_k = t \tag{4}$$

The key difference between Eqn. 1 and Eqn. 4 is that in Eqn. 4, there are $l_n, \ldots, l_k$ instead of k in Eqn. 1. The following Lemma enables us to eliminate this difference too.

Lemma 6 Eqn. 4 has integral solution $J_n, \ldots, J_k$ iff the following equation has an integral solution J

$$w^tP + Jl = t \tag{5}$$

where $l = \gcd(l_n, \ldots, l_k)$

Proof: Let $(l_n, \ldots, l_k) = l(l_n', \ldots, l_k')$ and $\gcd(l_n', \ldots, l_k') = 1$. It is easy to see that if Eqn. 4 has integral solution $J_n, \ldots, J_k$, then $J = l_n'J_n + \ldots + l_k'J_k$ is an integral solution to Eqn. 5.

Conversely, suppose Eqn. 5 has an integral solution J. Because $\gcd(l_n', \ldots, l_k') = 1$, there are integers $m_n, \ldots, m_k$ such that $m_nl_n' + \ldots + m_kl_k' = 1$. Therefore, $Jm_nl_n' + \ldots + Jm_kl_k' = J$ and Eqn. 5 becomes

$$w^tP + l(Jm_nl_n' + \ldots + Jm_kl_k') = t \tag{6}$$

Therefore, $J_n = Jm_n, \ldots, J_k = Jm_k$ is an integral solution to Eqn. 4. ∎

We then have the following analogue (proof omitted for brevity) of Th. 1

Theorem 4 The extrinsic iteration interval δ of any systolic array is l.

We can use the technique of Th. 2 to merge δ neighboring processors within a parallelepiped and derive for any systolic array a 100% efficient array.

Matrix Multiplication Example

For the Kung-Leiserson matrix multiplication array, $\delta = 3$, which is a prime number. All possible parallelepipeds which we can merge have a 1×3 aspect ratio. Furthermore, we have the following:

$$D = \begin{pmatrix} D_p \\ D_t \end{pmatrix} = \begin{pmatrix} 0 & 1 & -1 \\ 1 & 0 & -1 \\ 1 & 1 & 1 \end{pmatrix} = \begin{pmatrix} 1 & 0 & 0 \\ 0 & 1 & 0 \\ 1 & 1 & 3 \end{pmatrix} \begin{pmatrix} 0 & 1 & -1 \\ 1 & 0 & -1 \\ 0 & 0 & 1 \end{pmatrix} \tag{7}$$

Hence, we have $w^t = (1\ 1)$, $g = 1$ and the following two equations.

$$\begin{aligned} x + y &= c_1 \\ x + y &= 3c_2 \end{aligned}$$

where $\gcd(c_1, c_2) = 1$ and $\gcd(c_1, 3) = \gcd(c_2, 3) = 1$.

There are many ways to pick the basis vectors. For example, if we choose $c_1 = c_2 = 1$, we have $\nu_1 = (1\ 0)^t$ and $\nu_2 = (4\ -1)$ as the solutions to the above two equations respectively. Also is $\nu_1 = (0\,1)$ and $\nu_2 = (-1\,4)$. That is, we can cluster the 3 neighboring processors vertically or horizontally. Further, if we choose $c_1 = 2$ and $c_2 = 1$, we get $\nu_1 = (1\ 1)$ and $\nu_2 = (1\ 2)$ as a solution. ν_1 stands for the main diagonal. It should be noted that $(1\ -1)$ is not a solution to the first equation, because if it were, c_1 must be 0, hence $\gcd(c_1, 3) = 3$ which does not satisfy the condition as stated above. In fact, in the array, all processors along the direction $(1\ -1)$ are active or inactive at the same time.

6 Discussion and Conclusions

Recently, there has been some work in the context of merging processors of systolic arrays derived by conventional linear transformation to yield a fully efficient systolic array as reported by Bu and Deprettere [BDD90] and also by Clauss et al. [CMP90]. Both of them adopt the same approach, namely, selecting $\delta - 1$ vectors $\lambda_1, \ldots, \lambda_{\delta-1}$ in the problem domain and merging processors $p_0, p_0 + A\lambda_1, \ldots, p_0 + A\lambda_{\delta-1}$ (where A is the allocation matrix). We call this kind of approaches "enumeration approach". Our work differs from these approaches in two aspects.

First, due to the selection of vectors $\lambda_1, \ldots, \lambda_{\delta-1}$, it is very possible that $p_0, p_0 + A\lambda_1, \ldots, p_0 + A\lambda_{\delta-1}$ don't form a parallelepiped and may form a cluster of arbitrary shape. Although the resultant array is still regular (because the original array is regular), it is difficult (if not impossible) to come up with an **explicit** allocation function for the final array. Moreover it is not clear how the dense array constraint can be satisfied. In contrast, our approach is constructive and guarantees that an explicit allocation function satisfying the dense array constraint can be found.

Second, the enumeration approaches can only be applied to systolic arrays derived by the conventional linear transformation. In contrast, by studying the activation patterns of the RIA that is implemented by an arbitrary systolic array, our approach can be applied to any (pure) systolic array. Furthermore, the method can be extended to piece-wise systolic arrays as follows. A piece-wise array consists of a constant number of pieces of pure systolic arrays. We can thus apply our technique to each piece of the array and adjust the connections between boundary processors accordingly. This corresponds to using piece-wise quasi-linear functions, and is especially useful when dealing with ingenuous arrays designed by ad-hoc manner, outside a standard synthesis methodology. Such arrays are most likely to be piece-wise systolic, and our approach can be used to improve the efficiency of such arrays.

The results presented here generalize a previous result reported elsewhere [ZR91]. In that paper we show that there always exist some directions along which one can always merge δ neighboring processors to obtain a fully efficient array for any systolic array derived by the standard linear transformation. It is thus an instance of a parallelepiped, where the factorization of δ is $\delta = 11\ldots1\delta$ (i.e. all except one of the factors are one, and the last one is δ itself).

Our work reported here raises an interesting question regarding the cost (and hence optimality) of systolic arrays derived by linear transformations. Traditionally, the two cost measures that have been used are the computation time and the number of processors. However, by using the results in this paper, one can always reduce the processor count by a factor of δ. Thus, the "raw" processor count by itself is not an accurate measure. This corresponds to the volume v of a convex polyhedron (the domain of computation, $\mathcal{D}$), under the transformation to space-time, T. Except for two-dimensional recurrences, this is not a linear function and hence the optimal solution can be obtained only be enumeration. It would be interesting to investigate how such methods [WD89] can be adapted to use the new cost function which is $v/|T|$.

References

[Ban88] Utpal Banerjee. *Dependence Analysis for Supercomputing.* Kluwer Academic

Publishers, 1988.

[BDD90] J. Bu, E. F. Deprettere, and P. Dewilde. A design methodology for fixed-size systolic arrays. In S. Y. Kung and E. Swartzlander, editors, *International Conference on Application Specific Array Processing*, pages 591–602, Princeton, New Jersey, Sept 1990. IEEE Computer Society.

[CMP90] P. Clauss, C. Mongenet, and G. R. Perrin. Calculus of space-optimal mappings of systolic algorithms on processor arrays. In S. Y. Kung and E. Swartzlander, editors, *International Conference on Application Specific Array Processing*, pages 4–18, Princeton, New Jersey, Sept 1990. IEEE Computer Society.

[CS84] Peter R. Cappello and Kenneth Steiglitz. Unifying VLSI designs with linear transformations of space-time. *Advances in Computing Research*, 2:23–65, 1984.

[KL80] H. T. Kung and C. E. Leiserson. *Algorithms for VLSI Processor Arrays*, chapter 8.3, 'Introduction to VLSI Systems,' Mead, C. and Conway, L., pages 271–292. Addison-Wesley, Reading, Ma, 1980.

[KMW67] R. M. Karp, R. E. Miller, and S. Winograd. The organization of computations for uniform recurrence equations. *JACM*, 14(3):563–590, July 1967.

[LW85] G. J. Li and B. W. Wah. Design of optimal systolic arrays. *IEEE Transactions on Computers*, C-35(1):66–77, 1985.

[Mol83] D. I. Moldovan. On the design of algorithms for VLSI systolic arrays. *Proceedings of the IEEE*, 71(1):113–120, January 1983.

[Qui87] Patrice Quinton. *The Systematic Design of Systolic Arrays*, chapter 9, Automata Networks in Computer Science, pages 229–260. Princeton University Press, 1987. Preliminary versions appear as IRISA Tech Reports 193 and 216, 1983.

[Rao85] Sailesh Rao. *Regular Iterative Algorithms and their Implementations on Processor Arrays*. PhD thesis, Stanford University, Information Systems Lab., Stanford, Ca, October 1985.

[RF90] Sanjay V. Rajopadhye and Richard M. Fujimoto. Synthesizing systolic arrays from recurrence equations. *Parallel Computing*, 14:163–189, June 1990.

[RK86] Sailesh Rao and Thomas Kailath. What is a systolic algorithm. In *Proceedings, Highly Parallel Signal Processing Architectures*, pages 34–48, Los Angeles, Ca, Jan 1986. SPIE.

[Sch88] A. Schrijver. *Theory of Integer and Linear Programming*. John Wiley and Sons, 1988.

[Thi89] Lothar Thiele. On the design of picecwise regular processor arrays. In *International Symposium on Circuits and Systems*, pages 2239–2542. IEEE CAS, IEEE Press, 1989.

[WD89] Jiwan Wong and Jean-Marc Delosme. Optimization of the processor count for systolic arrays. Technical Report YALEU-DCS-RR-697, Computer Science Dept. Yale University, May 1989.

[ZR91] Xiaoxiong Zhong and Sanjay V. Rajopadhye. Synthesizing efficient systolic arrays. In *International Conference on Acoustics, Speech and Signal Processing*, pages accepted, Toronto, Canada, May 1991. IEEE.

[ZWR90] Xiaoxiong Zhong, Ivan M. Wong, and Sanjay V. Rajopadhye. Bounds on the solution space in systolic array design. Technical Report CIS-TR-90-10, University of Oregon, Computer Science Department, 120 Deschutes Hall, Eugene OR 97403-1202, April 1990. (submitted to Journal of VLSI Signal Processing; preliminary results reported in VLSI Signal Processing IV, San Diego CA, Nov 1990).

Affine Timings
for Systems of Affine Recurrence Equations †

C. MONGENET
Université Louis Pasteur, Département d'Informatique
7 rue René Descartes, 67084 STRASBOURG, FRANCE

Abstract. — This paper is devoted to the problem of the existence of affine timings for problems defined by Systems of Affine Recurrence Equations. After a first analysis, such problems may have no affine timing not because the problem is uncomputable but only because of the initial system of equations. This system can induce dependencies organized in an inappropiate way. We give conditions for a dependency to be well-organized in such a way that an affine timing may exist. When a dependency does not satisfy these conditions, we describe how to transform it in order to meet the conditions. A problem defined by a system of equations is analyzed by a step-by-step examination of its dependencies. For each dependency organized in an inappropriate way, a transformation is applied. The whole transformation process yields to the determination of a new equivalent system of equations from which an affine timing can usually be computed. Many practical problems need such transformations. We illustrate this transformation process on the Algebraic Path Problem.

Keywords. — systems of affine recurrence equations, affine timing functions, mapping, systolic arrays, processor arrays.

Introduction

Many research efforts have focused on methods to map problems on systolic arrays ([KUN 82]) and on processor arrays architectures ([MOL 83], [QUI 84], [DEI 85], [MON 85], [MOF 87], [FFW 87], [MOP 87], [YAC 88]).

The first approaches mainly dealt with systems of Uniform Recurrence Equations introduced by KARP, MILLER and WINOGRAD ([KMW 67]). However this restrictive class of equations cannot naturally express lots of problems. So, researchers now focus on a larger class of equations : the Systems of Affine Recurrence Equations (SARE in the following) ([CLP 88], [QVD 89]).

† This work has been supported by the Laboratoire d'Informatique de Besançon and by the French Coordinated Research Program C^3 of the CNRS.

The goal of these different methods is the same. Using the information given by the dependency analysis of the problem, they determine first a *timing* or *scheduling* of the elementary calculation points of this problem, then an *allocation* of these calculations on a set of processors. The timing is generally defined by an *affine timing function*.

This paper is concerned with the problem of the affine timing determination.

A problem defined by a given initial SARE can have no parallel solutions because the dependencies between the calculations are such that no possible affine scheduling of these calculations exists. This can be intrinsically related to the problem and the interdependencies between the calculations. But experience proves that in some cases, the lack of affine timing is not due to strong interdependencies but only to one ill-conditioned dependency. That is to say, the initial SARE is not appropriate. In such cases, a transformation of this initial SARE into an equivalent one can then yield to the existence of affine schedulings. This could be achieved by applying formal methods of recurrence equations transformations.

Instead of this formal approach, our objective is to include the problem of SARE transformation in our method for the mapping of SARE onto regular arrays ([MCP 91]). We shall use its general geometrical framework to discuss the existence of affine timings and the necessity of dependency transformations when no such timings exist. We propose to analyze individually each dependency in order to determine if it is ill-conditioned. When it is, i.e. when no affine timing exists according to this dependency, we present domain transformations in order to get a timing. Since each dependency of the problem will be examined separately, the existence of a timing for the whole problem will not be known and guarantied before the complete examination of all the dependencies. In practice, this approach is very fruitful for lots of problems (Gauss-Jordan, Algebraic Path Problem, etc) where the initial SAREs contain ill-conditioned dependencies. After application of the process described here, a well-conditioned description of the problem, i.e. an equivalent SARE, is obtained from which we easily deduce an affine timing.

We recall the different steps of our mapping method. We first define the notion of affine timing. We emphasize the interest of constant timing, i.e. timing independent on the problem size. The dependency analysis is then realized in terms of two classes of vectors : the *generating vectors* and the *inductive vectors*. Using this information, we first present the condition for an affine timing to be valid according to one dependency. We differentiate two validity conditions or constraints : the weak condition and the stronger one where broadcast is removed. When one of these validity conditions is not satisfied we describe how to transform the domain and the current dependency in order to get a valid timing according to this dependency.

This paper is organized as follows. Section 1 presents the concept of SARE. The notion of affine timing and the way we realize the dependency analysis are described in section 2. Section 3 focuses on the conditions for the existence of a valid affine timing according to one dependency and presents the appropriate domain transformation to be applied in case of non-existence of a timing. Finally section 4 illustrates these results on a classical problem, the Algebraic Path Problem ([ROT 87], [DEL 88], [BE-all 89]). We show why the initial problem specification does not yield to any affine timing. The transformations described in this paper are then applied on this problem and the resulting specification can be scheduled.

1. Problems definition

We consider problems that can be expressed in terms of a system of affine recurrence equations. Each equation of the system is of the form :

$$X_u[z] = f_u\Big(\ldots, X_v[\rho_{u,v,p}(z)], \ldots\Big) \qquad z \in D_{u,p} \tag{E}$$

where

• X_u and X_v are variable names. X_u is a result variable (possibly an intermediary one) whose items $X_u[z]$, $z \in D_{u,p}$, are calculated in terms of argument variables X_v appearing in the right part of (E). X_v is either a data or a result variable (possibly X_u). The variables X_u and X_v are indexed with integral indexes z and $z' = \rho_{u,v,p}(z)$ of constant dimension. If X_v is a result variable it must be fully indexed, i.e. $\rho_{u,v,p}(z)$ and z must be of the same dimension n.

• $D_{u,p} \subset \mathbb{Z}^n$ is a convex bounded polyhedron or a union of convex bounded polyhedra associated with equation (E). It is called the *index domain*. It defines the set of integral coordinates points on which the equation (E) is defined. Each point z of $D_{u,p}$ corresponds to the calculation of one item $X_u[z]$. The domains of all the equations (E) of the system are of the same dimension. If several equations (E) define the same result variable X_u, their domains must be disjoint to avoid a result to be defined twice. In practice we have $n = 2$ or 3 to get by projection of the domain a linear or bidimensional array. We call *domain* of the problem the union of the $D_{u,p}$.

• $p \in \mathbb{Z}^q$ is the *size parameter*. It characterizes the bounds of the index domain $D_{u,p}$.

• f_u is any function. We assume its complexity is O(1).

• the ... express that they are other arguments of the same form as $X_v[\rho_{u,v,p}(z)]$.

• $\rho_{u,v,p}$ is an affine function from $\mathbb{Z}^n \to \mathbb{Z}^{n'}$ ($n' \leq n$). It is called the *index mapping*. It defines the item of variable X_v used in equation (E) to calculate the item $X_u[z]$. It is of the form :

$$\rho_{u,v,p}(z) = A(z) + B(p) + C$$

The initialization of the recurrence defined by equation (E) has to be specified. This is done by the *initialization* or *input equations*. There is at least one such equation associated with each result X_u of the problem. It is of the form :

$$X_u[z_o] = d \qquad z_o \in D_{u_o,p} \subset \mathbb{Z}^n \tag{Init}$$

where

• z_o corresponds to the initial step of the recurrence. In practice the initialization domain $D_{u_o,p}$ is adjacent to the calculation domain $D_{u,p}$.

• d is a data of the problem. It is either a constant or a data variable over a set of integral indexes.

From such a problem definition we analyze the dependencies in order to characterize conditions for the existence of affine timings associated with each dependency of the problem. The dependency analysis is realized in terms of 2 classes of vectors : the *generating vectors* denoted Φ and the *inductive vectors* denoted Ψ.

2. Affine timing and dependency analysis

The dependency analysis information is used to determine the possible timing functions and at a later stage (not discussed in this paper) the allocation of the calculations to the processors.

In this paper we only consider affine timing functions. The timing function defines, for any point z of the domain, its execution instant $t(z)$ in the following way :

DEFINITION 1. — *An* affine timing function *associated with a problem is a function of the form :*

$$\begin{array}{rlcl} t: & D \subset \mathbb{Z}^n & \rightarrow & \mathbb{N} \\ & z & \rightarrow & t(z) = \lambda \cdot z + \alpha \end{array}$$

where $\lambda = (\lambda_1, \ldots, \lambda_n)$ *and* $\lambda_1, \ldots, \lambda_n, \alpha$ *are integral constants.*

The vector λ is orthogonal to hyperplanes of $\mathbb{Z}^n$ such that all the points belonging to one of them have the same execution instant. We call these hyperplanes the *timing surfaces.*

We now introduce the notion of efficient timing. Let z_1 and z_2 be any two points of the domain. An efficient timing function should guaranty that the difference between their execution instants $t(z_1) - t(z_2)$ is independent on the parameter size p of the problem. Hence if this difference depends on p , the delay between the execution instants of z_1 and z_2 would increase when p increases. That is why the only interesting timing functions are the constant timings defined by :

DEFINITION 2. — *An affine timing is* constant *if and only if it is independent on the size parameter p of the problem.*

From now one we shall only consider constant affine timing functions.

Because of the dependencies between elementary calculations of a problem, a timing must satisfy some conditions. We call them the *causal constraints.* They express that an elementary calculation point z_1 which uses as data a result item calculated by a point z_0 must be executed *after* z_0. Therefore these constraints restrict the timing and are only associated with the result variables whose items are used as data.

The problem also contains dependencies which are associated with data variables. These dependencies do not impose any causal constraints : any scheduling of all the calculations using a same data is valid.

To take into account the different incidence on the timing of the result and data variables, we analyze their dependencies differently.

2.1. Dependencies for a data

A data variable X_v does not impose any intrinsic causal constraint. Since a data variable does not need to be fully indexed we use $\zeta_0 \in \mathbb{Z}^{n'}$ as its index. The only information we need to know about X_v is on which subset of the domain D any given data item $X_v[\zeta_0]$ is used. This set of points is called a *utilization set* and is defined in the following way :

DEFINITION 3. — *The* utilization set *associated with a given data item $X_v[\zeta_0]$ that occurs in equation (E) defining the variable X_u is*

$$\mathrm{Util}_E(X_v, \zeta_0) = \{z \in D_{u,p} \mid \rho_{u,v,p}(z) = \zeta_0\}$$

Since the index mappings $\rho_{u,v,p}$ are affine functions, the solutions of any $\rho_{u,v,p}(z) = \zeta_0$ form an affine subspace of $\mathbb{Z}^n$. $\mathrm{Util}_E(X_v, \zeta_0)$ is then the intersection of this affine subspace with the domain $D_{u,p}$ which is a convex polyhedron or a union of convex bounded polyhedra. Therefore we have the following property :

PROPERTY 1. — *Every utilization set* $\mathrm{Util}_E(X_v, \zeta_0)$ *is a convex polyhedron or a union of convex bounded polyhedra.*

We introduce the notion of *generating vectors* to characterize these utilization sets.

DEFINITION 4. — *We call* generating vectors *associated with a data variable X_v that occurs in equation* (E) *the m vectors of a basis of* $\mathrm{Util}_E(X_v, \zeta_0)$ *for any ζ_0. We denote them $\Phi_{E,X_v,i} (i = 1 \dots m)$.*

Note that if for a given equation (E), the utilization set $\mathrm{Util}_E(X_v, \zeta_0)$ is of dimension 0, it is reduced to one point. This means that a given data item $X_v(\zeta_0)$ is used only on one calculation point z. Such a situation is characterized by a null generating vector $\Phi_{E,X_v} = 0$.

2.2. Dependencies for a result

We consider now the case where the variable X_v appearing in the recurrence equation (E) is a result. Recall that since X_v is a result variable, it is fully indexed. We use $z_0 \in \mathbb{Z}^n$ as an index of X_v.

The recurrence equation $X_u[z] = f_u(\dots, X_v[\rho_{u,v,p}(z)], \dots)$ expresses that the computation of variable X_u at point z causally depends on the computation of variable X_v at point $\rho_{u,v,p}(z)$, i.e. the computation of X_u at point z must occur *after* the one of X_v at point $\rho_{u,v,p}(z)$. For a given result item $X_v[z_0]$ with $z_0 = \rho_{u,v,p}(z)$ we need to characterize the set of all the points z depending causally on point z_0, i.e. using the result provided by z_0. This set is called a *reception set* and defined by :

DEFINITION 5. — *The* reception set *associated with a given result item $X_v[z_0]$ that occurs in equation (E) defining the variable X_u is*

$$\mathrm{Rec}_E(X_v, z_0) = \{z \in D_{u,p} \mid \rho_{u,v,p}(z) = z_0\}$$

Similarly to property 1 we have :

PROPERTY 2. — *Every reception set* $\mathrm{Rec}_E(X_v, z_0)$ *is a convex bounded polyhedron or a union of convex bounded polyhedra.*

These reception sets are defined in the same way as the utilization sets. However they have different conceptual meanings. A utilization set is associated with a data while a reception set is associated with a result. There is an intrinsic causal dependency underlying the notion of reception set which does not exist in the notion of utilization set. We will represent this causal dependency by *inductive vectors* defined in the following way :

DEFINITION 6. — *We call* inductive vector *associated with a result variable X_v that occurs in equation (E), any vector denoted by $\Psi_{X_v,\rho_{u,v,p}(z)}$ and defined by :*

$$\Psi_{X_v,\rho_{u,v,p}(z)} = z - \rho_{u,v,p}(z)$$

A reception set is then characterized by its basis of generating vectors (in the same way as a utilization set) and by the inductive vectors. These inductive vectors connect a point $z_0 = \rho_{u,v,p}(z)$ computing a result item $X_v[z_0]$ to all the points z, $z \in \mathrm{Rec}_E(X_v, z_0)$, that causally depend on z_0, i.e. using the result provided by z_0 for their own computation. Notice that there are as many sets of inductive vectors associated with a given result X_v as there are occurrences of X_v in the right side of any equation (E).

All the points $z_0 = \rho_{u,v,p}(z)$ which are origin of a vector of a given set of inductive vectors form a set. This set is called the *emission set* and defined by :

DEFINITION 7. — *The* emission set *associated with a given result variable X_v that occurs in equation (E) defining the variable X_u is*

$$\mathrm{Emit}_E(X_v) = \{z_0 \in D \mid \exists z \in D_{u,p} \textit{ such that } \rho_{u,v,p}(z) = z_0\} = \rho_{u,v,p}(D_{u,p})$$

Since $\mathrm{Emit}_E(X_v)$ is the image by the affine function $\rho_{u,v,p}$ of a convex bounded polyhedron or of the union of convex bounded polyhedra, we have the following property :

PROPERTY 3. — *Every emission set* $\mathrm{Emit}_E(X_v)$ *is a convex bounded polyhedron or a union of convex bounded polyhedra.*

3. Validity conditions for the existence of affine timings

We now use the data dependencies information to determine the validity conditions on the timing regarding to one dependency associated with a result variable.

We distinguish two validity conditions :
• the weak validity condition. It characterizes timings which are valid according to one dependency, i.e. timings that only satisfy the causal constraints.
• the stronger validity condition. It characterizes timings which are not only valid but which also remove the broadcast.

3.1. Weak validity condition and existence of constant affine timings

PROPERTY 4. Weak validity condition. — *An affine timing is* valid according to a given dependency associated with a result variable *if and only if it satisfies*

$$\lambda \cdot \Psi > 0$$

for all the inductive vectors Ψ characterizing this dependency.

DEMONSTRATION : A causal constraint expresses that the point z depending causally on a point z_0 must be executed after z, i.e. $t(z) > t(z_0)$. This is equivalent to $\lambda.(z-z_0) > 0$. By definition 6, $z-z_0 = \Psi$. The validity condition is therefore $\lambda.\Psi > 0$. □

This validity condition can be satisfied or not depending on the respective organization of the reception and emission sets. It expresses the existence of affine timings valid according to one dependency. Let us now discuss this problem in the context of the only efficient timings : the constant ones.

Let (S) be the affine subspace of equation $\rho_{u,v,p}(z) = z_0$. By definition of a reception set, $\mathrm{Rec}_E(X_v, z_0)$ is included in (S). Two cases can occur for the corresponding emission point z_0 : either z_0 belongs to (S) or it does not. Let us study this two cases regarding to the existence of constant affine timings.

3.1.1. $z_0 \in (S)$

In this case all the corresponding inductive vectors Ψ_{X_v,z_0} are colinear and belong to the subspace (S). These vectors associated to one point z_0 can be organized in 2 different ways :

• either they all belong to the same open half-subspace. Therefore we have the following property :

PROPERTY 5. — *When an emission point z_0 belongs to the subspace (S) of equation $\rho_{u,v,p}(z) = z_0$ and has all its inductive vectors Ψ_{X_v,z_0} in the same open half-subspace, these vectors define a* cone *whose origin is z_0. It is denoted $C(\Psi_{X_v,z_0})$.*

The existence of a timing for this dependency will depend on the way all the cones (for the different emission points z_0) are organized. This situation is a particular case of the problem discussed in section 3.1.2.

• or they belong to the whole subspace. In this case the validity condition $\lambda \cdot \Psi > 0$ can not be satisfied since there exist opposite inductive vectors. This situation does not allow any affine timing as stated by the following property.

Property 6. — *When an emission point z_0 belongs to the subspace (S) of equation $\rho_{u,v,p}(z) = z_0$ and is such that its inductive vectors Ψ_{X_v,z_0} describe the whole subspace, there is no affine timing.*

This situation is characterized by the fact the reception set is divided into 2 disjoints subsets. By translating one of them as it is presented on figure 1, we transform the problem into the first situation where all the vectors Ψ_{X_v,z_0} belong to the same open half-subspace. The existence of a timing will eventually be possible. Such a situation is illustrated on the example of the Algebraic Path Problem presented in section 4.

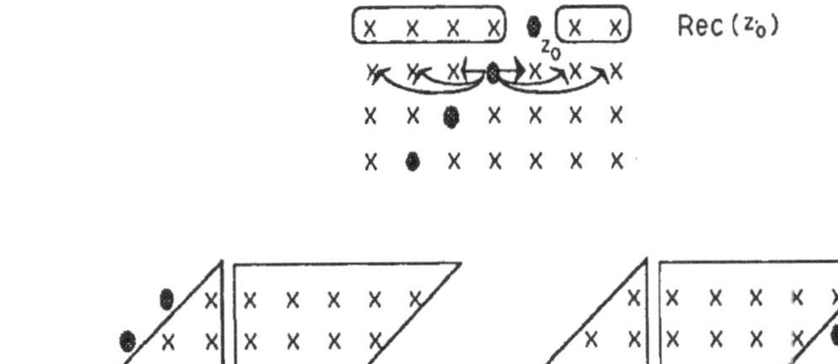

Figure 1 : transformation on a reception set when no affine timing exists

3.1.2. $z_0 \notin (S)$

In this case all the inductive vectors Ψ_{X_v,z_0} have the same origin z_0 and their extremities belong to the union of convex bounded polyhedra $\text{Rec}_E(X_v, z_0)$. We therefore deduce the following property :

Property 7. — *When an emission point z_0 does not belong to the subspace of equation $\rho_{u,v,p}(z) = z_0$ the set of inductive vectors Ψ_{X_v,z_0} associated with z_0 define a* cone *whose origin is z_0. It is denoted $C(\Psi_{X_v,z_0})$.*

We now consider the union of all these cones.

Definition 8. — *We call* extremal cone *C_p associated with a dependency on the result variable X_v the convex hull of the union of all the Ψ-cones $C(\Psi_{X_v,z_0})$, $z_0 \in \text{Emit}_E(X_v)$ where p is the parameter size.*

PROPERTY 8. — *Consider one dependency on a result variable of a problem characterized by an extremal cone C_p. A constant affine timing valid according to this dependency exists if and only if the extremal cone C_p is strictly included in a half-subspace of $\mathbf{R}^n$.*

DEMONSTRATION : If C_p is strictly included in a half-subspace of $\mathbf{R}^n$, then its complementary cone defined by $\{\mu \mid \mu \cdot \Psi < 0, \forall \Psi \in C_p\}$ is not empty. Since it contains the vectors μ such that $\mu \cdot \Psi < 0$, any vector $\lambda = -\mu$ satisfies $\lambda \cdot \Psi > 0, \forall \Psi \in C_p$. Therefore these vectors λ define affine timings. Among them, there exist constant ones. Vice versa, if C_p is not strictly included in a half-subspace of $\mathbf{R}^n$ (i.e. its either the half-subspace or the whole subspace) its complementary cone is empty and no affine timing exists. □

When no constant affine timing exists, domain transformations are necessary. Translations will not solve the problem. We need to apply transformations such as foldings. This point is not studied in this paper. We now examine the following question : When a constant affine timing does exist, under which conditions does a constant broadcast-removing (abbreviated BR in the following) affine timing exist?

3.2. Stronger validity condition and existence of constant BR affine timings

On many parallel architectures, the broadcast of data is not easily or efficiently implementable. We may wish to remove such a broadcast. This can be realized by restricting the timings to timings which do not allow broadcast. We call such timings the Broascast Removing timings (BR timings). They are easily characterized by the following property for problems whose reception sets are of dimension 1, i.e. described by only one generating vector Φ. In the general case, when the reception sets are of dimension greater than 1, a transformation is first necessary to partly remove the broadcast. This point is discussed in [MCP 90].

PROPERTY 9. Stronger validity condition. — *A BR affine timing associated with a given dependency characterized by a set of extremal vectors Ψ_{X_v,ex_i} and one generating vector Φ must satisfy*

$$\lambda \cdot \Psi_{ex_i} > 0 \quad \forall i \in \mathbf{N}$$
$$\lambda \cdot \Phi \neq 0$$

DEMONSTRATION : Since the extremal cone C_p is characterized by a finite set of extremal vectors Ψ_{ex_i} any inductive vector Ψ is equal to $\sum_i \alpha_i \Psi_{ex_i}$ with $\alpha_i \geq 0$. Therefore the condition $\lambda \cdot \Psi > 0, \quad \forall \Psi$ is equivalent to $\lambda \cdot \Psi_{ex_i} > 0, \quad \forall \Psi_{ex_i}$. Moreover the generating vector Φ directs lines whose points use the same item of X_v. To avoid a broadcast, these points must belong to different timing surfaces, i.e. $\lambda \cdot \Phi \neq 0$. □

PROPERTY 10. — *Consider one dependency on a result variable of a problem characterized by an extremal cone C_p strictly included in a half-subspace of $\mathbf{R}^n$. There exists a constant broadcast-removing affine timing valid according to this dependency if and only if $\lim_{p \to \infty} C_p$ is strictly included in a half-subspace of $\mathbf{R}^n$.*

DEMONSTRATION : Let us consider the complementary cone $\overline{C_p}$ of C_p defined by $\overline{C_p} = \{\mu \mid \mu \cdot \Psi < 0, \forall \Psi \in C_p\}$. If $\lim_{p\to\infty} C_p$ is not strictly included in a half-subspace of $\mathbf{R}^n$, i.e. if it is the whole half-subspace, $\lim_{p\to\infty} \overline{C_p}$ is reduced to a half-subspace. Let us show that this half-subspace only contains timings creating broadcast.

Since the extremal cone C_p is strictly included in a half-subspace of $\mathbf{R}^n$, by property 8 there exist constant affine timings. By definition , any reception set is included in a subspace of $\mathbf{Z}^n$. Therefore the set of all vectors ν orthogonal to a reception set define a subspace of $\mathbf{Z}^n$. These vectors are characterized by $\nu \cdot \Phi = 0$ for any generating vector Φ of the reception set. Moreover half of this subspace of vectors ν is characterized by $\nu \cdot \Psi > 0$. Therefore these vectors ν characterize affine timings. Since they satisfy $\nu \cdot \Phi = 0$, these timings create broadcast. Hence we have the following : whenever affine timings exist, the half-subspace of broadcast-creating timings defines solutions. Therefore if $\lim_{p\to\infty} \overline{C_p}$ is reduced to a half-subspace, this subspace is necessarily the set of the broadcast-creating timings. In this case there is no broadcast-removing timings.

Vice-versa, if $\lim_{p\to\infty} C_p$ is strictly included in a half-subspace, $\lim_{p\to\infty} \overline{C_p}$ is not reduced to the half-subspace of broadcast-creating timings. Therefore in this case, there exist constant broadcast-removing timings in the cone $\lim_{p\to\infty} \overline{C_p}$. □

REMARK : When the extremal cone C_p is strictly included in a half-subspace of $\mathbf{R}^n$, it is characterized by a finite set of extremal vectors $\Psi_{X_v, ex_i}, i \in \mathbf{N}$. Hence property 8 can be expressed by the following : there exists a constant broadcast-removing affine timing valid according to this dependency if and only if every pair of extremal vectors $(\Psi_{ex_1}, \Psi_{ex_2})$ is such that $\lim_{p\to\infty} \widehat{\Psi_{ex_1}, \Psi_{ex_2}} < 180°$ where $\widehat{\Psi_{ex_1}, \Psi_{ex_2}}$ denotes the angle of the two vectors.

When no constant BR affine timing exists, it is because the position of the emission set regarding to the reception set yields to a too wide extremal cone when the parameter size p increases. We can reduce this width by changing the relative positions of the emission and reception sets. This is realized by applying a transformation such as a translation directed by the generating vector on part of the reception set. An example of such a situation is presented on figure 2. This problem of non existence of a constant BR affine timing also occurs in the Algebraic Path Problem presented in section 4.

4. Example : the Algebraic Path Problem

Let us consider a weighted oriented graph $G = (V, E, w)$ where V is a finite set of n vertices, E is the set of edges and w is a function which associates a weight with each edge. These weights are defined in an $n \times n$ matrix A associated with G. $A = (a_{ij})$ is defined by $a_{ij} = w(i,j)$ if $(i,j) \in E$, $a_{ij} = 0$ otherwise.

The Algebraic Path Problem (APP) consists in finding, for all pairs of vertices (i,j), the quantities $d_{ij} = \oplus_{p \in M_{ij}} w(p)$ where M_{ij} denotes the set of all paths from i to j. The weight of a path $p \in M_{ij}$ is defined by $w(p) = \otimes_{e \in p} w(e)$.

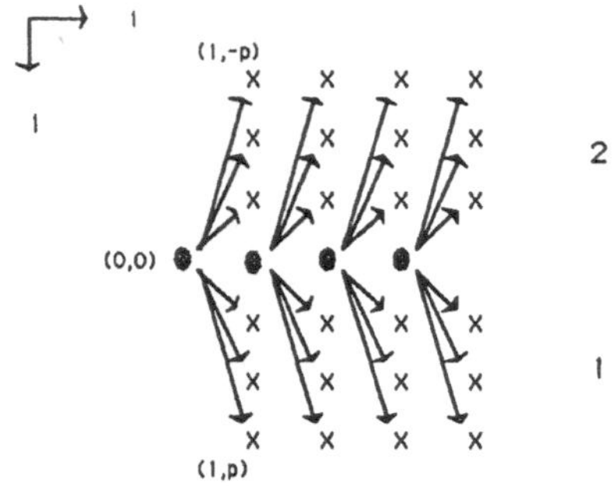

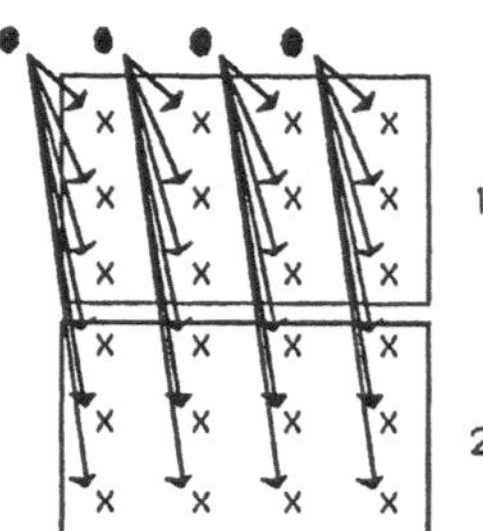

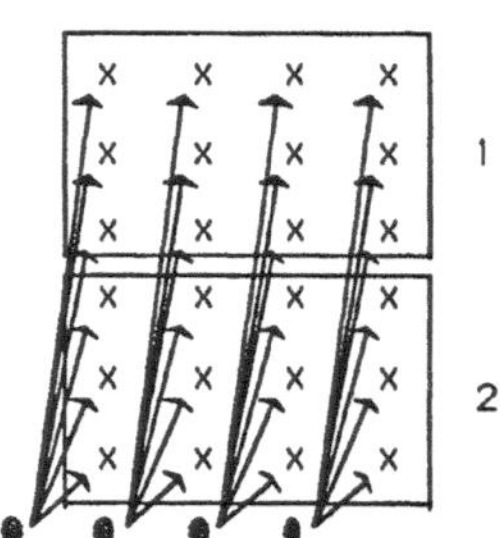

Figure 2 : transformation on a reception set when no constant BR affine timing exists

We denote $M_{ij}^{(k)}$ the set of all paths from i to j whose only intermediate vertices v are such that $1 \le v \le k$. When computing the recurrence $a_{ij}^{(k)} = \oplus_{p \in M_{ij}^{k}} w(p)$, we obtain $d_{ij} = a_{ij}^{(n)}$. The problem is defined by the following system of recurrence equations :

$$
\begin{array}{lll}
(1) & a(k,k,k) & = a(k,k,k-1) \quad 1 \le k \le n \\
(2) & a(i,j,k) & = a(i,j,k-1) \otimes a(k,j,k) \quad 1 \le i,k \le n,\ i \ne k,\ j = k \\
(3) & a(i,j,k) & = a(i,j,k-1) \otimes a(i,k,k) \quad 1 \le j,k \le n,\ j \ne k,\ i = k \\
(4) & a(i,j,k) & = a(i,j,k-1) \oplus a(i,k,k) \otimes a(k,j,k-1) \\
 & & \qquad 1 \le i,j,k \le n,\ i \ne k,\ j \ne k
\end{array}
$$

This problem is characterized by 4 subdomains respectively associated with the 4 equations :

$$
\begin{aligned}
D_1 &= \{(k,k,k) \in \mathbb{Z}^3 \mid 1 \le k \le n\} \\
D_2 &= \{(i,k,k) \in \mathbb{Z}^3 \mid 1 \le i,k \le n,\ i \ne k\} \\
D_3 &= \{(k,j,k) \in \mathbb{Z}^3 \mid 1 \le j,k \le n,\ j \ne k\} \\
D_4 &= \{(i,j,k) \in \mathbb{Z}^3 \mid 1 \le i,j,k \le n,\ i \ne k,\ j \ne k\}
\end{aligned}
$$

Notice that D_1 is a convex domain while D_2 and D_3 are the union of 2 convex disjoints subsets and D_4 the union of 4 subsets. They are presented on figure 3.

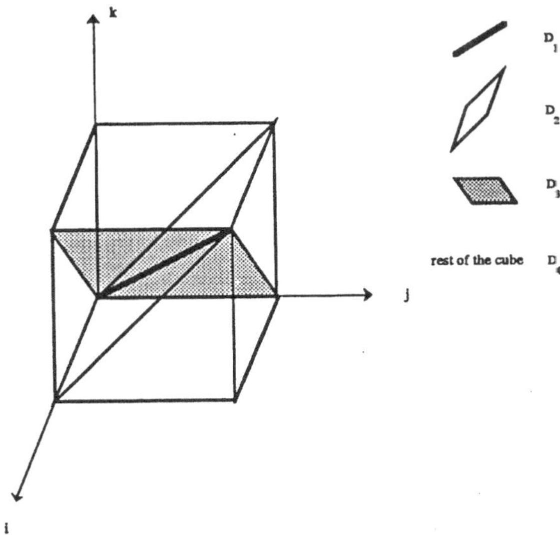

Figure 3 : domain of the Algebraic Path problem

On all four domains, we have the same dependency expressed by a constant inductive vector $\Psi_1 = (0,0,1)$. For each emission point $z_0 = (i_0, j_0, k_0)$ the corresponding reception set is reduced to one point $z = (i_0, j_0, k_0+1)$. Its generating vector is therefore the null vector.

On D_2 the use of $a(k,j,k)$ determines a reception set $\text{Rec}_{(2)}\ (a, z_0) = \{(i, k_0, k_0) \mid 1 \le i \le n,\ i \ne k_0\}$ characterized by one generating vector $\Phi_{(2)} = (1,0,0)$ and a set of inductive vectors $\Psi_{(2)}(z_0) = (i-k_0, 0, 0)$ with $1 \le i \le n, i \ne k_0$. The emission set is $\text{Emit}_{(2)} = \{(k_0, k_0, k_0) \mid 1 \le k_0 \le n\} = D_1$.

On D_3 the use of $a(i,k,k)$ determines a reception set $\text{Rec}_{(3)}\ (a, z_0) = \{(k_0, j, k_0) \mid 1 \le j \le n,\ j \ne k_0\}$ characterized by one generating vector $\Phi_{(3)} = (0,1,0)$ and a set of inductive vectors $\Psi_{(3)}(z_0) = (0, j-k_0, 0)$ with $1 \le j \le n, j \ne k_0$. The emission set is $\text{Emit}_{(3)} = \{(k_0, k_0, k_0) \mid 1 \le k_0 \le n\} = D_1$.

On D_4 the use of $a(i,k,k)$ determines a reception set $\text{Rec}_{(4),1}\ (a, z_0) = \{(i_0, j, k_0) \mid 1 \le j \le n,\ j \ne k_0\}$ characterized by one generating vector $\Phi_{(4),1} = (0,1,0)$ and a set of inductive vectors $\Psi_{(4),1}(z_0) = (0, j-k_0, 0)$ with $1 \le j \le n, j \ne k_0$. The emission set is $\text{Emit}_{(4),1} = \{(i_0, k_0, k_0) \mid 1 \le i_0, k_0 \le n, i_0 \ne k_0\} = D_2$.

Finally on D_4 too, the use of $a(k,j,k-1)$ determines a reception set $\text{Rec}_{(4),2}\ (a, z_0) = \{(i, j_0, i_0) \mid 1 \le i \le n,\ i \ne i_0\}$ characterized by one generating vector $\Phi_{(4),2} = (1,0,0)$

and a set of inductive vectors $\Psi_{(4),2}(z_0) = (i - i_0, 0, 1)$ with $1 \leq i \leq n, i \neq i_0$. The emission set is $\text{Emit}_{(4),2} = \{(i_0, j_0, i_0 - 1) \mid 1 \leq i_0, j_0 \leq n, i_0 \neq j_0\}$.

The 3 dependencies defined by $\Psi_{(2)}$, $\Psi_{(3)}$ and $\Psi_{(4),1}$ are characterized by the fact that their inductive and generating vectors are colinear. Moreover their inductive vectors define a line, i.e. there are opposite inductive vectors such as $\Psi_{(2)}(n,n,n) = (1-n,0,0)$ for $z = (1,n,n)$ and $\Psi_{(2)}(1,1,1) = (n-1,0,0)$ for $z = (n,1,1)$. This situation corresponds to case 3.1.1 mentioned above and forbids any affine timing. Translations on the respective domains are necessary to transform the sets of inductive vectors into cones strictly included into a half-line and therefore allow eventually a timing. These translations are presented on figure 4.

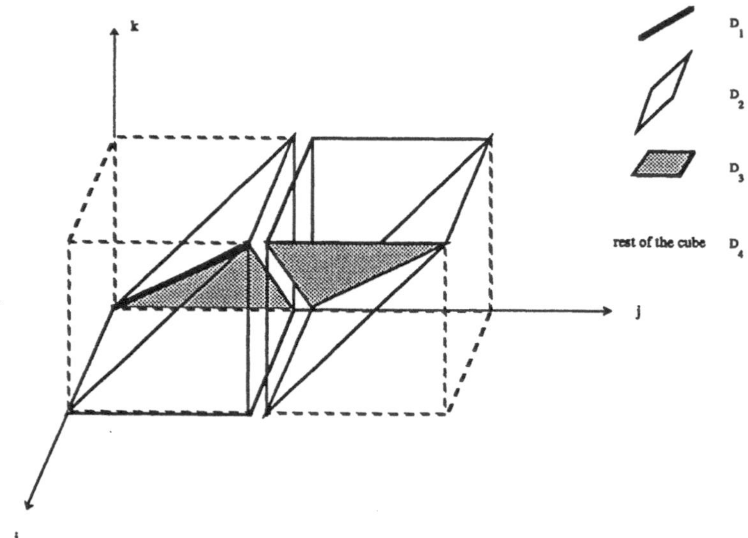

Figure 4 : first transformation on the domain of the APP

The last dependency defined by $\Psi_{(4),2}$ is characterized by a cone of inductive vectors associated with each emission point z_0 : $C(\Psi_{(4),2}, z_0) = \{(i - i_0, 0, 1) \mid 1 \leq i \leq n, i \neq i_0\}$. The extremal cone C_p, convex hull of the cones $C(\Psi_{(4),2}, z_0)$, $z_0 \in \text{Emit}_{(4),2}$ is strictly included in a half-subspace of $\mathbf{R}^3$ and characterized by 2 extremal vectors $\Psi_{ex_1} = (1-n,0,1)$ and $\Psi_{ex_2} = (n-1,0,1)$. By property 8 there exists a constant affine timing for this dependency. However $\lim_{n\to\infty} \widehat{\Psi_{ex_1}, \Psi_{ex_2}} = 180°$ and by property 10 there is no constant BR affine timing. When applying a translation on domain D_4 as it is presented on figure 5, the angle $\widehat{\Psi_{ex_1}, \Psi_{ex_2}}$ is reduced to less than 45° and a constant BR affine timing now exists for this dependency.

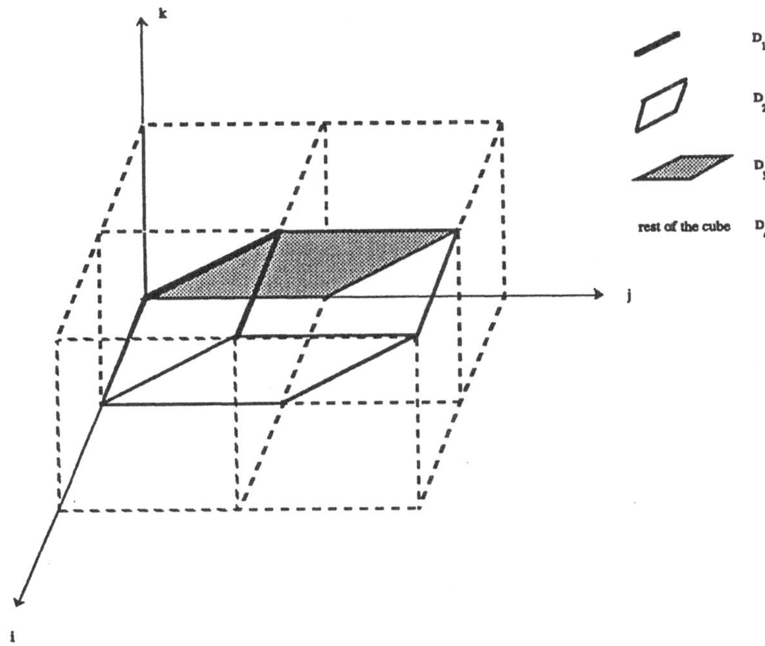

Figure 5 : second transformation on the domain of the APP

After all the transformations mentioned above, the dependencies are now expressed by :

$$\Psi_1 = (0,0,1)$$
$$\Psi_{(2)}(z_0) = (i-k_0,0,0) \quad for \quad k_0+1 \le i \le n+k_0$$
$$\Psi_{(3)}(z_0) = (0,j-k_0,0) \quad for \quad k_0+1 \le j \le n+k_0$$
$$\Psi_{(4),1}(z_0) = (0,j-k_0,0) \quad for \quad k_0+1 \le j \le n+k_0$$
$$\Psi_{(4),2}(z_0) = (i-i_0,0,1) \quad for \quad i_0+1 \le i \le n+i_0$$

The extremal vectors are then

$$\Psi_{(2),ex_1} = (1,0,0) \qquad \Psi_{(2),ex_2} = (n,0,0)$$
$$\Psi_{(3),ex_1} = (0,1,0) \qquad \Psi_{(3),ex_2} = (0,n,0)$$
$$\Psi_{(4),1,ex_1} = (0,1,0) \qquad \Psi_{(4),1,ex_2} = (0,n,0)$$
$$\Psi_{(4),2,ex_1} = (1,0,1) \qquad \Psi_{(4),2,ex_2} = (n,0,1)$$

The successive examination of each dependency has yield to some transformations of the problem. The existence of a global affine timing can now be tackled with the new values of the inductive vectors.Using these vectors, any affine timing $\lambda = (\lambda_1, \lambda_2, \lambda_3)$ must then verify $\lambda_1, \lambda_2, \lambda_3 > 0$. Since the generating vectors are $\Phi_{(2)} = \Phi_{(4),2} = (1,0,0)$ and $\Phi_{(3)} = \Phi_{(4),1} = (0,1,0)$, all the affine timing are broadcast removing timings. The most efficient affine timing is then defined by $\lambda = (1,1,1)$. The corresponding computation time is $5n-4$. The initial specification of the APP did not allow any affine timing. The appropriate transformations on the different dependencies now yield to affine timings.

Conclusion

We have presented a geometrical framework to analyze the existence of affine timing functions for a given SARE. Using a dependency analysis in terms of generating and inductive vectors, we express conditions for the existence of affine timings valid according to one dependency. We characterize different validity conditions : the weak condition and the broadcast-removing condition. We emphasize the notions of constant and broadcast-removing timings. When these conditions are not satisfied by an ill-conditioned dependency, transformations are proposed in order to get an affine timing valid according to this dependency. Notice that there is not a unique transformation (for example the translation on figure 2 can be applied either on set 1 or on set 2). The different possibilities should be tried to give the largest chances to find an affine timing for the whole problem since they are not necessarily all compatible.

The process presented here is a step-by-step method. It does not guaranty the existence of an affine timing for the whole problem which is not possible since the computability of such parameterized SAREs has been proved undecidable ([SAQ 90]).

However it is very fruitful in practice where the initial SAREs of many problems often present ill-conditioned dependencies. The results presented here allow to determine a new specification of these problems without any ill-conditioned dependencies. These results have been illustrated on the example of the Algebraic Path Problem.

Bibliography

[BE-all89] BENAINI A., QUINTON P., ROBERT Y., SAOUTER Y., TOURANCHEAU B. — Synthesis of a new Systolic Architecture for the Algebraic Path Problem, *IRISA Research Report,No1094*, 1989.

[CLP 88] CLAUSS Ph., PERRIN G.R. — Synthesis of Process Arrays, *CONPAR'88, Manchester, G.B.*, 1988.

[DEI 85] DELOSME J.M., IPSEN I.C.F. — An Illustration of a Methodology for the Construction of Efficient Systolic Architectures in VLSI, *Sd. Int. Symposium on VSLI Technology, Systems and Applications, Taipei, Taiwan, R.O.C.*, 1985, p. 268–273.

[DEL 88] DELOSME J.M. — A Parallel Algorithm for the Algebraic Path Problem, *Int. Workshop on Parallel and Distributed Algorithms, M. Cosnard et al. editors, North-Holland*, 1988.

[FFW 87] FORTES J.A.B., FU K.S., WAH B.W. — Systematic Approaches to the Design of Algorithmically Specified Systolic Arrays, *Int. Conf. on Acoustics, Speech and Signal Processing*, 1987.

[KMW 67] KARP R.M., MILLER R.E., WINOGRAD S. — The Organization of Computations for Uniform Recurrence Equations, *JACM*, t. **14,3**, 1967.

[KUN 82] KUNG H.T. — Why systolic architectures ?, *Computer*, t. **15-1**, 1982, p. 37–46.

[MCP 90] MONGENET C., CLAUSS Ph., PERRIN G.R. — Geometrical Tools to map Systems of Affine Recurrence Equations on Regular Arrays, *Research Report, LIB, Université de Franche-Comté*, 1990.

[MCP 91] MONGENET C., CLAUSS Ph., PERRIN G.R. — A Geometrical Cding to Compile Affine Recurrence Equations on Regular Arrays, *Fifth Int. Parallel Processing Symposium, Anaheim, California*, 1991.

[MOF 86] MOLDOVAN D.I., FORTES J.A.B. — Partitioning and Mapping Algorithms into Fixed Size Systolic Arrays, *IEEE Transactions on Computers*, t. **35-1**, 1986, p. 1–12.

[MOL 83] MOLDOVAN D.I. — On the Design of Algorithms for VLSI Systolic Arrays, *Proc IEEE conf.*, t. **71-1**, 1983, p. 113–120.

[MON 85] MONGENET C. — Une Méthode de Conception d'Algorithmes Systoliques, Résultats Théoriques et réalisation, *Thèse INPL, Nancy*, 1985.

[MOP 87] MONGENET C., PERRIN G.R. — Synthesis of Systolic Arrays for Inductive Problems, *Conf. PARLE, LNCS 259*, 1987.

[QUI 84] QUINTON P. — Automatic Synthesis of Systolic Arrays from Uniform Recurrence Equations, *Proc. IEEE 11th Int. Symp. on Computer Architecture, Ann Arbor, MI, USA*, 1984, p. 208–214.

[QVD 89] QUINTON P., VAN DONGEN V. — The Mapping of Linear Recurrence Equations on Regular Arrays, *The Journal of VLSI Signal Processing*, t. **1**, 1989, p. 95–113.

[ROT 87] ROBERT Y., TRYSTRAM D. — Systolic Solution of the Algebraic Problem, *Int. Workshop on Systolic Arrays, Oxford, Adam-Hilger*, 1987, p. 171–180.

[SAQ 90] SAOUTER Y., QUINTON P. — Computability of Recurrence Equations, *TR-1203, IRISA, Rennes*, 1990.

[YAC 88] YAACOBI Y., CAPPELLO P.R. — Scheduling a System of Affine Recurrence Equations onto a Systolic Array, *Int. Conf. on Systolic Arrays, San Diego, USA*, 1988, p. 373–382.

On the Computational Complexity of Optimal Sorting Network Verification

Ian Parberry*

Department of Computer Science

The Pennsylvania State University

Abstract

A *sorting network* is a combinational circuit for sorting, constructed from comparison-swap units. The depth of such a circuit is a measure of its running time. It is reasonable to hypothesize that only the fastest (that is, the shallowest) networks are likely to be fabricated. It is shown that the problem of verifying that a given sorting network actually sorts is Co-$\mathcal{NP}$ complete even for sorting networks of depth only $4\lceil\log n\rceil + O(1)$ greater than optimal. This is shallower than previous depth bounds by a factor of two.

1 Introduction

A *comparator network* is a combinational circuit constructed from comparison-swap units called *comparators*. A *sorting network* is a comparator network which sorts. The *size* of a comparator network is the number of comparators used. The *depth* is the number of layers of comparators, where each layer receives input only from the layers above it. Comparator networks can be fabricated relatively easily using VLSI techniques. It would be useful to be able to verify whether a given sorting network actually works. It is well known that in order to test whether a given n-input comparator network is a sorting network, it is sufficient to check that it sorts the $2^n - n - 1$ nonsorted zero-one inputs (which we will call bit-strings). This observation is called the *zero-one principle.*

Comparator networks which sort all but a single nonsorted bit-string are known. That is, for all nonsorted sequences of n bits x, there exists an n-input comparator network which sorts all

*Research supported by NSF Grant CCR-8801659. Author's current address: Department of Computer Sciences, P.O. Box 13886, University of North Texas, Denton, TX. 76203-3886, U.S.A. Electronic mail: ian@dept.csci.unt.edu.

bit-strings except x. These are called ***single exception sorting networks.*** Chung and Ravikumar [5] give a recursive construction of an n-input single exception sorting network of polynomial size and depth. They further deduce in [6] that the sorting network verification problem is Co-$\mathcal{NP}$ complete. Parberry [16] gave a non-recursive construction for a single exception sorting network of depth $D(n-1)+2\lceil\log(n-1)\rceil+2$, where $D(n)$ is the minimum depth of an n-input sorting network, and deduced, using the construction of Chung and Ravikumar [6], that the problem of verifying sorting networks of depth $2D(n)+6\lceil\log n\rceil+O(1)$ is Co-$\mathcal{NP}$ complete. We will show that the sorting network verification problem remains Co-$\mathcal{NP}$ complete even for sorting networks of depth $D(n)+4\lceil\log n\rceil+O(1)$.

The remainder of this paper is divided into six sections. The first section contains a more formal definition of a sorting network, and briefly describes some standard results. The second section contains a proof that a modified version of the satisfiability problem is $\mathcal{NP}$ complete. The third section contains a sketch of the reduction from that problem to the sorting network verification problem. The fourth section contains the details of the construction of an important component used in that reduction — a comparator network that sorts all except a specific set of inputs. The construction of this component uses the single exception sorting network of Parberry [16]. A slightly improved single exception sorting network is given in the fifth section of this paper. The sixth section contains details on how to reduce the depth of the construction to give the required result.

Let $\mathbf{N}$ denote the natural numbers, and $\mathbf{B}$ denote the Boolean set $\{0,1\}$. Members of $\mathbf{B}^n$ (the set of n-tuples of bits) will be called ***bit-strings.*** We will use the standard regular-expression notation to describe certain sets of bit-strings, for example, 0^n1^m denotes a single bit-string consisting of n ones followed by m zeros, and $(00\cup 11)^n$ denotes the set of n pairs of bits, where each pair is either 00 or 11, that is,

$$\{x_1y_1\cdots x_ny_n \mid x_i=y_i\in\mathbf{B} \text{ for } 1\le i\le n\}.$$

If A and B are sets, $A\backslash B$ denotes $\{x \mid x\in A, \text{ but } x\notin B\}$.

2 Sorting Networks

One of the early investigations into parallel sorting concerned the ***Bose-Nelson sorting problem,*** named by Floyd and Knuth [9], after Bose and Nelson [4]. The problem involves sorting n values by using a sequence of ***oblivious*** in-situ comparison-and-swap operations; that is, a sequence of comparisons between the ith and jth value, where i and j are independent of the values being sorted. The obliviousness property allows the following elegant hardware interpretation of the problem. Suppose that we are given a basic unit of hardware called a ***comparator.*** A comparator

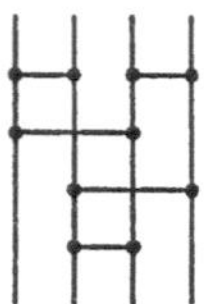

Figure 1: A 4-input sorting network of depth 3 and size 5.

takes two values as input and outputs them in ascending order. A *comparator network* consists of n parallel *channels*, which can be thought of as wires carrying values, to which comparators are attached. The network is divided into a finite number of *levels*. Each level consists of one or more comparators. Each comparator is attached to two channels. At most one comparator is placed on any channel at each level. Channels, in our diagrams, will be drawn as vertical lines, and comparators as horizontal lines with heavy dots emphasizing the connection-points. Levels are numbered vertically from top-to-bottom, and channels are numbered horizontally from left-to-right. Level 0 will be used to denote the inputs.

Let C be a comparator network. We define the value carried by channel i of C at level j on input $x = (x_1, \ldots, x_n)$, written $V(C, x, i, j)$, as follows. $V(C, x, i, 0) = x_i$ and for $j > 0$:

- If there is a comparator between channels i and $k > i$ at level j, then

$$V(C, x, i, j) = \min(V(C, x, i, j-1), V(C, x, k, j-1)).$$

- If there is a comparator between channels i and $k < i$ at level j, then

$$V(C, x, i, j) = \max(V(C, x, i, j-1), V(C, x, k, j-1)).$$

- Otherwise, $V(C, x, i, j) = V(C, x, i, j-1)$.

The output of an n-input, d-level comparator network C on input x is

$$V(C, x) = (V(C, x, 1, d), V(C, x, 2, d), \ldots, V(C, x, n, d)).$$

If for all inputs x, $V(C, x)$ is in nondecreasing order from left to right, then C is called a *sorting network*.

For example, Figure 1 shows a 4-input sorting network. The comparators on the first level compare the values in pairs. The second layer of comparators determines the maximum and minimum values: the minimum of the two minima is the minimum overall, and the maximum of the two maxima is the maximum overall. The third layer puts the remaining two values into the correct order.

Each level of a comparator network consists of a set of independent comparisons which may be performed in parallel. The number of levels is thus a reasonable measure of parallel time. This is called the *depth* of the network. Another resource of interest is its *size*, which is defined to be the number of comparators used. We will call an n-input sorting network *optimal* if it is $\mathcal{P}$-uniform (that is, there is an algorithm which outputs a description of the sorting network, on input n, in time polynomial in n), and has depth $O(\log n)$. Optimal sorting networks have size $O(n \log n)$.

There have been a number of recent results on optimal sorting networks. For a survey of some of these results, see Parberry [13, 14]. Sorting networks of optimal depth are known for $n \leq 10$ (Parberry [15]) and with optimal size for $n \leq 8$ (Knuth [11]). For all practical purposes, the best sorting networks are constructed using the recursive technique of Batcher [3], which gives depth $O(\log^2 n)$ and size $O(n \log^2 n)$, although some small improvement in the lower-order terms of the size have been made by Drysdale [7] and Van Voorhis [19] in exchange for a large increase in depth. Ajtai, Komlós and Szemerédi [1, 2] have demonstrated that asymptotically optimal (logarithmic depth and log-linear size) sorting networks can be constructed; however their method remains impractical for reasonable values of n despite the efforts of Paterson [17], since the constant multiples involved are extremely large.

We will make use of two standard results. Firstly, if we allow channels to be permuted obliviously between each layer, then the model is the same, in the sense that we can remove the permutations in polynomial time in a manner that affects neither the size of the network, the depth of the network, nor its ability to sort. This result is due to Floyd and Knuth [8]. Therefore, the sorting network verification problems with and without permutations allowed are polynomial-time equivalent. The inclusion of such permutations will simplify the presentation of our results since it will allow us to physically group together logically related channels. The second standard result is the *zero-one* principle, which states that a comparator network is a sorting network *iff* it sorts all bit-strings. In the light of this result, we will throughout this paper consider only zero-one inputs. These results are discussed at great length in Knuth [11], and Parberry [13, 16, 14]. We assume that the reader is familiar with the techniques and terminology of the theory of $\mathcal{NP}$ completeness. The reader who is not should consult a standard reference, such as Garey and Johnson [10].

More formally, the sorting network verification problem can be stated as a decision problem as follows:

Sorting Network Verification (NONSORT)

INSTANCE: A comparator network C.

QUESTION: Is there an input which C does not sort?

The number of comparators in C is a reasonable measure of input size for any sorting network verification program. Furthermore, the number of inputs to C is a valid measure of input size for any program which verifies optimal sorting networks. Clearly NONSORT $\in \mathcal{NP}$, since if we are given a comparator network C and an input x, the output of C on x can be determined in time polynomial in the number of comparators in C. It remains to show that NONSORT is $\mathcal{NP}$ hard. This will imply that the original sorting network verification problem is Co-$\mathcal{NP}$ complete.

3 One-in-Three 3SAT

The following set-theoretic problem was shown to be $\mathcal{NP}$ complete by Schaefer [18].

One-in-Three 3SAT (T3SAT)

INSTANCE: Sets $S_1, \ldots, S_n$ with $|S_i| \leq 3$ for $1 \leq i \leq n$.

QUESTION: Does there exist a set S such that $|S \cap S_i| = 1$ for $1 \leq i \leq n$?

Garey and Johnson [10] list Schaefer's result in a slightly different form which more clearly illustrates that it is a restricted case of the Satisfiability Problem. Define a *clause* over a set V to be a subset of $V \times V \times V$. A set $S \subseteq V$ is said to *satisfy* a clause (v_1, v_2, v_3) iff $|\{i \mid v_i \in S\}| = 1$. If C is a list of clauses over V, then $S \subseteq V$ is said to be a *satisfying assignment* for C iff S satisfies all clauses in C. Intuitively, the elements of V are variables, and S is a set of variables to be assigned the value `true`. A clause represents a ternary Boolean function which is `true` iff exactly one of its inputs is `true`. A list of clauses represents the conjunction of a list of these functions.

Modified One-in-Three 3SAT (M3SAT)

INSTANCE: A list of clauses C over a set V.

QUESTION: Is there a satisfying assignment for C?

Lemma 3.1 *M3SAT is $\mathcal{NP}$ complete.*

PROOF: Obviously, M3SAT $\in \mathcal{NP}$. It is easy to show that T3SAT $\leq_m^p$ M3SAT. □

We will find it useful to consider a restricted version of M3SAT.

Balanced One-in-Three 3SAT (B3SAT)

INSTANCE: A set of variables V and a list of clauses C over V in which every variable appears exactly three times.

QUESTION: Is there a satisfying assignment for C?

Note that each instance of B3SAT with n clauses must have n variables. (Since there are n clauses, there are $3n$ instances of variables. Since every variable has exactly 3 instances, there must be n variables.) Also, every *satisfiable* instance of B3SAT with n clauses must have n divisible by 3. (Let V be a variable-set, and $C = (C_1, \ldots, C_n)$ be a list of clauses over V. Suppose $S \subseteq V$, where $|S| = m$. Then C contains $3m$ instances of variables that are members of S. But if S is a satisfying assignment, since there are n clauses, each of which must contain exactly one instance of a variable in S, there must be n instances of variables that are members of S. Therefore $n = 3m$.)

Theorem 3.2 *B3SAT is $\mathcal{NP}$ complete.*

PROOF: Clearly B3SAT $\in \mathcal{NP}$. By Lemma 3.1, it suffices to show that M3SAT $\leq_m^p$ B3SAT.

Suppose $C_1, \ldots, C_n$ is an instance of M3SAT over variable-set V. The corresponding instance of B3SAT is constructed as follows. For every clause C_i, there are three new clauses $(a_{i,1}, a_{i,2}, a_{i,3})$, $(b_{i,1}, b_{i,2}, b_{i,3})$, $(c_{i,1}, c_{i,2}, c_{i,3})$ called *structural enforcers*. For each variable $v \in V$, let

$$X_v = \{a_{i,j}, b_{i,j}, c_{i,j} \mid v \text{ is the } j\text{th variable in } C_i\}.$$

Suppose

$$X_v = \{a_{i_1,j_1}, b_{i_1,j_1}, c_{i_1,j_1}, \ldots, a_{i_m,j_m}, b_{i_m,j_m}, c_{i_m,j_m}\}$$

for some $m \leq n$. Define

$$I_v = \{(a_{i_k,j_k}, b_{i_k,j_k}, c_{i_k,j_k}) \mid 1 \leq k \leq m\}.$$

If $k = 1$, define $E_v = I_v$, otherwise define

$$\begin{aligned} E_v \ = \ & \{(b_{i_1,j_1}, c_{i_1,j_1}, a_{i_2,j_2}), (b_{i_2,j_2}, c_{i_2,j_2}, a_{i_3,j_3}), \\ & \ldots, (a_{i_{m-1},j_{m-1}}, b_{i_{m-1},j_{m-1}}, c_{i_m,j_m}), (a_{i_m,j_m}, b_{i_m,j_m}, a_{i_1,j_1})\} \end{aligned}$$

For each $(p, q, r) \in I_v$ and each $(p, q, r) \in E_v$ there are three clauses (p, x, y), (q, x, y), and (r, x, y), called *equality enforcers*, where x and y are new variables not previously used. It is clear that this transformation can be computed in time polynomial in n.

The new instance of M3SAT consisting of the structural enforcers and the equality enforcers is actually an instance of B3SAT, that is, that every variable appears in exactly three clauses.

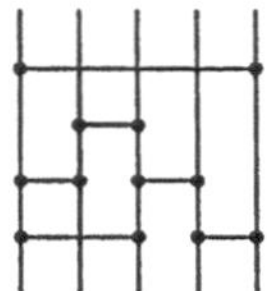

Figure 2: A variable component.

Each of the $a_{i,j}$, $b_{i,j}$, and $c_{i,j}$ variables occurs in exactly one structural enforcer, and two equality enforcers (one corresponding to an element of I_v, and one corresponding to an element of E_v, for some variable v). The extra variables in the equality enforcers also appear in exactly three clauses, by construction.

We claim that $C_1, \ldots, C_n$ is satisfiable iff there is a satisfying assignment for the structural enforcers and the equality enforcers. Clearly if S is a satisfying assignment for $C_1, \ldots, C_n$, then $\cup_{v \in S} X_v$ satisfies every structural enforcer and every equality enforcer. Conversely, suppose that S satisfies the structural enforcers and equality enforcers. Since the equality enforcers corresponding to members of I_v are satisfied by S, then for all variables $v \in V$, either $X_v \cap S = \{\}$ or $X_v \subseteq S$. Thus if we set $T = \{v \mid X_v \subseteq S\}$, then T satisfies $C_1, \ldots, C_n$. □

4 The Reduction

In order to show that NONSORT is $\mathcal{NP}$ complete, it is sufficient to show that B3SAT $\leq_m^p$ NONSORT. Suppose we are given an instance of B3SAT, that is, a list of clauses $C = (C_1, \ldots, C_n)$ over a set of variables $V = \{v_1, \ldots, v_n\}$ such that every variable in V appears exactly three times in C. We will construct a comparator network with $5n$ inputs. An input $x = (x_1, \ldots, x_{5n})$ to the comparator network is said to *correspond to assignment* S for C iff for all $1 \leq i \leq n$, $x \in (0^5 \cup 1^5)^n$, and $v_i \in S$ iff $x_{5i-4} = 1$. Our comparator network will sort only inputs that do not correspond to satisfying assignments for C, that is, inputs that do not correspond to any assignment, and inputs that correspond to nonsatisfying assignments. Therefore, it will be a sorting network iff C is not satisfiable.

For each variable $v \in V$, we have a *variable component*, consisting of five channels and six comparators, of depth three (see Figure 2). For each clause C_i we have a *clause component*, consisting of three channels and three comparators, of depth three (see Figure 3). These components are connected as follows.

Figure 3: A clause component.

The $5n$ inputs are divided into quintuples and put into n variable components, one for each variable. The center three outputs of each variable component are put into the inputs of the appropriate clause components. Specifically, suppose $C_i = (v_{i,1}, v_{i,2}, v_{i,3})$, for $1 \leq i \leq n$ where $v_{i,1}, v_{i,2}, v_{i,3} \in V$. Let $c(i,j)$ denote the number of previous occurrences of $v_{i,j}$ in $C_1, \ldots, C_i$, that is, $c(i,j) = |\{v_{k,l} \mid k < i \text{ or } (k = i \text{ and } l < j)\}| + 1$, for $1 \leq i \leq n$, $1 \leq j \leq 3$. Note that $0 \leq c(i,j) \leq 2$. Then for each clause $C_i = (v_{i,1}, v_{i,2}, v_{i,3})$, put the $(c(i,k)+2)$th output of the variable component corresponding to $v_{i,k}$ into the kth input of the clause component corresponding to C_i, for $1 \leq k \leq 3$. The first two outputs of the clause components are routed to the far right, and the last output of the clause components is routed to the far left. All of the channels are then put into a special component called a *selector*, which will sort every input except those which correspond to satisfying assignments.

The centre three outputs of a variable component corresponding to variable $v \in V$ are copies of variable v which are to be used in the clauses in which v appears. It can be demonstrated (by case analysis of the 32 different input strings) that if the first and last outputs of a variable component are zero, then all outputs are zero, and if the first and last outputs of a variable component are one, then all outputs are one. Thus the centre three outputs of a variable component carry a valid truth assignment for the variable v if the first and last output of the variable component are identical. The output of a clause component is 001 iff exactly one of its input channels carries a one (since each clause component is actually a three-input sorting network).

Suppose the input to the comparator network corresponds to a satisfying assignment S. Since there is exactly one variable from each clause in S, the first n inputs to the selector will be one, and the last $2n$ inputs to the selector will be zero. The remaining $2n$ inputs will consist of n pairs, each of which is either 00 or 11. Since variable appears in exactly three clauses, S must contain $n/3$ variables. Therefore $2n/3$ pairs will be 00, and $n/3$ pairs will be 11. That is, the input to the selector will be a member of the set $1^n(\mathrm{B}^{2n}_{2n/3} \cap (00 \cup 11)^n)0^{2n}$, where $\mathrm{B}^n_k = \{x \in \mathrm{B}^n \mid x \text{ has exactly } k \text{ ones}\}$. Conversely, if the input to the comparator network does not correspond to satisfying assignment, then it is clear that the input to the selector will not be of this form. The above reduction can be carried out in time polynomial in n provided there exist $\mathcal{P}$-uniform selectors. The construction of

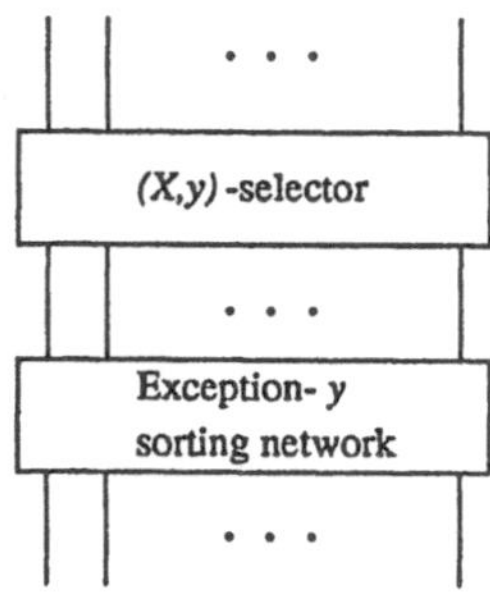

Figure 4: An exception-X sorting network.

the selector is the subject of the next section.

5 Selective Sorting Networks

The *selector* used in the previous section was a comparator network which selectively sorts all bit-strings except for members of a certain, slightly obscure set. We use the generic term *selective sorting network* for such a comparator network. If the comparator network sorts all members of $\mathbf{B}^n \backslash X$, than it is called an *exception-X sorting network.*

Comparator networks which sort all but a single bit-string are known. That is, for all nonsorted $x \in \mathbf{B}^n$, there exists an n-input comparator network which sorts all members of $\mathbf{B}^n$ except x. These are called *single exception sorting networks.* Chung and Ravikumar [5] give a recursive construction of an n-input single exception sorting network of polynomial size and depth. Parberry [16] gives a non-recursive construction for a single exception sorting network of depth $D(n-1)+2\lceil\log(n-1)\rceil+2$, where $D(n)$ is the minimum depth of an optimal n-input sorting network. We wish to find a particular selective sorting network with $5n$ inputs, where n is a multiple of three, and an exception set of size

$$\binom{n}{n/3}.$$

If $X \subseteq \mathbf{B}^n$, and $y \in \mathbf{B}^n$ is nonsorted, a comparator network C is an (X,y)*-selector* iff for all $x \in \mathbf{B}^n$, the output of C on input x is y iff $x \in X$.

Theorem 5.1 *If $X \subseteq \mathbf{B}^n$, and there exists an (X,y)-selector of depth $D_X(n)$, then there exists an exception-X sorting network of depth $D_X(n) + D(n) + 2\lceil\log n\rceil + 2$.*

PROOF: Suppose $X \subseteq \mathbf{B}^n$, and there exists a comparator network C of depth $D_X(n)$ and nonsorted

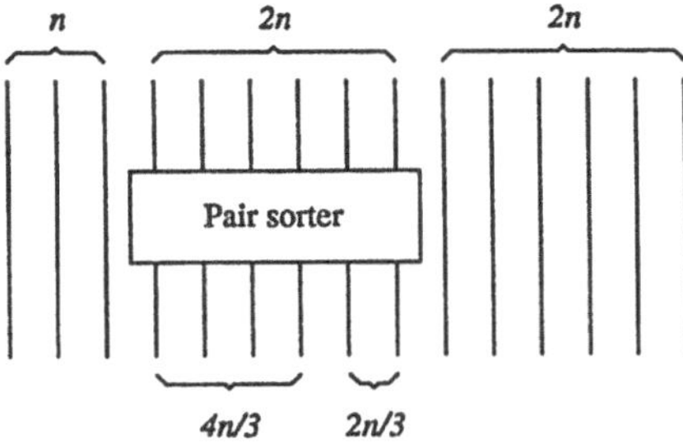

Figure 5: An X-selector.

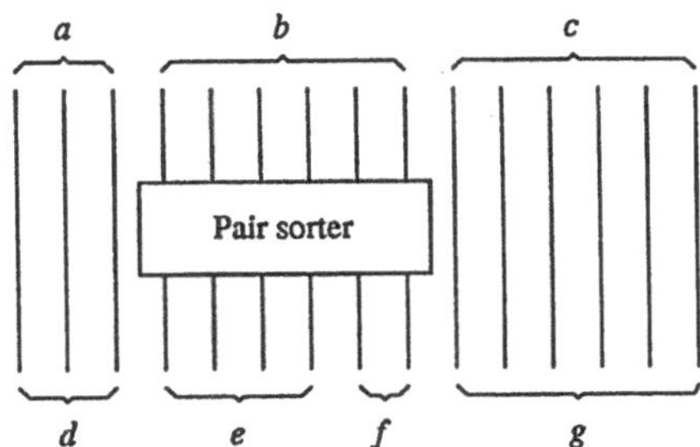

Figure 6: An X-selector labelled for the proof of Theorem 5.2.

$y \in \mathbf{B}^n$ such that for all $x \in \mathbf{B}^n$, the output of C on input x is y iff $x \in X$. Then an exception-X sorting network is constructed by composing C and an exception-y sorting network as in Figure 4. The total depth is bounded above by the depth of C plus the depth of the exception-y sorting network. Parberry [16] gives a construction for exception-y sorting networks for all nonsorted y of depth $D(n) + 2\lceil \log n \rceil + 2$. □

The exception-set from the previous section is the set of bit-strings of the following form: n ones, followed by n pairs of bits, $2n/3$ of which are 00 and $n/3$ of which are 11, followed by $2n$ zeros. That is,

$$X = 1^n(\mathbf{B}^{2n}_{2n/3} \cap (00 \cup 11)^n)0^{2n}.$$

Theorem 5.2 *Suppose $n \in \mathbf{N}$, and $X = 1^n(\mathbf{B}^{2n}_{2n/3} \cap (00 \cup 11)^n)0^{2n}$. There exists an exception-X sorting network of depth $D(n) + D(5n) + 2\lceil \log(5n) \rceil + 3$.*

PROOF: Let $X = 1^n(\mathbf{B}^{2n}_{2n/3} \cap (00 \cup 11)^n)0^{2n} \subseteq \mathbf{B}^{5n}_{5n/3}$. By Theorem 5.1, we can build an exception-

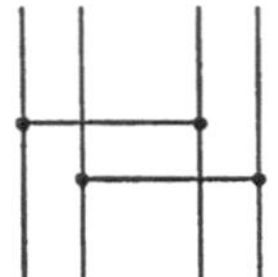

Figure 7: A pair comparator.

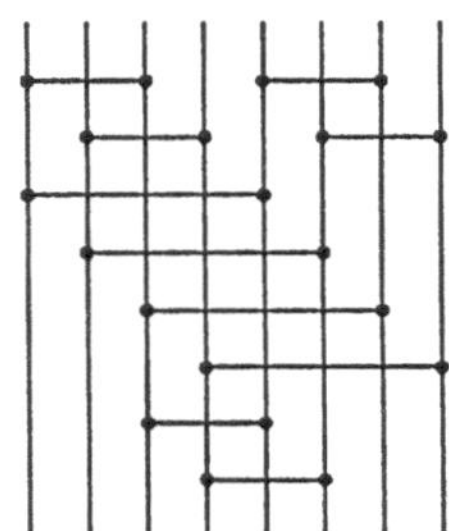

Figure 8: A 4-pair sorter constructed from Figure 1.

X sorting network from an (X, y)-selector. An (X, y)-selector with $y = 1^n 0^{4n/3} 1^{2n/3} 0^{2n}$ can be constructed as in Figure 5, by leaving the first n and the last $2n$ channels alone, and placing a $2n$-input *pair sorter* on the remaining $2n$ channels. A *pair sorter* is a comparator network which has an even number of inputs. The inputs are treated as bit-pairs. Each bit-pair is sorted, and the sorted bit-pairs are then sorted into lexicographic order, that is, the output of the pair sorter is a member of $(00)^*(01)^*(11)^*$. We will return to the construction of the pair sorter later in this proof.

Suppose the input to the network shown in Figure 5 is $abc \in \mathbf{B}^{5n}_{5n/3}$, where $a \in \mathbf{B}^n$, $b, c \in \mathbf{B}^{2n}$, and its output is $defg \in \mathbf{B}^{5n}_{5n/3}$, where $d \in \mathbf{B}^n$, $e \in \mathbf{B}^{4n/3}$, $f \in \mathbf{B}^{2n/3}$, and $g \in \mathbf{B}^{2n}$ (see Figure 6). We claim that $defg = 1^n 0^{4n/3} 1^{2n/3} 0^{2n}$ iff $abc \in X$. Suppose $abc \in X$. then $b \in \mathbf{B}^{2n}_{2n/3} \cap (00 \cup 11)^n$. Since $b \in (00 \cup 11)^n$, the output of the pair sorter, ef, is a sorted sequence of bits. Since $b \in \mathbf{B}^{3n}_{2n/3}$, $ef = 0^{4n/3} 1^{2n/3}$. Therefore, since $d = a$ and $g = c$, $defg = 1^n 0^{4n/3} 1^{2n/3} 0^{2n}$, as claimed. Conversely, suppose that $abc \notin X$. We claim that $defg \neq 1^n 0^{4n/3} 1^{2n/3} 0^{2n}$, that is, either $d \neq 1^n$, $e \neq 0^{4n/3}$, $f \neq 1^{2n/3}$, or $g \neq 0^{2n}$. Since $abc \notin X$, either $a \neq 1^n$, (in which case there is a zero in d), $c \neq 0^{2n}$, (in which case there is a one in g), or $b \notin \mathbf{B}^{2n}_{2n/3} \cap (00 \cup 11)^n$. In the latter case, suppose $b \in \mathbf{B}^{2n}_m$. If $m = 2n/3$, then $b \notin (00 \cup 11)^n$, and so there is a one in e. If $m < 2n/3$, then there is a zero in f. If $m > 2n/3$, then there is a one in e. In all cases, $defg \neq 1^n 0^{4n/3} 1^{2n/3} 0^{2n}$, as claimed.

It is clear that the depth of our $5n$-input (X, y)-selector is equal to the depth of a $2n$-input pair sorter. This pair-sorter is constructed as follows. The pairs are sorted with a single layer of comparators, one per pair. A pair of pairs can be sorted into lexicographic order by comparing the first element of the first pair with the first element of the second pair, and simultaneously comparing the second element of the first pair with the second element of the second pair (see Figure 7). Therefore n pairs can be sorted into lexicographic using a comparator network obtained from an n-input sorting network by doubling all the channels, and replacing every comparator with a pair-comparator (for example, see Figure 8). Thus the depth of the pair sorter is $D(n) + 1$. Therefore by Theorem 5.1, the depth of our $15n$-input exception-X sorting network is $D(n) + D(5n) + 2\lceil\log(5n)\rceil + 3$. □

Theorem 5.3 *NONSORT is $\mathcal{NP}$ complete even for n-input sorting networks of depth* $2D(n) + 2\lceil\log n\rceil + 9$.

PROOF: The reduction is as described in Section 4, using the selector from Theorem 5.2. The depth of an n-input comparator network constructed using this technique is bounded above by 3 for the variable components, plus 3 for the clause components, plus $2D(n) + 2\lceil\log n\rceil + 3$ for the selector, a total of $2D(n) + 2\lceil\log n\rceil + 9$. □

6 Improved Single Exception Sorting Networks

In the construction of the selector in the previous section, we used the single exception sorting network from Parberry [16], which has depth $D(n-1)+2\lceil\log(n-1)\rceil+2$. It is possible to improve that result by a small constant. Suppose $n \in \mathrm{N}$, and $1 \leq k < n$. A better single exception sorting network with exception 1^k0^{n-k} is constructed as follows (see Figure 9). In [16], a single exception sorting network with this exception is called a *canonical k-ones single exception sorting network*. The first k inputs are put into a *min network*. The leftmost output of this network is the minimum of its inputs. The last $k-1$ outputs of this network, and the remaining $n-k$ channels are sorted together. The leftmost channel, and the leftmost $n-k-1$ outputs of the sorting network are put into an *insertion network*. This network takes as input a single value followed by a sorted sequence, and it inserts the new value into the correct place in the sequence. It is straightforward to recursively construct n-input min and insertion networks of depth $\lceil\log n\rceil$.

It is easy to see that this construction gives a single exception sorting network. Suppose the input to the network is 1^k0^{n-k}. Then the leftmost output of the min network is 1, and the output of the sorting network is $0^{n-k}1^{k-1}$, and hence the output of the insertion network is $0^{n-k-1}101^{k-1}$,

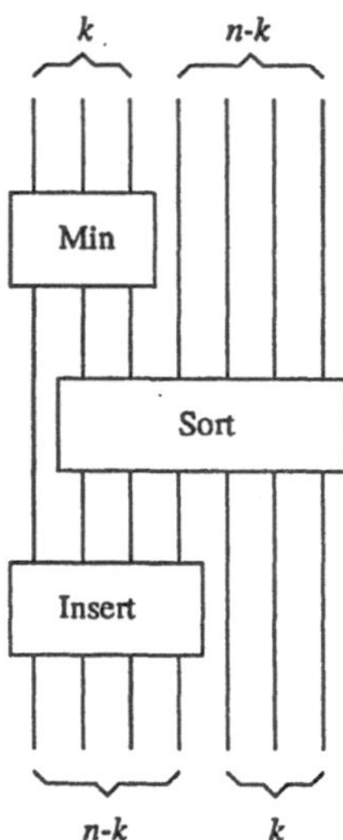

Figure 9: A single exception sorting network.

which is not sorted. Now suppose the input to the network is not 1^k0^{n-k}. In particular, suppose it is $ab \neq 1^k0^{n-k}$, where $a \in \mathrm{B}^k$, $b \in \mathrm{B}^{n-k}$. Then either $a \neq 1^k$ or $b \neq 0^{n-k}$. In the former case, the leftmost output of the min network is 0, hence the values on the channels immediately after the sorting network are sorted, and they remain sorted through the rest of the network. If $a = 1^k$ and $b \neq 0^{n-k}$, then b contains at least one 1, and so the insertion network carries the 1 from the leftmost channel into the correct place.

Theorem 6.1 *For all $n > 1$ and all nonsorted bit-strings x, there exists an n-input comparator network of depth $D(n-1) + 2\lceil\log n\rceil - 1$. which sorts all bit-strings except x.*

Proof: If $D(n)$ is the depth of the optimal n-input sorting network, then the depth of the canonical k-ones single exception sorting network shown in Figure 9 is

$$D(n-1) + \lceil\log\lceil n/2\rceil\rceil + \lceil\log\lfloor n/2\rfloor\rceil \leq D(n-1) + 2\lceil\log n\rceil - 1.$$

If x is an arbitrary nonsorted bit-string with k ones, then a comparator network can be constructed from the canonical k-ones single exception sorting network using the technique of Theorem 7 of [16]. □

This new construction can be used to improve slightly on the results in [16], and to improve slightly on the bound in Theorem 5.3.

Theorem 6.2 *NONSORT is $\mathcal{NP}$ complete even for n-input sorting networks of depth* $2D(n) + 2\lceil\log n\rceil + 6$.

PROOF: Use Theorem 6.1 in the construction of Theorem 5.3. □

7 Improved Selectors

The selective sorting network described in Section 5 was developed using general techniques which will work for any exception set of the appropriate form. However, the exception set that appears in the reduction of Section 4 has additional special properties which allow a reduction in the depth bound.

Let us re-draw the selector using the canonical single exception sorting network from Section 6 (see Figure 10). The selector consists of the pair sorter of depth $D(n)$, a sorting network of depth $D(5n)$, and a min network and an insertion network of combined depth $2\lceil\log(5n)\rceil - 1$. Thus the total depth of the $5n$-input selector is $2D(5n) + 2\lceil\log(5n)\rceil - 1$. We route the last $2n/3$ outputs from the pair sorter to the right of the first $4n/3$ outputs, so that the exception becomes $1^{5n/3}0^{10n/3}$ instead of $1^n0^{4n/3}1^{2n/3}0^{2n}$.

However, we have not used the fact that the output of the pair sorter is sorted in pairs. Thus there is no need for the sorting network in the part of the selector corresponding to the single exception sorting network. If we sort the first n and last $2n$ values in parallel with the pair sorter, then all we need to do is merge the sorted pairs in two groups after the pair sorter, and merge these with the outputs of the sorters (see Figure 11).

Parberry [12] gives a construction for an n-input pair merger (which is called an *alternating merging network* in that reference), of depth $\lceil\log n\rceil$. Batcher [3] gives a construction for an n-input merging network of depth $\lceil\log n\rceil + 1$. Therefore the new $5n$-input selector has depth bounded above by $D(2n)$ for the sorters, plus $\lceil\log(4n/3)\rceil \leq \lceil\log(5n)\rceil - 1$ for the pair mergers, plus $\lceil\log(5n)\rceil + 1$ for each of two mergers, plus $\lceil\log(5n)\rceil$ for the inserter, giving a total of $D(2n) + 4\lceil\log(5n)\rceil + 1$.

Therefore, we obtain the main result of this paper:

Theorem 7.1 *NONSORT is $\mathcal{NP}$ complete even for n-input sorting networks of depth* $D(n) + 4\lceil\log n\rceil + 7$.

PROOF: The proof is similar to that of Theorem 5.3, substituting the more efficient selector described in this section. The depth in this case is bounded above by 3 for the variable components, 3 for the clause components, plus $D(2n/5) + 4\lceil\log n\rceil + 1$ for the selector. □

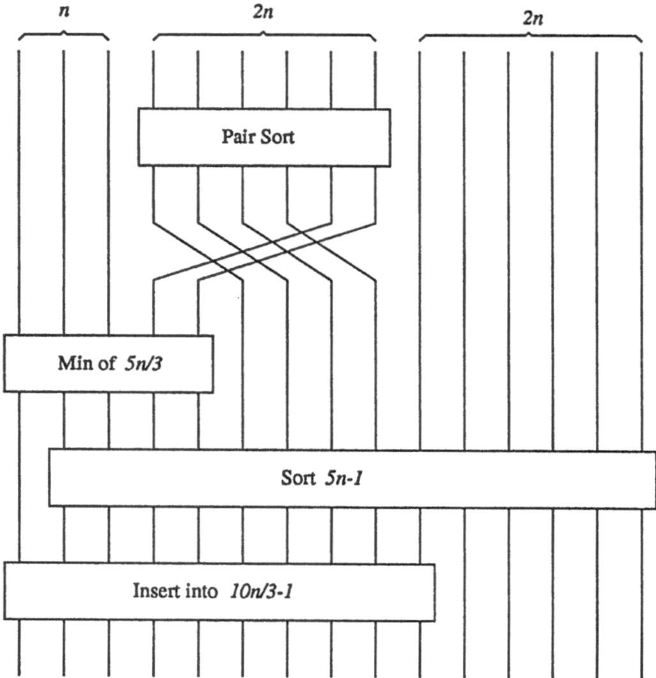

Figure 10: Details of the selector construction.

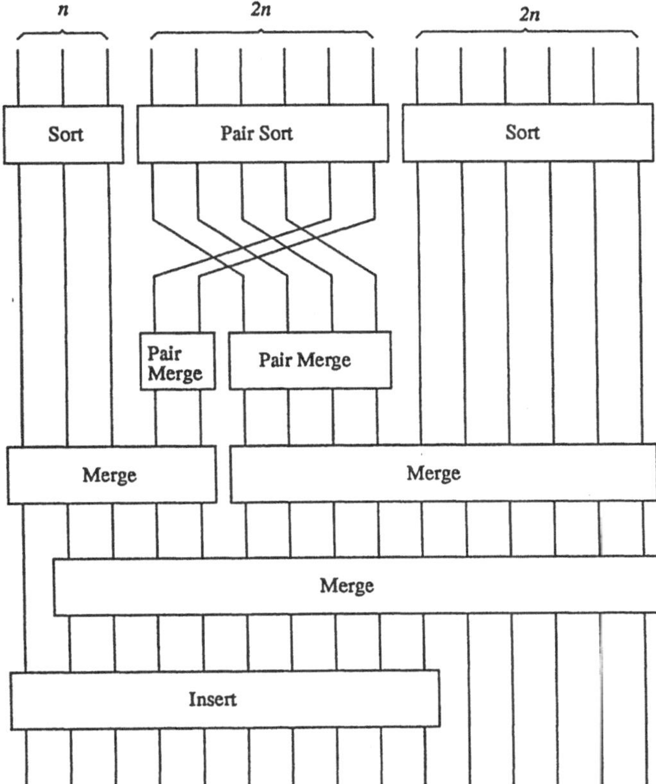

Figure 11: Details of the modified selector construction.

8 Conclusion and Open Problems

We have shown that sorting network verification is intractable even for sorting networks of depth $D(n)+4\lceil\log n\rceil+7$, where $D(n)$ is the depth of an optimal n-input sorting network. This is smaller by a factor of two than previous results. Our result is fairly strong, given the current state of knowledge about $D(n)$, which is that $\lceil\log n\rceil \leq D(n) < 6100\lceil\log n\rceil$ (the lower bound is obvious, and the upper bound is from Paterson [17]). We conjecture that sorting network verification remains intractable even for the shallowest sorting networks, that is, sorting networks of depth $D(n)$.

It is also interesting to consider the depth of single exception sorting networks, since their existence implies an exponential time lower bound for deterministic and probabilistic verification algorithms based on the zero-one principle (Parberry [16]). If $S(n)$ is the minimum depth of an n-input single exception sorting network, we know that $D(n)-1 \leq S(n) \leq D(n)+2\lceil\log n\rceil-1$, where $D(n)$ is the minimum depth of an n-input sorting network. We conjecture that $S(n)=D(n)$.

It should be noted that it is an open problem as to whether the result of Theorem 7.1 is better than that of Theorem 6.2. The former is better than the latter iff $D(n) \geq 2\lceil\log n\rceil-1$, which is the case for large enough n (Yao [20]).

References

[1] M. Ajtai, J. Komlós, and E. Szemerédi. An $O(n.\log n)$ sorting network. *Proc. 15th Ann. ACM Symp. on Theory of Computing*, pages 1–9, April 1983.

[2] M. Ajtai, J. Komlós, and E. Szemerédi. Sorting in $c\log n$ parallel steps. *Combinatorica*, 3:1–48, 1983.

[3] K. E. Batcher. Sorting networks and their applications. In *Proc. AFIPS Spring Joint Computer Conference*, volume 32, pages 307–314, April 1968.

[4] R. C. Bose and R. J. Nelson. A sorting problem. *J. Assoc. Comput. Mach.*, 9:282–296, 1962.

[5] M. Chung and B. Ravikumar. On the size of test sets for sorting and related problems. In *Proc. 1987 International Conference on Parallel Processing*. Penn State Press, August 1987.

[6] M. J. Chung and B. Ravikumar. Strong nondeterministic Turing reduction — a technique for proving intractability. *J. Comput. System Sci.*, 39(1):2–20, 1989.

[7] R. L. Drysdale. Sorting networks which generalize batcher's odd-even merge. Honors Paper, Knox College, May 1973.

[8] R. W. Floyd and D. E. Knuth. Improved constructions for the Bose-Nelson sorting problem (preliminary report). *Notices of the AMS*, 14:283, 1967.

[9] R. W. Floyd and D. E. Knuth. The Bose-Nelson sorting problem. In J. N. Srivastava, editor, *A Survey of Combinatorial Theory*. North-Holland, 1973.

[10] M. R. Garey and D. S. Johnson. *Computers and Intractability: A Guide to the Theory of NP-Completeness*. W. H. Freeman, 1979.

[11] D. E. Knuth. *Sorting and Searching*, volume 3 of *The Art of Computer Programming*. Addison-Wesley, 1973.

[12] I. Parberry. The alternating sorting network. Technical Report CS-87-26, Dept. of Computer Science, Penn. State Univ., September 1987.

[13] I. Parberry. *Parallel Complexity Theory*. Research Notes in Theoretical Computer Science. Pitman Publishing, London, 1987.

[14] I. Parberry. Sorting networks. Technical Report CS-88-08, Dept. of Computer Science, Penn. State Univ., March 1988.

[15] I. Parberry. A computer-assisted optimal depth lower bound for sorting networks with nine inputs. In *Proceedings of Supercomputing '89*, pages 152–161, 1989.

[16] I. Parberry. Single-exception sorting networks and the computational complexity of optimal sorting network verification. *Mathematical Systems Theory*, 23:81–93, 1990.

[17] M. S. Paterson. Improved sorting networks with $O(\log n)$ depth. *Algorithmica*, 5(4):75–92, 1990.

[18] T. J. Schaefer. The complexity of satisfiability problems. In *Proc. 10th Annual ACM Symposium on Theory of Computing*, pages 216–226. Association for Computing Machinery, 1978.

[19] D. C. Van Voorhis. An economical construction for sorting networks. In *Proc. AFIPS National Computer Conference*, volume 43, pages 921–926, 1974.

[20] A. Yao. Bounds on selection networks. *SIAM Journal on Computing*, 9, 1980.

Managing a Parallel Heap Efficiently

Sajal K. Das[1] and Wen-Bing Horng

Center for Research in Parallel and Distributed Computing
Department of Computer Science
University of North Texas
Denton, TX 76203-3886, USA

Abstract

We design a cost-optimal algorithm for managing a parallel heap on an exclusive-read and exclusive-write (EREW), parallel random access machine (PRAM) model. This is an improvement in space and time over the one recently proposed by Deo and Prasad [4]. Our approach efficiently employs p processors in the range $1 \leq p \leq n$, where n is the maximum number of items in a parallel heap. It is assumed that a *delete-think-insert cycle* is repeatedly performed, and each processor requires an arbitrary but the same amount of time (called the *think time*) for processing an item which in turn generates at most α (a constant) new items. The time required for deleting p items of the highest priority from the parallel heap is $O(1)$, while that for inserting at most αp new items is $O(\log n)$. With or without incorporating the think time, the speedup of our algorithm is proved to be linear, i.e. $O(p)$. Using a global, working data structure for each level of the heap, it is shown that the additional memory space required for our parallel heap is much less than that for the existing one [4]. Furthermore, the proposed algorithm retains the strict priority ordering of a sequential heap.

Index Terms: Algorithm analysis, data structure, EREW PRAM, heap, parallel algorithm, priority queue, optimal speedup.

1 Introduction

This paper concerning parallel heap data structure is an outcome of our research on parallel Branch-and-Bound (B&B) algorithms. In a B&B algorithm [6], a priority queue is often used to store the live nodes of a state space tree. Since the insertion and deletion operations on a heap can be performed efficiently, this data structure is used for implementing a priority queue. In the sequential version of a B&B algorithm, a *delete-think-insert cycle* (called an *iteration*) is repeatedly performed. At the beginning of each iteration, an item of the highest priority is deleted from the top of the heap and some processing (or thinking) is done on that item which possibly generates new items. These items are then inserted back into the heap. On the other hand, in a parallel B&B algorithm, p items (where p is the number of processors) of the highest priority

[1]This work was in part supported by a Junior Faculty Summer Research Fellowship from the University of North Texas at Denton.

are deleted at each iteration for the think phase to start. Therefore, managing heaps in parallel is an important problem.

Existing literature on parallel B&B algorithms mostly deal with upper bounds on the speedup or conditions for anomaly to occur [10-12, 15]. Although it is apparent that p items of the highest priority can be deleted from the priority queue in constant time, it is not clear how to efficiently insert the generated items. So various mechanisms have been proposed with an attempt to parallelize heaps for shared-memory parallel computers [2, 4, 7, 13, 14]. Table 1 summarizes the performance of these mechanisms on various machine models. For a comprehensive review, readers may refer to Horng and Das [5]. Quinn and Yoo [13] presented the *software pipelining* mechanism to delete an item from a heap in $O(1)$ time using $\lceil \log n \rceil$ processors, where n stands for the maximum number of items in the heap. Jones [7] proposed a *concurrent skew heap*, in which heap operations are performed in $O(\log n)$ amortized time. Biswas and Browne [2] also presented a concurrent data structure for heaps. Rao and Kumar [14] developed a practical mechanism by introducing a top-down insertion on a heap. Since at most $O(\log n)$ processors are efficiently utilized in these *concurrent heaps*, a maximum of $O(\log n)$ speedup is attainable. For increased parallelism, two other concurrent heaps — called *pipelined* and *partitioned heaps* — are proposed in [5].

Recently, Deo and Prasad [4] presented a new data structure, called the *parallel heap*, which achieves linear speedup for an n-item binary heap using p processors in the range $1 \leq p \leq n$. However, their approach suffers from several limitations as outlined here.

1. The available processors are partitioned into *general* and *maintenance processors*, which alternatively remain idle during the insert-delete and think phases.

2. At each delete-think-insert cycle, only r processors participate simultaneously in the think phase, where $r < p$ is the number of items in a heap-node.

3. The working memory space for managing the parallel heap is $O(n)$, a large value.

4. Linear speedup is not achieved if the think time is larger than $O(\log p)$.

In this paper, we propose an efficient algorithm for managing a parallel heap on an exclusive-read and exclusive-write (EREW), parallel random access machine (PRAM) model [8]. Our approach overcomes the limitations cited above, and includes improvement in time and space. A parallel heap of n items is represented as a complete k-ary (for $k \geq 2$) tree, each node of which contains x items, for $1 \leq x \leq p$. It is assumed that a delete-think-insert cycle is repeatedly performed, a processing item generates at most a constant number (α) of new items, and each processor requires the same amount of think time, which is arbitrary. Based on the number of processors available, the algorithm is presented into three distinct cases by applying the variants of a single technique. In each of these cases, p processors (for $1 \leq p \leq n$) are effectively utilized such that p items of the highest priority are deleted from the parallel heap in $O(1)$ time, and at most αp new items are inserted in $O(\log n)$ time. With or without incorporating the think time, a linear speedup is guaranteed, and thus the heap-management algorithm is cost-optimal. Furthermore, the strict priority ordering of a sequential heap is retained.

Unlike the algorithm due to Deo and Prasad [4], all p processors in our approach are treated as the general processors during the think phase; while during the insert-delete

Table 1: Summary of concurrent and parallel heap (priority queue) algorithms

Researchers	Model	# of Proc.	Insert_1	Delete_1	N	D	Insert_p	Delete_p	Type	Locking
Quinn, Yoo	MIMD-TC	$\lceil \log n \rceil$	–	$O(1)$	1	1	–	$O(p)$	Concurrent (Software Pipelining)	No
Biswas, Browne	MIMD-TC	$[1, n]$	$O(\log n)$	$O(\log n)$	1	1	$O(p + \log n)$	$O(p + \log n)$	Concurrent	Yes
Rao, Kumar	MIMD-TC	$[1, n]$	$O(\log n)$	$O(\log n)$	1	1	$O(p + \log n)$	$O(p + \log n)$	Concurrent	Yes
Jones	MIMD-TC	$[1, n]$	$O(\log n)$	$O(\log n)$	1	1	$O(p + \log n)$	$O(p + \log n)$	Concurrent (Skew Heap)	Yes
Horng, Das	MIMD-TC	p	$O(n/p + \log p)$	–	1	p	$O(p)$	$O(1)$	Concurrent (Pipelined Heap)	Yes
Horng, Das	MIMD-TC	$[1, n]$	$O(\log n)$	$O(\log n)$	1	p	$O(p + \log n)$	$O(p + \log n)$	Concurrent (Partitioned Heap)	Yes
Deo, Prasad	EREW PRAM	$[1, \log n]$	$O(\log n)$	$O(\log n)$	1	1	$O(p + \log n)$	$O(p + \log n)$	Concurrent (same as [14])	Yes
Deo, Prasad	EREW PRAM	$(\log n, n]$	$O(\log r)$	$O(1)$	r	r	$O(\log n)$	$O(1)$	Parallel	No
This paper	EREW PRAM	$[1, \log n)$	$O(\log n)$	$O(\log n)$	1	p	$O(\log n)$	$O(1)$	Parallel	No
This paper	EREW PRAM	$[\log n, \sqrt{n}]$	$O(\log n)$	$O(1)$	q	p	$O(\log n)$	$O(1)$	Parallel	No
This paper	EREW PRAM	$(\sqrt{n}, n]$	$O(\log n)$	$O(1)$	p	p	$O(\log n)$	$O(1)$	Parallel	No

[Note]: The number of items in a pipelined heap is $n \leq \lceil p^2 / \lceil \log(p+1) \rceil \rceil$.

MIMD-TC : multiple-instruction stream and multiple-data stream, tightly coupled multiprocessors

Insert_p : time required for inserting p items into the heap

Delete_p : time required for deleting p highest-priority items from the heap

N : number of items in a heap-node

D : number of items deleted at a time

Type : classification of the heap

Locking : whether employing semaphore or locking mechanism

phase, they act as the maintenance processors for maintaining the heap property at each level. The working memory space is reduced by exploiting the fact that at each level of a parallel heap, at most one heap-node is accessed at a time. Hence rather than allocating data fields to each node for inserted items or for book-keeping, we associate a global data structure with each level.

2 Preliminaries

A *heap* of n items can be represented by a complete k-ary (for $k \geq 2$) tree of depth $d = \lceil \log_k(n(k-1)+1) \rceil - 1$. The root is assumed to be at level 0. The *min-heap* (respectively, *max-heap*) *property* is one where the value of the item at any node is no greater (respectively, less) than those of the items at each of its children. In this paper, a heap always means a k-ary min-heap. For details on heap operations, refer to [2, 5].

A heap of maximum size n can be conveniently implemented by an array HEAP[0 .. $n-1$] such that the root (i.e., node 0) of the heap occupies location 0. The k children of node i in location i occupy consecutive locations $ki+1, \ldots, k(i+1)$, whereas the parent of node i is at location $\lceil \frac{i}{k} \rceil - 1$. Associated with a heap are the data fields LAST and TARGET. Let u be the current size of the heap. Then LAST $= (u-1)$ is the index of the last non-empty node of the heap, and TARGET $= u$ is the index of the first empty node of the heap. Nodes with index being equal to LAST and TARGET are called the *last node* and the *target node*, respectively. The path from the root of a heap to a target node is called the *insertion path* of the target node [14]. Suppose a target node, $TARGET$, is at level t. Let $FIRST(t) = (k^t - 1)/(k-1)$ be the index of the first (leftmost) node at level t, and let $DP = TARGET - FIRST(t) \geq 0$. Then, as shown in Horng and Das [5], the insertion path IP of the target node can be represented by the sequence of digits obtained by representing DP in radix k. That is, $DP = (IP)_k = (e_1 e_2 \cdots e_t)_k$.

An extremal case of insertion paths is that the root has its insertion path IP equal to $NULL$ with the length $|IP| = 0$. Figure 1 shows an example of 3-ary min-heap of twelve items with the insertion path from the root to the target node 12. In this figure a circle represents a node, the upper half of the circle contains the node number (or array index), the lower half contains the value of the item at that node, and the number outside the node is its path number. Here, we have $k = 3$, $d = 2$, $LAST = 11$, $TARGET = 12$, $FIRST = 4$. The path number of the target node is 022 while the insertion path $IP =$ 22. Note that $DP = TARGET - FIRST = 8 = (22)_3 = (IP)_k$.

3 Parallel Heap Algorithm

Analogous to a sequential heap, a *parallel heap* of n items, with each node containing x (for $1 \leq x \leq p$) items and thus having $m = \lceil \frac{n}{x} \rceil$ nodes, can be represented by a complete k-ary tree of depth $d = \lceil \log_k(m(k-1)+1) \rceil - 1$. The root is assumed to be at level 0, and thus the total number of levels is $d+1$. The *parallel heap property* is that the value of the items at any node of a parallel heap are no greater (or less) than those of the items at each of its children.

The parallel heap algorithm proposed in this paper can effectively utilize p processors

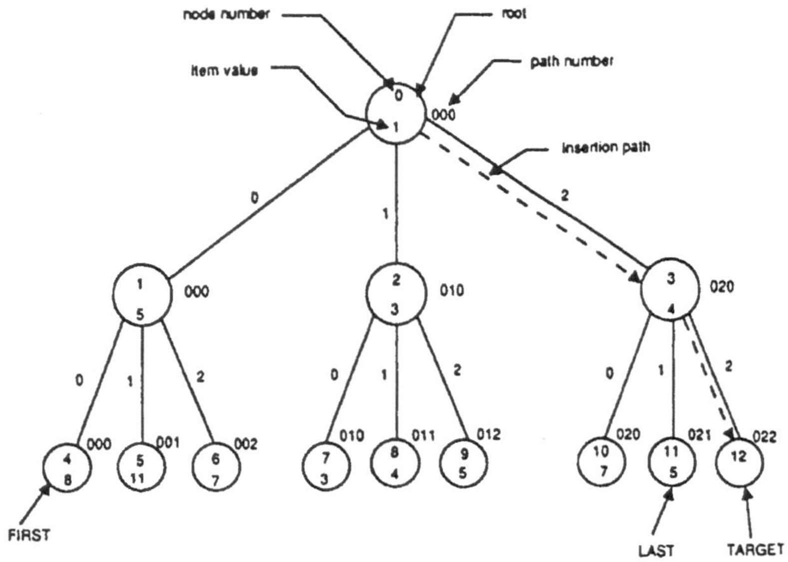

Figure 1: A 3-ary heap of 12 items

in the range $1 \leq p \leq n$ and achieves optimal speedup. The employed processors act both as the general and maintenance processors during different phases. For example, during the think phase, all p processors are treated as general processors to perform the thinking process, while during the insert-delete phase, they are considered as maintenance processors to maintain the heap property at each level. Each level of the heap is associated with $\lceil x/\log_k x \rceil$ processors for maintenance operations when $x \geq 2$. For the special case of $x = 1$, only one processor is associated with each level. Since the number of items in a node and the number of maintenance processors assigned to a level of a k-ary heap vary depending on the total number of processors available, we divide the algorithm into three distinct cases by applying appropriate variants of a single technique:

- Case I: $(nk - n)^{1/(1+\log k)} < p \leq n$, and $x = p$
- Case II: $\lceil \log_k(n(k-1)+1) \rceil \leq p \leq (nk-n)^{1/(1+\log k)}$, and $x = q$
- Case III: $1 \leq p < \lceil \log_k(n(k-1)+1) \rceil$ and $x = 1$,

where q is a function of n and p and $1 \leq q < p$. In Case III, since the number of processors is less than the number of levels, each of p levels of the heap is associated with only one maintenance processor at any pipeline cycle (described in Section 3.2).

3.1 Generic Parallel Heap Data Structures

The data structures maintained by a parallel heap consist of: (i) data structures for the heap itself and a heap-node, (ii) data structure for a level of the heap, and (iii) other data fields. Figure 2 shows a layout of a parallel heap data structure with three levels. These data structures are elaborated in the following.

(i) Let each node of a parallel heap contains x items, where $1 \leq x \leq p$. Then the parallel heap of at most n items can be represented as an array PHEAP[0 .. $m-1$], where $m = \lceil \frac{n}{x} \rceil$ is the number of nodes, each of which is a record of the form:

```
type NODE = record
          ITEM : array [0 .. x - 1] of item_type; /* items */
          #ITEMS : 0 .. x; /* currently */
          EXPECTED : 0 .. x; /* eventually */
     end_record;
var  PHEAP : array [0 .. m - 1] of NODE; /* parallel heap */
```

The array ITEM[0 .. $x-1$] stores items at a node, or the ***substitute items*** for an iteration of the delete-reheapification at that node. The field #ITEMS indicates how many items are currently in the node, and EXPECTED shows how many are eventually expected to be in it. An item which should be in the node but not yet available is called an ***expected item***. In the following, we use #ITEMS(i) and EXPECTED(i) to denote the values of the fields #ITEMS and EXPECTED of node i, respectively, and use ITEM(j, i) to denote the value of the field ITEM[j] of node i.

(ii) A data structure, called BLOCK, for each level of the heap is used for maintaining the parallel heap property. It is defined as follows.

```
type BLOCK = record
          WNODE : 0 .. m - 1; /* node to be processed */
          WTYPE : 0 .. α;
               /* 0 for delete-reheapification,
                  i for the ith insert-reheapification, where 1 ≤ i ≤ α */
          INSERT_ITEM : array [0 .. x - 1] of item_type; /* items to be inserted */
          LENGTH : 0 .. x - 1; /* number of inserted items */
          TARGET : 0 .. m - 1; /* index of the target node */
          INSERTION_PATH : 0 .. m - 1; /* insertion path */
     end_record;
var  LEVEL : array [0 .. d] of BLOCK; /* d is the depth of the parallel heap */
```

Like a ***context block*** in operating systems, the data structure BLOCK for each level of the heap stores all the required information for a node at that level which needs to perform an iteration of the insert- or delete-reheapification. The field WNODE is the index of the node to be processed at that level. If WNODE $= -1$, then no operations are performed at that level during that pipelined cycle. The field WTYPE indicates what kind of reheapification needs to be performed at node WNODE. If WTYPE $= 0$, then an iteration of delete-reheapification is performed. On the other hand, if WTYPE ≥ 1, an insert-reheapification process is performed. The array INSERT_ITEM[0 .. $x-1$] contains the items to be inserted during that insert-reheapification. LENGTH indicates how many items are to be inserted during that insert-reheapification, and TARGET is the index of the node where these items should be inserted eventually. INSERTION_PATH is a nonnegative integer representing the remaining path from WNODE to TARGET.

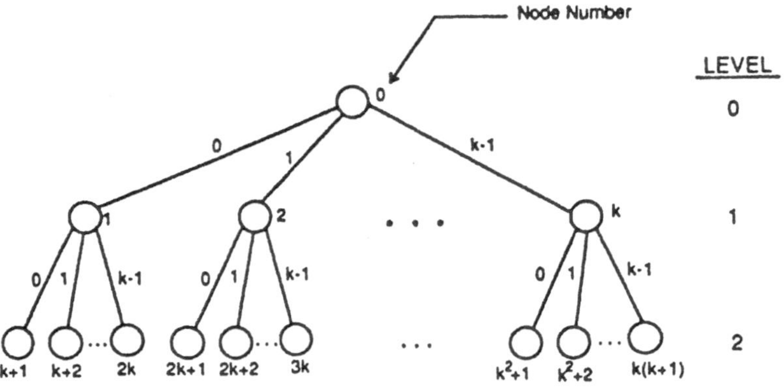

(a) A k-ary heap of three levels

ITEM [0 .. x -1]

ITEMS

EXPECTED

(b) Data Structure for a node

WNODE WTYPE

INSERT_ITEM [0 .. x -1]

LENGTH

TARGET

INSERTION_PATH

(c) Data Structure for a level

READY [0 .. p -1]

NEW_ITEM [0 .. $\alpha \cdot p$ -1]

WANTED_BY [0 .. $x \cdot (y+1)$ -1]

LAST

(d) Other data fields

Figure 2: Data structures of a parallel heap

The array LEVEL contains $d+1$ data structures of type BLOCK such that level l of the heap is associated with LEVEL[l]. In the following, WNODE(l), WTYPE(l), LENGTH(l), TARGET(l), and INSERTION_PATH(l) denote the values of their corresponding data fields at level l, respectively. Also INSERT_ITEM(j,l) denotes the value of INSERTION_ITEM[j] at level l.

(iii) The remaining data fields associated with a parallel heap are three arrays READY, NEW_ITEM, and WANTED_BY, and an integer variable LAST. Assuming $y = \lceil \frac{p}{x} \rceil$, the declarations are:

```
var  READY : array [0 .. p − 1] of item_type;
     NEW_ITEM : array [0 .. α * p − 1] of item_type;
     WANTED_BY : array [0 .. x * (y + 1) − 1] of −1 .. p − 1;
     LAST : 0 .. m − 1; /* index of the last node */
```

The array READY contains p items (if the parallel heap is not empty) of the highest priority since the previous think phase is complete. The array NEW_ITEM is used to store at most αp new items produced by the processing items of the current think phase. Processors access the READY array simultaneously, each P_i reading the item READY[i], for $0 \leq i \leq p$, in constant time for the next think phase to process. After this think phase is complete, processor P_i puts its new generated items in consecutive locations $\alpha i, \alpha i+1, \cdots, \alpha(i+1)-1$ of the array NEW_ITEM. The field LAST is the index of the first node whose EXPECTED value is less than p. The array WANTED_BY is used by maintenance processors to check whether to move items from the bottom of the heap to the array NEW_ITEM. Initially, the field WANTED_BY[i], for $0 \leq i < x(y+1)$, is set to -1 to indicate that no movement of items is required. The array WANTED_BY is divided into $y+1$ subarrays such that the subarray WANTED_BY[$x(y-i)$ .. $x(y-i+1)-1$] is for node LAST$-i$, for $0 \leq i \leq y$. If the total number, say w, of new generated items after a think phase is less than p, then the additional $d_w = p - w$ items (called the *wanted items*) will be moved from the bottom of the heap to the subarray NEW_ITEM[w .. $p-1$] as follows. There are d_w processors allocated, one-to-one, to the last d_w items (including expected ones) in the parallel heap such that the last item (i.e., the item ITEM[EXPECTED-1] in node LAST) will be moved to the field NEW_ITEM[$p-1$], the last but one item to NEW_ITEM[$p-2$], and so on. If these wanted items are already in their target nodes, then all of them are moved to the subarray NEW_ITEM[w .. $p-1$] at the same time. If any one of these wanted items is still somewhere else along its insertion path, then its corresponding WANTED_BY fields will be set to the index of the subarray NEW_ITEM[w .. $p-1$] indicating where it should be moved to. The fields #ITEMS and EXPECTED of the nodes which move wanted items and/or is going to move expected items to NEW_ITEM array are updated appropriately. The data field LAST is updated accordingly. At the next iteration of the insert-reheapification, the maintenance processors at each level of the heap will first check the WANTED_BY array. If any element WANTED_BY[j] is set to value $g \neq -1$, then the corresponding wanted item at that level is moved to the NEW_ITEM[g], and WANTED_BY[j] is reset to -1.

3.2 Generic Parallel Heap Operations

The parallel heap algorithm presented in this paper is based on the framework due to Deo and Prasad [4]. In our approach, we also use the parallel prefix algorithm due to Ladner and Fischer [9] for packing the array NEW_ITEM, the optimal parallel mergesort due to Cole [3] for sorting new generated items, and the adaptive bitonic merging algorithm due to Bilardi and Nicolau [1] for maintaining the parallel heap property at each level of the heap. On an EREW PRAM model, each of the parallel prefix and bitonic merge algorithms on input of size $O(x)$ require $O(\log x)$ using $\lceil x/\log x \rceil$ processors; while Cole's mergesort has $O(\log x)$ time complexity for sorting $O(x)$ items employing x processors.

In order to maintain the parallel heap property, two generic operations — *delete-reheapification* and *insert-reheapification* — are performed. Since each iteration of the parallel reheapification is performed in a pipelined fashion, at most one node may violate the parallel heap property at each level. Such a state is called a *partial parallel heap.* In the following we briefly outline parallel reheapification procedures.

An *iteration* of the *parallel delete-reheapification* at level l is performed if BLOCK[l]. WTYPE = 0. Let i be the value of BLOCK[l].WNODE. If node i satisfies the parallel heap property, the current reheapification terminates. Otherwise, merge the substitute items in PHEAP[i].ITEM and the items in its children, move the items in the merger to their appropriate places, and reset the data fields of the next lower level. A formal algorithm, called Parallel_Delete_Reheapification, for an iteration of the delete-reheapification is presented in [5], which uses a subprocedure Parallel_Bitonic_Merge.

An *iteration* of the *parallel insert-reheapification* at level l is performed if BLOCK[l]. WTYPE $\geq$ 1. Let i be the value of BLOCK[l].WNODE. The procedure Parallel_Insert _Reheapification first moves the wanted items to the array NEW_ITEM, if necessary. Then, it merges these leftover inserted items and those in PHEAP[i].ITEM, moves items in the merger to their appropriate places, and resets the data fields of the next lower level. The details of this algorithm utilizing a subprocedure Parallel_Prefix for packing is presented in [5].

The maintenance processors at each level l of a partial parallel heap performs either an iteration of the insert-reheapification or delete-reheapification on node WNODE(l), according to the value of WTYPE(l). One such iteration is called a *pipeline cycle*, which is described as the following procedure.

```
procedure Pipeline_Cycle(l);
begin
    if (BLOCK[l].WTYPE = 0) then Parallel_Delete_Reheapification(l)
    else if (BLOCK[l].WTYPE ≥ 1) then Parallel_Insert_Reheapification(l)
end; /* Pipeline_Cycle */
```

Analyzing the subprocedures Parallel_Bitonic_Merge and Parallel_Prefix, an iteration of the parallel delete-reheapification (or insert-reheapification) as well as a pipeline cycle require $O(\log x)$ time. It is to be noted that a pipeline cycle defined in this paper is different from that in [4], where one iteration of the delete-reheapification and two iterations of the insert-reheapification are included in one pipeline cycle.

3.3 Case I: For p Processors, where $(nk-n)^{1/(1+\log k)} < p \leq n$

As mentioned earlier, we divide our algorithm into three distinct cases based on the number of processors available. Since we apply variants of the same technique for these three cases, for brevity, we present only Cases I and II in the following. Interested readers may find the third case in [5].

In Case I, $(nk-n)^{1/(1+\log k)} < p \leq n$, $x = p$, $m = \lceil \frac{n}{p} \rceil$, and $d = \lceil \log_k(m(k-1)+1) \rceil - 1$. Each level in the heap is associated with $\lceil p/\log p \rceil$ processors. The parallel heap algorithm divides a delete-think-insert cycle into two phases: ***think phase*** and ***delete-insert phase*** as discussed in the following.

THINK PHASE:

During the think phase, all the available p processors are switched to serve as general processors. Each processor P_i, for $0 \leq i < p$, performs the thinking process by accessing item READY[i]. After this think phase is complete, processor P_i generates at most α new items and places them in the subarray NEW_ITEM[αi .. $\alpha(i+1)-1$]. There are at most αp new items in the array NEW_ITEM.

INSERT-DELETE PHASE:

All p processors are switched back to serve as maintenance processors during this phase which includes the following five steps.

Step 1: *Packing new generated items*

Since some processors may generate less than α new items, the parallel prefix algorithm [9] is employed to pack these new items and to calculate the total number of such items produced by using the maintenance processors at level 0 (the root) of the heap. While packing new generated items, the maintenance processors at other levels perform one pipeline cycle at the same time.

Step 2: *Moving the wanted items to the array NEW_ITEM, if necessary*

After the new items are packed, the maintenance processors at level 0 check whether the total number of new generated items, say w, is less than p. If it is the case, all p processors are employed in moving $d_w = p - w$ items to the subarray NEW_ITEM[w .. $p-1$] if they are at the bottom of the heap, or set their WANTED_BY fields if they are still in their insertion paths. If any field of the array WANTED_BY is set, then one pipeline cycle is performed on the heap to move all the wanted items to the array NEW_ITEM, and w is updated to be the total number of items in NEW_ITEM.

Step 3: *Sorting the array NEW_ITEM*

If there are no items in NEW_ITEM, the algorithm terminates. Otherwise, the p available processors perform Cole's (optimal) parallel mergesort on NEW_ITEM.

Step 4: *Filling out the array READY*

After sorting the new items, all processors are switched back to serve as maintenance processors, and one pipeline cycle is issued at each level of the heap. Meanwhile, the maintenance processors at level 0 use bitonic merging algorithm to merge the items in the root and the first p items in NEW_ITEM if $w \geq p$ (or all the items in NEW_ITEM if $w < p$). Then move the first p items of the merger to array READY and move the remaining ones to the root. Also both the fields WNODE and WTYPE of BLOCK[0] are set to 0 to indicate that heap-node 0 needs to perform an iteration of the delete-reheapification. After the current pipeline cycle is complete, one more pipeline cycle is issued on the heap so that, at this time the maintenance processors at level 0 perform an iteration of the delete-reheapification.

Step 5: *Inserting the remaining items in the array NEW_ITEM*

Let g be the number of remaining items in the array NEW_ITEM, where $0 \le g \le p(\alpha - 1)$. If $g = 0$, no further operation is needed since the heap has satisfied the partial parallel heap property at the root. Otherwise, let $d_p = p -$ EXPECTED(LAST). Then g can be expressed as $g = u + h \times p + v$, where $0 \le h < \alpha - 1$, $0 \le v < p$, and

$$u = \begin{cases} g & \text{if } g < d_p \\ d_p & \text{otherwise.} \end{cases}$$

This means that we need one (the first) iteration of insert-reheapification for inserting u new items, h (if $h \neq 0$) iterations for inserting p new items each, and one (the last, if $v \neq 0$) iteration for inserting v new items. Thus, at most α iterations of insert-reheapification process are required at the root. That is, at most α pipeline cycles will be issued on the heap. During each of these iterations, the maintenance processors at each level (other than level 0) perform their required operations, while the maintenance processors for level 0 will (i) set the field INSERTION_PATH according to the value LAST (as described in Section 2), (ii) increment the field LAST by one, (iii) update the fields WNODE and WTYPE to 0 and its corresponding iteration i, respectively, (iv) merge the items in the root and those in the array NEW_ITEM required by its corresponding iteration, and (v) follow the usual operation of an insert-reheapification process. After all such iterations are complete, the heap satisfies the partial parallel heap property and it starts the next think phase.

3.3.1 Time Complexity

Recall that one pipeline cycle requires $O(\log p)$ time in this case. Step 1 uses a parallel prefix algorithm to pack new items which requires $T_1 = O(\log p)$ time. In Step 2, the time required to move the wanted items from the bottom of the heap to the array NEW_ITEM and/or to set the fields WANTED_BY, and to perform one more pipeline cycle to move the wanted items to the array NEW_ITEM (if necessary) is given by $T_2 = O(1) + O(\log p)$. Step 3 performs parallel mergesort on the array NEW_ITEM in $T_3 = O(\log p)$ time. In Step 4, the time required by the root to merge the items in the subarray NEW_ITEM[0 .. $\beta - 1$] and those in the root, to move items to READY and the root, and to perform one pipeline cycle is $T_4 = 4 \times O(\log p)$. In Step 5, at most α pipeline cycles are issued for the root to perform an insert-reheapification process and the worst-case time is $T_5 = \alpha \times O(\log p)$ time. Therefore, the overall (parallel) time for an insert-delete phase to delete p items and to insert at most αp items is $T_{par} = T_1 + T_2 + T_3 + T_4 + T_5 = O(\log p)$, since α is a constant. Thus, the speedup in this case is $T_{seq}/T_{par} = O(p)$, where $T_{seq} = O(p \log n)$ is the time required for performing $O(p)$ sequential heap operations, and $\log p = O(\log n)$, for $(nk - n)^{1/(1+\log k)} < p \le n$.

Assuming the time required for performing a think phase to be T_{think}, the total time for a delete-think-insert cycle is $O(\log p) + T_{think}$. Obviously, the speedup for this case is also $O(p)$. Therefore, with or without incorporating the think time, our parallel heap algorithm for Case I achieves linear speedup (within a constant factor).

3.3.2 Space Complexity

The memory space required for the data structure PHEAP is $m(p+2) = \lceil \frac{n}{p} \rceil (p+2)$, that for LEVEL is $(d+1)(p+5) = \lceil \log_k(m(k-1)+1) \rceil (p+5)$, and that for other data fields is $(\alpha+2)p+1$. Thus, the additional memory space (other than that required for storing the heap) is approximately $2(n/p) + (p+5)\log_k(n/p) + (\alpha+3)p$. For the special case when $k = \alpha = 2$, the number p of processors can be represented as $p = n^{\epsilon}$, where $1/2 < \epsilon < 1$. Then the additional memory space in our algorithm is approximately $O(n^{\epsilon}\log n)$, whereas that required in [4] is $O(n)$. In general, since p is much less than n, it is clear that our algorithm can save much more additional memory space than that in [4].

3.4 Case II: For p Processors, where $\lceil \log_k(n(k-1)+1) \rceil \leq p \leq (nk-n)^{1/(1+\log k)}$

In this case, $x = q$, $m = \lceil \frac{n}{q} \rceil$, and $d = \lceil \log_k(m(k-1)+1) \rceil - 1$, where q is a function of n and p, and $1 \leq q < p$. Each level is associated with $\lceil q/\log q \rceil$ (or 1 if $q = 1$) processors. Thus the inequality $p \geq (d+1)\lceil q/\log q \rceil$ must hold, which yields $q \approx (p\log q)/(d+1)$. The algorithm for this case is similar to that in Case I except that there are at most $\lceil \frac{p}{q} \rceil$ iterations of the delete-reheapification for filling out the READY array and at most $\lceil (\alpha-1)p/q \rceil + 1$ iterations of the insert-reheapification for inserting the remaining items in the NEW_ITEM array to the heap. Since the think phase and Steps 1, 2, and 3 of the insert-delete phase are similar to those in Case I, in the following we describe only Steps 4 and 5 of the insert-delete phase.

INSERT-DELETE PHASE:

Step 4: *Filling out the array READY*

Let $m_w = \min(p, w)$ and $m_w = u \times q + v$, where $0 \leq u < \lceil m_w/q \rceil$ and $0 \leq v < q$. This indicates that there are u iterations of moving q items and one (the last, if $v \neq 0$) iteration of moving v items from NEW_ITEM to READY. Each of these iterations is performed as follows. At iteration i, for $1 \leq i \leq u$, the maintenance processors at the root use bitonic merging algorithm to merge the items in the root and those in the subarray NEW_ITEM$[(i-1)q \,..\, iq-1]$. At the same time, one pipeline cycle is performed at other levels of the heap. Then, move the first q items out of the merger to READY$[(i-1)q \,..\, iq-1]$, and move the remaining items to the root. Also, update WNODE and WTYPE of level 0 to zero to indicate that node 0 needs to perform an iteration of the delete-reheapification. After the current pipeline cycle is complete, one more such cycle is issued to the heap so that the root performs the delete-reheapification at this time. If $v \neq 0$, the $(u+1)$-st iteration is performed as follows. While the maintenance processors at other levels of the heap perform one pipeline cycle, the v items in NEW_ITEM$[uq \,..\, m_w-1]$ are merged with those in the root. Out of the merger, the first v items are placed into the subarray READY$[uq \,..\, m_w - 1]$, the remaining items are moved to the root, and WNODE and WTYPE of level 0 are set to zero. After the current pipeline cycle is complete, one more pipeline cycle is issued to the heap as in the previous iteration.

Step 5: *Inserting the remaining items in the array NEW_ITEM*

If $w \leq p$, no insert-reheapification is needed since at this point, no items are left in NEW_ITEM. Otherwise, let $g = w - p$ be the number of remaining items in NEW_ITEM, where $0 \leq g \leq (\alpha - 1)p$. Denoting $u = q -$ EXPECTED(LAST), we get $g = u + h \times q + v$, where $0 < u \leq q$, $0 \leq h < \lceil (\alpha - 1)p/q \rceil$, and $0 \leq v < q$. Then follow the same method as in Case I to perform at most $h + 2$ iterations of the insert-reheapification and start the next think phase.

3.4.1 Time Complexity

In this case, one pipeline cycle requires $O(\log q)$ time. Therefore, the time required for each of Steps 1 and 3 is $O(\log p)$, and that required for Step 2 is $O(\log q)$. Step 4 requires $2(u+1) \times O(\log q) = O(p/q) \times O(\log q) = O(\log n)$ time. Step 5 performs at most $h + 2$ pipeline cycles and it requires $(h + 2) \times O(\log q) = O(p/q) \times O(\log q) = O(\log n)$ time. Therefore, the overall (parallel) time for an insert-delete phase to delete p items and to insert at most αp items is $O(\log n)$. Following the same approach as in Case I, with or without incorporating the think time, the speedup for this range of processors is still $O(p)$.

3.4.2 Space Complexity

The memory space required for data structure PHEAP is $m(q + 2) = \lceil \frac{n}{q} \rceil (q + 2)$, that for LEVEL is $(d + 1)(q + 5) = \lceil \log_k (m(k - 1) + 1) \rceil (q + 5)$, and that for other data fields is $(\alpha + 2)p + 1$. Therefore, the additional memory space is approximately $2(n/q) + (q + 5)\log_k(n/q) + q + (\alpha + 2)p$. For the special case when $k = \alpha = 2$, the number of processors p is in the range $\lceil \log(n + 1) \rceil \leq p \leq \sqrt{n}$. Then the additional memory space in our algorithm is $O((n/q) + q \log(n/q) + p)$. Clearly, on an average, our algorithm for this range of processors saves much more additional memory space than that required in [4], which is always $O(n)$.

3.4.3 Calculation of q

As mentioned earlier, each node of the heap in Case II contains q items, where q is a function of n and p. According to Section 3.4, the following inequality holds:

$$p \geq \lceil q / \log q \rceil \times \lceil \log_k (m(k - 1) + 1) \rceil,$$

from which an approximate equation can be derived:

$$q \approx p \log q / \lceil \log_k (m(k - 1) + 1) \rceil.$$

Now the approximate value of q can be calculated by the following algorithm.

Table 2: Performance of parallel binary heaps

Researchers	Das & Horng		Deo & Prasad	
Complexity	Space	Time	Space	Time
Case I	$O(n^{\epsilon}\log n)$, $\frac{1}{2} < \epsilon \leq 1$	$O(T + \log n)$	$O(n)$	$O(\frac{p}{r}(T + \log r))$
Case II	$O(p + \frac{n}{q} + q\log\frac{n}{q})$	$O(T + \log n)$	$O(n)$	$O(\frac{p}{r}(T + \log r))$
Case III	$O(\log n)$	$O(T + \log n)$	$O(n)$	$O(T + \log n)$

Algorithm: *Computing the value of q*
Input: n, p, and k.
Output: q.

1. $q' \leftarrow k$;
   repeat $q \leftarrow q'$; $q' \leftarrow p\log q/\lceil\log_k(m(k-1)+1)\rceil$ **until** $|q'-q| < 1$;
   $q \leftarrow \lceil q' \rceil$;
2. $m = \lceil n/q \rceil$; $l \leftarrow \lceil\log_k(m(k-1)+1)\rceil$; $s \leftarrow \lceil q/\log q \rceil$;
   if $s \times l \leq p$ **then return**(q);
3. $s \leftarrow \lceil p/l \rceil$;
   repeat $q \leftarrow q'$; $q' \leftarrow s\log q$ **until** $|q'-q| < 1$;
   $q \leftarrow \lceil q' - 1 \rceil$;
   goto Step 2.

3.5 Case III: For p Processors, where $1 \leq p < \lceil\log_k(n(k-1)+1)\rceil$

For the number of processors in this range, each node of the parallel heap contains only one item. Consequently, $m = n$, and $d = \lceil\log_k(n(k-1)+1)\rceil - 1$. Also, the data structures of the parallel heap and an iteration of the insert- or delete-reheapification can be simplified. Since $p \leq d$, instead of statically allocating a certain number of processors to each level as in the previous cases, each of p processors is dynamically assigned to a level of the parallel heap as the maintenance processor which performs the insert- or delete-reheapification. Due to space limitation, the details are omitted here and readers may refer to [5]. The performance of our algorithm for this case are summarized below. One pipeline cycle requires $O(1)$ time, and a linear speedup is achieved with or without incorporating the think time. The additional memory space required is approximately $5\log_k n + (\alpha+2)p$. For the special case when $k = \alpha = 2$, p is in the range $1 \leq p < \lceil\log(n+1)\rceil$ and the additional space required by our algorithm is $5\log n + 4p = O(\log n)$ while that space in [4] or [14] is $2n + \log n = O(n)$.

Table 2 provides a comparison of a special case of our algorithm on binary heaps (i.e., for $k = \alpha = 2$) with the approach due to Deo and Prasad. In this table, T stands for T_{think}, the quantity q is the number of items in a node for Case II of our algorithm, and r is the number of items in a node in Cases I and II of [4].

4 Proofs of Correctness

It can be shown that our parallel heap algorithm is deadlock-free, starvation-free, and provides mutual exclusion on each heap-node [5]. The correctness of the proposed algorithm is presented in Theorem 1. Before proving it, we establish four lemmas which can be proved from the generic parallel heap operations. As mentioned, a partial parallel heap stands for a k-ary partial heap with each node containing x items, where $1 \leq x \leq p$.

Lemma 1: For a partial parallel heap, the items in each node of the heap and those in INSERT_ITEM at each level are in sorted order.

Lemma 2: For a partial parallel heap, after an iteration of the insert-reheapification on a node i at level l, node i still satisfies the parallel heap property.

Lemma 3: For a partial parallel heap, after an iteration of an iteration of the delete-reheapification on a node i is complete, the child which is not a leaf node and has the smaller last item among the children of node i may violate the parallel heap property, while nodes i and the other children still satisfy this property.

Lemma 4: For a partial parallel heap, only those nodes which need to perform the delete-reheapification may violate the parallel heap property.

Theorem 1: After each insert-delete phase in a partial parallel heap, the values of the items in the root are less than or equal to those of the items in its descendants, and the items in the array READY are in sorted order with the values of these items being less than or equal to those of the items in the root of the heap.

Proof (by induction): The induction parameter s in this theorem is the iteration number of delete-think-insert cycle applied so far.

Induction Basis: At the beginning of the first insert-delete phase (i.e., $s = 1$), only one item is in the heap. After this phase is complete, the item is moved to the array READY and the heap becomes empty. Clearly, the theorem holds for this case.

Induction Hypothesis: Assume that at the end of i-th iteration (i.e., $s = i$) of delete-think-insert cycle, the theorem holds true.

Induction Step: Now, let us consider the beginning of the $(i + 1)$-th iteration of insert-delete phase. By definition, before an iteration of the delete-reheapification is introduced, the first x items of the smallest values from the merger of items in array NEW_ITEM and those in the root, are moved to array READY. Thus, after performing all delete-reheapification iterations in an insert-delete phase on the root, the items in READY are in sorted order; and their values are no more than those of the items in the heap and the remaining ones in NEW_ITEM. Essentially, we have a partial parallel heap with the root-node satisfying the parallel heap property, thereby implying that the items in the root contain the smallest values of the entire heap. □

Example: Let us illustrate the performance of our algorithm and compare it with Deo-and-Prasad's algorithm for $n = 2^{32}$ and $p = 2^{13}$. Note that the given values of n and p correspond to Case II of our approach. The results are given in Table 3.

We observe the following facts from Table 3: (i) the number of items in a node in our algorithm is almost 1.7 times larger than that in a node in the algorithm due to Deo and Prasad, and (ii) in our algorithm 8180 processors are allocated as the maintenance processors during the insert-delete phase and all of 8192 processors are switched back

Table 3: Comparison of two algorithms when $n = 2^{32}$ and $p = 2^{13}$

Algorithm	#items per node	#nodes in heap	#levels of heap	#processors per level	#maintenance processors	#general processors
Das and Horng	5029	854041	20	409	8180	8192
Deo and Prasad	2951	1455480	21	257	5397	2951

to serve as the general processors during the think phase. However, in the algorithm presented in [4], only 5397 processors are dedicated to serve as the maintenance processors and 2951 processors act as the general processors. Therefore, approximately 66% and 34% of the available processors are idle during the think phase and insert-delete phase, respectively.

5 Concluding Remarks

The contribution of this paper is to develop a cost-optimal algorithm for managing a parallel heap, which is also an improvement over the parallel (binary) heap proposed in [4], for an EREW PRAM model. Our algorithm efficiently employs p processors in the range $1 \leq p \leq n$. We represent a parallel heap as a complete k-ary tree with each node containing x items (for $1 \leq x \leq p$) depending on the number of processors available. A delete-think-insert cycle is repeatedly performed, assuming that the think time is identical for each processor. Also, it is assumed that each processing item generates at most α (a constant) new items. The proposed algorithm is divided into three distinct cases, each of which utilizes an appropriate variant of a single strategy. The p processors are efficiently utilized such that the time required for deleting p items with the highest priority from the heap is $O(1)$, while that for inserting at most αp new items is $O(\log n)$. Though our algorithm originates from the idea developed by Deo and Prasad [4], the design strategy and data structures differ in several ways. Some of the salient features and major advantages of our approach are highlighted below.

1. At each delete-think-insert cycle, p items are deleted for the think phase to start.

2. Rather than partitioning p processors into two disjoint sets for distinct usages, namely *general* and *maintenance* processors for the think and insert-delete phases, we use all of them efficiently for both purposes.

3. Unlike in [4], where all the working data structures are stored in each heap-node, we propose a global data structure associated with each level of the heap. This leads to a reduction of extra memory space for implementing our algorithm.

4. Variants of the same algorithmic strategy are applied for three different ranges of available processors between 1 and n, whereas the authors in [4] apply the mechanism proposed in [14] for the number of processors in the range $1 \leq p \leq \log n$.

5. With or without incorporating arbitrary think time, our algorithm is proved to achieve optimal speedup.

6. The number of items stored in a heap-node is greater than or equal to that in [4], and it depends on the available range of processors.

As part of future work, we intend to implement the parallel heap algorithm on commercial shared memory parallel computers, in order to verify how closely its actual performance follows the theoretical analyses developed here. Since this algorithm allows simultaneous deletion of p items from the heap, we are currently investigating how to apply it for estimating the speedup of parallel algorithms based on the branch-and-bound strategy. Our preliminary observations are encouraging, which will be reported in a forthcoming paper.

References

[1] G. Bilardi and A. Nicolau, "Adaptive Bitonic Sorting: An Optimal Algorithm for Shared-Memory Machines," *SIAM J. Comput.*, vol. 18, no. 2, pp. 216-228, Apr. 1989.

[2] J. Biswas and J. C. Browne, "Simultaneous Update of Priority Structures," *Proc. Int. Conf. Parallel Process.*, 1987, pp. 124-131.

[3] R. Cole, "Parallel Merge Sort," *SIAM J. Comput.*, vol. 17, no. 4, pp. 770-785, Aug. 1988.

[4] N. Deo and S. Prasad, "Parallel Heap," *Proc. Int. Conf. Parallel Process.*, vol. III, 1990, pp. 169-172.

[5] W.-B. Horng and S. K. Das, *Heaps — Concurrency and Parallelism*, Tech. Rep. #N-90-003, Dept. Comput. Sci., Univ. North Texas, Denton, Mar. 1990.

[6] E. Horowitz and S. Sahni, *Fundamentals of Computer Algorithms*, Computer Science Press, Rockville, Maryland, 1978.

[7] D. W. Jones, "Concurrent Operations on Priority Queues," *Commun. ACM*, vol. 32, no. 1, pp. 132-137, Jan. 1989.

[8] R. M. Karp and V. Ramachandran, "Parallel Algorithms for Shared-Memory Machines," in *Handbook of Theoretical Computer Science, Volume A: Algorithms and Complexity* (J. van Leeuwen, Ed.), MIT Press, Cambridge, MA, 1990, pp. 869-941.

[9] R. E. Ladner and M. J. Fischer, "Parallel Prefix Computation," *J. ACM*, vol. 27, no. 4, pp. 831-838, Oct. 1980.

[10] T.-W. Lai and S. Sahni, "Anomalies in Parallel Branch-and-Bound Algorithms," *Commun. ACM*, vol. 27, no. 6, pp. 594-602, June 1984.

[11] T.-W. Lai and A. Sprague, "Performance of Parallel Branch-and-Bound Algorithms," *IEEE Trans. Comput.*, vol. C-34, no. 10, pp. 962-964, Oct. 1985.

[12] G.-J. Li and B. W. Wah, "Coping with Anomalies in Parallel Branch-and-Bound Algorithms," *IEEE Trans. Comput.*, vol. C-35, no. 6, pp. 568-573, June 1986.

[13] M. J. Quinn and Y. B. Yoo, "Data Structures for the Efficient Solution of Graph Theoretic Problems on Tightly-Coupled MIMD Computers," *Proc. Int. Conf. Parallel Process.*, 1984, pp. 431-438.

[14] V. N. Rao and V. Kumar, "Concurrent Access of Priority Queues," *IEEE Trans. Comput.*, vol. 37, no. 12, pp. 1657-1665, Dec. 1988.

[15] B. Wah and Y. Ma, "MANIP — A Parallel Computer System for Implementing Branch-and-Bound Algorithm," *Proc. 8th Annu. Symp. Comput. Archi.*, 1982, pp. 239-262.

Parallel complexity in the design and analysis of concurrent systems

Carme Àlvarez* José L. Balcázar* Joaquim Gabarró* Miklós Sántha+

Dep. de Llenguatges i Sistemes Informatics
Universitat Politècnica de Catalunya
Pau Gargallo 5, 08028 Barcelona
Spain

CNRS - LRI
Université Paris-Sud
91405 Orsay
France

Keywords: Petri nets; partially commutative monoids; CCS; PRAM algorithms; boolean circuits; P-completeness.

Abstract: We study the parallel complexity of three problems on concurrency: decision of firing sequences for Petri nets, trace equivalence for partially commutative monoids, and strong bisimilarity in finite transition systems. We show that the first two problems can be efficiently parallelized, allowing logarithmic time Parallel RAM algorithms and even constant time unbounded fan-in circuits with threshold gates. However, lower bounds imply that they cannot be solved in constant time by a PRAM algorithm. On the other hand, strong bisimilarity in finite labelled transition systems can be classified as P-complete; as a consequence, algorithms for automated analysis of finite state systems based on bisimulation seem to be inherently sequential in the following sense: the design of an efficient parallel algorithm to solve any of these problems will require an exceedingly hard algorithmic breakthrough.

1. Introduction

Given the intrinsic difficulty of designing large software systems, it is natural that software tools would be designed to help in performing this task. The possibility of formalizing both specifications and implementations in the same, or in a closely related, formal language yields the potential of automated analysis, allowing for early checking of correctness and provably correct prototypes.

The design of correct concurrent programs is even more difficult than in the sequential case, and their verification using formal systems may give rise to formidable computational problems. For instance, the study of the correctness and liveness properties of mutual exclusion algorithms for just two processes already requires resorting to computerized analysis [27]; if more processes are considered, the state space soon becomes intractable.

One reason to develop concurrent programs stems from the fact that important advantages can be gained from the use of massive parallelism. In view of the large number of

* Research supported by the ESPRIT II Basic Research Actions Program of the EC under contract No. 3075 (project ALCOM).

+ Research supported by the Programme MERCURE of the DCSTD of the Ministère Français des Affaires Etrangères and the DGICYT of the Ministerio de Educación y Ciencia de España. This research was performed while visiting the Dep. de Llenguatges i Sistemes Informàtics of the Universitat Politècnica de Catalunya.

parallel algorithms discovered in recent years (see [10] and [16]), it might be hoped that one such application would be the study of concurrent systems, and that algorithms running on highly parallel machines could perform automated analysis of large concurrent programs substantially faster than sequential algorithms. Such a behaviour corresponds to a running time roughly logarithmic in the size of the state space (assumed finite); and being able to tackle problems of relevant size corresponds to algorithms that use a large but feasible number of processors (cf. the definition of the class *NC* below).

One of the first models issued to study concurrent systems was the Petri net model. A Petri net consists of two different kinds of objects: places and transitions. Places serve to model pre and post conditions and transitions model events. A transition needs to satisfy some conditions to be fired, and its firing changes the valuations on the places (see below for exact definitions). The net evolves firing transitions sequentially and the behaviour of the whole system is described by the set of all possible firing sequences.

For each fixed Petri net, we exhibit an *NC* algorithm to decide very efficiently in parallel whether a given sequence of transitions is a firing sequence. We also discuss some lower bounds on the parallel time necessary to solve this problem.

In a monoprocessor environment, a concurrent system is fully described by the set of all sequential evolutions. A possible evolution of the system is described by a finite word called trace. Let us consider with more attention the sequencing of two events x and y in a trace $w = \cdots xy \cdots$. Let $w' = \cdots yx \cdots$ be the trace obtained from w by commuting the order of the events x and y. We have two different possibilities:

- The events x, y are independent from each other. In this case the order of execution is irrelevant and x, y commute. Then the traces w and w' correspond to the same parallel behaviour.
- The execution of x modifies the environment of y. Then these events are in conflict. The trace w' represents a behaviour different from w.

A basic question is: given two traces, do they model the same concurrent behaviour? A way to deal with this approach is to consider partially commutative monoids. This framework has been fully developed by Mazurkiewicz [18]. A mathematical characterization of trace equivalence was found in [8], and it can be used to find a fast sequential algorithm. Here we prove the existence of an *NC* algorithm, and discuss also some lower bounds.

A capability that seems natural to expect from software tools for aiding the design of concurrent systems is to be able to decide some form of equivalence of finite state systems. Indeed, this problem plays a fundamental role in the study of concurrent systems, and has been widely studied both from theoretical and practical points of view. Milner specifies in [21] a complete set of axioms for proving equivalence of finite state agents. Kanellakis and Smolka consider in [15] efficient sequential algorithms to solve this problem. On the more practical side, the prototype named Concurrency Workbench, implemented in Standard ML, has been used by Walker [27] for undertaking the automated analysis of mutual exclusion algorithms via finite state systems, using the fact that the state space of all these algorithms is finite.

Until now, the analysis of concurrent systems by means of bisimulation techniques has been based on sequential algorithms. A natural question to ask is: do the automatic bisimulation techniques admit fast parallel algorithms?

In this paper we give a strong evidence that unfortunately the answer to the above question is negative. More precisely, we prove that deciding bisimulation in finite transition systems is a P-complete problem. P-complete problems have efficient sequential algorithms but it is widely believed that they do not admit fast parallel ones.

In fact, this concept plays a role analogous to the notion of NP-complete problems. These are problems that can be solved by an exponentially slow exhaustive search, and they inherently seem to require superpolynomial time algorithms. The NP-completeness of a problem implies that success in designing a polynomial time sequential algorithm for it is highly unlikely.

Analogously, P-complete problems are identified as inherently sequential problems: if there are any problems in P that do not admit efficient parallel algorithms, then all P-complete problems are among them. Conversely stated, if a parallel algorithm is found for a P-complete problem which uses a feasible (e.g. polynomial) amount of hardware and runs in polylogarithmic time, then all problems solvable in polynomial time have also such feasible and very fast parallel algorithms. However, strong research in the area during several years has failed to produce such an algorithm for any of the well studied P-complete problems. Thus, the design of a parallel algorithm with these characteristics for a P-complete problem would require a breakthrough in Algorithmics. Actually the conjecture of many researchers in the field is that such an algorithm does not exist at all. Surveys of P-complete problems have appeared in [13] and [22].

2. Preliminaries

2.1 Sequential and parallel complexity classes. For the formal study of the possible existence of parallel algorithms we will consider two main complexity classes: P and NC. We will mention also some interesting subclasses of NC. The class P models problems with *efficient sequential algorithms*; the class NC models problems with *fast parallel algorithms*, using a feasible number of processors. Each of these classes has many characterizations that support this description.

• By definition, the class P contains the problems for which a polynomial time sequential algorithm exists. This can be formalized by considering an abstract model of sequential computation for which "time" is a well-defined notion. Polynomial time RAM algorithms (a model quite close to a real computer [1]), polynomial time Turing machines ([1], [4]), or even polynomial size uniform circuits (see below) are all suitable for this purpose, and give equivalent definitions of the class P.

• The class NC formalizes the concept of efficiently parallelizable problems: it contains those problems for which a parallel algorithm can be designed that runs in polylogarithmic parallel time and uses a feasible (i.e. polynomial) amount of hardware. There are many characterizations of this class. Consider for instance Parallel RAM (PRAM) machines, which are one of the basic abstract models of parallel computers [26]. NC can be defined as the class of all the problems that can be solved in a PRAM within $O(\log^k n)$ time for constant k and using polynomially many processors.

For theoretical analysis sometimes *unbounded fan-in boolean circuits* are preferable [7]. A boolean circuit is a directed, acyclic, labelled graph in which the nodes of indegree zero are the inputs, the nodes of indegree 1 compute boolean negation, and the nodes of indegree 2 or more compute either boolean conjunction or disjunction of all their inputs, according to their respective label. The nodes of outdegree zero are the output nodes. The *size* of a circuit is the number of its nodes; the *depth* is the length of the longest path from an input to an output. The nodes in a circuit are called also *gates*. A boolean circuit computes a boolean function by substituting values for the inputs, evaluating all the nodes, and collecting values at the output nodes. Binary inputs and outputs might be binary encodings of other objects assuming some simple coding scheme.

To use boolean circuits to solve problems, we have to select a different circuit for each input length; but such a selection might be very hard to compute. Here we will explicitly rule out those families of circuits for which this selection is indeed hard, and will restrict ourselves to uniform families. A family of circuits is *uniform* if basic facts about the connection of the gates can be answered in deterministic logarithmic time, or equivalently can be expressed in an extended version of first order logic (see [6] for precise definitions).

It is well known that in many aspects PRAMs and uniform unbounded fan-in circuits are equivalent [26], with bounds on number of processors corresponding to bounds on the size of the circuit, and bounds on the PRAM time corresponding to bounds on the depth of the circuit. Thus NC is formed by the problems solvable by polylogarithmic depth, polynomial size uniform circuits.

NC has some interesting subclasses. In particular, AC^0 contains the problems solvable by unbounded fan-in uniform circuits of constant depth and polynomial size, or equivalently solvable by a PRAM in constant time with a feasible (i.e. polynomial) number of processors; and AC^1 contains the problems solved by unbounded fan-in uniform circuits of logarithmic depth and polynomial size, which corresponds to logarithmic time in a PRAM with again a feasible number of processors. AC^0 contains some problems with long history, for instance the addition of two integer numbers.

Lying between AC^0 and AC^1 is the class TC^0, defined by uniform constant depth polynomial size circuits that are allowed to use threshold gates. This class can be motivated by the growing of a complexity theory of neural networks [23], and is important for tight analysis of the complexity of certain problems; it also contains very natural and interesting problems such as the multiplication of two integer numbers [7].

Since threshold gates can simulate AND and OR gates we have that $AC^0 \subset TC^0$, but these two classes do not coincide: Ajtai [2] and Furst, Saxe and Sipser [9] proved that the inclusion was strict. This was shown by proving that the majority problem, coded as the set $MAJ = \{w \in (0+1)^* \mid |w|_1 \geq |w|_0\}$, cannot be solved by a constant depth polynomial size circuit having only AND and OR gates. The proof does not require any uniformity condition on the AC^0 circuits.

To compare and classify problems in P we use the *constant depth reducibility* [7]. A function f is constant depth reducible to g, denoted here as $f \leq_{cd} g$, if there is a family of circuits which compute f with polynomial size, constant depth, and oracle gates for g. The cost and depth of an oracle gate is 1. It can be easily shown that AC^0, TC^0,

AC^1, and AC are closed under this reducibility; e.g., if $g \in TC^0$ and $f \leq_{cd} g$, then $f \in TC^0$.

A problem S is *P-complete* under $\leq_{cd}$ reductions if $S \in P$ and every problem in P is $\leq_{cd}$-reducible to S. It can be shown that this reducibility is transitive, and therefore to prove that a problem in P is P-complete, it is enough to prove that some other complete problem in P is reducible to it. There are several standard P-complete problems which are natural candidates for the reduction. One of these is the Circuit Value Problem *CVP*. The input to this problem is pair formed by a circuit and an input to the circuit. The problem consists of computing the output of the circuit on the given input. When suitable additional hypotheses are assumed on the given circuit, we obtain variants of this problem that still are P-complete. In order to prove our results we consider one of these variants: the evaluation problem for *monotone alternating circuits*. Such a circuit is divided into levels, so that the inputs to a gate at a given level are all outputs of gates from the immediately preceeding level. The circuit does not contain negation gates; instead it receives each input together with its negation. All gates in the same level are of the same type, and the levels alternate between AND and OR levels. Figure 1 gives us an example of a monotone alternating circuit.

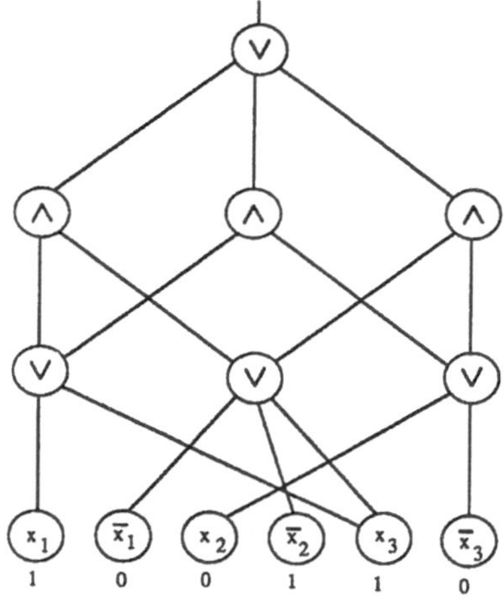

Fig. 1 A monotone alternating boolean circuit C

The following is known [13]:

Theorem 1: The Monotone Alternating Circuit Value Problem *MACVP* is P-complete.
Input: An encoding of a monotone alternating circuit c with one output, together with boolean input values $x_1, \overline{x}_1, \ldots, x_n, \overline{x}_n$.
Output: The value of c on these input values.

Let us end here our complexity-theoretic notions and go on to introduce the problems whose complexity will be classified. The notations introduced here will be necessary for later description of parallel algorithms.

2.2 Petri nets. The Petri net model was one of the first models introduced to describe concurrent processes with distributed control [25]. Formally a Petri net is a tuple $N = \langle P, T, F, M_0 \rangle$ where:

1. The set $P = \{p_1, \ldots, p_r\}$ is called the *set of places*. During the evolution of the net, a place p contains a number of tokens denoted as $M(p)$ and called its marking. Such a marking models some local aspect of the system with global state $M = (M(p_1), \ldots M(p_r))$.
2. The *set of transitions* is $T = \{t_1, \ldots, t_s\}$. Transitions model the events of N and every sequential behaviour is represented by a word $w \in T^*$.
3. The *flow function* $F : \{(P \times T) \cup (T \times P)\} \to \mathbb{N}$ connects between them places and transitions. The value of F fixes the precondition to be fulfilled in order to fire a transition t in a marking M. The *firing rule* is:

$$\forall p \in P : M(p) \geq F(p, t).$$

   Additionally, F gives us the new marking M' reached after the firing of t in M, denoted as $M[t\rangle M'$, and defined by:

$$\forall p \in P : M'(p) = M(p) - F(p, t) + F(t, p).$$

4. $M_0 : P \to \mathbb{N}$ is the initial marking.

We denote by $\Delta(p, t)$ the variation on the number of tokens in a place p when t is fired, $\Delta(p, t) = F(t, p) - F(p, t)$. Then $M'(p) = M(p) + \Delta(p, t)$. The firing rule can be extended from transitions to words $w \in T^*$ as usual and the whole sequential behaviour of the net N is described by the set of *firing sequences* which is:

$$S_N(M_0) = \{w \in T^* \mid \exists M : M_0[w\rangle M\}.$$

Our first main result in the next section will classify the problem for Petri nets defined as follows:

Problem 2: Fixed a Petri net $N = \langle P, T, F, M_0 \rangle$, the membership problem for firing sequences on this net N, denoted as *N-PETRI-FIRING* is:

Input: $w \in T^*$.

Question: $w \in S_N(M_0)$?.

2.3 Partially commutative monoids. Another way to model concurrent systems is with concurrent alphabets and partially commutative monoids [18]. We call ***concurrent alphabet*** a pair $(\Sigma, \sim)$ where $\Sigma = \{x_1, \ldots x_s\}$ is a finite alphabet denoting the set of events and $\sim$ is a symmetric and irreflexive binary relation on Σ called the ***commutation relation***. The complementary notion is also useful: the ***conflict relation*** is defined as $\Sigma \times \Sigma \setminus \sim$. To describe equivalent behaviours in Σ^* we introduce the congruence generated by the commutation relations (i.e. if x and y commute we consider the relation $xy \sim yx$) and we denote as $w \overset{*}{\sim} w'$ the equivalence given by this congruence. The quotient monoid $\Sigma^*/\sim$ is called ***partially commutative monoid*** and its elements are called ***traces***. If w and w' are equivalent then they model two sequential evolutions corresponding to a unique parallel behaviour. To study this equivalence we need the projection function over a subset Δ of Σ denoted as $\pi_\Delta : \Sigma^* \longrightarrow \Delta^*$. This function is defined as $\pi_\Delta(x) = x$ if $x \in \Delta$ and $\pi_\Delta(x) = \lambda$ otherwise. The trace equivalence $w \overset{*}{\sim} w'$ has been characterized in [8] in the following way:

1. for every event x in Σ we have $\pi_{\{x\}}(w) = \pi_{\{x\}}(w')$ and
2. for every pair (x, y) of different events in conflict we have $\pi_{\{x,y\}}(w) = \pi_{\{x,y\}}(w')$.

In our main results we will consider the following problem:

Problem 3: Fixed a concurrent alphabet $(\Sigma, \sim)$, the trace equivalence problem, denoted as $(\Sigma, \sim)$-*TRACE-EQUIVALENCE*, is

Input: A string $w\$w'$ where $w, w' \in \Sigma^*$ and $\$ \notin \Sigma$.

Question: It is true that $w \overset{*}{\sim} w'$?.

2.4 Finite transition systems. Concurrent systems can be analyzed also by means of transition systems [17]. Recall that a ***finite labelled transition system*** (FLTS for short) is a triple $M = \langle Q, \Sigma, T \rangle$, where Q is a finite set of states (or processes), Σ is a finite alphabet of actions and $T \subseteq Q \times \Sigma \times Q$ is the set of transitions. A transition $(q, x, q') \in T$ has label x and is denoted by $q \xrightarrow{x} q'$. Given two states p and q, the idea of having the same behaviour is formalized by the notion of strong bisimulation [24] (see also [20]).

A relation $S \subseteq Q \times Q$ is a ***strong bisimulation*** if $(p, q) \in S$ implies, for all $x \in \Sigma$, the following bisimilarity conditions:

(i) whenever $p \xrightarrow{x} p'$, then for some q', $q \xrightarrow{x} q'$ and $(p', q') \in S$,

(ii) whenever $q \xrightarrow{x} q'$, then for some p', $p \xrightarrow{x} p'$ and $(p', q') \in S$.

The ***strong bisimilarity*** relation $\sim$ is defined as the union of all strong bisimulations, that is

$$\sim \; = \bigcup \{ S \mid S \text{ is a strong bisimulation } \}$$

Notice that the strong bisimilarity relation is also a strong bisimulation.

Other relationships such as bisimulation and observational equivalence can be defined in similar ways, using "invisible actions" [20]. It is not difficult to see that the decisional problems for these notions are equivalent to the decision of strong bisimulations.

We will prove the P-completeness of the the following problem:

Problem 4: The problem *STRONG-BISIMILARITY* is

Input: An encoding of a finite transition system with two selected states p^* and q^*.

Question: Are p^* and q^* strongly bisimilar?.

3. Main results

3.1 Petri net firing. Fixed a Petri net $N = \langle P, T, F, M_0 \rangle$ we would like to study the complexity of *N-PETRI-FIRING* problem. We start with an intuitive massivelly parallel algorithm able to solve this problem. After, we will consider some tight bounds.

Proposition 5: Given a Petri net N, the decision problem *N-PETRI-FIRING* belongs to NC.

Proof. Given a Petri net N and a sequence of transitions $w = x_1 \ldots x_i \ldots x_n$ it is easy to prove that w is a firing sequence iff the following holds:

- To fire the transition x_1 the following property has to be satisfied:

$$\forall p \in P : M_0(p) \geq F(p, x_1).$$

- To fire the transition x_i $(1 < i \leq n)$ it is necessary to fulfil two conditions. First the prefix $x_1 \ldots x_{i-1}$ is a firing sequence. And second, all the places have to contain enough tokens to enable x_i. Both conditions can be expressed together as

$$\forall 1 < i \leq n \quad \forall p \in P : M_0(p) + \sum_{t \in T} \Delta(p, t) \cdot |x_1 \ldots x_{i-1}|_t \geq F(p, x_i).$$

These conditions can be easily verified in parallel. To do this we associate a processor to every transition x_i of the input string. The processor i will operate fundamentally with transition x_i. The NC program solving this problem is given in the program "N-Petri-Firing". ∎

To obtain a tight upper bound we can express the *N-PETRI-FIRING* problem in terms of first order logic enlarged with majority quantifiers [3]. Considering Immerman's work [14], this formalism can be transformed into parallel programs running over PRAM machines enlarged with threshold operations. In our case these programs have constant time. We also give a lower bound by showing that *N-PETRI-FIRING* problem is equivalent to the *MAJ* problem under constant depth reductions. Hence this problem cannot be solved in constant time by a standard PRAM with a polynomial number of processors.

---

for $1 \leq i \leq n$ **do in parallel**
  for $1 \leq j \leq s$ **do**
    (* by prefix sum techniques processor i compute $count_i[j]$ *)
    $count_i[j] := |x_1 \ldots x_i|_{t_j}$
  end for;
  for $1 \leq k \leq r$ **do**
    $delta_i[k] := count_i[1] \cdot \Delta(p_k, t_1) + \ldots + count_i[s] \cdot \Delta(p_k, t_s);$
  end for;
  if $i = 0$
    then $enabled_i := \bigwedge_{1 \leq k \leq r} (M_0[k] \geq F(p_k, x_i))$
    else $enabled_i := \bigwedge_{1 \leq k \leq r} (M_0[k] + delta_{i-1}[k] \geq F(p_k, x_i))$
  end if;
end parallel for;
(* by recursive folding all the processors help to compute the result *)
$N\text{-}Petri\text{-}Firing := \bigwedge_{1 \leq i \leq n} enabled_i$

Program. N-Petri-Firing

Proposition 6: The *N-PETRI-FIRING* problem belongs to TC^0. Moreover, a lower bound complexity is fixed by the following two assertions:

1. Fixed a Petri net N we have *N-PETRI-FIRING*$\leq_{cd}$ *MAJ*.
2. There exists a Petri net N such that *MAJ* $\leq_{cd}$ *N-PETRI-FIRING*.

From 1 and 2 we conclude that fixed a Petri net N the firing cannot solved in constant time by a PRAM with a polynomial number of processors.

For a detailed proof of this proposition see [3].

3.2 Trace equivalence. Fixed a concurrent alphabet $(\Sigma, \sim)$ we would like to study the complexity of the $(\Sigma, \sim)$-*TRACE-EQUIVALENCE* problem. As we have done above, first we will propose an intuitive massively parallel algorithm to solve this problem, and second we will consider some tight bounds.

Proposition 7: Given a concurrent alphabet $(\Sigma, \sim)$, the decision problem $(\Sigma, \sim)$-*TRACE-EQUIVALENCE* belongs to NC.

Proof. Given w and w' it is easy to prove that $w \overset{*}{\sim} w'$ iff the two conditions given by [8] are satisfied. These conditions can be verified in parallel.

For the first condition, i.e., every letter x of the alphabet Σ appears in w and in w' the same number of times, we use masking and prefix sum techniques as we propose in the program "Equal-Length". And for the second one, the letters of every pair in conflict appear in w and in w' following the same order, is verified also using masking and prefix sum techniques and as many processors as $\max\{|w|, |w'|\}$. For every pair (x, y) in conflict, the processor i verifies that the ith letter of $\Pi_{\{x,y\}}(w)$ is equal to $\Pi_{\{x,y\}}(w')$. The program "Equal-Conflicts" verifies this condition. ∎

```
equal-length := TRUE;
for x ∈ Σ do
   (* by masking and prefix sum techniques compute *)
   l1 := |Π_x(w)|;
   l2 := |Π_x(w')|;
   equal-length := equal-length ∧ (l1 = l2);
end do
```

Program Equal-Length

We can obtain a tight upper bound on the complexity of $(\Sigma, \sim)$-*TRACE-EQUIVALENCE* by expressing it in terms of first order logic enlarged with majority quantifiers. We also have a lower bound of this problem. It can be seen that $(\Sigma, \sim)$-*TRACE-EQUIVALENCE* cannot be solved by a PRAM in constant time because it is equivalent to the MAJ problem under a constant depth reduction which increases the computation time only with a constant.

Proposition 8: The $(\Sigma, \sim)$-*TRACE-EQUIVALENCE* problem belongs to TC^0. Moreover, a lower bound complexity is fixed by the following two assertions:

1. Fixed a concurrent alphabet $(\Sigma, \sim)$ we have

$$(\Sigma, \sim)\text{-}\mathit{TRACE\text{-}EQUIVALENCE} \leq \mathit{MAJ}.$$

2. There exists a concurrent alphabet $(\Sigma, \sim)$ such that

$$\mathit{MAJ} \leq (\Sigma, \sim)\text{-}\mathit{TRACE\text{-}EQUIVALENCE}.$$

From 1 and 2 we conclude that the trace equivalence on the partially commutative monoid generated by $(\Sigma, \sim)$ cannot be solved in constant time by a PRAM with a polynomial number of processors.

```
equal-conflicts := TRUE;
for every pair (x,y) in conflict do
  (* by masking and prefix sum techniques compute *)
  k1 := |Π{x,y}(w)|; k2 := |Π{x,y}(w')|;
  if k1 = k2
    then
      for 1 ≤ i ≤ k1 do in parallel
        u := letter i^th of Π{x,y}(w); v := letter i^th of Π{x,y}(w');
        test_i := (u = v)
      end parallel for;
      (* by prefix sum techniques *)
      equal-conflicts := equal-conflicts ∧ ⋀_{1≤i≤k1} test_i
    else
      equal-conflicts := FALSE
    end if
end for
```

Program Equal-Conflicts

For a detailed proof of this proposition see [3].

3.3 Bisimulations. In contrast with these problems allowing very fast and feasible parallel algorithms we prove next that the *STRONG-BISIMILARITY* problem is P-complete

It is well known that strong bisimilarity in a LFTS is a polynomial time decidable property [20]. To see this, it suffices to construct $\sim$ as intersection of the sequence of relations $\equiv_0, \equiv_1, \ldots,$ which are defined by induction as follows:

(i) For every $(p,q) \in Q \times Q$, $p \equiv_0 q$,

(ii) $p \equiv_{i+1} q$ if for every $x \in \Sigma$,

whenever $p \xrightarrow{x} p'$, then for some q', $q \xrightarrow{x} q'$ and $p' \equiv_i q'$;

whenever $q \xrightarrow{x} q'$, then for some p', $p \xrightarrow{x} p'$ and $p' \equiv_i q'$.

It is easy to see that these relations can be constructed in polynomial time. This is because $\sim$ coincides with $\equiv_k$ where k is the number of states in the finite transition system. More efficient algorithms to solve this problem have been considered in [15]. The P completeness of the *STRONG-BISIMILARITY* problem will follow from the following lemma. A more detailed proof appears in [5].

Lemma 9: *MACVP* can be reduced to *STRONG-BISIMILARITY.*

Proof. We will transform an arbitrary instance of the circuit value problem for monotone alternating circuits *MACVP* into an instance of *STRONG-BISIMILARITY* in three steps.

• We define the k-alternating pattern A_k. Figure 2 shows A_4. This is a circuit of height k, where every level has two gates, one valuated to 0 and other valuated to 1. It is easy to check then that in A_k the following two conditions are satisfied:

- every OR gate has an input valuated to 0,
- every AND gate has an input valuated to 1.

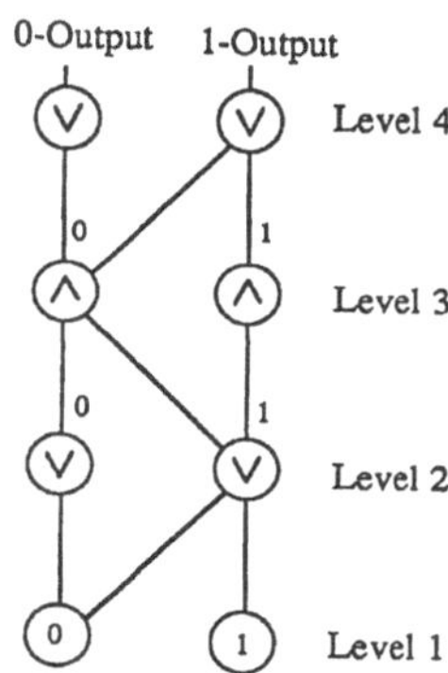

Fig. 2 The 4-alternating pattern A_4

• We couple the k-alternating pattern A_k with the circuit C to get a new circuit C'. Figure 3 shows the circuit C' constructed from the example of Figure 1. The circuit C' satisfies the following three properties:

- every OR gate has at least an input valuated to 0,
- every AND gate has at least an input valuated to 1,
- every gate of C' evaluates to the same value as the corresponding gate in A_k or C.

• We now transform the circuit C' into a FLTS M over a one letter alphabet. M contains a state corresponding to each gate of C'. These states are called ***ordinary states***. In addition M contains $n+1$ ***auxiliary states***, associated with the $n+1$ inputs of C' which evaluate to 1 (the n inputs of C of value 1, and the constant 1 input of A_k). We say that these auxiliary states are on level 0. Figure 4 shows M in our example.

By induction it can be shown that the circuit C evaluates to 1 with the given input values if and only if states p^* and q^* in M are strongly bisimilar.

As a consequence of the precedent lemma we obtain our announced result.

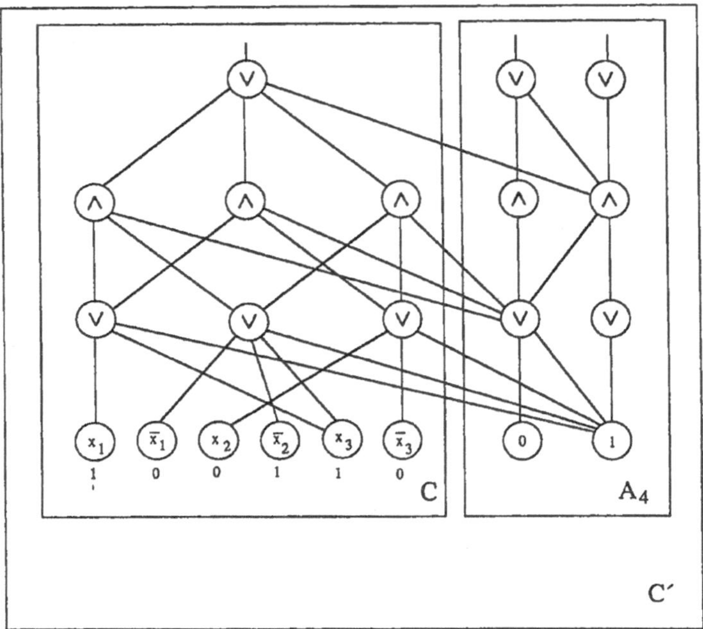

Fig. 3 The coupling of C and A_4 into C'

Theorem 10: The *STRONG-BISIMILARITY* problem is P-complete.

Other results related to this one can be obtained by the same proof idea. The properties named Observation Equivalence and Observation Congruence are defined in [19]. We can state:

Theorem 11: The problem of deciding Observation Equivalence and the problem of deciding Observation Congruence of two states in a LFTS are both P-complete.

4. Extensions

We have presented a quite precise classification of three problems on concurrency. These were the decision of firing sequences for Petri nets, the trace equivalence for partially commutative monoids, and the strong bisimulation decision problem for finite transition systems. These classifications give us hints about the complexity of massively parallel algorithms to solve them: the first two have such algorithms but however the third one cannot have such an algorithm unless all problems in P do.

Now we want to complete the discussion by raising some questions. For the bisimulation problem, our version of the statement requires the system to be part of the input. This

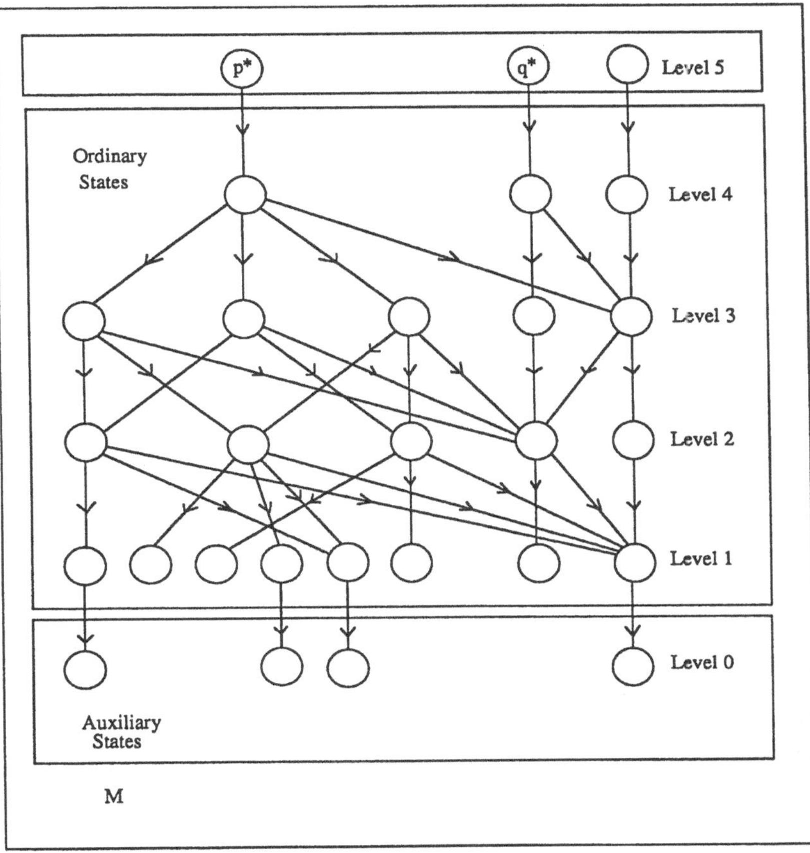

Fig. 4 The transition system M corresponding to C'

is necessary since, the system being finite, if we fix it then we obtain only a finite number of possible pairs of states, and therefore the problem can be solved in constant time.

On the contrary, in our first two problems the devices (i.e. Petri nets and concurrent alphabets) are independent of the input, and the proofs rely strongly of this fact. It is interesting to see what happens when the description of the device is added as a part of the input. For comparison, recall that context-free parsing can be done efficiently in parallel [16]; however when a coding of the grammar is added as a part of the input the complexity grows up substantially, becoming P-complete [11]. Let us briefly describe the properties of our first two problems assuming that the devices are part of the input. No proofs will be provided here.

Let us consider first the problem of general trace equivalence for partially commutative monoids. Inputs are a concurrent alphabet $(\Sigma, \sim)$, and words w and w' in Σ^*; the problem is to decide whether w and w' are equivalent. Using more complex arguments, we can prove that this problem is in uniform TC^0, and thus has the same complexity as the problem for fixed monoid; it therefore can be solved by fast parallel algorithms.

In the same way we can consider a more general version of membership for Petri net firing sequences, where the net is a part of the input. Our way of computing the variation of tokens in each place due to a prefix of the trace relies on the fact that the net is fixed, and therefore the numbers of places and transitions are constants. But if the net is part of the input, these expressions are sums of nonconstant numbers, and we must resort to a multiple addition. It is known that this problem belongs to a uniform version of TC^0 [6]. Thus we obtain that the problem belongs to uniform TC^0.

Finally, let us present some additional considerations regarding the bisimulation problem. Since it is so relevant to the design of concurrent systems, our negative P-completeness result calls for new concepts of equivalence that might be of practical value, yet testable by fast parallel algorithms. On the other hand, from the standpoint of a developer of a concurrent system, another relevant issue is whether interaction with a software tool might be more efficient than completely automatic equivalence testing.

It is well known from the study of NP problems that in many cases "verifying" is easier than "computing". This is also true in our case; indeed, the problem of whether a given relation is a bisimulation is in NC. This opens a possible way to partially overcome the P-completeness obstacle. The idea would be to design concurrent systems in an interactive way through a sequence of stepwise refinements, e.g. in the line of [12], in such a way that at every step the designer keeps direct intuition of how to transform the precedent bisimulation to obtain a new one. He then can guess the result and verify it. Perhaps only in some rare cases the designer will need to compute the whole bisimulation, and if this case is infrequent enough he would accept such a long computational process.

References

[1] Aho, A., Hopcroft, J., Ullman, J.: *The Design and Analysis of Computer Algorithms.* Addison-Wesley (1975).

[2] Ajtai, M.: Σ_1^1-formulae on finite structures. *Ann. Pure Appl. Logic* **24**, 1–48 (1983).

[3] Àlvarez, C., Gabarró, J.: The parallel complexity of two problems on concurrency. To appear at IPL.

[4] Balcázar, J.L., Díaz, J., Gabarró, J.: *Structural Complexity I.* Springer Verlag EATCS Monographs in Theoretical Computer Science, v. 11 (1988).

[5] Balcázar, J., Gabarró, J., Sántha, M.: Deciding bisimilarity is P-complete. Report LSI-90-25, Universitat Politècnica de Catalunya. Submitted for publication.

[6] Barrington, D.M., Immerman, N., Straubing, H.: On uniformity within NC^1. *J. Comp. Syst. Sci.* **41**,274–306 (1990).

[7] Chandra, A.K., Stockmeyer, L., Vishkin, U.: Constant depth reducibility. *SIAM J. Comput.* **13**, 2, 423–439 (1984).

[8] Cori, R., Perrin, D.: Automates et commutations partielles. *RAIRO Inf. Theor.* **19**, 21–31 (1985).

[9] Furst, M., Saxe, J.B., Sipser, M.: Parity, circuits and the polynomial time hierarchy. *Math. Syst. Theory* **17**, 13–27 (1984).

[10] Gibbons, A., Rytter, W.: *Efficient Parallel Algorithms*. Cambridge University Press (1988).

[11] Goldschlager, L.: ϵ-Productions in context-free grammars. *Acta Informatica* **16**, 303–308 (1981).

[12] He Jifeng: Process Simulation and Refinement. *Formal Aspects of Computing* **1**, 229–241 (1989).

[13] Hoover, H.J., Ruzzo, W.L.: A Compendium of Problems Complete for P. Manuscript (1984).

[14] Immerman, N.: Expressibility and parallel complexity. *SIAM J. Comput.* **18**, 3, 625–638 (1989).

[15] Kanellakis, P.C., Smolka, S. A.: CCS expressions, finite state processes, and three problems of equivalence. *Information and Computation* **86**, 202–241 (1990).

[16] Karp, R., Ramachandran, V.: Parallel Algorithms for Shared Memory Machines. In: *Handbook of Theoretical Computer Science*, (vol A), 869–941, editor Jan Van Leeuwen, Elsevier (1990).

[17] Keller, R.M.: Formal Verification of Parallel Programs. *Comm. ACM*, **19**, 7, 371–384 (1976).

[18] Mazurkiewicz, A: Basic notions of trace theory. Springer Verlag Lecture Notes in Computer Science 354, 285–363 (1989).

[19] Milner, R.: *A Calculus of Communicating Systems*. Springer Verlag Lecture Notes in Computer Science 92 (1980).

[20] Milner, R.: *Communication and Concurrency*. Prentice Hall (1989).

[21] Milner, R.: A Complete Axiomatization for Observation Congruence of Finite-State Behaviours. *Information and Computation*.

[22] Miyano, S., Shiraishi, S., Shoudai, T.: A list of P-complete problems. Technical Report RIFIS-TR-CS-17, Kyushu University 33, 1989.

[23] Parberry, I.: A primer on the complexity theory of neural networks. In: *Formal techniques in artificial intelligence*, R.B. Banerji (editor), North-Holland (1990).

[24] Park, D.: Concurrency and Automata on Infinite Sequences. Springer Verlag Lecture Notes in Computer Science 104, 168–183 (1981).

[25] Peterson, J.L.: *Petri net theory and the modeling of systems*. Prentice-Hall (1981).

[26] Stockmeyer, L., Vishkin, U.: Simulation of parallel random access machines by circuits. *SIAM J. Comput.* **13**, 2, 409–422 (1984).

[27] Walker, D.J.: Automated Analysis of Mutual Exclusion Algorithms using CCS. *Formal Aspects of Computing*, 1, 273–292 (1989).

FORK
A High–Level Language for *PRAM*s

T. Hagerup[1], A. Schmitt[2], H. Seidl[2]

Fachbereich Informatik, Universität des Saarlandes
Im Stadtwald, 6600 Saarbrücken, Germany

Abstract

We present a new programming language designed to allow the convenient expression of algorithms for a parallel random access machine (*PRAM*). The language attempts to satisfy two potentially conflicting goals: On the one hand, it should be simple and clear enough to serve as a vehicle for human-to-human communication of algorithmic ideas. On the other hand, it should be automatically translatable to efficient machine (i.e., *PRAM*) code, and it should allow precise statements to be made about the amount of resources (primarily time) consumed by a given program. In the sequential setting, both objectives are reasonably well met by the Algol-like languages, e.g., with the *RAM* as the underlying machine model, but we are not aware of any language that allows a satisfactory expression of typical *PRAM* algorithms. Our contribution should be seen as a modest attempt to fill this gap.

1 Introduction

A *PRAM* is a parallel machine whose main components are a set of processors and a global memory. Although every real machine is finite, we consider an ideal *PRAM* to have a countably infinite number of both processors and global memory cells, of which only a finite number is used in any finite computation. Both the processors and the global memory cells are numbered consecutively starting at 0; the number of a processor is called its *processor number* or its *index*, and the number of a memory cell is, as usual, also known as its address. Each processor has an infinite local memory and a local program counter. All processors are controlled by the same global clock and execute precisely one instruction in each clock cycle. A *PRAM* may hence also be characterized as a synchronous shared-memory MIMD (multiple-instruction multiple-data) machine.

The set of instructions available to each processor is a superset of those found in a standard *RAM* (see, e.g., [2]). The additional instructions not present in a *RAM* are an instruction LOADINDEX to load the index of the executing processor into a cell in its local memory and instructions READ and WRITE to copy the contents of a given global memory cell to a given cell in the local memory of the executing processor, and vice versa. All processors can access a global memory cell in the same step, with some restrictions concerning concurrent access by several processors to the same cell (see Section 2.5).

Among researchers working on the development of concrete algorithms, the *PRAM* is one of the most popular models of parallel computation, and the number of published *PRAM* algorithms is large and steadily growing. This is due mainly to the convenient and very powerful mechanism for inter-processor communication provided by the global memory. Curiously, there is no standard *PRAM* programming language, and each researcher, in so far as he wants to provide a formal description of his algorithms, develops his own notation from scratch. The disadvantages of this are evident:

[1] supported by Deutsche Forschungsgemeinschaft, SFB 124, TP B2

[2] supported by Deutsche Forschungsgemeinschaft, SFB 124, TP C1

1. At least potentially, difficulties of communication are aggravated by the lack of a common language;

2. The same or very similar definitions are repeated again and again, resulting in a waste of human time and journal space;

3. Since the designer of an algorithm is more interested in the algorithm than in the notation used to describe it, any language fragments that he may introduce are not likely to be up to current standards in programming language design.

In the wider area of parallel computing in general, much effort has gone into the development of adequate programming languages. Most of these languages, however, are intended to be used with loosely coupled multiprocessor systems consisting of autonomous computers, each with its own clock, that run mainly independently, but occasionally exchange messages. The facilities provided for interprocessor communication and synchronization are therefore based on concepts such as message exchange (Ada [16]; OCCAM [15]; Concurrent C [10]) or protected shared variables (Concurrent Pascal [13]). In particular, a global memory simultaneously accessible to all processors is not supported by such languages, and it can be simulated only with an unacceptably high overhead. While such languages may be excellent tools in the area of distributed computing, they are not suited to the description of *PRAM* algorithms.

Before we go on to discuss other languages more similar in spirit to ours, we describe what we consider to be important features of such languages. Most obviously, they must offer a way to state that certain operations can be executed in parallel. Secondly, we want to write programs for a shared–memory machine. Therefore, the language should distinguish between shared variables, which exist only once and can be accessed by a certain group of processors, and private variables, of which there might be several instances, each residing in a different processor's private memory and possibly having a different value.

Also, the machine facilities of synchronous access to shared data, should be reflected in the language. Finally, program constructs like recursion, which are well suited for writing clear and well structured sequential programs, should be allowed to be freely combined with parallelism. Recursion is characterized by a subdivision of a given problem into a set of subproblems that can be solved independently and possibly in parallel. Each subproblem may again be worked on by several processors. Therefore, the programming language should provide the programmer with a means of generating independently working subgroups of synchronously running processors. Since the efficiency of many algorithms relies on a subtle distribution of processors over tasks, an explicit method should be available to assign processors to newly created subgroups.

A frequently used tool for indicating parallelly executable program sections is a *for* loop where all loop iterations are supposed to be executed in parallel. Such a construct is, e.g., used in extensions of sequential imperative languages like FORCE [14] and *ParC* [4]. Also textbooks about *PRAM* algorithms, e.g. [3, 11], usually employ some Algol–style notation together with a statement like **for i:=1 to n pardo ... endpardo** .

A different approach is taken in the language GATT [7]. In GATT all processors are started simultaneously at the beginning. During procedure calls subgroups can be formed to solve designated subproblems. However, since GATT is designed for describing efficient algorithms on processor networks, GATT lacks the concept of shared variables. Instead, every variable has to reside in the private memory of one of the processors.

For *PRAM*s, there are various examples of descriptions of recursive algorithms using an informal group concept, e.g., see [3, sect. 4.6.3, p. 101], [8, 6]. An attempt to formulate a recursive *PRAM* algorithm more precisely is made in [5]. Corresponding to the machine-level *fork* instruction of [9], a *fork* statement is introduced, which allows a given group of synchronously working processors to be divided into subgroups. This *fork* statement gave the name to our language.

The present paper embeds the *fork* statement suggested in [5] into a complete programming language. In detail, the contributions of *FORK* are the following:

- It adds a *start* construct, which allows a set of new processors with indices in a specified range to be started.

- It makes precise the extent to which the semantics guarantees synchronous program execution (and hence synchronous data access).
- Besides the implicit synchronization at the beginning of every statement, as proposed in [5], it introduces implicit splitting into subgroups at every branching point of the program where the branch taken by a processor depends on its private variables.

It is argued that the available program constructs can be freely nested. In particular, iteration and recursion are compatible both with the starting of new processors and the forking of subprocesses.

The paper is organized as follows. In Section 2 we explain the mechanism of synchronism of *FORK* together with the new constructs in *FORK* for maintaining parallelism. Moreover, we introduce the three basic concepts of *FORK*, namely the concepts of a "logical processor", of a "group" of logical processors and of "synchronous program execution". These basic concepts are used in Section 3, which gives the semantics of the main constructs of *FORK* in an informal way. A complete description of *FORK* can be found in [12]. Section 4 concludes with some hints on how programs of the proposed language can be compiled to efficiently running *PRAM* machine code.

It should be emphasized that although our language design aims to satisfy the needs of theoreticians, we want to provide a practical language. The language *FORK* was developed in close connection with a research project at the Saarbrücken Computer Science Department that in detail explores the possibilities of constructing a *PRAM* [1] and is going to build a prototype. We plan to write a compiler for our language that produces code for this physical machine.

Both a formal semantics of *FORK* and a more precise description of a compiler for *FORK* are in preparation.

2 An overview on the programming language *FORK*

Parallelism in *FORK* is controlled by two orthogonal instructions, namely **start** [<*expr*>..<*expr*>] and **fork** [<*expr*>..<*expr*>]. The *start* instruction can be used to readjust the group of processors available to the *PRAM* for program execution, whereas the *fork* instruction leaves the number of processors unchanged, but creates independently operating subgroups for distinct subtasks and allows for a precise distribution of the available processors among them. The effect of these instructions together with *FORK*'s concept of synchronous program execution will be explained by the examples below.

2.1 Creating new processors: The *start* statement

A basic concept of *FORK* is a *logical processor*. Logical processors are meant to be mapped onto the physical processors provided by the hardware architecture. However, the number of actually working logical processors may vary during program execution; also, the number of logical processors may exceed the number of physically available processors. Therefore, these two kinds of processors should not be confused. In the sequel, if we loosely speak of "processors" we always mean "logical processors". If we mean physical processors we will state so explicitly.

Every (logical) processor p owns a distinguished integer constant $\sharp$ whose value is referred to as the *processor number* of p. Also, it may have other private constants, private types, and private variables which are only accessible by itself. Objects declared as *shared* by a group of processors can be accessed by all processors of the given group.

As a first example consider the following problem. Assume that we are given a forest F with nodes $1,\ldots,N$. F is described by an array A of N integers, where $A[i] = i$ if i is a root of F, and $A[i]$ contains the father of i otherwise. For an integer constant N, the following program computes an array R of N integers such that $R[i]$ contains the root of the tree to which i belongs in F.

As in PASCAL, the integer constant N, the loop variable t, and the arrays A and R must be declared in the surrounding context; in *FORK* this declaration indicates whether variables are *shared* (as in the example) or *private* and hence only accessible to the individual processor itself.

```
(1)  ...
(2)  shared const N = ...;
(3)  shared var t : integer;
(4)  shared var A : array [1 .. N] of integer;
(5)  shared var R : array [1 .. N] of integer;
(6)  ...
(7)  start [1..N]
(8)      R[♯] := A[♯] ;
(9)      for t := 1 to log(N) do
(10)     /* log(N) denotes ⌈log2(N)⌉ */
(11)         R[♯] := R[R[♯]]
(12)     enddo
(13) endstart
(14) ...
```

Initially there is just one processor with processor number 0. The instruction *start* [1..N] in line (7) starts processors with processor numbers $1, \ldots, N$. The corresponding instruction *endstart* stops these processors again and reestablishes the former processors. Hence the sequence of an instruction *start* [1..N] immediately followed by an instruction *start* [1..M] does not start NM processors but only M processors. An occurrence of *endstart* finishes the phase where M processors were running and again there are N processors with numbers $1, \ldots, N$.

At the machine level every instruction consumes exactly one time unit. However, the semantics of a high-level program should be independent of the special features of the translation schemes. Therefore, it should be left unspecified how many time units are precisely consumed by, e.g., an assignment statement of *FORK*.

For this reason a notion of synchronous program execution is needed which only depends on the program text itself. Again, the semantic notion of "synchronous program execution" should not be confused with the notion of a global clock of a physical *PRAM*. For example, the underlying hardware may allow different processors to execute different instructions within the same clock cycle, whereas our notion of synchronism does *not* allow for a synchronous execution of different statements. Being synchronous is a property of a set of processors. It implies that all processors within this set are at the same program point. This means that they not only execute the same statement within the same loop within the same procedure. It also means that the "history" of the recursive call to that procedure and the number of iterations within the same loop agree. There is no explicit synchronization mechanism in *FORK*. Implicit synchronization in *FORK* is done statement by statement. At the end of each statement there is an (implicit) *synchronization point*. This means that if a set of processors synchronously executes a statement sequence

<statement>$_1$; *<statement>*$_2$

the processors of this set first synchronously execute *<statement>*$_1$. When all processors of this set have finished the execution of *<statement>*$_1$ they synchronously execute *<statement>*$_2$. Note that within the execution of *<statement>*$_1$ different processors may reach different program points. Thus they may become asynchronous in between.

FORK is well structured; there are no *gotos*. Hence implicit synchronization points cannot be circumvened. Nontermination caused by infinite waiting for deviating processors is therefore not possible.

In the given example all the processors execute the same code. According to our convention they execute statement by statement synchronously. Hence, first every processor copies the value of $A[\sharp]$ to $R[\sharp]$. Recall that the constant $\sharp$ is distinct for every processor. Then all processors assign 1 to the variable t, followed by the execution of line (11). Then they assign 2 to t, and so forth. Since the upper bound for t depends on shared data only (namely on N), all processors finish the *for* loop at the same time.

Observe here that the synchronous execution of an assignment statement is subdivided into three synchronously executed steps: first, the right-hand side is evaluated; secondly, the variable corresponding to the left-hand side is determined; finally, the value of the right hand side is assigned to the variable described by the left-hand side.

In our example, in line (11), first the value of $R[R[\sharp]]$ is computed in parallel for every processor, secondly, the variable $R[\sharp]$ is determined, which receives its new value in step three.

2.2 Forming groups of processors: The *fork* statement

FORK allows free combination of parallelism and recursion. This gives rise to the second basic concept of *FORK*: a *group*. Groups are formed by a (possibly empty) set of processors. Shared variables are always shared relative to a group of processors, meaning that they can be accessed by processors within this group but not by processors from the outside.

Groups can be divided into subgroups. This is done by the *fork* construct of *FORK*. The most recently established groups are called *leaf groups*. Leaf groups play a special role in *FORK*. As a minimum, the processors within one leaf group work synchronously. Also, if new shared objects are declared, they are established as shared relative to the leaf group executing this declaration.

Every group has a group number. The group number of the most recently created group can be accessed by its members through the distinguished private integer constant @. Clearly, the values of @ are equal throughout that group. Initially, there is just one group with group number 0 which consists of the processor with processor number 0.

As an example, consider the following generic divide-and-conquer algorithm *DC*. *DC* has a recursion parameter N describing the maximal number of processors available to solve the given problem and additional parameters containing the necessary data, which for simplicity are indicated by .... Assuming that the problem size is reduced to its square root at every recursion step, *DC* may be programmed as follows:

```
(1)  procedure DC(shared const N: integer; ... );
(2)  ...
(3)  if trivial(N)
(4)      then conquer( ... )
(5)      else
(6)          fork [0 .. sqrt(N)-1]
(7)          @ = ♯ div sqrt(N) ;
(8)          ♯ = ♯ mod sqrt(N) ;
(9)              DC(sqrt(N), ... ) /* sqrt(N) denotes ⌈√N⌉ */
(10)         endfork;
(11)         combine( ... )
(12) endif;
(13) ...
```

When a leaf group reaches the *fork* instruction in line (6), a set of subgroups with group numbers $0, \ldots, sqrt(N) - 1$ is created. These newly created groups are leaf groups during the execution of the rest of the *fork* statement, which, in the example, consists of line (9). Observe that procedure calls inside a *fork* may allocate distinct instances of the same shared variable for each of the new leaf groups.

Executing the right-hand side of line (7), every processor determines the leaf group to which it will belong.

In order to make the call to a recursive procedure simpler it may be reasonable for a processor to receive a new processor number w.r.t. the chosen leaf group. In the example this new number is computed in line (8).

When the new leaf groups have been formed, the existing processors have been distributed among these groups, and the processor numbers have been redefined, the leaf groups independently execute the statement list inside the *fork* construct. In the example this consists just of a recursive call to *DC*. Clearly, the parameters of this recursive call which contain the specification of the subproblem in general depend on the value of the constant @ of its associated leaf group.

When the statements inside a *fork* statement are finished the leaf groups disappear — in the example at line (10). The original group is reestablished as a leaf group, and all the processors continue to synchronously execute the next statement (11).

2.3 Why no *pardo* statement?

There is no *pardo* statement in *FORK*. This choice was motivated by the observation that in general *pardo* is used simply in the sense of our *start*. A difference occurs for nested *pardos*. Consider the program segment

```
for i := 1 to n pardo        (1)
   for j := 1 to m pardo     (2)
      op(i,j)                (3)
   endpardo                  (4)
endpardo                     (5)
```

Using a similar semantics as for the *start* instruction in *FORK*, the second *pardo* simply would overwrite the first one, which means that on the whole only m processors execute line (3); moreover, the value of i in line (3) would no longer be defined. This is not the intended meaning.

Instead, two nested *pardos* as in lines (1) and (2) are meant to start nm processors indexed by pairs (i, j). Precisely, a *pardo* statement of the form

for i := <*expr*>$_1$ **to** <*expr*>$_2$ **pardo** <*statement*> **endpardo**

where the expressions <*expr*>$_1$ and <*expr*>$_2$ and the statement <*statement*> do not use any private objects, can be simulated as follows:

```
begin
  /* declare two new auxiliary constants in order
     to avoid double evaluation of the
     expressions <expr>1 and <expr>2
  */
  shared const a1 = <expr>1;
  shared const a2 = <expr>2;

  /* start a2−a1+1 new processors ... */
  start[a1 .. a2]
    /*... and distribute them among a2−a1+1 new groups */
    fork[a1 .. a2]
      @ = ♮;
      ♮ = 0;
      /* each leaf group creates a new variable i and
         initilizes it with the group number
      */
      begin
        shared var i : integer;

        i := @;
        <statement>
      end
    endfork
  endstart
end /* of the pardo simulation */
```

In order to avoid redundancies we decided not to include the *pardo* construct in *FORK*.

On the other hand one may argue that the *fork* construct as provided by *FORK* is overly complicated. Using the very simple *pardo* would suffice in every relevant situation. Using *pardo* a generic divide-and-conquer algorithm may look as follows:

```
procedure DC(shared const N: integer; ... );
...
if trivial(N)
    then conquer( ... )
    else
        for i := 1 to sqrt(N) pardo
            DC(sqrt(N), ... )
        endpardo;
        combine( ... )
endif;
...
```

In the *pardo* version of DC beginning with one processor, successively more and more processors are started. In particular, every subtask is always supplied with one processor to solve it. Opposed to that, in the *fork* version the leaf group of processors is successively subdivided and distributed among the subtasks. The leaf group calling *DC* does not necessarily form a contiguous interval. Hence there might be subtasks which receive an empty set of processors and thus are not executed at all. In fact, this capability is essentially exploited in the order-chaining algorithm of [5]. This algorithm is not easily expressible using *pardos*. This was one of the reasons for introducing the *fork* construct.

2.4 Forming subgroups implicitly: The *if* statement

So far we have not explained what happens if the processors of a given leaf group synchronously arrive at a conditional branching point within the program. As an example, assume that for some algorithm the processors $1, \ldots, N$ are conceptually organized in the form of a tree of height $\log(N)$. At time t, a processor should execute a procedure *op1*(#) if its height in the tree is at most t, and another procedure *op2*(#) otherwise. The corresponding piece of program may look like:

```
(1) shared var t : integer;
(2) ...
(3) for t := 1 to log(N) do
(4)     if height(#) <= t
(5)       then op1(#)
(6)       else op2(#)
(7)     endif
(8) enddo
(9) ...
```

For every t the condition of line (4) may evaluate to *true* for some processors, and to *false* for others. Moreover, the evaluation of both *op1*(#) and *op2*(#) may introduce local shared variables, which are distinct even if they have the same names. Therefore, every *if–then–else* statement whose condition depends on private variables implicitly introduces two new leaf groups of processors, namely those that evaluate the condition to *true* and those that evaluate it to *false*. Both groups receive the group number of their father group, i.e. the private constants @ are not redefined.

Clearly, within each new leaf group every processor is at the same program point. Hence, they in fact can work synchronously as demanded by *FORK*'s group concept. As soon as the two leaf groups have finished the *then* and the *else* parts, respectively, (i.e., at the instruction *endif*) the original leaf group is reestablished and the synchronous execution proceeds with the next statement. *Case* statements and loops are treated analogously.

In the above example the condition of the *for* loop in line (3) depends only on the shared variable t. Therefore, the present leaf group is not subdivided into subgroups after line (3). However, this subdivision occurs after line (4). The two groups for the *then* and the *else* parts execute lines (5) and (6) in parallel, each group internally synchronously but asynchronously w.r.t. the processors of the other group. Line (7) reestablishes the original leaf group, which in return synchronously executes the next round of the loop, and so on.

The fact that we implicitly form subgroups at branching points whose conditions depend on private data allows for an unrestricted nesting of *if*s, loops, procedure calls and *fork*s.

Observe that at every program point the system of groups and subgroups containing a given processor forms a hierarchy. Corresponding to that hierarchy, the shared variables can be organized in a tree–like fashion. Each node corresponds to a group in the hierarchy and contains the shared variables relative to that group. For a processor of a leaf group all those variables are relevant that are situated on the path from this leaf group to the root. Along this path, the ordinary scoping rules hold.

For returning results at the end of a *fork* or for exchanging data between different leaf groups of processors it is necessary also to have at least in some cases a synchronous access to data shared between different subgroups of processors.

Consider the following example:

```
(1)  ...
(2)  shared var A : array [0..N−1] of integer;
(3)  shared var i : integer;
(4)  ...
(5)  fork [0..N−1]
(6)  @ = ... ;
(7)  # = ... ;
(8)      for i := 0 to N−1 do
(9)          A[ (@+1) mod N ] := result(i,@,A)
(10)     enddo
(11) endfork
(12) ...
```

In this example the array A is used as a mail box for communication between the groups $1, \ldots, N$. The loop index i and the limits of the *for* loop of lines (8) and (9) are shared not only by the processors within every leaf group, but also between all groups generated in line (5). If (as in the example) the loop condition depends only on variables shared by all the existing groups, the semantics of *FORK* guarantees that the loop is executed synchronously throughout those groups.

Hence, the results computed in round i are available to all groups in round $i+1$. The general rule by which every processor (and hence also the programmer) can determine the largest surrounding group within which it runs synchronously is described in detail in Section 3.2.2.

2.5 How to solve read and write conflicts

So far we have explained the activation of processors and the generation of subgroups. We left open what happens when several processors access the same shared variable synchronously. In this case, failure or success and the effect of the succeeding access is determined according to an initially fixed regime for solving access conflicts. Most *PRAM* models allow common read operations. However, we do not restrict ourselves to such a model. The semantics of a *FORK* program also may be determined, e.g., w.r.t. an exclusive read regime where synchronous read accesses of more than one processor to the same shared variable leads to program abortion. Also, several regimes for solving write conflicts are possible. For example, we may fix a regime where common writes are allowed provided all processors assign the same value to the shared variable. In this case, the *for* loop in the example above is executed successfully, whereas if we fix a regime where common writes are forbidden, a *for* loop with a shared loop parameter causes a failure of program execution.

As another example consider a regime where common writes are allowed, and the result is determined according to the processor numbers of the involved processors, e.g., the processor with the smallest number wins. This regime works fine if the processors only access shared data within the present leaf group. If processors synchronously write to variables declared in a larger group g they may solve the write conflict according to their processor numbers relative to that group g. Consider the following example:

```
(1)  ...
(2)  start[0 .. 2*N-1]
(3)      ...
(4)      shared var A : array[0 .. N-1] of integer;
(5)      ...
(6)      fork[0 .. 1]
(7)      @ = # div N;
(8)      # = # mod N;
(9)          A[#] := result(A,@,#)
(10)     endfork
(11) endstart
(12) ...
```

The shared variable A is declared for some group g having according to line (2) processors numbered $0, \ldots, 2 * N - 1$. For $n \in \{0, \ldots, N-1\}$ there are processors $p_{n,0}$ and $p_{n,1}$ of processor number n in the first and the second leaf group, respectively, that want to assign a value to $A[n]$ in line (9). This conflict is solved according to the processor numbers of $p_{n,0}$ and $p_{n,1}$ relative to group g, i.e. n and $N + n$, respectively. Hence, in line (9), the result of processor $p_{n,0}$ is stored in $A[n]$.

Observe that this scheme fails if another *start* occurs inside the *fork* statement because the original processor numbers can no longer be determined. In this case program execution fails.

In any case, the semantics of a *FORK* program is determined by the regime for solving read and write conflicts. Hence, *FORK* is intended to give the syntax and a *scheme* for the semantics of *FORK* programs.

2.6 Input/Output in *FORK*

There are at least two natural choices for designing input and output facilities for *FORK*. First, one may provide shared input/output facilities. These can be realized by (one–way infinite) shared arrays. Within this framework, synchronous I/O is realized by synchronous access to the corresponding array where conflicts are solved according to the same regime as with other accesses to shared variables (see Section 2.5).

As a second possibility one may provide private input/output facilities. These can be realized by equipping every logical processor with private input streams from which it can read, and private output streams to which it can write. The latter is the straightforward extension of PASCAL's I/O mechanism to the case of several logical processors.

Which of these choices is more suitable depends on the computing environment, e.g., the available hardware, the capabilities of an operating system and the applications. Therefore, input functions and output procedures are not described in this first draft on *FORK*.

3 The semantics of *FORK*

In this section we give a description of the constructs of the *PRAM* language *FORK*. As usual, nonterminals are enclosed in angled brackets, and we use them also for denoting arbitrary elements of the corresponding syntactical category, e.g., an arbitrary statement may be addressed by *<statement>*. The empty word is denoted by ε.

Programs are executed by processors which are organized in a group hierarchy. For each construct of *FORK* we have to explain how the execution of this construct affects the group hierarchy, the synchronism among the processors, and the scopes of objects.

We use the following terminology. A *group hierarchy* H is a finite rooted tree whose nodes are labeled by sets of processors. A node of H is called a *group (in H)*. Assume G is a group. A *subgroup of* G is a node in the subtree with root G. A *leaf group of* G is a leaf of this tree. A processor p *is contained in a group* G if it is contained in (the label of) a leaf group of G.

In the sequel we define inductively the actual group hierarchy and the notion of a *maximally synchronous* group w.r.t. this hierarchy. According to the inductive definition all processors in a maximally synchronous group are at the same program point. Also, each leaf group is a subgroup of a maximally synchronous group. A group G is called *synchronous* if it is a subgroup of a maximally synchronous group. If we loosely speak of a synchronous group G executing, e.g., a statement, we always mean that all the processors within G synchronously execute this statement.

Before program execution, the group hierarchy H consists just of one group numbered 0 which contains a single processor also numbered 0. This group is maximally synchronous. At the end of program execution we again have this hierarchy H.

Declaration sequences and statement sequences are executed by maximally synchronous groups (w.r.t. the actual group hierarchy). In the sequel G always denotes such a maximally synchronous group w.r.t. the actual group hierarchy and H the subtree with root G. The execution of declarations and statements may change the group hierarchy and the synchronism — but only within the subtree H. Therefore, we only describe these changes.

3.1 *Declarations*

In order to have e.g. dynamic arrays, the values of constants may not be computable at compile time and hence have to be determined at runtime.

Assume G executes a declaration sequence

<declaration> ; *<decls>* .

Then G first executes the declaration *<declaration>*. During the execution of *<declaration>* the synchronism among the executing processors and the group hierarchy with root G may change. However, at the end of *<declaration>*, H is reestablished, and G is maximally synchronous again w.r.t. the actual group hierarchy. Now G starts executing *<decls>*. As in PASCAL, there are constant, type, variable, function and procedure declarations. Constant and variable declarations additionally have to determine the *access type* of the newly created data objects, i.e., whether they are *private* or *shared*. If a data object is declared private, a distinct instance of the object is created for every processor executing the declaration. If a data object is declared shared, each leaf group executing this declaration receives a distinct instance of this object, which is accessible by all the processors of this leaf group.

Also, the return values of functions and the formal parameters of procedures and functions are treated differently to PASCAL:

- In a function declaration we have to specify whether the return value should be treated as a private or as a shared value. If it is declared shared then it is shared relative to the leaf group which called the function.
- Our language uses *const* parameters instead of PASCAL's *value* parameters.
- Every formal parameter has to be declared as private or as shared.

3.2 *Statements*

Assume G is a maximally synchronous group executing a statement sequence

<statement> ; *<stats>* .

Then G first executes the statement *<statement>*. During the execution of *<statement>* the synchronism among the executing processors and the group hierarchy H with root G may change. However, at the end of *<statement>*, H is reestablished, and G is maximally synchronous again w.r.t. the actual group hierarchy. Then G starts executing *<stats>*.

Our language supports 6 kinds of statements:

1. assignments
2. branching statements (*if* and *case* statement)
3. loop statements (*while*, *repeat* and *for* statement)
4. procedure calls
5. activation of new processors (*start* statement)
6. splitting of groups into subgroups (*fork* statement)

The statements of types 1—4 are similar to their counterparts in PASCAL. But due to the fact that there are usually several processors executing such a statement synchronously there are some differences in the semantics.

The *start* statement allows the activation of new processors by need. If an algorithm needs a certain number of processors these processors are activated via *start*. When the algorithm has terminated, the new processors are deactivated and the computation continues with the processors that were active before the *start*.

The *fork* statement does not change the number of active processors, but refines their division into subgroups.

3.2.1 *The assignment statement*

Assume an assignment statement of the form

<*expr*>$_1$:= <*expr*>$_2$

is executed. First all processors of G synchronously evaluate the two expressions (see Section 3.3). Each processor of G evaluates the left–hand side expression to a private or shared variable. Then the processors of G synchronously bind the value of the right–hand side expression to this variable. The effect of this is defined as follows:

1. if the variable is private, then after the assignment it contains the value written by the processor.
2. if the variable is shared, then the regime for solving write conflicts (see Section 2.5) determines the success or failure of the assignment, and, in case of success, the value to which the variable is bound.

Example Assume that four processors numbered 0 — 3 synchronously execute the assignment

$x := @ + \sharp$

where x is a variable of type integer. The value of x after the assignment depends on whether x is declared as private or as shared. We list some cases below.

1. x is a private variable. In this case there are four distinct incarnations of x. Then after the assignment x contains for each processor the value of the expression $@ + \sharp$.
2. the four processors form a leaf group and the variable x is shared relative to that leaf group. In this case the four processors write to the same variable and this write conflict is solved according to the regime for solving write conflicts (see Section 2.5).
3. the four processors form two different leaf groups (i.e. processors 0 and 1 are in the first one, 2 and 3 in the second one). Each leaf group has a distinct instance of the variable x. Then processors 0 and 1 write to the same variable and so do processors 2 and 3. The regime for solving write conflicts determines the value of the (two distinct) variables x after the assignment.

Above we have described the assignment of basic values. Our language supports assignment of structured values (e.g., $a := b$, where a and b are arrays of the same type), which is carried out component by component synchronously.

3.2.2 *The if statement*

Assume an *if*–statement

if <*expr*> **then** <*stats*>$_1$ **else** <*stats*>$_2$ **endif**

is executed. Then all processors of G first evaluate the expression <*expr*> synchronously. Depending on the result of this evaluation the processors synchronously execute the statements of the *then* or of the *else* part, respectively. Since different processors may evaluate <*expr*> to different values, we cannot make sure that all of them continue to work synchronously. The group hierarchies and the new maximally synchronous groups executing the *then* and the *else* part (if present), respectively, are determined according to the constants, variables and functions on which the expression <*expr*> "depends". Precisely, we say <*expr*> *depends* on a variable x if x occurs in <*expr*> outside any actual parameter of a function call. The case of constants and functions is analogous. We have to treat two cases:

1. *the expression does not depend on any private variables, constants, or functions.*

   In this case we do not change the group hierarchy H. Only the synchronism among the processors may change. Consider a processor p of G. We choose as the maximally synchronous group containing p the maximal group g_p of H which satisfies the following conditions:

   - g_p is a subgroup (not necessarily a proper one) of G.
   - p is contained in g_p.
   - The condition <*expr*> does not depend on any shared variables or other shared data relative to a proper subgroup of g_p. Note that the return value of a shared function is a shared datum only relative to a leaf group (see Section 3.3).

   Under these conditions all processors in g_p evaluate <*expr*> to the same value. Therefore all these processors choose the same branch of the *if* statement and hence are at the same program point.

   Note that a processor outside of g_p runs asynchronously with the processor p even if it evaluates expression <*expr*> to the same value.

   Example Assume that during program execution we have obtained the following group hierarchy.

   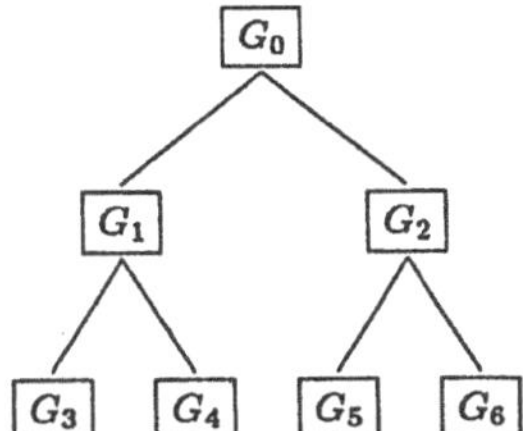

   and that all processors execute synchronously the *if* statement

   if $x = 5$ **then** S_1 **else** S_2 **endif**

   where an instance of x is a shared variable relative to G_1, and another instance of x is a shared variable relative to G_2. Then the processors of group G_3 work synchronously with the processors of group G_4 and the same holds for groups G_5 and G_6, respectively. The processors of group G_4 work asynchronously with the processors of group G_6 even if the two instances of variable x contain the same value. □

   When a processor of G has finished the execution of <*stats*> it waits until the other processors of G have finished their statement sequences. When all processors of G have arrived at the end of the *if* statement, G becomes maximally synchronous again. This means that even if two processors of G work asynchronously inside the *if* statement they become synchronous again after the *if* statement.

2. *The expression depends on private variables, constants, or functions*

   In this case we cannot even be certain that all processors inside the same leaf group evaluate <*expr*> to the same value. In this case both the group hierarchy H and the synchronism are changed as follows.

   Each leaf group of G generates two new leaf groups: The first one contains all processors of the leaf group that evaluate <*expr*> to *true*, and the second one contains the rest. All new leaf groups obtain the group number of their father group. The new leaf groups become maximally synchronous. Thus, the processors inside the same new subgroup work synchronously, while the processors of different subgroups work asynchronously.

   Again, when a processor reaches the end of the *if* statement, it waits until the other processors reach this point. When all the newly generated leaf groups have terminated the execution of the *if* statement, i.e., when all processors have reached *endif*, the leaf groups are removed. The original group hierarchy H is reestablished and G is again maximally synchronous.

3.2.3 *The while statement*

A *while* statement

```
while <expr> do <stats> enddo
```

is semantically equivalent to

```
if <expr> then
  <stats> ;
  while <expr> do <stats> enddo
endif
```

Thus, its semantics is determined by the semantics of *if—then—else* and recursion. The semantics of other kinds of loops, e.g., *for* or *repeat* loops, may be reduced to the semantics of the *while* loop.

3.2.4 *Procedure calls*

Syntax and semantics of procedure calls are similar to those of PASCAL. There are some slight differences due to the fact that processors can access two different kinds of objects: shared and private objects. This imposes some restrictions on the actual parameters of a procedure call. One only has to be careful of the access types of formal and actual parameters:

1. If a formal parameter is a shared–var–parameter then the corresponding actual parameter has to be a variable which does not depend on any private object, e.g., ar[♯] is not allowed as an actual shared var parameter even if ar is a shared array.
2. If a formal parameter is a private–var–parameter then the corresponding actual parameter can be a private or a shared variable.
3. If a formal parameter is a shared–const–parameter then the corresponding actual parameter may denote a private or a shared value.
4. If a formal parameter is a private–const–paramenter then the corresponding actual parameter may denote a private or a shared value.

The case where a formal shared–var–parameter is bound to a private variable is explicitly excluded since it allows some processor to modify the contents of this private variable, which may belong to a different processor. In contrast, in the case of a formal shared–const–parameter the value of that formal parameter during the execution of the procedure is determined by the regime for solving write conflicts. This is done in the same way as when determining the value of a shared variable after the assignment of a private value (see Section 3.2.1).

Assume the maximally synchronous group G executes a procedure call

$$\langle name\rangle(\langle expr\rangle_1,\ldots,\langle expr\rangle_n)\ .$$

First the actual parameters $\langle expr\rangle_1,\ldots,\langle expr\rangle_n$ are synchronously evaluated and bound to the corresponding formal parameters synchronously from left to right. Then G synchronously executes the procedure block, i.e. the declarations and the statements of that procedure.

3.2.5 *The start statement*

The *start statement* is used for activating new processors. The new processors are not allowed to access the private objects of the old processors. Therefore, statements inside the *start* statement should not refer to any private types, variables or constants, except @ and ♯, that are declared outside the *start* statement.

When the processors of G execute the statement

start [<*expr*>$_1$..<*expr*>$_2$] <*stats*> **endstart**

they first evaluate the range <*expr*>$_1$ .. <*expr*>$_2$. This range must not depend on any private data. Thus, the evaluation gives for all processors in the same leaf group g the same range $v_{g,1},\ldots,v_{g,2}$. At every leaf group g of G a new leaf group is added which contains $v_{g,2}-v_{g,1}+1$ new processors numbered with the elements of $\{v_{g,1},\ldots,v_{g,2}\}$. The group numbers of the new leaf groups are the same as for their father groups. G remains maximally synchronous. Now G executes <*stats*>, which means that the new processors of the new leaf groups execute <*stats*> synchronously. When G reaches *endstart*, the leaf groups are removed, i.e. the original hierarchy H is reestablished.

3.2.6 *The fork statement*

The *fork statement* is used to generate several new leaf groups explicitly. The new leaf groups obtain new group numbers and the processors inside the new leaf groups are renumbered.

Assume the processors of G execute the statement

fork[<*expr*>$_1$ .. <*expr*>$_2$]
@ = <*expr*>$_3$;
♯ = <*expr*>$_4$;
<*stats*>
endfork

First each processor p of G evaluates the expressions <*expr*>$_1,\ldots,$<*expr*>$_4$. The expressions <*expr*>$_1$ and <*expr*>$_2$ must not depend on any private data. Thus, all processors within the same leaf group g of G evaluate <*expr*>$_1$ and <*expr*>$_2$ to the same values $v_{g,1}$ and $v_{g,2}$ respectively. At every leaf group g of G $v_{g,2}-v_{g,1}+1$ new leaf groups are added which are numbered with the elements of $\{v_{g,1},\ldots,v_{g,2}\}$. The new leaf group with number i is labeled by the subset of those processors of g which evaluate <*expr*>$_3$ to i. Each of these processors obtains the value of <*expr*>$_4$ as its new processor number. G remains maximally synchronous and executes <*stats*>. When G reaches *endfork*, the new leaf groups are removed, i.e., the original group hierarchy is reestablished.

3.3 *Expressions*

Syntax and semantics of expressions are similar to those of PASCAL. There are just two new predefined constants: the processor number ♯ and the group number @, which both are private constants of type integer. Every processor of the maximally synchronous group G evaluates the expression and returns a value. The return value of a shared function is determined according to the choosen regime for solving write conflicts seperately for each leaf group of G and is treated as a shared object of this leaf group. Note that the evaluation of expressions may cause read conflicts, which are solved according to the choosen regime for solving read conflicts.

Example

```
...
shared var a: array[1 .. 10] of integer;
...
... := a[3]+3
...
```

Determining the variables corresponding to the subexpression a[3] does not cause a read conflict, whereas determining the values of these variables may cause read conflicts. □

4 Implementation

In this section we sketch some ideas showing that programs of *FORK* can not only be translated to semantically equivalent *PRAM* code, but also to code that runs efficiently. These considerations are supposed to be useful both to theoreticians and to compiler writers, who may have different realizations of *PRAMs* available, possibly without powerful operating systems for memory management and processor allocation.

Our basic idea for compiling *FORK* is to extend the usual stack-based implementation of recursive procedure calls of, e.g., the P-machine [17] by a corresponding regime for the shared data structures, a synchronization mechanism, and a management of group and processor numbers. Hence, here we address only the following issues:

- creating new subgroups;
- synchronization;
- starting new processors with *start*.

4.1 Creating new subgroups

The variables which are shared relative to some group have to be placed into some portion of the shared memory of the *PRAM*, which is reserved for this group. Therefore, the crucial point in creating new subgroups is the question of how the subgroups obtain distinct portions of shared memory. There are (at least) two ways to do this with little computational overhead:

1. by address arithmetic, as suggested in [5],
2. taking into account that in practice the available shared memory always is finite, by equally subdividing the remaining free space among the newly created subgroups.

The first method corresponds to an addressing scheme where the remaining storage is viewed as a two–dimensional matrix. Its rows are indexed by the group numbers, whereas the column index gives the address of a storage cell relative to a given group. For the second method the role of rows and columns are simply exchanged. In both cases splitting into subgroups can be executed in constant time. Also, the addresses in the physical shared memory can be computed from the (virtual) addresses corresponding to the shared memory of a subgroup in constant time. This memory allocation scheme is well suited to group hierarchies with balanced space requirements. It may lead to an exponential waste of space in other cases. Consider the following while loop, whose condition depends on private variables:

```
while cond(#) do          (1)
      work(#)             (2)
enddo                     (3)
```

Whenever the group of synchronously working processors executes line (1), it is subdivided into two groups, one consisting of the processors that no longer satisfy *cond*(#), and one consisting of the remaining processors. Hence, the first group needs no shared memory space (besides perhaps some constant amount for organizational reasons); a fair subdivision into two equal portions would unnecessarily halve the space available to the second group to execute *work*(#) of line (2).

However, there is an immediate optimization to the above storage distribution scheme: we attach only a fixed constant amount of space to groups of which it is known at compile time that they do not need new shared memory space, and subdivide the remaining space equally among the other subgroups.

This optimization clearly can be performed automatically for loops as in the given example, but also for one-sided *if*s, or *if*s where one alternative does not involve blocks with a non–empty declaration part.

4.2 Synchronization

In order to reestablish a group g, the runtime system has to determine when all the subgroups of group g have finished. This is the termination detection problem for subgroups.

If all processors run synchronously, no explicit synchronization is necessary. In the general case where the subgroups of group g run asynchronously, there are the following possibilities for implementing termination detection:

1. Use of special hardware support such as a unit-time *fetch&add* operation which allows the processors within a group to simultaneously add an integer to a shared variable.

2. Static analysis (possibly assisted by user annotation); most of the *PRAM* algorithms published in the literature are of such a simple and regular structure that the relative times of execution sequences can be determined in this way.

3. Use of a termination detection algorithm at runtime. The latter is always possible; however, complicated programs cheating a static analyzer will be punished by an extra loss of efficiency.

4.3 Starting new processors

The following method which is analogous to the storage distribution scheme works only for concurrent-read machines with a finite number of processors. Before running the program, all processors are started, all of them with processor number $\sharp = 0$. If in a subsequent *start* statement of the program fewer processors are started than what are physically available in the leaf group executing this statement, then several physical processors may remain "identical", i.e., receive the same new processor number. These identical processors elect a "leader". All of them execute the program but only the leader is allowed to perform write operations to shared variables. Consider the following example. Assume that we are given 512 physical processors.

```
start [0 .. 127]                    (1)
    if ♯ < 64 then                  (2)
        start [0 .. 212]            (3)
            compute(212)            (4)
        endstart                    (5)
    endif                           (6)
endstart                            (7)
```

Before line (1), all the 512 physical processors are started. After line (1), there are always four processors having the same processor number $\sharp$. Having executed the condition of line (2) all the processors whose processor number is less than 64 enter the *then* part: these are 256. All of them are available for the *start* instruction of line (3), where they receive the new numbers $0, \ldots, 212$. Two physical processors are assigned to each of the first 43 logical processors whereas one physical processor is assigned to each of the remaining 170 logical processors. The original processor identities are put onto the private system stack. When the *endstart* in line (5) is reached, the processors reestablish their former processor numbers.

If more processors are started than physically available in the present group, then every processor within that group has to simulate an appropriate subset of the newly started processors.

In both cases *start* can be executed in constant time by every leaf group consisting of a contiguous interval of processors: this is the case, e.g., for *starts* occuring in the statement sequence of the toplevel block. Thus, a programming style is encouraged where the logical processors necessary for program execution are either started at the beginning, i.e., before splitting the initial group into subgroups, or are started in a balanced way by contiguous groups.

To maximally exploit the resources of the given *PRAM* architecture, a programmer may wish to write programs which use different algorithms for different numbers of physically available processors. Therefore, a (shared) system constant of type integer should be provided, whose value is the number of physical processors available on the given *PRAM*. This allows programs to adopt themselves to the underlying hardware.

References

[1] F. Abolhassan, J. Keller, and W.J. Paul. On physical realizations of the theoretical PRAM model. Technical Report 21/1990, Universität des Saarlandes, SFB 124, 1990.

[2] A.V. Aho, J.E. Hopcroft, and J.D. Ullman. *The Design and Analysis of Computer Algorithms.* Addison–Wesley, Reading Massachusetts, 1974.

[3] S.G. Akl. *The Design and Analysis of Parallel Algorithms.* Prentice Hall, 1989.

[4] Y. Ben–Asher, D.G. Feitelson, and L. Rudolph. ParC — An extension of C for shared memory parallel processing. Technical report, The Hebrew University of Jerusalem, 1990.

[5] P.C.P. Bhatt, K. Diks, T. Hagerup, V.C. Prasad, S. Saxena, and T. Radzik. Improved deterministic parallel integer sorting. *Information and Computation,* to appear.

[6] A. Borodin and J.E. Hopcroft. Routing, merging and sorting on parallel models of computation. *J. Comp. Sys. Sci. 30*, pages 130 – 145, 1985.

[7] M. Dietzfelbinger and F. Meyer auf der Heide (ed.). Das GATT-Manual. In: Analyse paralleler Algorithmen unter dem Aspekt der Implementierbarkeit auf verschiedenen parallelen Rechenmodellen. Technical report, Universität Dortmund, 1989.

[8] F.E. Fich, P. Ragde, and A. Widgerson. Simulations among concurrent-write PRAMs. *Algorithmica 3*, pages 43 – 51, 1988.

[9] S. Fortune and J. Wyllie. Parallelism in random access machines. In *10th ACM Symposium on Theory of Computing,* pages 114–118, 1978.

[10] N.H. Gehani and W.D. Roome. Concurrent C. In N.H. Gehani and A.D. McGettric, editors, *Concurrent Programming,* pages 112–141. Addison Wesley, 1988.

[11] A. Gibbons and W. Rytter. *Efficient Parallel Algorithms.* Cambridge University Press, 1988.

[12] T. Hagerup, A. Schmitt, and H. Seidl. FORK — A high–level language for PRAMs. Technical Report 22/1990, Universität des Saarlandes, SFB 124, 1990.

[13] P.B. Hansen. The programming language Concurrent Pascal. *IEEE Transactions on Software Engeneering 1(2),* pages 199–207, June 1975.

[14] H.F. Jordan. Structuring parallel algorithms in a MIMD, shared memory environment. *Parallel Comp. 3*, pages 93–110, 1986.

[15] Inmos Ltd. *OCCAM Programming Manual.* Prentice Hall, New Jersey, 1984.

[16] United States Department of Defense. Reference manual for the Ada programming language. ANSI/MIL-STD-1815A-1983.

[17] St. Pemberton and M. Daniels. *Pascal implementation: The P4 compiler.* Ellis Horwood, 1982.

NEURAL NETWORK-BASED DECISION MAKING FOR LARGE INCOMPLETE DATABASES

A. R. Hurson, B. Jin and S. H. Pakzad
Department of Electrical and Computer Engineering
The Pennsylvania State University
University Park, PA 16802

Abstract

As an extension to the relational algebra, maybe algebra operations have been proposed to handle incomplete information. Such a set of operations allows the user to investigate the potential set of data values (i.e. tuples) to draw his/her own conclusions. However, maybe algebra operations could return nonrelevant data, generate low quality results, and offer low physical performance. Hence, it is appropriate to design a scheme to investigate the results generated by the maybe operations, in order to improve the data quality and performance of large databases. Such a mechanism should be dynamic to adjust itself according to the user's query and the characteristics of the underlying databases. In this paper, an artificial neural network-based decision support system for handling large databases containing incomplete information is proposed. It is a subsystem which learns and constructs a knowledge base to filter out the data that is not of any importance to the user. The network accomplishes the decision-making task in a massively parallel manner. This paper also discusses the implementation of the decision-making network based on the VLSI design of a Basic Neural Unit (BNU). Using a weight-centered design principle, BNU can be expanded and reconfigured to satisfy the requirements of the underlying environment.

I. Introduction

Missing information presents a problem for any data model. Such a data value can be interpreted as the "value at present unknown". The inclusion of missing information provides unknown data for the attributes in a database. Within the scope of relational model, a partial tuple is the one that contains one or more incomplete (null) data. A tuple is said to be total if it contains no null values. Similarly, a relation is total if all of its tuples are total. As the relational database model has matured, researchers have examined the question of how to handle missing data [2,4,5,6,8,9,14,15]. The extensions of relational algebra have been addressed in [4,8,9,15]. This study is based on Codd's maybe algebra[4] which adopts a three-valued logic for handling partial relations. Such an extended capability allows the user to probe the database for potential data relationships that might be useful but can not be retrieved by true relational algebra.

While such a set of maybe operations brings a better information utilization, it also has drawbacks.

For example, maybe join operator has the potential to return nonrelevant data relationships which are not defined according to the semantics of the underlying database. Moreover, it might generate extremely large result relations, which could cause drastic degradation in performance if most of the resultant tuples are not of interest to the user. [8] shows that the loss in physical performance is paralleled by a loss in logical performance. Zaniolo's generalized join [19] removes the join over two nulls, but it still has the potential to produce extremely large relations.

What is needed then is a dynamic mechanism which functions as a filter with learning capability to adjust itself according to the specific characteristics of the underlying database and the requirement of the user's query. To fulfill the requirement of such a dynamic mechanism, we propose a neural network-based decision support system. The neural network-based approach has advantages over conventional knowledge-based systems, due to its strong dynamic and self-adaptive characteristics [7]. In a neural network, learning algorithm sets the connection strengths to their appropriate values based on valid input/output pairs (i.e. training pairs) [1]. In this way, the network learns and constructs a knowledge base to filter out the useless information. This paper will explore these points and will show how a knowledge acquisition model can be built for the decision-making network.

The dynamic nature of the databases and user's query require expandability and reconfigurability at the underlying decision-making network. Such a requirement creates a major problem for the traditional neuron-centered design techniques. In this paper, a weight-centered design principle is used instead, which yields a high parallelism and reconfigurability for the constructed networks. A Basic Neural Unit (BNU), based on the weight-centered approach, is designed to be used as basic building block in the construction of an expandable and reconfigurable decision-making network.

This paper is organized as follows: Section 2 overviews the concept of neural network-based decision support system. The VLSI design of BNU and the construction of a decision-making network using BNUs are discussed in Section 3. Section 4 presents the simulation results to analyze the performance of the proposed decision support system. Finally, Section 5 concludes the paper.

II. A Neural Network-Based Decision Support System

The maybe operations could offer a low-level logical and physical performance. To improve such a behavior, it is desired to build an intelligent mechanism which can make appropriate judgments over the resultant tuples, i.e., filtering out low quality and erroneous data and providing higher quality result to the user. To achieve such a goal, we propose a neural network-based decision support system, as shown in Figure 1. This system learns, adjusts and adapts itself to make certain decisions by applying its learned knowledge to the real database environment.

The neural network-based decision support system is composed of four modules: i) decision-making network; ii) knowledge acquisition; iii) decision-controlled buffer; and iv) user-system interface.

2.1 Adaptive Learning and Decision-Making Network

This module is a three layer neural network, an input layer, a hidden layer, and an output layer (Figure 2). The decision-making network is characterized by a set of neurons, pattern interconnection, propagation rule, output function, and learning rule. It is a fully connected network, i.e., all input units are connected to all hidden neuron-units, and all hidden neuron-units are connected to all output neuron-units.

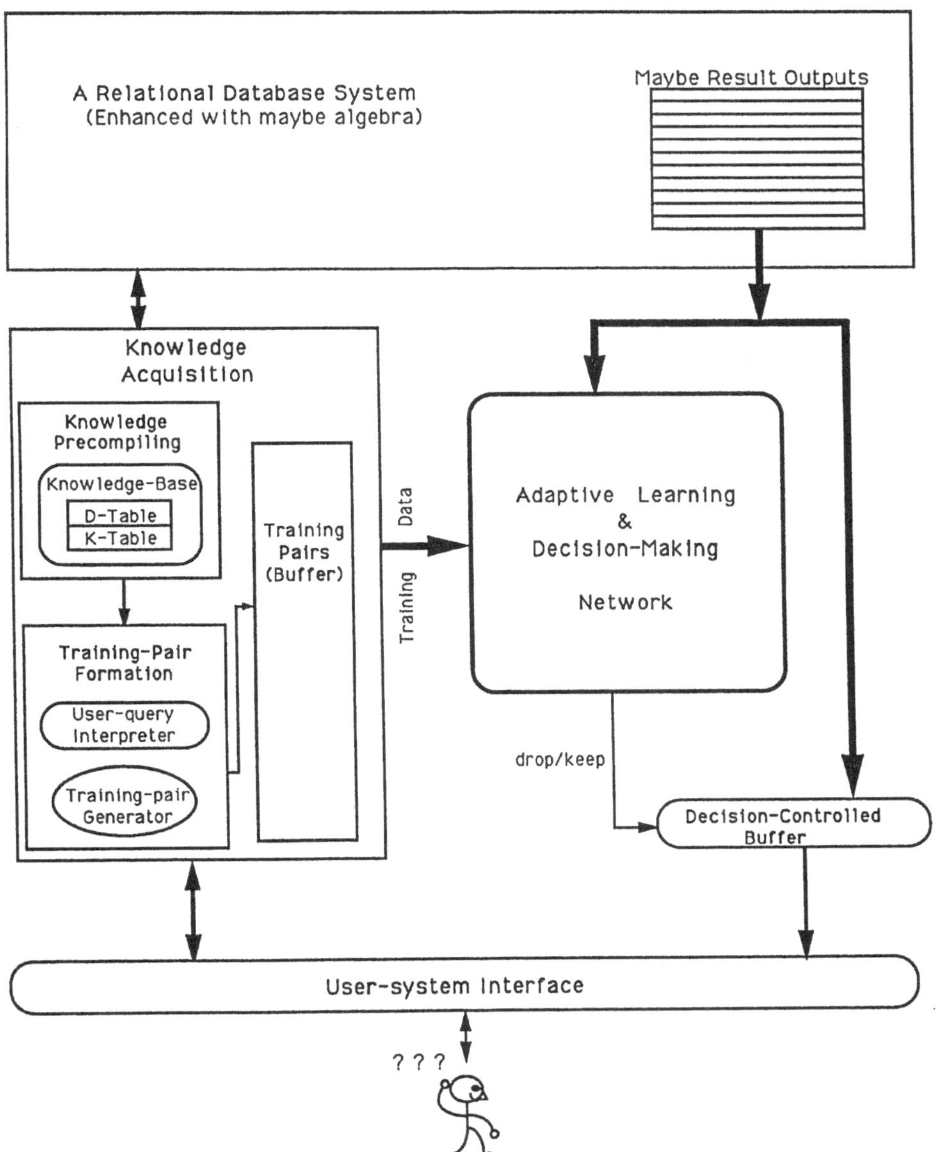

Figure 1. A neural network-based decision support system

No computation function is performed by the units of the input layer. They simply distribute an input pattern to the hidden layer. Input is a binary bit pattern, representing the resultant tuples generated by maybe operations, such as maybe join. Each input unit can take a value of either "1" (meaning known attribute value) or "0" (meaning unknown or null attribute value). The number of input units in the decision-making network varies with the number of attributes in the resultant tuple. The proposed decision-making network is designed to have a reconfigurable architecture in both software simulation and hardware implementation [7]. Hidden neuron-units compose a layer of abstraction, pulling features from the input pattern[11,13,16,18]. In the decision-making network, the number of hidden units is also adjustable [7]. The decision responses are shown in the output layer. It represents the classification associated with each input pattern. Classifications (or network outputs) are binary and can be interpreted as categories. The leftmost output unit, in Figure 2, represents the "keep" decision, which means that the associated input tuple should be passed to the user for its high quality information; while the rightmost unit represents the "drop" decision due to the low quality and/or erroneous information. Besides these two categories, we can define certain intermediate categories representing some degree of data quality. This will provide the user a more flexible "keep"/"drop" range. It can be recognized as a flow-controlled filter which controls the quantity (as well as the quality) of the filtered tuples. Hence the number of output neuron-units should also be adjustable.

The decision-making network functions in either of the two operation modes, learning mode or decision-making mode. When the network works in the learning mode, it receives a number of training pairs from the knowledge acquisition module. Each training pair consists of a known input pattern and a desired output pattern. Using supervised learning algorithm of the "generalized delta rule", the network adjusts its weights so that it can respond to the input patterns as closely to their respective desired responses as possible. Once the network is trained, it is ready to switch to the decision-making mode. Working in the decision-making mode, the network receives input patterns generated by maybe operations, and responds to the closest classification (e.g. "keep", "drop", etc.). Generalization is considered to be successful if the network responds correctly, with high probability, to input patterns that were not included in the training set [11]. The decision made by the network is transferred to decision-controlled buffer as a control signal.

2.2 Knowledge Acquisition

The goal of knowledge acquisition in the decision support system is to gather a set of training pairs for the decision-making network. In the knowledge acquisition module, knowledge can be obtained from the database semantics, i.e. data dependency relationships and key attributes. Decision-making process requires to understand the semantics of the underlying database as well as the requirements of the user's query. We propose a two-level hierarchical knowledge acquisition module for the decision support system (Figure 1). In the first level, knowledge can be acquired by extracting relationships that are implicitly/explicitly defined in a database system, independent of the user's query. This can be done by a knowledge precompiling unit. It provides long-term knowledge with a strong portability. However, to form the training pairs proper information should be also extracted from the user's query. This part of the knowledge can be extracted by a user-query interpreter unit in the second level of knowledge acquisition. In this level, knowledge acquisition takes place after a query is submitted. The acquired knowledge, referred to as short-term knowledge, differs from query to query and can not be shared. Since we assume that the knowledge-base is constructed in advance, only the speed of the short-term knowledge acquisition could affect the performance of knowledge acquisition module in the real-time situation.

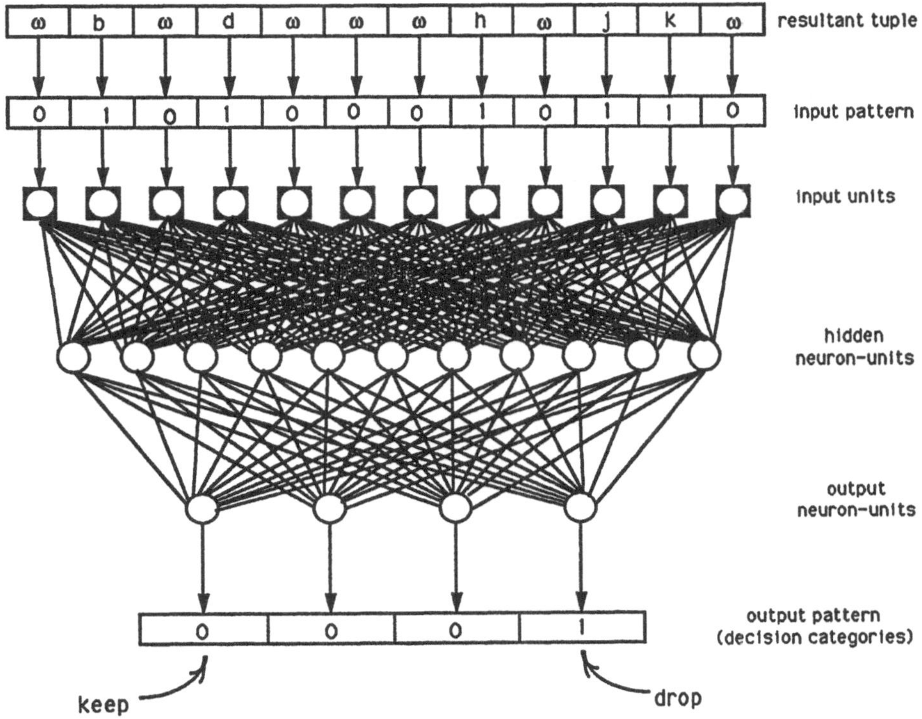

Figure 2. A neural natwork-based decision-making network

Level 1: Knowledge Precompiling

In the decision support system, accumulating useful information directly from the underlying relational database system and reorganizing it for the formation of training pairs, is referred to as knowledge precompiling. Two concepts in a relational database system form the theoretical foundation for knowledge precompiling: *data dependencies* and *key attributes* . The knowledge base consists of a D-table and a K-table. D-table stores the knowledge captured by analyzing the data dependency relationships. A K-table is constructed as part of the knowledge base to store the knowledge acquired from various kinds of key attributes, i.e., *candidate keys* (including *primary keys, alternate keys*) and *foreign keys* [17].

Level 2: Training-Pair Formation

The second level of the knowledge acquisition is the training-pair formation. Each time a query is submitted to the system, a set of training pairs ought to be generated based on the user's query. A user-query interpreter is designed to analyze and recognize the set of query dependent information needed to generate the training pairs. Consequently, the long term and short term knowledge are used, along with a set of heuristic criteria, to form the training pairs. This set of heuristic rules is determined by a careful analysis of the user's requirements and the semantics of the database, such as:

- If a resultant tuple has null values for all the candidate keys, then the decision should be "drop".
- If a resultant tuple has null value for the primary key attribute, but with one of the alternate keys defined, then the decision should be "keep".
- If, in the resultant tuple, attribute A functionally determines attribute B, and A is null, then the tuple should be "dropped".
- Under the previous dependency condition, if A is defined, but B is null, then this tuple should be "kept".
-
-
-

These guiding criteria should be enforced by the system designer. Moreover, it should be flexible enough to satisfy the requirements of the ever changing environment. Finally, all generated training pairs are stored in a training-pair buffer to be sent to the decision-making network.

2.3 Decision-Controlled Buffer

Decision-controlled buffer is a delay device which simply holds the resultant tuples and waits for the output from the decision-making network. By the time when a decision arrives, the buffer will either pass a resultant tuple to the user (if a "keep" decision was made) or filter out the tuple (if a "drop" decision was made). The delay time would depend on the execution speed of the decision-making network.

2.4 User-System Interface

It functions as an interface between the user and the decision support system. It provides a convenient communication environment for both the user and the system. It explains the result to the user, and more importantly, links the user to the knowledge acquisition module through a dialogue system.

III. Implementation of Decision-making Network Using BNU Chips

Software simulations have been conducted to demonstrate the feasibility of the proposed decision support system, e.g., the decision-making network and the knowledge acquisition modules[7]. Simulations were run to investigate two performance parameters, the accuracy and the training time of the decision-making network. As a result, the design of such a decision support system was verified by the obtained average accuracy of network response between 70 percent to 97 percent. It was shown that the decision-making network converged within short training times. As part of the simulation, it was demonstrated that these two parameters are also a function of the number of hidden units[7]. While the simulation results have shown the feasibility of the proposed scheme, it did not give any insight regarding the parallel capability of the proposed decision-making network. To demonstrate the parallel capability of the proposed decision-making network, a Basic Neural Unit (BNU) was designed and fabricated. The BNU is used as the basic building block in the construction of a decision-making network. The effectiveness of parallel processing in decision support was exploited by hardware implementation of the decision-making network.

3.1 The VLSI Design of BNU(Basic Neural Unit) chip

<u>Neuron-centered Design v.s. Weight-centered Design</u>

The decision-making network is composed of a set of processing elements (PEs) and the synaptic weight interconnections. Each PE simulates a neuron which sums all the weighted inputs from other PEs (neurons). Typically, this is a neuron-centered PE, as shown in Figure 3(a). In neuron-centered design, once a PE or a set of PEs is implemented, it is usually difficult to expand the size of the network with the increased number of inputs. For example, by combining two networks of Figures 3(b) and 3(c), an expanded 9-input, 3-output, three layer network can be constructed, as shown in Figure 3(d). However, such a small scale expansion requires 36 additional interconnections. In VLSI implementation, the requirement for additional connections complicates the design strategy where strong modularity, expandability and reconfigurability are necessary.

Due to this restriction we propose a new approach, namely weight-centered design. Figure 4(a) shows a 4-input, 1-output weight-centered processing element. It consists of 4 weight units, 1 accumulator, and a nonlinear function unit. It is functionally equivalent to a 4-input, 1-output neuron-centered PE. Each weight unit stores a weight value. Similar to a neuron-centered PE where each weight value has an associated input, in a weight-centered PE each input should be received by its associated weight unit to form the weighted input value that is shifted down to the accumulator. The accumulator then sums all the weighted inputs and sends the result to a nonlinear function unit to generate the final output. In the proposed weight-centered PE, two ports are left open for the purpose of expansion and reconfiguration. Horizontally, inputs can be sent to more weight units to increase the number of output neurons. For example, by adding another set of weight units, the network of Figure 4(a) can be expanded to a 2-output processing unit (Figure 4(b)). In this way, the number of output neurons can be easily expanded and reconfigured. Vertically, weight-centered PEs can be stacked up to increase the number of input neurons. Figure 4(c) shows an 8-input, 1-output processing unit. Note that the nonlinear function unit is not a part of the basic building block, it is required only at the final output.

To form a multilayer decision-making network, weight units are divided into groups, called layer-

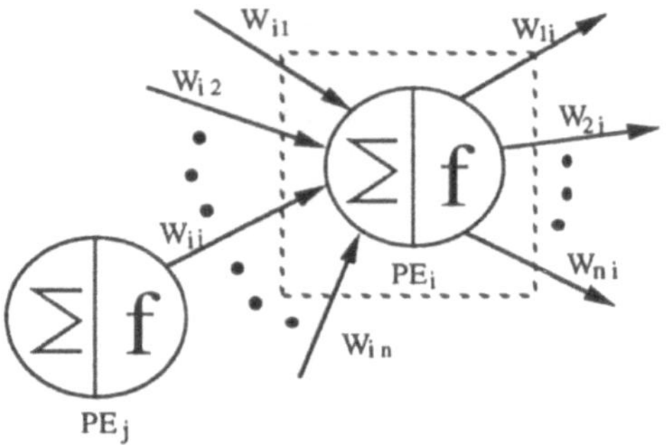

(a) A neuron-centered processing element

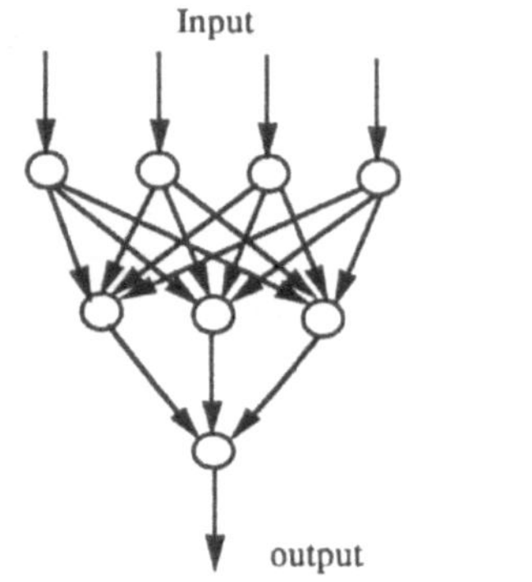

(b) A 4-input, 1-output three-layer network

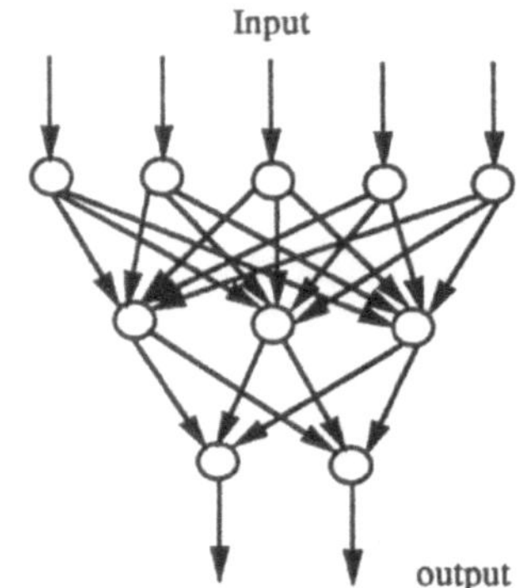

(c) A 5-iuput, 2-output three-layer network

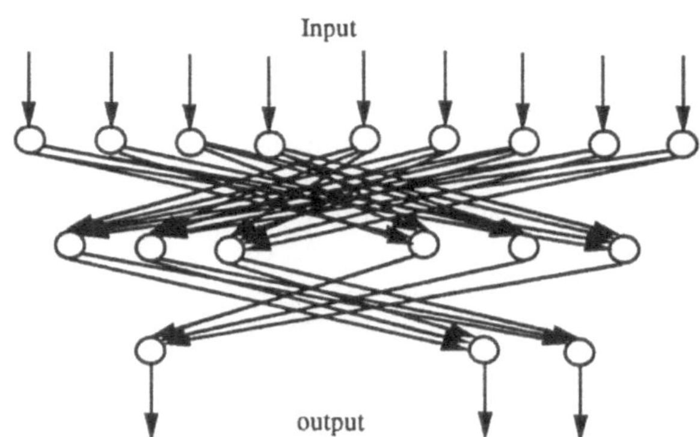

(d) A 9-input, 3-output three-layer network constructed by combining (b) and (c)

Figure 3. Neuron-centered design of decision-making networks

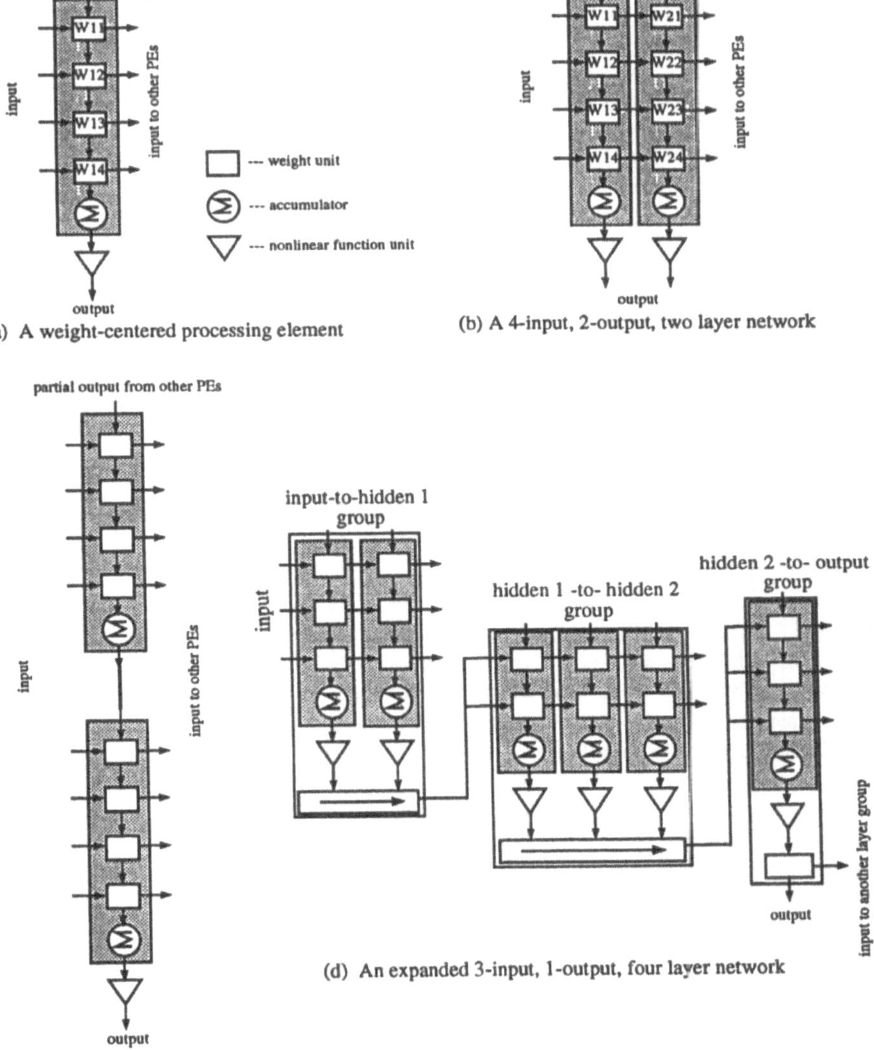

(a) A weight-centered processing element

(b) A 4-input, 2-output, two layer network

(c) An expanded 8-input, 1-output, two layer network

(d) An expanded 3-input, 1-output, four layer network

Figure 4. Weight-centered design of decision-making networks

to-layer board. Each group (board) represents weight units between each two adjacent layers. For example, Figure 4(d) is a four layer network constructed by using weight-centered design principle.

The Design of Basic Neural Unit (BNU)

Weight-centered design technique is promising due to its modularity and expandability. In contrast to the neuron-centered design, it simplifies the inter-neuron connections especially when expanding a network. Based on the concept of weight-centered design, a Basic Neural Unit (BNU) was proposed and fabricated. Besides the expandability and reconfigurability, embedding more weight units into a single silicon chip serves the main objective in designing the BNU. Due to the fabrication process available to us, a BNU consisting of 30 weight units, 10 inputs and 3 outputs was fabricated by the IBM corporation. A photomicrogragh of BNU chip is shown in Figure 5(a). BNU is packaged in a 120-pin package. The BNU chip has a maximum time delay of 153.6 ns to generate the partial output sums.

Figure 5(b) shows the circuit schematic of a BNU chip. Each BNU forms three identical and independent processing strings, i.e., three weight groups, running in a parallel fashion. W's represents the weight input. The 10 inputs are denoted by a's. b(k+1) is the output. And, b(k) represents the partial output coming from other BNU chip when more than one BNU are used in constructing a network. One weight group is functionally equivalent to a neuron-centered PE which has 10 inputs and 1 output. The BNU works in two functional modes. It first enters the initialization mode in which weight values are loaded into three weight groups, then it is switched to the execution mode in which three groups receive the same set of inputs, and generate three outputs. Figure 5(c) demonstrates the detail schematic of one weight group in a BNU chip. The 10 weight units are implemented as shift registers. At the top is a multiplexing register that can either load the weight values during the network initialization or form the WEIGHT ITERATION LOOP when the network is in the execution mode. Each group is augmented by two more multiplexing registers. One is used to receive the INPUT via an enable input ("en"). It either passes a weight value to the adder if the input value is "1" or rejects it (i.e. sends out a "0") if the input is "0". Another multiplexing register is used to loop back the partial sum of weighted inputs to form the ADDITION LOOP. With the help of these two registers, the adder generates a partial sum output of 10 weighted inputs. The latter register is also responsible for receiving partial output from another BNU and pass to the adder if more than one BNU are used in the network. It is noted that because of our limited fabrication process, the "a" values are passed in sequential fashion. Naturally, such approach has affected the execution time of the chip.

3.2 Constructing a Decision-making Network with BNU Chips

Varying physical characteristics of databases and dynamic nature of the user's queries require varying number of neurons on input, output and hidden layers. As a result, the supporting decision-making network should have the ability to be reconfigured to satisfy the requirements of the user's query. This can be accomplished by using BNU chips as basic building blocks in the construction of a decision-making network. On one hand a BNU itself forms a 10-input, 3-output, two layer network, on the other hand it can be connected with other BNUs to construct a larger network.

Each BNU chip is a 10-input, 3-output, two layer network. By adding another BNU chip along the vertical direction (i.e., the PARTIAL SUM OUTPUT of one BNU is connected to the PARTIAL SUM INPUT of another BNU), it can be extended to a 20-input, 3-output module. By the same token, attaching one BNU to another in the horizontal direction will extend the network to a 10-input, 6-output module. For example, Figure 6(a) depicts a 4 connected BNUs on the input-to-hidden layer board which forms 20

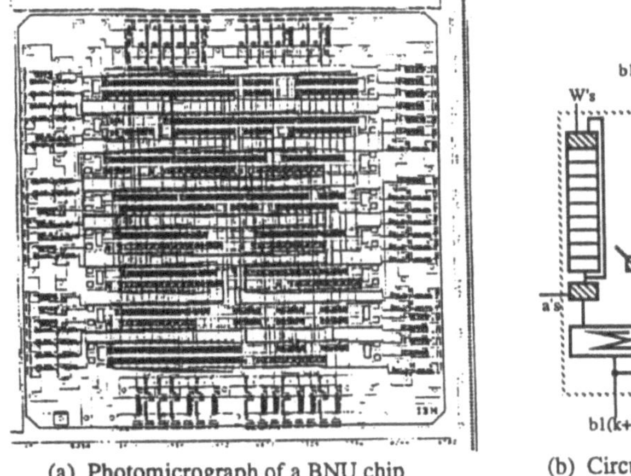

(a) Photomicrograph of a BNU chip

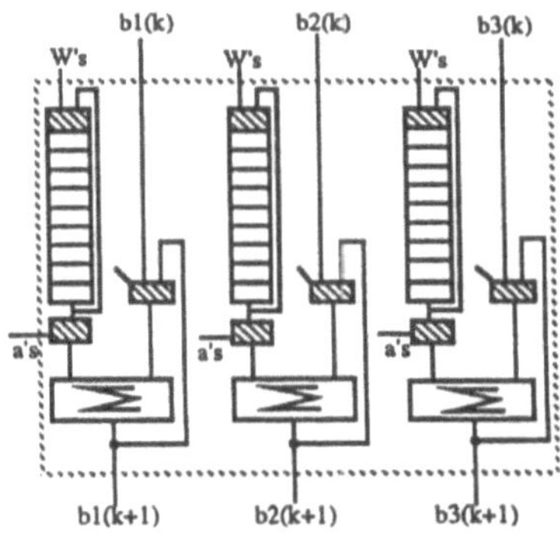

(b) Circuit schematic of a BNU chip

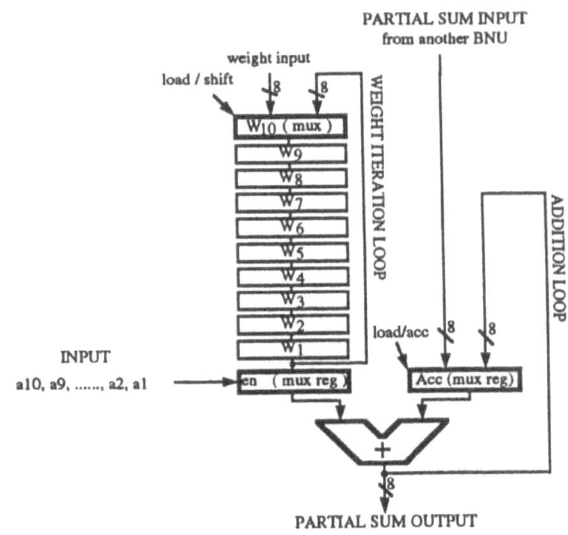

(c) single weight group in a BNU chop

Figure 5. VLSI design of a Basic Neural Unit (BNU) chip

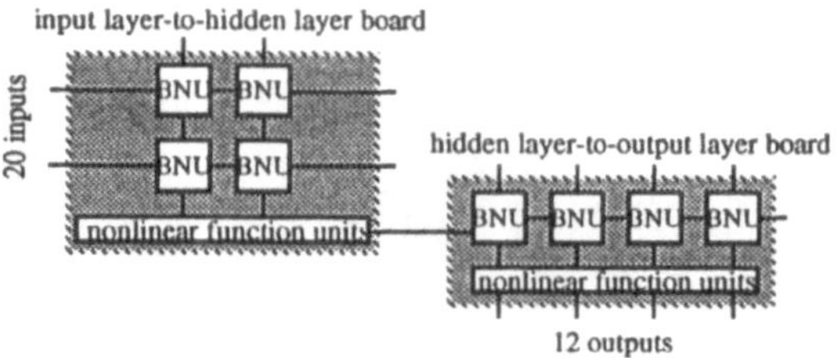

(a) A three layer network constructed by 8 BNU chips with 20 inputs and 12 outputs.

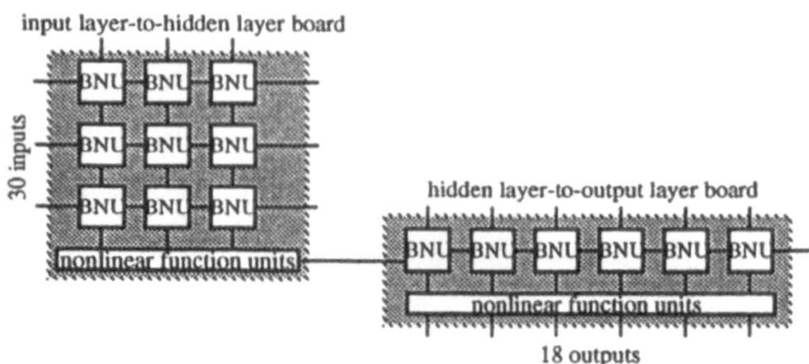

(b) A three layer network constructed by 18 BNU chips with up to 30 inputs and 18 outputs.

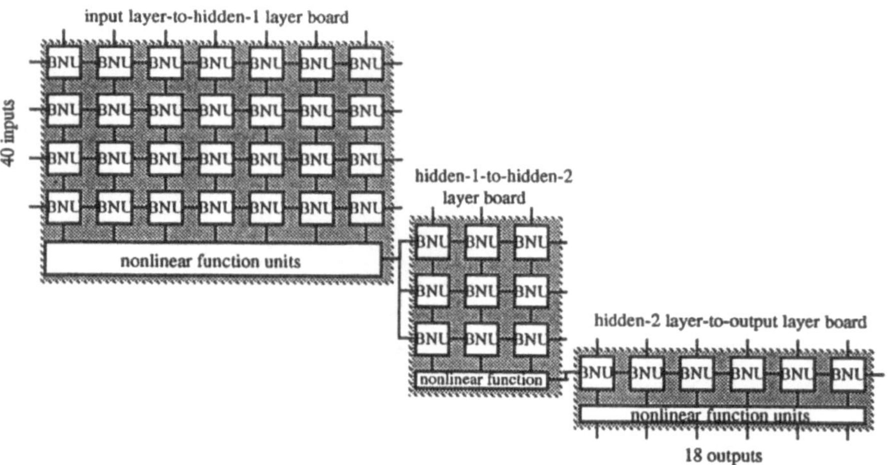

(c) A four layer network constructed by 48 BNU chips with up to 40 inputs and 18 outputs.

Figure 6. Constructing a decision-making network using BNU chips

inputs and 6 outputs. In general, adding a "row" of BNUs in the vertical direction of a network can increase the number of inputs by 10, while adding a "column" of BNUs in the horizontal direction can increase the number of outputs by 3. Figure 6(b) shows a 30-input, 9-output, two layer network on the input layer-to-hidden layer board.

The proposed BNU can be used to expand the number of layers in a multilayer network through the use of layer-to-layer board. In general, each two-layer network forms a group of BNUs which are mounted on a circuit board, called a layer-to-layer board. For example, in Figure 6(a) two layer-to-layer boards, an input-to-hidden layer board and a hidden-to-output layer board, are used to obtain a three layer network. When connecting the two, the number of outputs in the first group should match up with the number of inputs in the second group. Note that the second group has 10 inputs and 12 outputs. However, only 6 inputs are used to match the 6 outputs generated by the first group.

When expanding a multilayer network, the same design principle can be applied to an individual layer board. As an example, Figure 6(b) is an expanded three layer network based on the one shown in Figure 6(a). On the input-to-hidden layer board, a row of BNUs is added to increase the number of input neurons by 10. Adding another column of BNUs, the number of hidden neurons can be extended to 9. On the hidden later-to-output layer board, 2 more BNUs are used in order to form 18 outputs. Moreover, adding one more layer board will increase one layer in the decision-making network. Figure 6(c) demonstrates the construction of a four layer network consisting of 3 layer-to-layer boards. It can be easily observed that the network has maximum 40 input neurons, 21 neurons on the first hidden layer, 9 neurons on the second hidden layer, and 18 output neurons.

IV. Performance Analysis and Simulation Results

A number of simulation runs have been conducted to show the feasibility of the proposed decision support system. A generator was developed to generate two relations --- student relation and company relation. The student relation contained the social security number, class standing, major, area of interest, and GPA. The company relation was composed of company's name and it's hiring requirements (e.g., area of interest, major, minimum required GPA).

The simulation is based on the following assumptions:

1. The entire database resides in memory.
2. There are 138 majors offered.
3. There are approximately 600 areas of interest.
4. There are 500 tuples in the student relation, and 50 tuples in the company relation.

The simulation was run based on four different classes of queries, to represent a variety of join operations (i.e., join over single and multiple attributes, θ-join and attribute maybe join):

Query class 1 --- The first class of queries was performed using the company's hiring area of interest and a student's area of interest as the join attribute. The null value in the student relation simply meant that the student was not sure about his/her area of interest was.

Query class 2 --- The second class of queries concerns only those students who satisfied the minimum GPA requirement for hiring, i.e., a company's minimum GPA requirement. In this type of queries, the company's minimum GPA had a domain from 2 to 4.

Query class 3 --- The third class of queries was performed using area of interest, major, and the GPA as join attributes. Even if a student doesn't know his/her area of interest, the student's GPA and major might satisfy a company's criteria and therefore should be considered.

Query class 4 --- The fourth class of queries was based on the attribute maybe-join on major and area of interest.

The following table lists some information about the simulation results:

Query class	Total # of Resultant Tuples	# of training pairs Generated	# of Correct Decisions	# of Incorrect Decisions	Accuracy of Response
1	608	98	458	150	75.33%
2	412	76	357	55	86.65%
3	337	55	274	63	81.31%
4	153	28	114	39	74.51%

In the above table, the training pairs for each query class were generated by the knowledge acquisition module. As a verification of the performance of this module, the decision-making network achieved an accuracy of response between 74% to 86%. The higher accuracy obtained in query class 2 is mainly due to the high quality of training pairs. The issue of how to improve the quality of training pairs is under the consideration.

To investigate the performance improvement of the proposed BNU-based network against software simulator, the simulation was run based on the characteristics of fabricated BNU chip. Three layer decision-making network was used for all four query classes, while the size of the network varies among different classes, depending on the length of the resultant tuple. Figure 7(a) shows the simulation results. Note that the length of the resultant tuples of query class 1 is the longest among all. Therefore, class 1 requires the longest decision-making time. Moreover, the time difference between hardware and software implementation tends to be larger when the length of the resultant tuple is longer, i.e., the size of the decision-making network is larger. In another word, the length of the resultant tuple affects the decision-making time required by the software more than it affects the hardware. In a separate simulation run the accuracy of knowledge acquisition module was analyzed based on the size of the training-pair set for each query class (Figure 7(b)). We observed that the size is a function of the length of the resultant tuples. As a result, for query class 1 the module was able to generate more training pairs than the other classes of queries. This is due to the fact that for relations with larger length the module can extract more short-term and long-term knowledge. In addition, as can be observed the accuracy after a certain point, is a function of the number of training pairs. This can be interpreted that too many training pairs will bring too much noise which would degrade the accuracy of the network responses. On the other hand, fewer number of training pairs may decrease the generalization ability of the network which will also affect the accuracy.

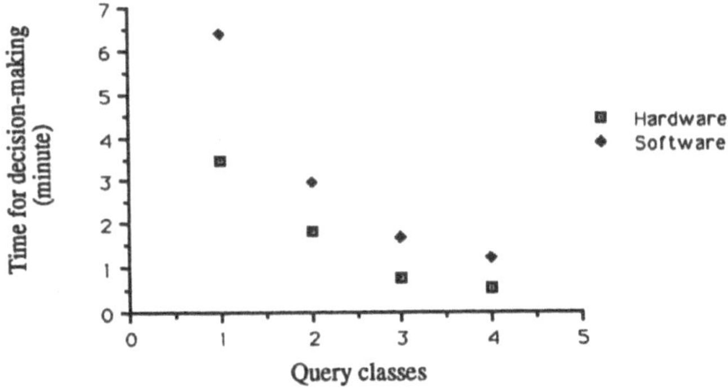

(a) Time required for decision-making

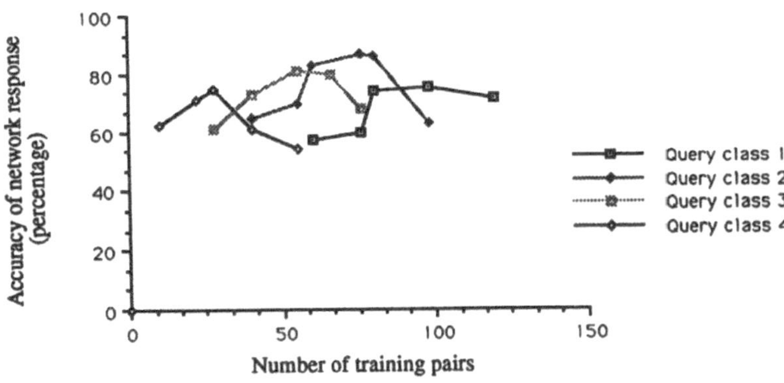

(b) Accuracy led by the size of the set of training pairs

Figure 7. Some simulation results and performance analysis

V. Conclusion

A neural network-based decision support system for handling large databases containing incomplete information is proposed. The major difficulty which hinders the use of maybe algebra operations is due to its tendency in generating large volume of low quality resultant data. It is the learning capability and strong dynamic self-adaptive nature of a neural network which make it effective to filter out low-quality tuples and extract higher-quality ones.

The design of the proposed decision support system has resulted in a model consists of four modules: adaptive learning and decision-making network (neural network), knowledge acquisition module, decision-controlled buffer, and user-system interface. The principle of operation of each module and the interrelationships among modules have been discussed. The VLSI implementation of a Basic Neural Unit based on the concept of weight-centered design was addressed. It was shown that BNU offers a high degree of modularity, expandability and reconfigurability if it is used as the basic building block of the proposed decision-making network. The execution time and accuracy of the proposed decision-making network based on the architectural features of the implemented BNU chip was simulated and analyzed. In addition, we have shown that these performance measures are directly related to the length of the input data.

References

[1] Below, R.K., "Designing Appropriate Learning Rules for Connectionist Systems," in Proceedings, IEEE first International Conference on Neural Networks, June 1987, pp. II479-II486.

[2] Biskup, J., "A Foundation of Codd's Relational Maybe-Operations," ACM Transaction Database Systems, Vol. 8, No. 4, 1983, pp. 608-636.

[3] Burr, D.J., "A Neural Network Digit Recognizer," in Proceedings of the IEEE Conference on Systems, Man, and Cybernetics, 1986, pp. 1621-1625.

[4] Codd, E.F., "Extending the Database Relational Model to Capture Meaning," ACM Transactions on Database Systems, Vol. 4, No. 4, 1979, pp. 397-434.

[5] Codd, E.F., "Missing Information (Applicable and Inapplicable) in Relational Databases," SIGMOD Record, Vol. 15, No. 4, 1986, pp. 53-78.

[6] Grant, J., "Null Values in a Relational Database," Information Processing Lett. 6 (1977) 156-157.

[7] Jin, B., and Hurson, A.R., "Neural Network-Based Decision Support For Incomplete Database Systems," Proceedings of Analysis of Neural Net Applications Conference, ANNA-91, May 1991, To appear.

[8] Hurson, A.R., Miller, L.L. and Pakzad, S.H., "Incomplete Information and the Join Operation in Database Machines," Proceedings of Fall Joint Computer Conference, 1987, pp. 436-443.

[9] Hurson, A.R., and Miller, L.L., "Database Machine Architecture for Supporting Incomplete Information," Journal of Computer System Science and Engineering, Vol. 2, No. 3, 1987, pp. 107-116.

[10] Jin, B., and Raggad, B., "A Reconfigurable Architecture for A VLSI Implementation of Artificial Neural Networks," Proceedings of 1990 International Neural Network Conference (INNC 90), Paris, July 1990, pp. 665-668.

[11] Kohonen, T., "State of the Art in Neural Computing," In Proceedings, IEEE First International Conference on Neural Networks, June 1987, pp. I79-I90.

[12] Lendaris, G.G., "Neural Networks, Potential Assistants To Knowledge Engineers", The Journal of Knowledge Engineering, Vol. 1, No. 3, Dec. 1988, pp 7 - 18.

[13] Lippermann, R.P., "An Introduction to Computing with Neural Nets," IEEE ASSP Magazine, April 1987, pp. 4-22.

[14] Lipski, W., "On Semantic Issues Connected with Incomplete Information Databases, ACM Transactions on Database Systems, Vol. 4, No. 3, 1979, pp. 262-296.

[15] Pakzad, S.H., Hurson, A.R., and Miller, L.L., "Maybe Algebra and Incomplete Data in Database Machine ASLM," Journal of Database Technology, 1990.

[16] Rumelhart, D.E., Hinton, G.E., and Williams, R.J., "Learning Internal Representations by Error Propagation," in Parallel Distributed Processing (PDP): Explorations in the Microstructure of Cognition, Vol. I: Foundations, MIT Press, Cambridge, Massachusetts, 1986, pp. 318-362.

[17] Ullman, J.D., Principles of Database Systems, 2nd Edition, Computer Science Press, Rockville, MD 1982.

[18] Widrow, B., Winter, R.G., and Baxter, R.A., "Learning Phenomena in Layered Neural Networks," in Proceedings, IEEE first International Conference on Neural Networks, June 1987, pp. II411-II429.

[19] Zaniolo, C., "Relational views in a database system: support for queries," IEEE COMPSAC, 1977, pp. 267-275.

An Optical Content-Addressable Parallel Processor for Fast Searching and Retrieving *

Ahmed Louri
Department of Electrical and Computer Engineering
The University of Arizona, Tucson, Arizona 85721

Associative processing based on content-addressable memories has been argued to be the natural solution for non-numerical information processing applications. Unfortunately, the implementation requirements of these architectures using conventional electronic technology have been very cost prohibitive, and therefore associative processors have not been realized. Instead, software methods that emulate the behavior of associative processing have been promoted and mapped onto conventional location-addressable systems. This however, does not bring about the natural parallelism of associative processing, namely the ability to access many data words simultaneously.

The inherently parallel nature and high speed of optics, combined with the recent technological advancements in optical logic, storage and interconnect devices are raising hopes for practical realization of highly parallel optical computing systems. This paper presents the principles of designing an optical content-addressable parallel processor, called OCAPP, for the efficient support of high speed symbolic computing. The architecture is designed to fully exploit the parallelism an high speed of optics. Several parallel algorithms are mapped onto OCAPP in bit-parallel as well as word-parallel fashion, resulting in efficient symbolic algorithms with execution times dependent only on the precision of the operands and not on the problem size. This makes OCAPP very suitable for applications where the number of data sets to be operated on is high e.g., massively parallel processing. A preliminary optical implementation of the architecture using currently available optical components is also presented.

1 Introduction

The "information explosion" seen in recent years has stimulated the development of computer-based information systems to assist in the creation, storage, modification, classification, and retrieval of mainly textual or symbolic data. For example, progress in database management systems, expert systems, and intelligent knowledge-based systems is increasing demand for symbolic information processing such as text editing, file processing, table sorting, searching, and retrieval. In fact a substantial proportion of the work-load of modern information processing systems involve searching and sorting symbolic data[1]. Nevertheless, a majority of today's computers are designed mainly for numerical computations, and suffer from a fundamental handicap, which stems from the principle of addressing the memory.

When a search for a value is made through a location-addressable memory, the entire memory may need to be searched one word at a time (if the data is not sorted in memory) which consumes a great deal of time. There is no logical reason why the search must be done sequential. The only reason stems from the fundamental handicap of separating processing and memory and addressing memory one word at a time. This fundamental flaw has forced system analysts and programmers to develop sophisticated software techniques for symbolic information processing such as hashing

*This research was supported by an NSF Grant No. MIP-8909216.

and indexing[2]. However, the implementation of such software techniques on location-addressed computers has lead to complex, expensive, and inefficient information processing systems.

Searching, retrieving, sorting and modifying symbolic data can be significantly improved by the use of content-addressable memory (CAM) instead of location-addressability. In a content-addressable memory data is addressed by its contents[2]. An associative processor is a parallel processing machine in which the data items are content-addressable with the added capability to write in parallel into words satisfying certain criterion. It may be that the entire contents of stored words may be changed or just a few bits of the words. Using this model, processing is carried out within the associative memory, without transfer to an independent processing unit. Since there is no addressing of data and no data movement, this implies the elimination of the fundamental von Neumann bottleneck encountered in conventional systems. Moreover, the amount of time required for searching, retrieving, and updating information is independent of the data set sizes.

However, this model of computing is not being largely used because of the difficulty and high cost of implementing it in conventional electronic technology. This can be seen from the following:

1. Each bit cell in an associative memory is much more complex and requires more circuitry than does a conventional cell. Even with the advent of VLSI technology, the single cell complexity still does not allow for the use of large associative memories.

2. The memory storage provides poor storage density compared with conventional memory.

3. The third major difficulty is the complexity of the interconnects. Recall that in order for all cells to compare their values to that of the comparand register, the control unit must broadcast the value to all cells involved in the comparison. However, using conventional technology, the time delays associated with the broadcasting function are very appreciable. Moreover, inter-cell interconnects become cumbersome for large array size.

4. The fourth difficulty is the lack of efficient means of implementing parallel access to the cells, namely parallel input and output.

There are two hypotheses underlying this paper:

1. that CAM-based processing provides a sound basis to uncover inherent parallelism in symbolic processing and information retrieval applications, and

2. that optics is, potentially, the ideal medium to exploit such parallelism by providing efficient implementation support for it.

2 Optical Content-Addressable Parallel Processor

Optical systems hold the promise for providing efficient support for future parallel processing systems. Optics advantages have been cited on numerous occasions[3, 4, 5, 6]. These include inherent parallelism, high spatial and temporal bandwidths, and non-interfering communications. For CAM-based processing, optics may be the ideal solution to the fundamental problems faced by electronic implementations, namely cell complexity, interconnects latency, difficulty of implementing information broadcasting and parallel access to the stored data. Optics can alleviate the cell complexity by migrating the implementation of wiring and logic into free-space. The multi-dimensional nature of optical systems allows for data storage and logic to be performed on two-dimensional planes while the third dimension can be used for interconnects. The high degree of connectivity available in free-space space-invariant optical systems (10^6 to 10^8), and the ease with which optical signals can be expanded (which allows for signal broadcasting) and combined (which allows for signal funneling) can also be exploited to solve the interconnects problems[7, 8]. Moreover, optical and electro-optical systems can offer a considerable storage capacity and parallel access than do pure electronic systems[9].

Figure 1 depicts a preliminary organizational structure for an optical content-addressable parallel processor called OCAPP. The architecture is organized in a modular fashion, and consists of a *selection unit*, a *match/compare unit*, a *response unit*, an *output unit*, and a *control unit*. The architecture is developed to meet four goals, namely: (1) exploitation of maximum parallelism; (2) amenability to optical implementation with existing devices; (3) modular design in that it can be scalable to bigger problems; and (4) ability to efficiently implement information retrieval, and symbolic computations. Moreover, the programming methodology for OCAPP is compatible with that of existing single-instruction multiple data (SIMD) systems. In what follows we describe the role of each unit. Detailed optical implementation of OCAPP will be presented in Sec.3.

The selection unit is schematically described in Fig.2. It is comprised of (1) a storage array of n words, each m bits long (in actuality, the storage array capacity is $n \times 2m$, since each bit position is comprised of a true bit w_{ij} and its complement $\bar{w}_{ij}$); and (2) word and bit-slice enable logic to enable/disable the words and/or the bit-slices that participate in the match operation, and reset the rest. It is assumed that the storage array can be loaded in parallel and (if need be) read in parallel.

The match/compare unit shown in Fig.3, contains a (1) $1 \times m$ interrogation register I; (2) logic hardware to perform parallel bitwise comparison between the bits of the interrogation register and the enabled bits of the storage array; (3) two $n \times 1$ working registers, G and L, which are used for magnitude comparisons (to be explained later); (4) a $n \times 1$ response register R for displaying the result of the comparison; and (5) a single indicator bit called the match detector MD, which indicates whether or not there is any matching words. This unit allows comparison of a single operand stored in the interrogation register and the words stored in the storage array. As such it is considered an SIMD (single-instruction-multiple data) unit. Bit position R_i of R is set to one when word W_i of the storage array matches the contents of I. The I register is a combination of the comparand register C and the mask register M as shown in table 1. As such, it holds the operand (depending on masking information if any) being searched for or being compared with. It is assumed that register I is available in dual-rail logic (both true and compliment bits available).

The response unit is responsible for selecting one or several matching words. It comprises several scratchpad registers and a priority circuit for selecting the first matching word. Depending on program control, the output of the response unit is routed either to the output unit for outputting the result or fed back to the selection unit for further processing of the matching words. All units are under the supervision of a conventional control unit with conventional storage (eg., a local RAM) which stores the program instruction. Its role is to load/unload the storage array, set/rest various registers such as the I, R, G and L of the match/compare unit, enable/disable memory words, perform conditional instructions, monitor the MD bit, and test program termination. In what follows, we describe the implementation of several parallel algorithms on the OCAPP in order to show its use and processing benefits.

Table 1: Formulation of the interrogation register

Search bit c_j	Mask bit m_j	Interrogation bits I_j $\bar{I}_j$
0	0	0 1
1	0	1 0
0	1	1 1
1	1	1 1 (no comparison is performed at this bit position)

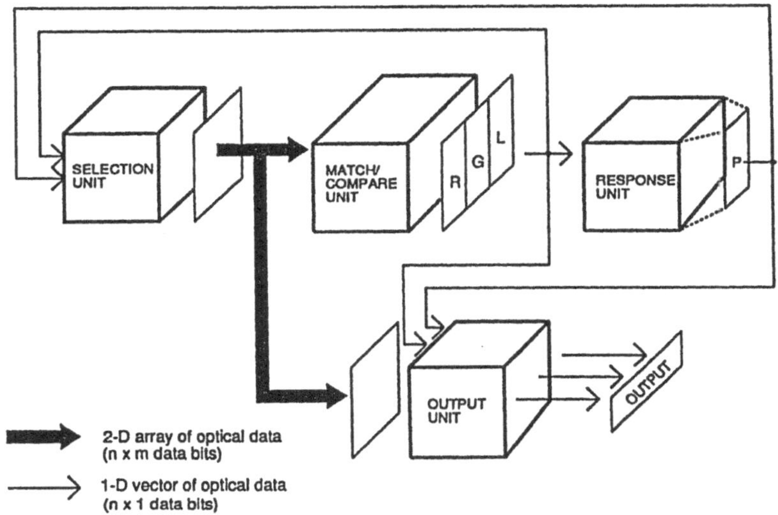

Figure 1 : A schematic organization of the proposed optical content-addressable parallel processor : OCAPP.

3 Parallel Search Algorithms on OCAPP

We classify search operations as basic and compound operations. A basic search operation is one which can be completed in one sweep over all the bit-slices of the storage array. It does not involve any feedback processing. A compound search operation requires a feedback from the response unit to the selection unit. As a consequence, it takes more than one sweep over the storage array to complete. Under basic search operations, we group the following operations:

- *Equivalence Search*: The equality search, the not-equal-to search, and the similarity search (search for a match within a masked field).
- *Threshold Search*: The smaller-than, the not-smaller-than, the greater-than, and the not-greater-than searches.
- *Extrema Search*: The greatest value search, and the smallest value search.

Compound search operations can be implemented in a series of basic search operations. Under the compound search, we group the following operations:

- *Adjacency Search*: Next-above search, and next-below search.
- *Between-Limits Search*: Search for words z, between two limits X and Y ($X < Y$): a) $X \leq z \leq Y$, b) $X < z \leq Y$, c) $X \leq z < Y$, and d) $X < z < Y$.

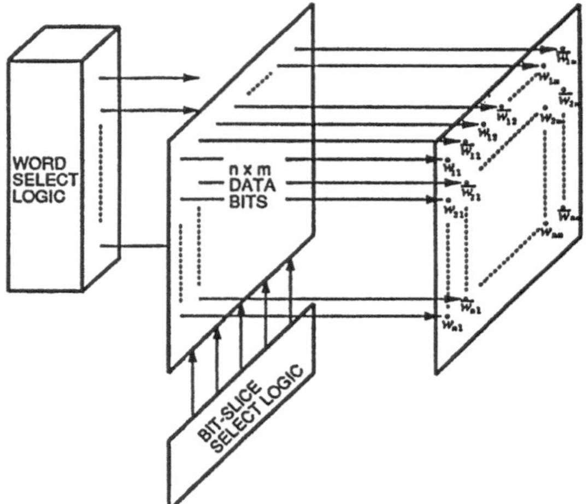

Figure 2 : Organization of the selection unit.

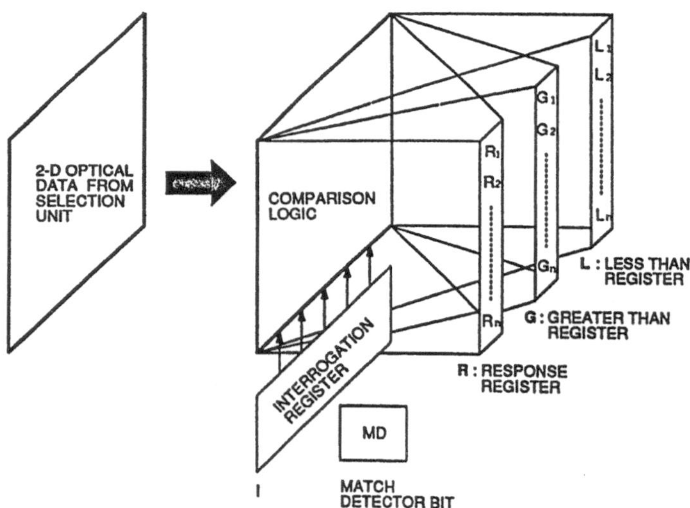

Figure 3 : Organization of the match/compare unit.

- *Outside-Limits Search*: Search for words z, outside two limits X and Y ($X < Y$): a) $X \leq z$ or $z \geq Y$, b) $X > z$ or $z \geq Y$, c) $X \geq z$ or $z > Y$, and d) $X > z$ or $z > Y$.
- *Ordered Retrievals (sorting)*: Ascending order retrieval, and descending order retrieval.

Of course many more compound search operations can be formulated using the basic search operations. The above search operations are the most frequently used in information retrieval applications.

3.1 Parallel Algorithms for Basic Search Operations on OCAPP

In what follows, we denote a memory word as $W_i = (w_{im}w_{im-1}\dots w_{i1})$ where w_{ij} is the jth bit cell of the word W_i. We denote the jth bit-slice by $B_j = (w_{1j}w_{2j}\dots w_{nj})$, which is made up of the jth bit of every word in the storage array. The interrogation and response registers are denoted by $I = I_m I_{m-1}\dots I_1$, and $R = R_1 R_2 \dots R_n$ respectively. The comparand word (search argument) and the mask register words are denoted by $C = (c_m c_{m-1}\dots c_1)$, and $M = (m_m m_{m-1}\dots m_1)$.

3.1.1 Equivalence Search

In this type of search, the memory is partitioned according to the magnitude of the search word C into two sets, namely, words which are equal to C and words which are not. The equality and masked search operations can be implemented by a bitwise match. For equality match all the bits of the search word need to be matched, whereas for the masked search, only a subset of the bits of the search word is compared with the respective bits of the memory words. For $m_j = 0$ means that c_j is not masked, while $m_j = 1$ means c_j is masked. These two search modes can be combined as shown in Table 1. Given an interrogation word I, a bit match denoted by b_{ij} on the jth cell of the ith word is given by:

$$b_{ij} = (I_j \wedge w_{ij}) \vee (\bar{I}_j \wedge \bar{w}_{ij}) \quad \text{(equivalence)} \tag{1}$$

where the symbols $\wedge$, $\vee$, and the bar ($^-$) denote the logical AND, logical OR and logical NOT respectively. Now the exact matching of memory word W_i with interrogation vector I requires the logical product of the bits b_{ij} for $j = 1, \dots, m$, therefore:

$$R_i = \bigwedge_{j=1}^{j=m} b_{ij} = b_{im} \wedge b_{im-1} \wedge \dots \wedge b_{i1}. \tag{2}$$

where $\bigwedge$ denotes a logical AND over all bits. The above equation indicates that matching words in memory will be flagged by having their corresponding R bit set to one, and all mismatches will have their R bits set to zero. Equations 1 and 2 are space-invariant and can be implemented in bit-parallel as well as word-parallel fashion. Therefore, all R_is for $i = 1, \dots, n$, are computed at the same time with a single access to the storage array.

Equivalence Search Algorithm:

1) *Initialization:*

   a) Load I (this will depend on the search word and the masking condition);

   b) Clear R (clear all bits of the R register);

2) *Perform comparison:*

   a) $b_{ij} = (I_j \wedge w_{ij}) \vee (\bar{I}_j \wedge \bar{w}_{ij})$;

   b) $R_i = \bigwedge_{j=1}^{j=m} b_{ij}$ for $i = 1, \dots, n$. ($R_i = 1$ if and only if W_i matches I).

3.1.2 Threshold Search

This mode of search partitions the memory according to the magnitude of the search word C into three sets, namely words which are equal to C, words which are less than C, and words which are greater than C. The result of the search is stored in the three registers of the response unit, namely R, G and L. Initially, all memory words are made active by making control registers RGL = 100. The memory is scanned from the most significant to the least significant bit position by enabling a single bit-slice at a time. When the comparand bit c_j is one, we select all active memory words with $w_{ij} = 0$ as "less than" by setting their corresponding bit position RGL = 001. These words are then disabled from further comparisons (the disabling process will be explained later). Similarly, when $c_j = 0$, we select all active memory words with $w_{ij} = 1$ as "greater than" by setting their corresponding bit position RGL = 010, and then disable them from further processing. At the end of the last bit position, words still in the state RGL = 100 are equal to the comparand, words in the state RGL = 010 are greater than the comparand, and words in the state RGL = 001 are less than the comparand. It is important to note that, even though we are scanning the memory from most significant bit to least significant bit, the search process can be terminated any time there are no matching words at a given bit position ($R_i = 0$ for all $i = 1, \ldots, n$). Such a condition is easily detectable by checking the MD bit. The detailed algorithm follows.

Threshold Search Algorithm:

1) *Initialization:*

   a) Load I (depending on the search word and masking condition);

   b) Enable memory words;

   c) Set R, clear G, clear L, set $j = m$ (the variable j is used by the control unit to scan the storage array);

2) *Perform Magnitude Search at bit-slice j:*

   a) $R_i = \bigwedge_{j=1}^{j=m} b_{ij}$, $G_i = \bigwedge_{j=1}^{j=m} I_i \wedge \bar{w}_{ij}$, $L_i = \bigwedge_{j=1}^{j=m} \bar{I}_i \wedge w_{ij}$ for $i = 1, \ldots, n$ (note that only the enabled bit-slice j determines the values of R_i, G_i and L_i, all other bit-slices are disabled at this time, and therefore have no influence);

   b) Test if MD = 1 (is there any words that match the I register at the current bit position j ?);

3) *If* $MD = 1$ *do :*

   a) Disable memory words whose corresponding bits in R are zero (memory words W_i with $R_i = 0$ have already been decided on);

   b) Decrement j: $j \leftarrow j - 1$, and test if $j = 0$?;

   c) If $j \neq 0$, go to step 2;

   d) If $j = 0$, go to step 4;

4) *If* $(MD = 0)$ *or* $(j = 0)$, then we are done and the search result is reported in R, G, and L.

The following example illustrates the algorithm for a magnitude search of 7 words, each 5 bits long:

Example 1: Threshold Search

Search word, S :	10110
Mask word:	00000
I register:	10110 (effective word search)

Memory word i	W_i	State of RGL at the end of the jth iteration				
		j = 5	j = 4	j = 3	j = 2	j = 1 (last iteration)
1	10111	100	100	100	100	010 $(W_1 > S)$
2	11000	100	010	010	010	010 $(W_2 > S)$
3	10010	100	100	001	001	001 $(W_3 < S)$
4	10110	100	100	100	100	100 $(W_4 = S)$
5	10101	100	100	100	001	001 $(W_5 < S)$
6	01101	001	001	001	001	001 $(W_6 < S)$
7	11101	100	010	010	010	010 $(W_7 > S)$

3.1.3 Extrema Search

This type of search refers to finding the maximum (or minimum) of a set of (or all) memory words. We consider first the search for maximum.

A. Maximum Search

To find the maximum, we scan memory words from the most to the least significant bit positions. As we scan the bit-slices, we determine if any of the enabled words have a one in the current bit position. If we find some, we disable all those words that do not have a one in this position. If none of the words at the current position possess a one, we do nothing. At any given time, all remaining candidates are equal as far as we have examined them, because for every bit position either everybody had a zero in that bit position, or whenever some words have ones, we disable the ones with zeros. Therefore, at bit position j, enabled words with $w_{ij} = 1$ are larger than enabled words with $w_{ij} = 0$. Since we are seeking the maximum, we disable the ones with $w_{ij} = 0$. This process is repeated until we exhaust all bit positions at which time the maximum word will be indicated by $R_i = 1$.

Algorithm for Finding the Maximum:

1) *Initialization:*
   a) Load I :$I \leftarrow 11\ldots11$ (I is loaded with all bits set to one);
   b) Clear MD, set $j = m$;
   c) Enable memory words, and set $R_i = 1$ for $i = 1, \ldots, n$;
2) *Perform equivalence search at bit-slice j:*
3) *Test if $MD = 1$* (is there any words with a one in the current bit position j ?);
4) *If $MD = 1$ do :*
   a) Disable all words which do not have a one in the current bit position (these words are indicated by $R_i = 0$).
   b) Clear R and MD;
   c) Go to step 5;
5) *Decrement j : $j \leftarrow j - 1$*, and test if $j = 0$?;
   a) If $j \neq 0$, go to step 2;
   b) If $j = 0$, output maximum value indicated by $R_i = 1$.

B. Minimum Search

The search for the minimum is very similar to the search for the maximum except that the I register is initially loaded with zeros and that if any enabled word has a zero in the current bit position (There exists a memory word W_i such that its corresponding $R_i = 1$), we disable the words with a one in the current bit position ($R_i = 0$). These words are bound to be greater than the minimum sought. The process is repeated until we exhaust all bits of the enabled words. The minimum value will also be indicated by a one in register R.

3.2 Parallel Algorithms for Compound Search Operations on OCAPP

Compound search operations such as the ones stated earlier can not be economically implemented by a single sweep over the memory words. We therefore choose to implement such operations as a series of basic searches. The rationale is to keep the architecture as simple as possible, and therefore making it highly amenable to optical implementation. Of course speed improvements can be gained by implementing these search operations as basic search, but the amount of logic circuits may be extensive.

3.2.1 Double Limits Search (Between and Outside Limits)

Given two numbers called HIGH and LOW, the double limits search consists of finding those words that are between this limits and/or words that are outside these limits. This gives rise to eight different searches which can be accomplished in a very similar manner. Let us consider the between limit search. Given the two numbers HIGH and LOW, we wish to find those words that are greater than LOW but less than HIGH namely, find all W_i such that $LOW < W_i < HIGH$. We can accomplish this search by using the magnitude comparison search as follows. First, we determine the words that are less than the comparand HIGH. These words will be indicated by a one in the L register. We then disable all other words except the ones that are less than HIGH, and perform another threshold search using the comparand LOW. After the second search, words that are less than HIGH and greater than LOW will be marked with a one in the G register, which could be routed to the output unit for outputting the search result.

3.2.2 Adjacency Search

To find the word that is next-above the comparand (the smallest word larger than the comparand), we search for all words that are larger than the comparand and then select their minimum. Similarly, to find the word that is next-below the comparand (the largest word smaller than the comparand), we search for all words less than the comparand and select their maximum. The search for the largest word smaller that the comparand (next-below search) can be carried out by a similar algorithm as the one above. In this case, step two of the next-above algorithm is replaced by a search for words that are less than the comparand, and step four is replaced by a maximum search.

3.2.3 Ordered Retrieval (Sorting)

The sorting or ordered retrieval of a set of data can be achieved by performing the extrema search repeatedly until all the data are retrieved. For the ascending order retrieval, we enable the memory words to be sorted, and determine their minimum (using the minimum search operation). We output the obtained minimum value and disable it from the storage array. We repeat these steps until we retrieve (in ascending order) all the enabled words. For descending order retrieval, we select the maximum value at each step.

4 Optical Implementation

In this section we identify the fundamental and basic operations required to implement the optical architecture, and describe possible optical components for achieving them. Detailed practical implementation issues and experimental setups will be the subject of a different publication.

4.1 Basic Operations and Hardware Components Required

An analysis of the conceptual OCAPP, the basic operations, and the algorithms reveals that in order to optically implement OCAPP, the following functions are required: (1) data is optical and must be available in dual-rail format (both the value and its complement is required); (2) parallel access for writing into and reading from the storage array as well as the various control registers; (3) disabling/enabling a memory word (or several memory words) based on certain criteria; (4) logical AND, and logical OR; (5) space-invariant optical transmission of information (one-to-one connections); (6) spreading a single bit (actually two bits due to the dual-rail format) of information to several spatial locations (one-to-many connection); (7) combining several bits of information into a single spatial location (many-to-one connection); and (8) dynamic routing of information (e.g., routing contents of register R to selection unit, or output unit, or response unit depending on the algorithm). The optical components required to accomplish the above operations can be divided into (1) logic elements, (2) storage elements, and (3) information transfer elements (or interconnects).

For optical logic and storage, many approaches are being investigated. One approach is the adaptation of the spatial light modulator (SLM) technology to optical logic[10]. Another approach for realizing optical components capable of performing logic, is to optimize the device from the beginning for digital operations. The recent emergence of the quantum-well self-electrooptic effect device (SEED) and its derivatives (S-SEED, T-SEED, D-SEED) is one such a product[11]. The SEED devices can be used to realize both logic operations such as NOR, OR, AND, NAND, etc. as well as for storage such as S-R latches[11]. Optical resonators are another family under this approach intended for optical logic[12]. Two similar bistable devices, etalons, and interference filters both based on the Fabry-Perot resonator are being actively pursued[13, 12]. All data movements and information transfer in OCAPP are space-invariant which may render their implementation easier. Classical optical components such as lenses, mirrors, beam splitters, holographic deflectors, and delay elements are most likely to be used for this purpose[14]. In addition, halfwave plates, shutters, and masks may be used for dynamic routing.

4.2 A Modular Implementation of OCAPP

In this paper, we present a modest design example of OCAPP, using existing optical hardware in order to highlight the potential implementation issues of a practicable realization. The implementation of this first version will make use of the SEED device operating as a NOR gate for optical logic, and of the S-SEED device operating as a S-R latch for storage[11]. The NOR gate is preferable to any other form of thresholding nonlinearity because it only requires distinguishing between the state where no light comes in and the state where light come in. Thus the NOR gate requires an SNR better than one only. In addition, a NOR function constitutes a complete logic set capable of implementing any boolean or arithmetic function[15]. The family of SEED devices seem to be easy to use, capable of high speed, low energy operation, and can be fabricated in 2-D format. Space-invariant optical interconnects, dynamic masking components, and beam spreading and combining devices are assumed for data routing[16].

The S-SEED device has two inputs, S, R, and two outputs Q and $\bar{Q}$. The state of the device is set by a pair of unequal signal beams labeled S (for setting the output $Q = 1$, $\bar{Q} = 0$) and R (for resetting the output $Q = 0$, $\bar{Q} = 1$). The device is set (Q = 1) when the power incident on the S input is much higher than the power incident on the R input. The state of the device is read by applying two equal-power (clock signal) beams to both inputs. During the setting of the device, the clock beams must be low, compared to the signal beams. The device holds its state when no clock signal is incident. Thus the device can operate as a latch. Moreover, during the application of the clock signal (the reading process) the state of the device is unaltered. As described earlier, the optical processor can be constructed from several units: the selection unit, the match/compare unit, the response unit, the output unit, and the control unit. In what follows, we describe the optical implementation (architectural rather than experimental setup) of each of these units. Moreover, the details in the routing and imaging paths such as lenses, holographic elements, masks, beam splitters, and polarizers have been omitted in this version to assist the reader's conceptual understanding of these configurations.

4.3 The Optical Selection Unit

The optical selection unit of Fig.4 is composed of a storage array which consists of a 2-D $n \times m$ array of clocked S-SEED devices (each entry in the array at position i, j has two incoming bits S, R and two outgoing bits w_{ij}, $\bar{w}_{ij}$), a $n \times 1$ word register A which serves at setting and resetting data words in the storage array, a $1 \times (m+1))$ bit-slice loading register B for loading a single bit-slice of the storage array. The first bit B_0 and its complement $\bar{B}_0$ are called set-E and reset-E respectively, since they are used for setting and resetting the $n \times 1$ enable register E which is used for the matching process (to be explained below). Memory words are disabled through the $n \times 1$ NOR gate array, representing the D register. The D register can be loaded from R, G or L registers. In addition, the E register can also be reset from the priority register P of the response unit (to be explained below).

A. Writing a Word/Bit-Slice into the Storage Array:

The storage array is assumed to be loaded in parallel at the beginning of the program. During program execution, the contents of the storage array can be altered by the use of the A and the B registers. To write a word in the storage array, say at word position i, the word is first written in the flip-flops of the B register. In the next clock cycle, the clock signals of B bits are pulsed high, and the contents of the B register is spread out vertically such that each bit B_j impinges on the set ports of the j-th column of the storage array. Next, bit A_i of A (corresponding to word position i) is pulsed high and spread out horizontally such that it impinges on the set/reset ports of the i-th row of the storage array. A one bit is written in bit position w_{ij} of the storage array if and only if a high A_i and a high B_j coincide at the set port of bit w_{ij}. Similarly, a zero bit is written in bit position w_{ij}, if and only if a high A_i and a high $\bar{B}_j$ coincide at the reset port of w_{ij}. This of course assumes that the set/reset thresholds of the S-SEED devices are so designed. Similar operations take place for writing a bit-slice in the j-th column of the storage array, with the exception of interchanging the roles of the B and A registers.

B. Enabling/Disabling Memory Words:

By enabling a memory word W_i, it is meant including it in the matching process. Similarly, by disabling it, it is meant excluding it from further matching operations. To allow a memory word W_i in participating in the matching process, its corresponding bit E_i in the E register must be set high. Similarly, to disallow a memory word w_i from participating in the matching process, its corresponding bit E_i in the E register must be made low. To enable/disable the entire memory words, the set-E/reset-E bits ($B_0/\bar{B}_0$) are spread out vertically and broadcast to all the set/reset ports of E. To selectively disable memory words whose R, or G or L bits are not asserted (R= 0, or G = 0 or L = 0), requires the routing of the appropriate register (R, G, or L) to the NOR gate array D. The output of D (which represents the complement of the routed register) is imaged onto the reset ports of register E. For example, to disable memory words whose R bits are not asserted (R = 0) from further matching operations, first contents of R is routed to D, which in turn image

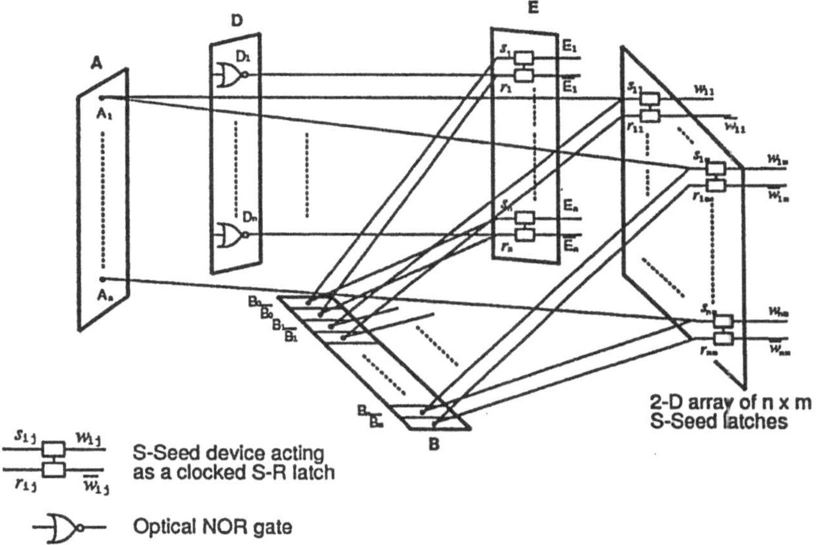

Figure 4 : Optical implementation of the selection unit.

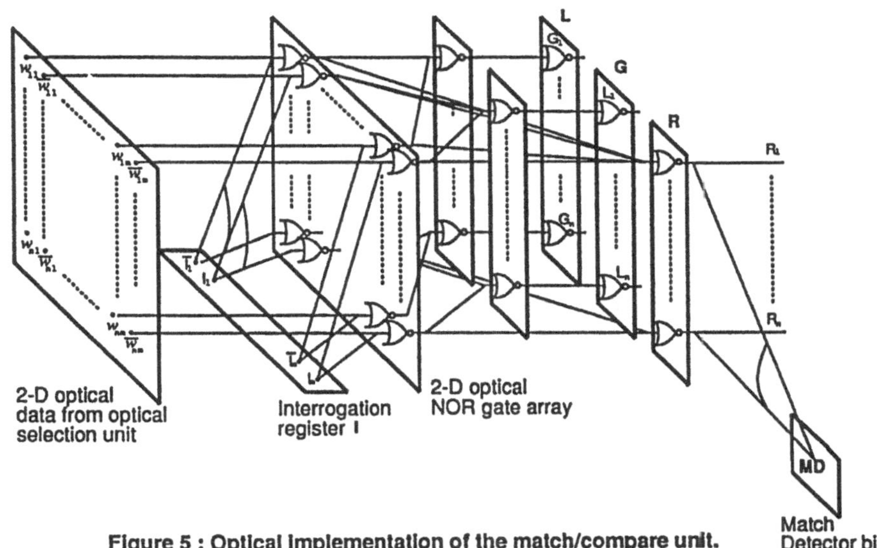

Figure 5 : Optical implementation of the match/compare unit.

the complement of R onto the reset ports of E. Thus a low bit R_i of the R register will disable the i-th bit of the E register, which in turn disables memory word W_i from participating in further comparisons. The role of the E register in the match operation is explained next.

4.4 The Optical Match/Compare Unit

This unit performs exact match, and magnitude comparison searches between the interrogation register I and words of the storage array. As shown in Fig.5, It contains several SEED arrays operating as NOR gate arrays, and three registers, namely the response register, R, the greater than register G, and the less than register L. Parallel comparison takes place between memory words emanating from the storage array and the interrogation register I. A match at bit w_{ij} is detected by an exclusive-and principle as indicated in Eq.1. For that, register I needs to be spread out vertically so that each bit I_j impinges on one port of the NOR gates of the j-th column of the array, while data bits w_{ij} impinge on the second port of the NOR gates of the same j-th column. Matches between I_j and w_{ij} are reported in bit R_i of the R register. Otherwise, the G and L registers indicate the relative magnitude. The contents of R , G, and L are routed to the response unit as well as fed back to the selection unit.

As stated above, the enable register E determines whether or not a memory word participates in the matching process. Thus, the word match condition of Eq.2 is rewritten as follows:

$$R_i = [(I_{m+1} \wedge E_i) \vee (\bar{I}_{m+1} \wedge \bar{E}_i)] \wedge [(I_m \wedge w_{im}) \vee (\bar{I}_m \wedge \bar{w}_{im})] \wedge, \ldots, \wedge [(I_1 \wedge w_{i1}) \vee (\bar{I}_1 \wedge \bar{w}_{i1})] \quad (3)$$

where bit I_{m+1} is set to one ($\bar{I}_{m+1} = 0$) during a match operation. It can be seen from Eq.3 that a memory word W_i will participate in the match process if and only if its enable bit E_i is set to one. The R register bits are logically ORed to form the Match Detector (MD) bit. The MD flip-flop indicates if there is any match between I and memory words.

The optical match/compare unit of Fig.5 consists of a single interrogation register I, and therefore allows comparison of one search argument with the words stored in the storage array. However, using the multi-dimensionality of optical systems, this unit can be extended to perform multiple search operations in a single step. That is, several search arguments are compared simultaneously with the words of the storage array. An extended MIMD match/compare unit would have a $k \times m$ two-dimensional array of k search arguments, a two-dimensional storage array of n words each m bits long, and a $m \times k$ two-dimensional response array as shown in Fig.6. Each response register R_l ($l = 1, \ldots, k$) would indicate the match between interrogation register I_l ($l = 1, \ldots, k$) and the words of the storage array. The two-dimensional match operation can be thought of as an optical binary matrix-matrix multiplication which can be implemented using several optical techniques [17].

4.5 The Optical Response Unit

The response unit, contains a combinational priority circuit, and a priority register P for indicating the first matching word in memory (It may also contain few scratchpad registers for temporary storage). The priority circuit allows only the first responder (the first memory word W_i whose R_i is one) to pass to the priority register P. The priority circuit can be implemented using several stages of the NOR gate arrays in the form of a binary tree with space-invariant interconnections between them[18]. Contents of the P register are routed to the output unit, and also fed back to the selection unit.

4.6 The Optical Output Unit

The output unit outputs memory word whose corresponding bit in the priority register P, R, G or L is set to one (Fig.7). These latter registers are routed to a $n \times 1$ NOR gate array, denoted by N in Fig.7, whose sole purpose is to invert their values. Each bit N_i of N is logically NORed with memory word W_i using a 2-D NOR gate array. Next, each column of the NOR gate array is logically ORed to form output bit O_i of the output register. This latter could be a photosensitive device which only detects the presence of light and outputs electrical signals, or a 1-D array of SEED devices acting as OR gates, and outputting optical signals. It should be noted that parallel readout of selected memory words is also achievable by replacing P with a 2-D output device and eliminating the OR function.

4.7 The Control Unit

OCAPP is under the control of a memory control unit which comprises a local memory for storing programs and a program sequencer for executing instructions that control the optical hardware such as the S-R latches, the NOR gate arrays, the routing shutters, and splitters, etc. The instruction set is composed of conventional assignment and conditional statements, and additional instructions required to implement associative parallel processing. This includes data movement between units, comparison operations, memory loading and unloading, monitoring the MD bit, etc. These additional instructions are very few in nature and are derived from the required fundamental operations described above. It should be noted that application programs for OCAPP can be written in conventional high-level languages such as Pascal or C, with few calls to external procedures which support parallel associative processing.

4.8 Estimated Execution time

An exact performance analysis of the proposed OCAPP including cost and power budget breakdown is currently not feasible due to the lack of optical S-R latches and thresholding devices with reasonable size (e.g., 500 × 500 gates), low operating energy, and fast response time. However, major efforts are being pursued[10, 11] in developing these devices in larger sizes. These efforts will soon culminate in the required devices and components for OCAPP as well as for other digital optical computing models.

We therefore try to theoretically estimate the execution time in terms of gate delay of the various basic search and arithmetic algorithms presented. This time does not include memory loading and unloading. We assume that the response times of the S-R latches (S-SEED) and that of the NOR gate arrays (SEED) are comparable and are both equal to T_s, and T_p is light propagation time through the processing loop (from the selection unit and back). It is assumed that the reading of memory words from the storage array and the I register is done at the same time, and takes one gate delay T_s; enabling/disabling of memory words is achieved in one gate delay; testing of the MD bit takes one gate delay, and the priority circuit takes $\log_2 n$ gate delays, where n is the number of words participating in the matching process.

Table 2 presents the estimated execution time of the algorithms presented. It can be seen that the execution time in equivalence search is a constant factor, and independent of the number of words in memory. The time in threshold search, double limits search, adjacency search, and extrema search, ordered retrieval (minimum time only) is proportional to the precision (number of bits) of the operands, and is independent of the number of words involved in the operation.

Note that the availability of the Match Detector (MD) bit provides major speed improvements to the above algorithms, in that certain conditions to terminate the computation as early as possible can be easily detected. Take for example, the threshold search algorithm. After the first comparison operation, if there are no words that match the comparand at that bit position (a condition that can be easily be detected by checking the MD bit), then all the words have been decided on in only $4T_s$ delay time, and the result is obtained in a much shorter time. The same considerations take place for double limits search, and adjacency search.

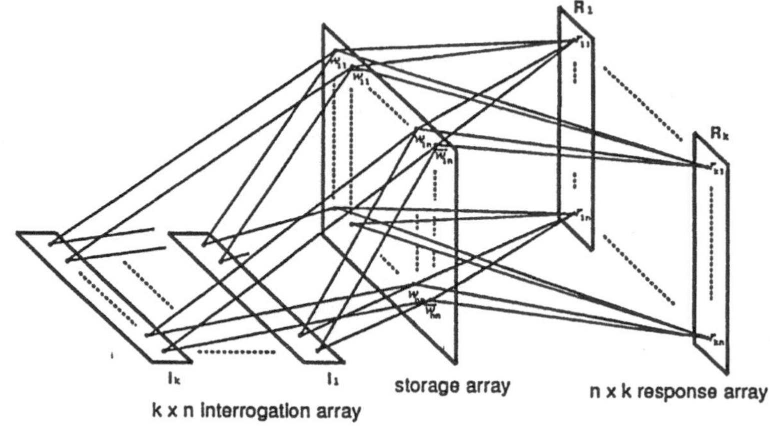

Register R_i indicates the match/mismatch of memory words with interrogation registers I_i(for i=1 to k).

Figure 6 : Optical implementation of a 2-D match/compare unit : the interrogation as well as the response registers of figure 5 are replaced by two-dimensional arrays of search arguments and response registers respectively.

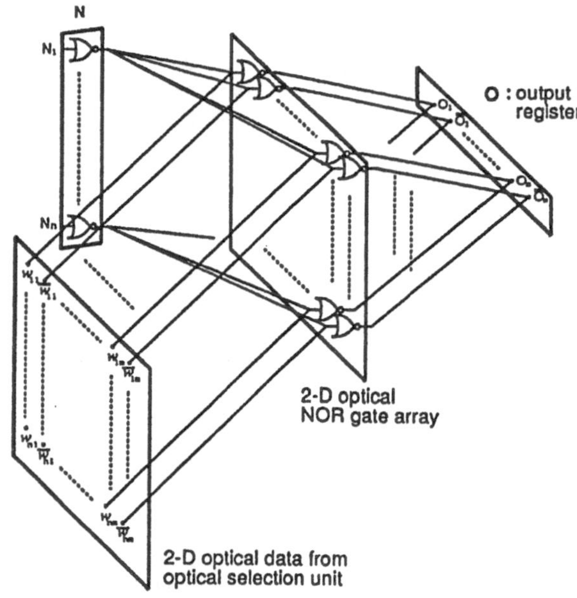

Figure 7 : Optical implementation of the output unit.

Table 2: Estimated execution time of the parallel algorithms on OCAPP

Search Algorithm	Minimun Execution Time	Maximum Execution Time
Equivalence Search	$3T_s$	$3T_s$
Threshold Search	$4T_s$	$(5T_s + T_p) \times m$
Minimum Search	$(6T_s + T_p) \times m$	$(6T_s + T_p) \times m$
Maximum Search	$(6T_s + T_p) \times m$	$(6T_s + T_p) \times m$
Double Limits Search	$10T_s + 2T_p$	$T_s + T_p + 2(5T_s + T_p) \times m$
Adjacency Search	$5T_s + T_p + (6T_s + T_p) \times m$	$T_s + T_p + (11T_s + 2T_p) \times m$
Ordered Retrieval Search	$((6T_s + T_p) \times m + T_p + 3T_s)$	$((6T_s + T_p) \times m + (\log_2 n)T_s + T_p + 3T_s) \times n$

The parameters m and n in the above table represent the word length and the number of operands respectively.

5 Conclusions

CAM-based processing has been argued to be the natural solution for non-numerical information processing applications. Unfortunately, the implementation requirements of these architectures using conventional electronic technology have been very cost prohibitive. This paper presented the principles and initial design concepts of an CAM-based parallel processing architecture that matches well with optics advantages, and therefore is highly amenable to optical implementation. The architecture relies heavily on the use of space-invariant interconnections, optical signal broadcasting and funneling (combining), and the simultaneous application of the same operation to many data points (SIMD mode of computing). The motivations behind this is the ease with which these operations can be realized with optics. A representative set of search algorithms have been presented to show the use and merits of the architecture. These algorithms are key components which occur in large computing tasks. It is important to note that these fundamental search algorithms are implemented on the optical architecture with an execution time independent of the problem size (the number of words to be processed). This indicates that the architecture would be best suited to applications where the number of data sets to be operated on is high. Some of the applications being investigated are: (1) real-time information retrieval, (2) database management, (3) knowledge-base and expert system implementation, and (4) list and string processing.

We presented a preliminary and simple version of an optical implementation of OCAPP. This version is only meant to show the feasibility of the architecture with existing optical devices. No optimization attempts were made. Nevertheless, this preliminary version reveals several key design issues that will determine the physical realization of such an optical architecture. Even if we assume the availability of optical nonlinear devices (latches, and NOR gates) in large sizes, the effective memory size will be critically determined by the beam spreading/combining optics, the contrast ratio and the fan-in/fan-out factors of the logic elements to be used. These practical implementations issues will be fully detailed in a follow-up paper.

References

[1] K. Hwang and D. Degroot, *Parallel Processing for Supercomputers and Artificial Intelligence*, McGraw-Hill, New York, 1988.

[2] T. Kohonen, *Content-addressable memories*, Springer-Verlag, 1980.

[3] A. A. Sawchuk and T. C. Stand, "Digital optical computing," *Proceedings of The IEEE*, vol. 72, no. 7, pp. 758–779, July 1984.

[4] W. T. Cathey, K. Wagner, and W. J. Miceli, "Digital computing with optics," *Proceedgins of the IEEE*, vol. 77, pp. 1558 - 1572, Oct. 1989.

[5] A. Louri, "A parallel architecture and algorithms for optical computing," *Optics Communications*, vol. 72, no. 1, pp. 27 – 37, July 1, 1989.

[6] A. Louri, "3-D optical architecture and data-parallel algorithms for massively parallel computing," *IEEE MICRO*, April 1991.

[7] B. K. Jenkins, P. Chavel, R. Forchheimer, A. A. Sawchuk, and T. C. Strand, "Architectural implications of a digital optical processor," *Applied Optics*, vol. 23, no. 19, , October 1984.

[8] J. W. Goodman, F. J. Leonberger, S. Y. Kung, and R. A. Athale, "Optical interconnections for VLSI systems," *Proceedings of the IEEE*, vol. 72, no. 7, pp. 850–866, July 1984.

[9] P. B. Berra, A. Ghafoor, M. Guizani, S. J. Marcinkowski, and P. A. Mitkas, "Optics and supercomputing," *Proceedings of the IEEE*, vol. 77, pp. 1797 – 1815, Dec. 1989.

[10] J. A. Neff, R. A. Athale, and S. H. Lee, "Two-dimensional spatial light modulators: a tutorial," *Proceedings of the IEEE*, vol. 78, pp. 836 – 855, May 1990.

[11] A. L. Lentine, H. S. Hinton, D. A. B. Miller, J. E. Henry, J. E. Cunningham, and L. M. F. Chirovsky, "Symmetric self-electrooptic effect device: optical set-reset latch, differential logic gate, and differential modulator/detector," *IEEE J. of Quantum Electron.*, vol. 25, pp. 1928 - 1936, Aug. 1989.

[12] J. L. Jewell, M. C. Rushford, and H. M. Gibbs, "Use of a single non-linear Fabry-Perot etalon as optical logic gate," *Appl. Phys. Lett.*, vol. 44, pp. 172 – 174, Jan. 1984.

[13] S. D. Smith, J. G. H. Mathew, M. R. Taghizadeth, A. C. Walker, B. S. Wherret, and A. Hendry, "Room temprature, visible wavelength optical bistability in ZnSe interference filters," *Optics Communications*, vol. 51, pp. 357 – 362, Oct. 1984.

[14] A. W. Lohmann, "What classical optics can do for the digital optical computer," *Applied Optics*, vol. 25, no. 10, pp. 1543 – 1549, 15 May 1986.

[15] A. Louri and K. Hwang, "A bit-plane architecture for optical computing with 2-d symbolic substitution algorithms," In *Proc. 15th Int'l. Symp. on Computer Arch.*, Honolulu, Hawaii, May 30 - June 4, 1988.

[16] K. Hwang and A. Louri, "Optical multiplication and division using modifed signed-digit symbolic substitution," *Optical Engineering, Special issue on Optical Computing*, vol. 28, no. 4, pp. 364 – 373, April 1989.

[17] R. A. Athale, "Optical matrix processors," *In Proc. SPIE, Optical and Hybrid Computing*, vol. 634, pp. 96 – 111, 1986.

[18] C. C. Foster, "Determination of priority in associative memories," *IEEE Transactions on Computers*, vol. C-17, pp. 788 – 789, Aug. 1968.

Towards an Efficient Hybrid Dataflow Architecture Model

Guang R. Gao† *Herbert H.J. Hum*‡ *Jean-Marc Monti*†

†McGill University
School of Computer Science
McConnell Engineering Building
3480 University St.
Montréal, Canada, H3A 2A7.

‡Centre de recherche informatique de Montréal
3744 Jean Brillant, Bureau 500
Montréal, Canada, H3T 1P1.

Abstract

The dataflow model and control-flow model are generally viewed as two extremes of computation models on which a spectrum of architectures are based.

In this paper, we present a hybrid architecture model which employs conventional architecture techniques to achieve fast pipelined operation, while exploiting fine-grain parallelism by data-driven instruction scheduling. A mechanism for supporting concurrent operations of multiple instruction threads on the hybrid architecture model is presented and a compiling paradigm for *dataflow software pipelining* which efficiently exploits loop parallelism in loops is outlined. Simulation results attest that hybrid evaluations can indeed be beneficial.

1 Introduction

There have been two basic models in computer architectures: (1) the von Neumann sequential control model; and (2) the data-driven distributed control model. The parallel architectures based on the von Neumann model are aimed at exploiting coarse-grain parallelism, while the traditional dataflow architecture model was conceptualized to handle fine-grain parallelism. For the past decade, researchers have been debating on which model is a "better" basis for future large-scale parallel computer systems.

The work described in this paper is based on our view that the two models are not orthogonal, and that a flexible architecture model can be developed by extending the dataflow model to allow hybrid dataflow and control-flow evaluation. As a result, the grain-size of the parallelism supported by such a hybrid parallel architecture model is "flexible" – compiler and hardware techniques can be combined to "tune" the effective grain size for the needs of efficient exploitation.

We believe that the dataflow model of computation offers a sound, simple, yet powerful model of parallel computation. In the dataflow programming and architectures, there is no notion of a single point or locus of control. Dataflow architectures have promised solutions addressing the two fundamental problems of von Neumann computers in multiprocessing: the memory latency and synchronization overhead [2]. However, we must not ignore the efficiency and simplicity of the instruction sequencing mechanism in von Neumann architecture models, as well as over 40 years of optimizations in the instruction execution mechanism. We have compared dataflow architectures designed as direct execution engines for dataflow graphs and those that perform the execution of dataflow graphs using features of conventional von Neumann computer architectures. We believe that the latter has the potential of offering a better performance/cost ratio.

In this paper, we develop a framework for a hybrid architecture model. The essential elements of such a framework must consist of both a simple architecture model and an effective compiling methodology which can structure code to expose parallelism for hybrid evaluation. Our application domain is general-purpose scientific computations where a program usually consists of a number of *code blocks* (or loops) which define the major array values for the computation. A compiling paradigm is established which exploits loop parallelism through *dataflow software pipelining*. We establish a set of basic results which show that the fine-grain parallelism in a loop can be fully exploited by a a simple scheduling scheme, achieving time and space efficiency simultaneously.

In Section 2, we describe one processing element of the McGill Dataflow Architecture Model (MDFA) which employs the *argument-fetching principle*. We then describe the basic architecture mechanisms for supporting concurrent (recursive) function invocations. For multiprocessing support, an efficient inter-processor synchronization and communication mechanism is described for sending and receiving data through an interconnection network. In Section 3, we present a hybrid evaluation model based on the MDFA which is structured around the notion of instruction threads, i.e., instructions in each thread are executed sequentially, while multiple (sequential) instruction threads can operate concurrently through a data-driven style instruction scheduling mechanism. In Section 4, we present the principle of dataflow software pipelining and the limited balancing technique – a technique that can be used to obtain instruction threads (or "macro" dataflow actors) for the hybrid model. Simulation results which attest to the need for hybrid evaluation are briefly discussed in Section 5. However, a major assumption made for the simulation runs has created a need to further extend the hybrid model so that executions can be performed more efficiently. We outline these other extensions in section 5.3. Finally, conclusions and future work are briefly outlined in the last section.

2 The McGill Dataflow Architecture Model

In this section, we introduce the operational model of the McGill Dataflow Architecture.

The operational semantics of dataflow programs for this model are defined in terms of familiar concepts used in conventional architectures. This architecture is based on the *argument-fetching* principle [5], where the instruction scheduling is decoupled from the main execution datapath, yielding a unique dataflow model which makes the extension to hybrid evaluation straightforward.

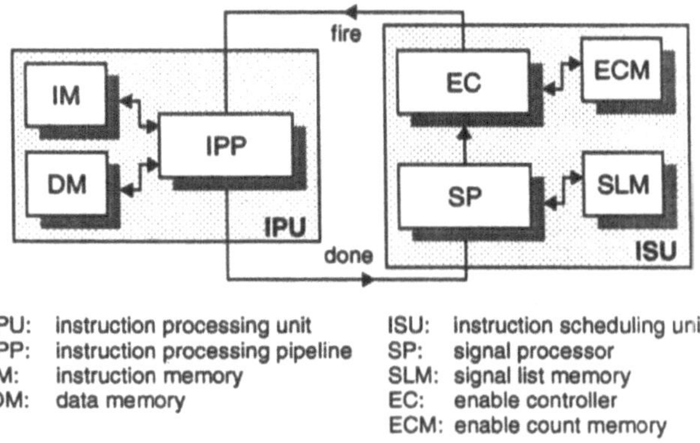

Figure 1: The McGill Dataflow Architecture Model.

2.1 The Architecture and the Program Tuples

The computation in our dataflow model is specified by a dataflow program and its operational semantics. A dataflow program is a tuple $\{P, S\}$, where the P portion, called P-code, is a set of 3-address instructions, and the S portion, called S-code, is a directed graph named the *signal graph.*

The architecture consists of an instruction processing unit (IPU) and an instruction scheduling unit (ISU) as shown in Figure 1. A three-address instruction (called a p-instruction or p-node) in the P-code is similar to that in a conventional architecture, and is stored in the IM. The operands and results of instructions are stored in the DM. For example, in a multiplication instruction $<$ times a b c $>$, the identifiers a and b are the memory addresses of the two operands and c is the address of the result. The storage model of the DM will be introduced shortly. Each p-instruction is uniquely identifiable by a p-instruction address in the IPU, pointing to its program memory in IM. In Figure 2, we show an example of a program tuple.

When a p-instruction address, say i, is presented to the "fire" input (we call this the *fire signal*), it is executed by the IPU in a fashion similar to that of any von Neumann style architecture, i.e., it goes through the usual stages of instruction fetch, operand fetch, execution, result store, etc., and accesses the appropriate memories accordingly. The major difference is that after the execution is completed, the IPU delivers the p-instruction address (in this case, i) to the "done" link as the *done signal.*

The scheduling of p-instruction execution is performed by the processing of S-code in the ISU. S-code consists of a set of nodes (named s-instructions or s-nodes), interconnected by directed arcs. Each s-node has a status field with the information required for scheduling: an *enable count*, indicating how many signals are yet to be received for the s-instruction to become enabled, and a *reset count* indicating the total number of signals it requires to become enabled again after the s-instruction is executed. The status field of the s-nodes are stored in the *enable count memory* (ECM) of the ISU. Each s-node is identified by an s-node address (or s-instruction address). For the purpose of this paper, we assume that there is a one-to-one correspondence between each s-instruction address

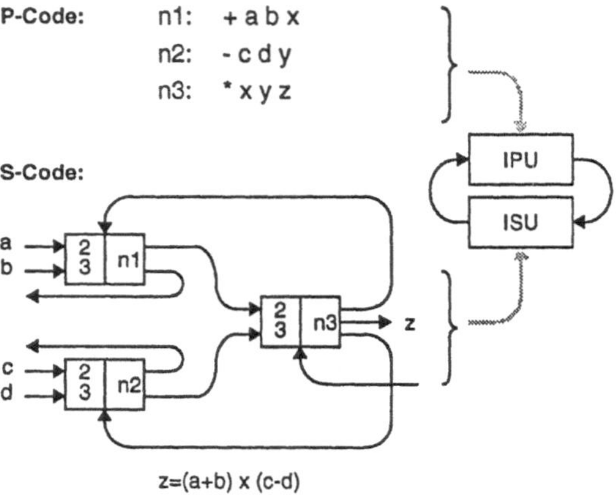

Figure 2: A Dataflow Program Tuple.

and each p-instruction address.

In reality, there are different ways the graph structure in S-code can be specified. For the purpose of this paper we assume that each s-node contains a list of s-node addresses as the destinations of its output arcs. This list is stored in the *signal list memory* (SLM) within the ISU. The principal function of S-code will be described shortly.

2.2 Run-time Storage Model

One of the important features of the architecture is its run-time storage model. In dataflow computations, it is perfectly legal (and also desirable) to have multiple function invocations active concurrently. Therefore, the stack model in a conventional architecture is not appropriate: there is no single point of control corresponding to the "top of stack".

A program tuple is often structured as multiple code modules, i.e., named function modules. Each function module corresponds to code generated from functions in a high level language program. A function module can be shared by many invocations of the function. To support multiple invocations, the memories in the IPU and the ISU are organized to support dynamic allocation of frames of locations, one for each invocation. A frame is also called a *function overlay* in [8]. Although an overlay appears to play a similar role to that of a stack frame in conventional processors, the run-time structure of the overlays in the memory is very different.

An overlay is composed of an IPU overlay and an ISU overlay. For each invocation of a function f, a data memory overlay is allocated to store the result values of each p-instruction composing the function body. The enable status fields of all s-nodes of a particular invocation of f are stored in f's corresponding ISU overlay in ECM. A value stored in a function overlay can be accessed through an address consisting of two parts: a base address of the overlay and an offset. For the purpose of this paper, we assume that the base address is generated and managed (together with memory overlays) by the run-time storage mechanism. This is done through the execution of *apply* and *return* operators for function invocations described in [8].

2.3 The Operational Model

Now, we can present the operational semantics of our dataflow model by the following *firing rules* of a dataflow program tuple:

- a p-instruction i is *enabled* if its S-code counterpart, an s-instruction i in S-code becomes *enabled*, i.e., it has received all the signals it needs;
- an enabled p-instruction may be selected to fire when a fire signal $< i, b >$ (where i is the p-instruction address and b is the base address of the corresponding function overlay) is generated;
- an enabled p-instruction i is processed (or *fired*) in the IPU by accessing the values stored in its function overlay using the base address b; a result value may be generated and stored in a location within the function overlay (note that in an unoptimized form, one location is allocated to each p-instruction);
- after the p-instruction processing is completed, a done signal $< i, b >$ is generated, signaling the completion of the firing of the p-instruction; and
- a scheduling phase (S-code processing) is activated by a done signal $< i, b >$; an s-instruction i is processed by sending signals to all s-instructions in its output signal list, and the status fields (referenced through the base address b) of these s-instructions are updated accordingly. If an s-node has received all the signals it needs, it will become enabled and its enable status will be reset accordingly.

The acute readers may recall the classical formulation of the firing rules based on token flow in a dataflow graph as described in [4]. Although our firing rules do not introduce the concept of tokens, they remain equivalent to the classical firing rules. It is beyond the scope of this paper to present a proof, but interested readers may convince themselves by considering the following hint: an actor in a dataflow graph can be mapped into a pair of p-instruction and s-instruction, and the classical firing rules can be implemented by the firing rules above.

Furthermore, it should be possible to extend the concept of well-behaved dataflow graphs [4] and its translation rules from high-level languages for this model.

2.4 Interprocessor Communication Supports

Thus far, we have only described one processing element of the MDFA. For multiprocessing support, we introduce an Interprocessor Communications Unit (ICU) [16]. Figure 3 shows a schematic block diagram of the ICU and the MDFA within a multiprocessor system.

In this section, we present an efficient interprocessor synchronization method to allow two data-dependent nodes residing in different processing elements (PEs) to synchronize their execution. Also, since dataflow architectures can effectively tolerate long latency memory operations, we have provided the system with a shared global memory, physically distributed among the PEs.

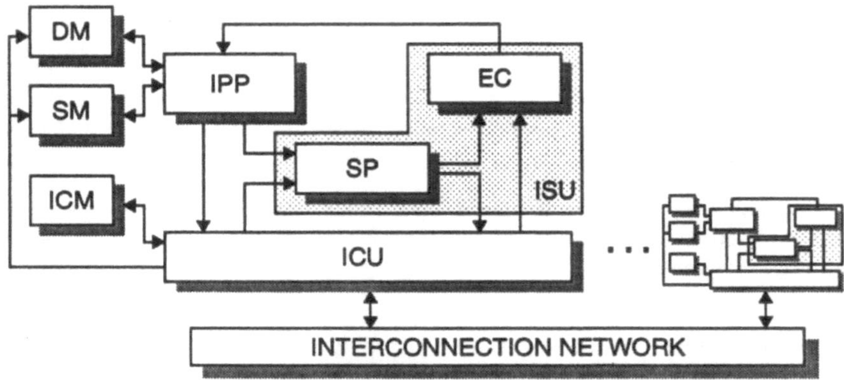

Figure 3: The MDFA in a multiprocessor context.

2.4.1 The Interprocessor Communication Unit

An important issue in the synchronization of remote events is the reduction of network traffic and one of the most effective ways of reducing network traffic is by pairing signals and data in the same packet whenever possible. Also, we believe that it is more efficient to implement interprocessor communications by using attributed elements in the signal list memory instead of special "send" and "receive" actors. Hence, in our scheme, we introduce one more kind of signal list elements: those tagged with *ip-count*. Regular signal list elements send their count signals to the EC, while the ones tagged with ip-count trigger the ICU to send interprocessor count signals. There are two kinds of ip-counts: *data-count* signals, which signifies that data has to be transmitted along with the signal, and *remote-count* which only convey the signal. The remote specifications of the target node as well as the type of packet that has to be sent onto the network are stored in the interprocessor communication memory (ICM). The following are the packet structures for the data-count and remote-count signals:

$$\langle \text{data-count}, PE_{rem}, \text{s-node}_{rem}, \text{dm\_add}_{rem}, \text{data\_value} \rangle$$
$$\langle \text{remote-count}, PE_{rem}, \text{s-node}_{rem} \rangle$$

where PE_{rem} is the remote PE address, s-node$_{rem}$ is the target node address (within PE_{rem}), dm_add$_{rem}$ is the address where the data_value has to be stored in the remote DM.

At the remote PE, reception of ip-signals is straightforward. For a data-count signal, the data conveyed in the packet is stored into DM before the count signal is sent to the EC unit while a remote-count ip-signal just sends a count signal to the EC.

2.5 A Summary of Distinct Features of the MDFA

The features of the MDFA in comparison with other proposed dataflow architecture models can be summarized as follows:

- *Elimination of instruction pipeline gaps due to operand matching.* The data-driven instruction scheduling mechanism is clearly separated from the instruction processing unit. Thus, the MDFA does not perform token "matching" in the critical instruction execution datapath, either explicitly using a *matching store* as in [1, 11] or implicitly at the frame slot with the direct token addressing scheme as in [18]. This eliminates pipeline "gaps" due to operand matching (in dyadic operators), provided that the ISU can handle the signals generated by IPU to "hide" the matching cost from the IPU. (The tradeoff will be elaborated later).

- *Avoiding token duplication.* In the MDFA, a result value never needs to be duplicated (copied) and routed to the input "arcs" of destination nodes. It is stored in the operand memory (only once when it is produced), and will be directly fetched when the subsequent instruction execution needs it as an operand – in a manner similar to any conventional architecture.

- *Efficient dataflow software pipelining of loops.* This is the subject of a more detailed discussion in later sections.

The MDFA supports multiple recursive function invocations, in this sense the model has the power of dynamic dataflow architecture models [1]. However, the MDFA treats loops differently from the loop unraveling scheme used in most dynamic dataflow models [1]. It retains the simplicity of the static dataflow model by limiting the number of concurrent activations of an instruction (to one instance per instruction) and multiple initiations of a loop body can be concurrently executed through dataflow software pipelining. In Section 4, we will comment on the relation between our scheme and the *loop bounding* scheme proposed as a practical realization for the loop unraveling scheme [3].

3 Hybrid Evaluation

In this section, we start with the operational model of the MDFA model and show how it can be easily extended to support sequential instruction execution.

3.1 Support of Sequential Instruction Execution on the MDFA Model

Although the processing of a p-instruction in the argument-fetching dataflow model is similar to that in a von Neumann architecture, there is one missing feature – the sequencing mechanism of instruction execution. In the control-flow model of computation, this mechanism implies that: 1) a program counter is maintained to contain the unique address of the currently executable instruction; 2) the program counter is updated during an instruction execution by default (i.e., for instructions other than branch instructions) – it will be incremented by one – automatically pointing to the next instruction in the sequence; and 3) only branch instructions may update the program counter with a destination address possibly different from the default value.

There is little doubt that the program counter based control mechanism is simple and effective for sequential instruction processing. It turns out that the extension of the basic MDFA to support hybrid evaluation is very simple. All that needs to be done is to provide an alternative for the instruction continuation, after the execution phase of a p-instruction is completed. That is, if desirable, the IPU can directly generate the next p-instruction address to be executed, instead of going through the scheduling phase in the ISU. Of course, this implies that some notion equivalent to the program counter of the control-flow model must be introduced in the dataflow model. In fact, we need a number of program counters: one for each instruction sequence.

To implement this hybrid model, each p-instruction is extended to carry an extra field – a tag field (called a *von Neumann bit*) which indicates whether the instruction is following a dataflow style scheduling or a von Neumann style scheduling. For example, a multiplication instruction now has the following format:

times a b c v-tag

where the *v-tag* is the tag field.

Depending on the value of the v-tag field, the instruction is either scheduled in dataflow mode (*D-mode*) or von Neumann mode (*V-mode*). If v-tag = 0 (dataflow mode), the done signal is generated as usual, i.e., it contains the address of the p-instruction so that it will be processed by the ISU. Otherwise, a new address is generated as the address of the next p-instruction to be fired, in a way similar to the update of the program counter in a von Neumann architecture. Note that this is different from the *repeat-on-input* mechanism in the $\mathcal{E}$psilon dataflow processor which was proposed to exploit the locality inherent in parameter duplication [10].

3.2 The Hybrid Operational Model

Since in the basic model, a p-instruction is processed by the IPU in a style similar to von Neumann style processing, the above hybrid evaluation model can be implemented in a straightforward fashion. The only change to be made is to allow a "short-cut" path from the done link to the fire link, thus allowing a V-mode p-instruction to directly "fire" the next instruction, bypassing the scheduling phase (ISU). In Figure 4, we illustrate how a sequence of dataflow nodes can be grouped into a "macro" dataflow node, and a short-cut signal mechanism is used for their sequencing.

The operational model for hybrid evaluation becomes:

- a p-instruction i is *enabled* if
  - its S-code counterpart, an s-instruction i in S-code becomes *enabled*, i.e., it has received all the signals it needs;
  - or, it is the next instruction in a sequence of V-mode p-instructions, and its predecessor p-instruction $(i-1)$ has completed its execution;
- An enabled p-instruction may be selected to fire when a fire signal $< i, b >$ (where i is the p-instruction address and b is the base address of the corresponding function overlay) is generated;

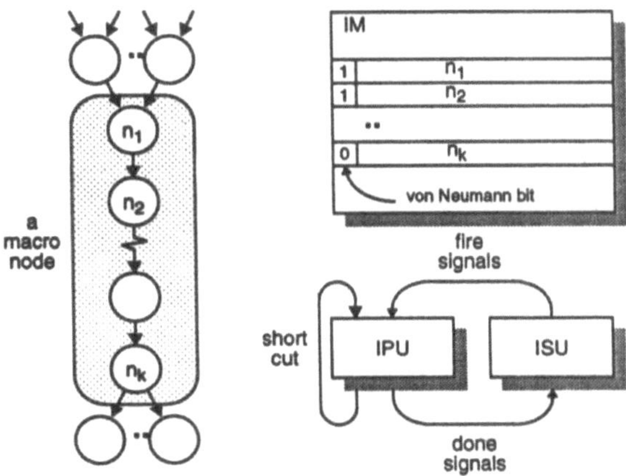

Figure 4: The Hybrid Evaluation Model.

- An enabled p-instruction i is processed (or *fired*) in the IPU by accessing the values stored in its function overlay using the base address b.
- After the p-instruction is fired and depending on its mode, one of the following happens:
  - for a V-mode instruction (vtag = 1), a fire signal is sent through the short-cut path, which contains the next p-instruction address; and
  - for a D-mode instruction, a done signal $< i, b >$ is generated, signaling the completion of the firing of the p-instruction and the start of the ISU scheduling phase below.
- a scheduling phase (S-code processing) is activated by a done signal $< i, b >$; an s-instruction i is processed by sending signals to all s-instructions in its output signal list, and the status fields (referenced through the base address b) of these s-instructions are updated accordingly. If an s-node has received all the signals it needs, it will become enabled and its enable status will be reset accordingly.

3.3 Features of the Hybrid MDFA Model

We briefly summarize the features of the hybrid MDFA model:

- *Generality*: The hybrid MDFA model supports both thread level and instruction level parallelism through efficient fine-grain synchronization. At any time, the IPU can execute several instructions in parallel: any instruction may be a D-mode instruction or a V-mode instruction, and the V-mode instructions may themselves come from different instruction threads. Thus our model is different from so-called "macro-dataflow" schemes where dataflow scheduling can only be done at the inter-procedural level [14]. It retains the advantage of dataflow models in terms of dealing

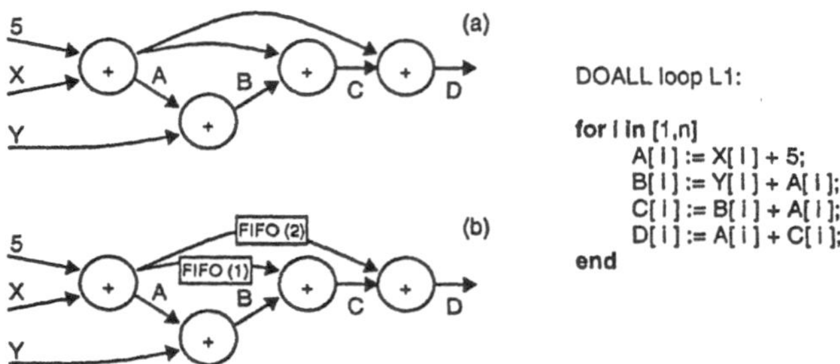

Figure 5: Example of Dataflow software pipelining

with the two fundamental issues of von Neumann multiprocessing as discussed in the introduction.

- *Flexibility*: There is no restrictions to the number of instruction threads which can be supported by this model. In fact, a variable number of instruction threads each with a different size can be active concurrently. This is different from some other multi-threaded architectures where the maximum number of threads are fixed *a priori* such as in the HEP [19].
- *Simplicity*: Under the hybrid MDFA model, any instruction in a program can be set to one of two modes, regardless of its function or type. This flexibility certainly makes the job of a compiler easier, since the mode control and the operation of an instruction become orthogonal. Note that ISU performs the fork/join operations of the threads implicitly through signal processing, while some other multi-threaded architectures may execute explicit fork and join instructions [17].

4 Dataflow Software Pipelining

In this section, we present a compiling paradigm which exploits loop parallelism through *dataflow software pipelining* using *limited balancing.* The limited balancing technique is one method which can be employed to obtain macro dataflow nodes for the hybrid architecture model.

4.1 Background

The principle of dataflow software pipelining can be explained using loop L1 as an example (in fig. 5). The loop is translated into a pipelined dataflow graph as shown in figure 5 (a). Successive waves of elements of the input arrays X and Y will be fetched and fed into the dataflow graph, so that the computation may proceed in a pipelined fashion. This is called *dataflow software pipelining*: the arcs drawn between actors correspond to addresses in stored dataflow machine code, not to the wired connections between logic elements.

Previous work in dataflow software pipelining are based on an idealized dataflow machine model. Here is a summary of some known results which are valid under the ideal

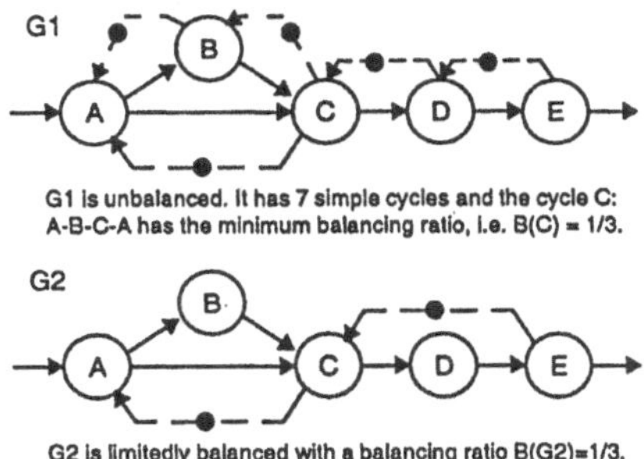

Figure 6: Balancing ratio for some example graphs.

model. A graph is (fully) *balanced* if every path between a pair of nodes contain exactly the same number of actors. To achieve maximum pipelining, a basic technique (called *balancing*) is used to transform an unbalanced dataflow graph into a (fully) balanced graph by introducing FIFO buffers on certain arcs.

For example, the dataflow graph in figure 5 (a) is not balanced We can introduce two FIFO buffers as shown in figure 5 (b) so the resultant graph will become fully balanced, and hence can be maximally pipelined in an ideal dataflow machine model.

Unfortunately, the previous work does not cover the case where dataflow graphs contain cycles. We will discuss an extension to include these loops in the next section. More about dataflow software pipelining can be found in [6].

4.2 Limited Balancing

A loop may contain loop-carried dependencies, in which cycles may exist with some back edges due to inter-iteration data dependencies.

Let us introduce the notion of *balancing ratio*, denoted by $B(G)$, to describe the degree of balancing of a dataflow graph G. We give an informal definition of the balancing ratio based on the observation of token and space duality in dataflow graph models suggested by Kung et. al. [15]. Let us augment G with *acknowledgment arcs* and assign one initial (dummy) token on each acknowledgment arc in G, denoting the fact that the corresponding forward (data) arc has one empty place – the SDFP is initially empty under our assumptions. Then our (augmented) SDFP is similar to Kung's *augmented dataflow graph* where the acknowledgment arcs correspond to his "reverse arcs". For any directed simple cycle C in G, [1] let the length of C (number of arcs) be K_C, and the number of dummy tokens be D_C. The ratio $\frac{D_C}{K_C}$, called the balancing ratio of C (denoted by $B(C)$) in this paper, determines the fastest rate at which the node in the cycle can be activated [15]. A graph is *(limitedly) balanced* with balancing ratio $B(G)$ if all directed cycles C_i in the graph have the same balancing ratio, i.e., $B(C_i) = B(G)$ for all i.

Figure 6 gives an example to illustrate the notion of limited balancing and the concept of balancing factor. It is beyond the scope of this paper to present the algorithm for limited

[1] Simple cycles are cycles in which each node appears at most once. For our purposes, only simple cycles are considered.

graph balancing. Readers may find a dedicated discussion in [9]. The basic intuition is that adding buffers as well as removing redundant signals (both data and acknowledgment arcs) may be required during the limited balancing process. In our simple example in figure 6, the nodes A, B, C can be grouped into a thread, so can D, E, F. This can have the potential of significantly reducing the number of signals to be handled by the ISU in dataflow software pipelining. Note that actors grouped into threads will form a macro dataflow actor for hybrid evaluation. We will demonstrate this advantage in the simulation section.

4.3 Related Work

There have been several different suggestions as how to compile von Neumann instruction threads for a hybrid computing model [13, 20]. In one interesting approach proposed by Iannucci [13], a compiler is responsible for partitioning program graphs into *scheduling quantums*. A methodology is outlined which will generate multiple scheduling quantums without deadlock. His method is tightly-coupled with the hybrid architecture described in his doctoral dissertation [13]. Another interesting approach proposed by Traub is to view partitioning of a program into threads as the central problem (or in his words: "first order business") of functional language compilation [20]. As a result of his "compiling-as-partitioning" strategy, one gains a much cleaner understanding of the relationships between lazy vs. lenient evaluation, as well as sequential vs. parallel execution. These techniques can also be applied to the architecture described in this paper.

5 Simulation Results

5.1 The Impact of ISU Signal Processing Capacity

In this section, we first present some simulation results to show that reducing the number of signals to be handled by the ISU is very important, thus motivating us to study the limited balancing technique.

In a balanced design of the MDFA, the ISU has to supply fire signals to the IPU just fast enough to keep the instruction execution pipelines operating continuously at full capacity. There could be cases in which the signal processing demand in the ISU cannot be satisfied fast enough, and this could cause a bottleneck.

Figure 7 shows the performance of Livermore Loop 7 using various machine configurations. When the signal processing capacity is one, the execution time of loop7 remains the same, no matter how many IPUs are used. In this case the computation is *ISU bound*, that is, the ISU is unable to satisfy the demand for manipulating incoming signals, causing a performance degradation. As the signal processing capacity of the ISU is increased, more enabled actors are presented to the IPUs per cycle, and thus, optimum processing is achieved, as shown by the leveling off of run times for the various numbers of IPUs modeled within a PE. From the shape of the curves we postulated that there is a relationship between signal processing capacity and number of IPUs needed to obtain optimum cost-effective program execution. Let us examine this phenomenon in more detail. To

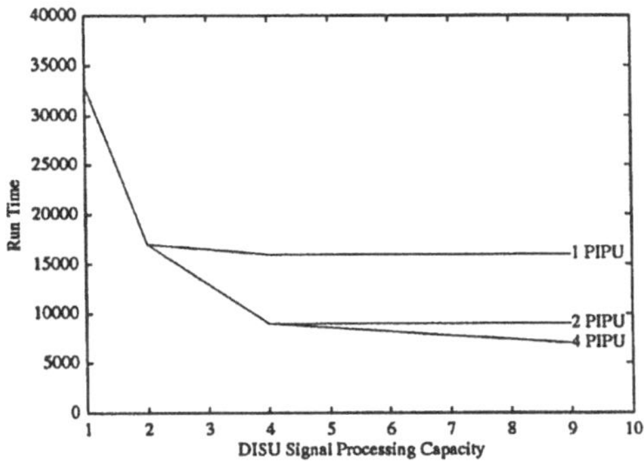

Figure 7: Varying machine configurations for loop7.

achieve such optimality, the following factors are important:

1. IPU capacity (denoted by P) must match the amount of parallelism within the program (*computational parallelism*), and

2. ISU signal capacity (denoted by C) must match the demand for processing the signals required to exploit the computational parallelism (*synchronization requirement*).

Our studies show that, even for a program having a computational parallelism sufficiently high to keep the IPU filled in an idealized ISU (one having infinite signal processing capacity), performance on a realistic machine might be far below that of the ideal value; the outcome depends upon the average number of signals S that are needed to fire an instruction. This number, called the *average signal density*, is given by the following formula:

$$S = \frac{TotalCountSignals}{TotalInstructionsExecuted}$$

So for a given machine configuration, the condition to keep the IPU pipeline usefully busy is:

$$\frac{C}{S} \geq P \tag{1}$$

Conceptually this means that to fully exploit the computational parallelism of a program, the ISU capacity should be at least equal to the product of the average signal density of the program and the number of IPUs of the given configuration. This is not surprising; *fine-grain parallelism has its price*, and for our architecture, the ISU pays.

Both the ISU capacity and IPU capacity can be taken as important parameters in compiler optimizations for the McGill dataflow architecture. In a machine in which signal traffic plays such an important role, the compiler should try to minimize the number of signals, while at the same time, trying to expose sufficient parallelism. This is one of the main motivations behind limited balancing.

IPU: 1, ISU Signal capacity: ∞, Pipeline Stages: 8 k: 8, n: 60, CB(G) = $\frac{1}{7.5}$						
Balance Factor (B)	$\frac{1}{2}$	$\frac{1}{3}$	$\frac{1}{4}$	$\frac{1}{5}$	$\frac{1}{6}$	$\frac{1}{7}$
Avg. Init. Cycle	60	60	60	60	60	66
Avg. Instr. Delay	11.08	11.14	6.34	3.68	1.73	1.28
Run Time	12,119	12,113	12,096	12,102	12,114	13,305
Utilization	99.1%	99.2%	99.3%	99.3%	99.2%	90.3%

Table 1: Simulation results for 1 IPU

5.2 The Effects of Limited Balancing

In this section, we demonstrate the effect of limited balancing. In each simulation run, the limited balancing has only been applied to one simple cycle of the graph in order to ease the job of manipulating the large number of signal arcs manually.

Table 1 shows the simulation results for the machine configuration of 1 pipeline (1 IPU) with 8 pipelined stages. The top row of the table indicates the associated machine configuration and the size of the graph n. The program graph does not have to be fully balanced to keep the IPU fully utilized. Instead, a *critical balancing ratio CB(G)* can be estimated based on the graph structure [7], which can be used to balance the graph. Each column of the table gives the corresponding results of applying a distinct balancing ratio (B) to the graph. The *Average Initiation Cycle* records the average pipelining period of each node in the graph; the average initiation rate of the graph is simply its reciprocal. The *Average Instruction Delay* records the average time delay of the enabled instruction waiting in the fire queue, which implicitly represents the amount of parallelism exposed in the graph that have not been fully utilized by the execution pipe.

Here is a summary of the major observations from table 1:

- In all simulation runs, we have observed that the loop initiation sequence will enter a steady state: a constant average initiate rate is reached.
- Once the average initiation rate *1/n* is achieved, the pipeline is maximally utilized.
- The critical balancing ratio for this loop is 1/7.5. Our simulation results are very close to this prediction. At B = 1/6, the pipeline is near fully utilized.
- The average instruction delay decreases gradually as the balancing ratio decreases. When the balancing ratio reaches the critical value, no excessive parallelism is exposed which may not be efficiently utilized, and the average instruction delay approaches its minimum.

The limited balancing scheme will reduce the amount of count signals to be handled by the ISU. In this case, if the limited balancing is applied to the entire graph, the total number of signal arcs can be reduced considerably.

5.3 Discussions

As we have shown, applying fine-grain synchronization to all actors in a dataflow graph does not necessarily yield optimal performance due to the fine-grain synchronization costs which must be absorbed by the synchronization mechanism in the underlying architecture. If there are enough parallelisms in the application, limiting the amount of exposed parallelism to the underlying machine by means of grouping actors into aggregates and then scheduling the aggregates themselves can lead to lower fine-grain synchronization costs and thus a more balanced use of the underlying processing and scheduling resources. This is fine, but in all the simulation runs, we have assumed that memory latencies are unity and as we all know, memory latencies in an actual implementation can never be kept fixed and low, especially not unity.

To address this problem, memory hierarchies must be investigated where fast and small memory like register files and conventional caches are utilized. However, in a multi-threaded architecture, lifetime analysis for register values are difficult, thus rendering registers less than effective. Moreover, the frequent switching between active threads wreaks havoc on the locality principle on which the conventional cache obtains its power, so caches are not ideal. In a paper appearing in the same conference [12], we introduce the *Register-Cache* mechanism which is a hybrid register file and cache mechanism. Register allocations are performed dynamically at run-time, so lifetime analysis of values are not necessary, and furthermore, all required values are loaded into the register-cache before a thread executes so that the execution pipe will see a fixed and low memory latency time. This register-cache mechanism is employed in the context of an extended MDFA model called the *Super-Actor Machine*, and readers are referred to [12] for more details.

6 Conclusion

We have presented an operational model for a hybrid dataflow and control-flow architecture. Based on the argument-fetching dataflow principles, we have demonstrated that a straightforward extension to the basic McGill dataflow architecture model can accommodate concurrent operations of multiple instruction threads. The flexibility of the new model not only resides in the support of hybrid evaluation, but also in exploring parallelism at any desirable level in a fine-grain fashion. A unique program mapping scheme for loops based dataflow software pipelining and the limited balancing technique was also presented. To take memory latencies into consideration, the hybrid MDFA model presented in this paper is further extended to be the *Super-Actor Machine* [12].

At McGill University, we are using the MDFA model and its variants as vehicles for studying a range of architectural issues for multi-threaded architectures, as well as compiling techniques for them. It is our plan to conduct a more extensive analysis and comparison of the advantages and disadvantages between the architectural models of the MDFA genre and other hybrid dataflow architecture models. We are also investigating the impacts of the hybrid model on language designs for multi-threaded computing.

7 Acknowledgment

We would like to thank the National Science and Engineering Research Council (NSERC) for their support of this work. Thanks also to the Bell Northern Research (BNR) for their support of research in parallel processing and dataflow.

We would also like to thank the members of the Advanced Architecture and Program Structures Group for joining us in the adventure of dataflow research. In particular, we thank René Tio, Robert Yates, Zaharias Paraskevas, Yue-Bong Wong, and Russel Olsen for working on the architecture, compiler, and simulation testbed for the McGill Dataflow Architecture. Without their support, the work described in this paper would not have been possible. Finally, we would like to thank J.B. Dennis for his valuable contributions.

References

[1] Arvind and D. E. Culler. Dataflow architectures. *Annual Reviews in Computer Science*, 1:225–253, 1986.

[2] Arvind and Robert A. Iannucci. Two fundamental issues in multiprocessing. In *Parallel Computing in Science and Engineering*, pages 61–88. Springer-Verlag, LNCS-295, 1987. *Proceedings of the 4th International DFVLR Seminar on Foundations of Engineering Sciences*, Bonn, June 1987.

[3] D. E. Culler. Managing parallelism and resources in scientific dataflow programs, Ph.D thesis. Technical Report TR-446, Laboratory for Computer Science, MIT, 1989.

[4] J. B. Dennis. First version of a data-flow procedure language. In *Proceedings of the Colloque sur la Programmation*, volume 19 of *Lecture Notes in Computer Science*, pages 362–376. Springler-Verlag, 1974.

[5] J. B. Dennis and G. R. Gao. An efficient pipelined dataflow processor architecture. In *Joint Conference on Supercomputing*, pages 368–373, Florida, November 1988. IEEE Computer Society and ACM SIGARCH.

[6] G. R. Gao. *A Code Mapping Scheme for Dataflow Software Pipelining.* Kluwer Academic Publishers, Boston, December 1990.

[7] G. R. Gao, H. H. J. Hum, and Y. B. Wong. An efficient scheme for fine-grain software pipelining. In *Proceedings of the CONPAR '90–VAPP IV Conference*, Zurich, September 1990.

[8] G. R. Gao, H. H. J. Hum, and Y. B. Wong. Parallel function invocation in a dynamic argument-fetching dataflow architecture. In *Proceedings of the PARBASE '90 Conference*, Miami Beach, FL, March 1990.

[9] G.R. Gao and Qi Ning. Loop storage optimization for dataflow machines. ACAPS Technical Memo 23, School of Computer Science, McGill University, Montréal, Qué., February 1991. In preparation.

[10] V. G. Grafe, G. S. Davidson, J. E. Hoch, and V. P. Holmes. The Epsilon dataflow processor. In *Proceedings of the 16th International Symposium on Computer Architecture*, pages 36–45, Israel, June 1989.

[11] J. R. Gurd, C. C. Kirkham, and I. Watson. The Manchester prototype dataflow computer. *Communications of the ACM*, 28(1):34–52, January 1985.

[12] H. H. J. Hum and G. R. Gao. A novel high-speed memory organization for fine-grain multi-thread computing. In *in the same Proceedings*, June 1991.

[13] R. A. Iannucci. Toward a dataflow/von Neumann hybrid architecture. In *Proceedings of the 15th Annual International Symposium on Computer Architecture*, pages 131–140. ACM, June 1988.

[14] D. Kuck, E. Davidson, D. Lawrie, and A. Sameh. Parallel supercomputing today and the cedar approach. *Science Magazine*, 231:967–974, February 1986.

[15] S. Y. Kung, S. C. Lo, and P. S. Lewis. Timing analysis and optimization of VLSI data flow arrays. In *Proceedings of the 1986 International Conference on Parallel Processing*, 1986.

[16] J.M. Monti. Interprocessor communication supports for a multiprocessor dataflow machine. Master's thesis, School of Computer Science, McGill University, Montréal, Qué, March 1991.

[17] R. Nikhil and Arvind. Can dataflow subsume von Neumann computing? In *Proceedings of the 16th International Symposium on Computer Architecture*, pages 262–272, Israel, 1989.

[18] G. M. Papadopoulos and D. E. Culler. Monsoon: An explicit token-store architecture. In *Proceedings of the Seventeenth Annual International Symposium of Computer Architecture, Seattle, Washington*, pages 82–91, 1990.

[19] Burton Smith. The architecture of HEP. In J. S. Kowalik, editor, *Parallem MIMD Computation: HEP Supercomputer and its Application*, pages 41–55. The MIT Press, 1985.

[20] K. R. Traub. Sequential implementation of lenient programming languages. Technical Report MIT/LCS/TR-417, Laboratory for Computer Science, MIT, 1988.

Data Flow Implementation of Generalized Guarded Commands

R. Govindarajan
VLSI Design Laboratory
McGill University
Montreal, H3A 2A7, Canada
govindr@pike.ee.mcgill.ca

Sheng Yu
Department of Computer Science
University of Western Ontario
London, N6A 5B7, Canada
syu@uwocsd.uwo.ca

Abstract

Earlier approaches to execute generalized alternative/repetitive commands of Communicating Sequential Processes (CSP) attempt the selection of guards in a sequential order. Also, these implementations are based on either shared memory or message passing multiprocessor systems, which exploit parallelism only among the processes of a CSP program. In contrast, we propose a data flow implementation for CSP with generalized guarded commands in which both inter-process and intra-process concurrencies are exploited. A significant feature of our implementation is that it attempts the selection of guards of a process in parallel. A simulated model empirically demonstrates correctness properties, namely 'safety' and 'liveness', of our implementation. The simulation experiments are also helpful in obtaining certain efficiency and fairness parameters of the implementation.

1 Introduction

Hoare's Communicating Sequential Processes (CSP) [14] has been widely accepted as a paradigm for programming parallel computation. But the initial definition of CSP does not allow output commands to appear in the alternative/repetitive commands. Not only is this a hindrance to the symmetry of the language, but this considerably constraints the expressibility [6, 7, 14, 15, 18]. This can be easily understood from the bounded buffer program or the dining philosopher program, where the inability to use output guards in the guarded commands necessitate additional signals. Following these arguments, the guarded commands of CSP have been generalized to allow even the output command to appear in them. While such a generalization is easy to perceive, an implementation is quite involved and requires reaching an agreement among the communicating processes. We explain this with the following example.

Process P1	Process P2	Process P3
... ...	... ...	... ...
*[true;P2!x1 --> S1	*[true;P1?y1 --> S3	*[true;P1!z1 --> S5
[] true;P3?x2 --> S2	[] true;P3!y2 --> S4	[] true;P2?z2 --> S6
]	]	]
... ...	... ...	... ...

The guarded statements in P1, P2 and P3 can be executed only if (i) both P1 and P2 agree to select the first guard (in their respective repetitive commands), or (ii) both P2 and P3 agree to select the second guard, or (iii) the processes P3 and P1 agree to select, respectively, the first and second guards. Let P1 and P2 make the agreement (or rendezvous). Then it is clear that P3 should not initiate a rendezvous with either P1 or P2. Thus the agreement is not only between the processes that rendezvous, P1 and P2 in the above example, but also with other processes, process P3 in this example, with which P1 and P2 can potentially communicate in the alternative/repetitive command. Thus a global agreement among mutually communicating processes is required to select a guard in the generalized guarded command.

Proposing an implementation for the generalized guarded command has been of constant interest to the research community [2, 3, 4, 7, 10, 18, 20]. Fujimoto's solution [10] was based on a shared memory model while the others employ a loosely-coupled architecture. A commonality observed in all these solutions is they are based on the conventional von Neumann framework, adopting either shared memory or message passing architecture. Also, the selection of guards in a particular process has so far been done sequentially. In contrast to the earlier proposals, we use data-driven evaluation [21] as the basic computation model for executing CSP programs and propose an implementation which parallelizes the selection of guards in a process. Attempting guards in parallel can significantly reduce the execution time of an alternative/repetitive command. It is important to note that our implementation does not sacrifice the semantics of the guarded commands. Other advantages of our implementation include exploiting fine-grain concurrency, allowing parallelism not only among the various processes of a program but also within a single process. Also, in our implementation, Processing Elements (PEs) do not busy wait for synchronization of guarded commands.

This paper is organized as follows. In the following section, we present the basic architecture and discuss the issues related to implementing generalized guarded commands. Section 3 describes the implementation scheme. A simulated model is developed to empirically demonstrate the correctness of our implementation. Further, certain performance parameters concerning efficiency and fairness of our implementation are also obtained from the simulation experiments. These results are reported in Section 4. Finally, we conclude by comparing our implementation with related works.

2 The Architecture and Related Issues

2.1 The Architecture

Data-driven evaluation has been chosen as the model of computation for our implementation for the following two reasons. Firstly, data flow model does not impose any order (other than what is dictated by data dependency) on the execution. The guards in a process can thus be concurrently attempted in a natural way. Further, fine grain asynchronous concurrency exploited by data flow machines ensures both inter- and intra-process parallelisms in a CSP program. The reader is referred to [1, 13, 21] for an introduction to data-driven evaluation.

Though the our implementation is suitable for any data flow computer, Manchester multi-ring data flow machine [5] is used as the base architecture in this paper. On top of the data flow architecture, we assume a shared memory module accessible to the Processing Elements (PEs) of the machine through an interconnection network. The architecture with the shared memory is shown in Fig.1. In order to increase the memory bandwidth — for obvious performance improvement reasons — the shared memory is low-order interleaved. The PEs are connected to the memory modules by means of a multi-stage interconnection network. Lastly, we assume the availability of a synchronization primitive such as the *fetch and increment* [11].

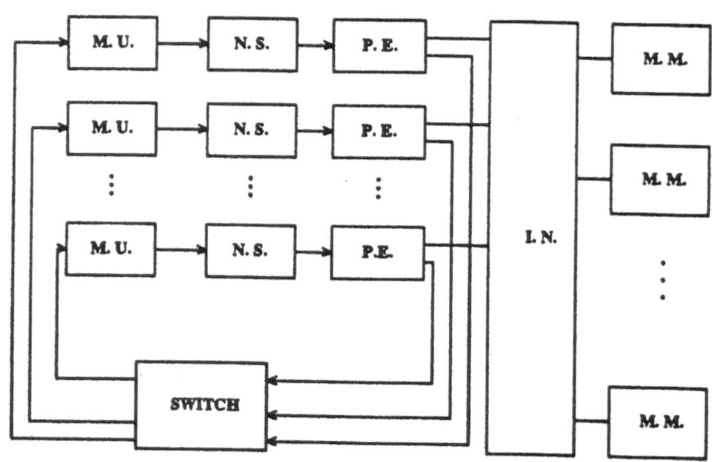

Fig. 1 A Multi-Ring Data Flow Machine

2.2 Implementing CSP

A CSP program consists of a set of communicating processes. We propose to execute a CSP program by converting each process into a data flow graph; the resulting graphs are executed on a data flow machine. The processes of a CSP program can be executed concurrently. A distributed execution of the data flow graph ensures both inter- and intra-process concurrencies. The implementation of the simple commands of CSP, namely skip, assignment, and parallel commands, is straightforward. Also, work on implementing CSP with constrained guarded commands (i.e., allowing only input commands in the guards) has already been reported [17]. Hence we restrict our attention to the implementation of the generalized guarded commands.

2.3 Notations

In a CSP program, the processes are given distinct process identifiers, called *process-ids*. Each invocation of an alternative or repetitive command is referred as a *transaction*. A unique identifier, called *trans-id*, is generated by concatenating the process identifier (of the process which invokes the transaction) and a sequence number. The *trans-ids* are totally ordered. In a transaction, there are a number of guarded commands, each guard having a distinct index. Consider a guarded statement $b; c \rightarrow S$ appearing in a process P_i, where b and c are the boolean and IO guards respectively. If c addresses the process P_j, then P_j is the communicating complement (or simply complement) of P_i and *vice-versa*.

As there can be many guarded commands in a transaction, there are many potential communicating pairs. So, implementing a generalized alternative or repetitive command is equivalent to reaching an agreement among the potential communicating processes. If processes P_i and P_j agree to communicate, then we say P_i and P_j rendezvous. When P_i and P_j rendezvous, it is implicit that the IO guards of them are complementary (that is, one is an input guard and the other an output guard).

2.4 Related Issues

In the first place, data flow implementation forbids the possibility of a process owning a Processing Element (PE). Also, since we allow parallel execution within a single process, various instructions (constructs) belonging to a process can reside on a number of PEs. As a consequence, a single (process) state cannot be assigned to a process. This is in contrast to the earlier implementations [2, 7, 10, 18] which rely heavily on the existence of a unique state for each process.

The execution of input/output commands in CSP warrants the synchronization of the communicating processes. If a PE executing a communication construct is allowed to 'busy wait', then this may lead to a deadlock situation: a situation where each PE executing a communication construct is waiting for synchronization, but the corresponding communicating complements are denied of PEs due to non-availability. Thus, to avoid deadlocks busy waiting in PEs should be prohibited. That is, a PE must be set free while the communication construct executed by it (PE) waits for synchronization. The details of the construct have to be stored in the shared memory to enable the execution of the construct later when its complement becomes ready.

3 The Implementation

Let us first consider the implementation of an alternative command; the same arguments can be extended to repetitive commands. Also, only those guards whose boolean components succeed are considered. In the following discussion, P_l refers the local process executing the guard g_l in transaction t_l. The process P_r addressed by the guard g_l is called the remote process.

Before we go into the details of the implementation, the basic principle is briefly described below. An alternative command of a process is converted into a data flow graph as shown in Fig.2. The execution proceeds by attempting the guards. The guard g_l of a process P_l is attempted by executing a *Tryguard* actor (to be explained subsequently). The actions performed by this actor are as follows:

1. First, the arrival of guard g_l is marked by writing an entry in a data structure stored in shared memory.
2. Then a matching guard for g_l is searched in the shared data structure.
3. A failure to find a matching guard results in termination of execution of the *Tryguard* actor. However, if a match is found, then the PE executing the *Tryguard* actor attempts to capture the respective processes (in the strict order of their *trans-ids*). This is done using the synchronization primitive *fetch and add.*
4. If the PE cannot capture either of the processes, then the execution of the *Tryguard* actor terminates. Otherwise (*i.e.*, when the PE captures both the processes), the rendezvous occurs.

It may be observed that execution of a *Tryguard* actor may or may not result in a rendezvous. However, when all the guards of a process fail to effect a rendezvous, the process gets into a state similar to the 'waiting' state of [10]. It is guaranteed that, eventually, some other process will find the 'waiting' process willing to rendezvous with it. At that time, the 'waiting' process is woken up by means of tokens.

3.1 Alternative Commands

We store the following variables in shared memory for each transaction t_l.

Committed (t_l): A boolean variable indicating whether transaction t_l is committed to some other transaction.

Excl (t_l): If a transaction t_l wants to rendezvous with t_r, then the PE attempting the rendezvous must acquire exclusive access to the *Committed* variables of both t_l and t_r. To accomplish this, the *Excl* flags are used. The rendezvous takes place only if the PE successfully acquires exclusive access to the *Committed* variables of t_l and t_r, and the *Committed* flags remaining *False* when the access was acquired. The capturing of local and remote processes goes by strict *trans-id* order to avoid cyclic dependencies and deadlocks.

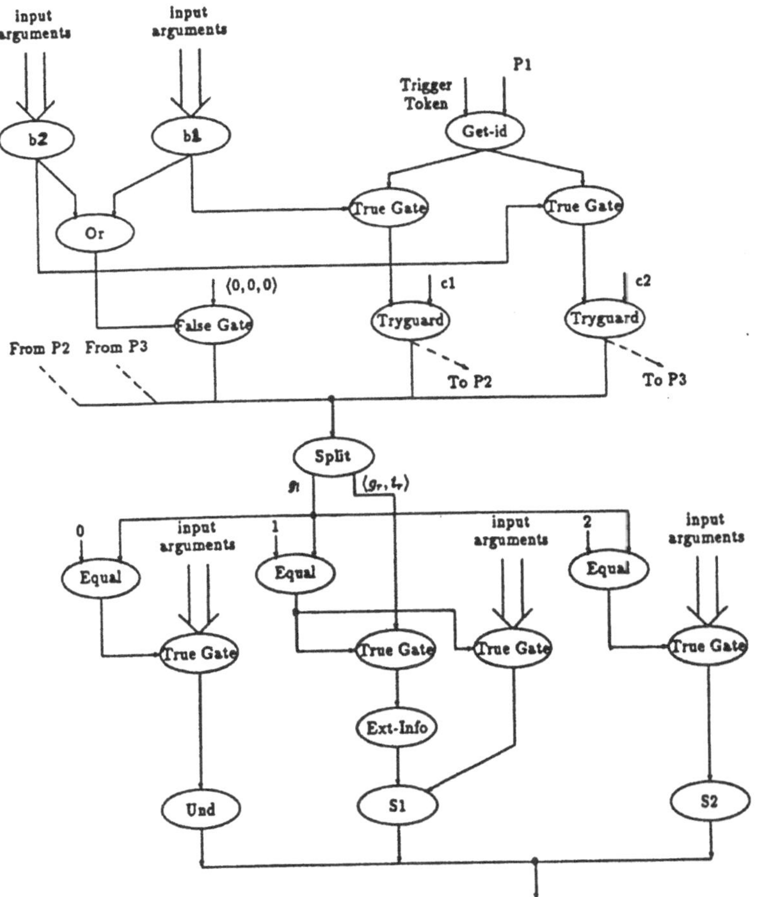

Fig.2 Data Flow Graph for $[b1; P2?x \rightarrow S1 [] b2; P3!y \rightarrow S2]$

G-list (t_l): *G-list* (t_l) is an array of linked lists indexed on the processes. For each process P_r, irrespective of whether P_l wishes to communicate to P_r or not, there is an entry *G-list* (t_l, P_r)[1]. Initially all entries of *G-list* (t_l) will contain nil pointers. When a guarded command g_l is attempted for execution, a new entry will be linked to *G-list* (t_l, P_r), where P_r is the remote process addressed by g_l. The entry is a tuple $\langle g_l, c\rangle$, where g_l is the guard index and c provides the communication details of g_l. Typically, c stores the remote process name, signal name, and a bit to identify whether the command is input or output. An entry $\langle g_l, c\rangle$ in *G-list* (t_l, P_r) indicates that the IO guard g_l in process P_l is willing to rendezvous with P_r, provided *Committed* (t_l) is *False*.

The following variables are associated with each process.

Seq-no (P_l): The integer variable *Seq-no* (P_l), initialized to zero, is used for generating a unique *trans-id* for each active transaction in process P_l.

Active-trans (P_l): This variable stores the *trans-id* of the active transaction in P_l. If no transaction is currently active, the variable stores a null value.

An alternative command can be executed by converting it into a data flow graph as shown in Fig. 2. Certain abstractions have been followed in the data flow graph for simplicity. For example, a set of input arguments is passed to a guarded statement through a single *True* gate. In practice a number of *True* gates have to be used for this purpose. Also, we have assumed unlimited fanout for each data flow actor. A few new data flow actors have been introduced in Fig.2. These data flow actors with their respective input and output arguments are shown in Fig. 3. To understand their operational semantics, we need to know about the following two procedures.

Store (P_r, t_l, g_l,c): This procedure links the entry $\langle g_l, c\rangle$ to *G-list* (t_l, P_r).

Checkguard (g_l, P_r, c): This procedure is responsible for searching a 'matching' and 'compatible' entry for g_l in *G-list* (t_r), where t_r is active transaction in P_r. The 'matching' condition ensures the remote guard is a potential communicating complement. Two entries are 'compatible' if the IO commands specified in them are complementary in nature. When this procedure succeeds (in finding the matching and compatible guard), it outputs g_r, the guard index in the remote transaction. A failure is indicated by outputting 0.

The operational semantics of the new data flow actors used in our implementation is presented below.

Get-id: The invocation of an alternative command commences with this actor. When the *Get-id* actor is invoked from a process P_l, the PE executing this actor fetches the *Seq-no* (P_l) and increments its contents by one using the synchronization primitive *fetch and increment*. A unique *trans-id* is generated by concatenating P_l with the fetched *Seq-no*.

[1]The arguments for indexing *G-list* (t_l) on *process-ids*, rather than on guard indices are: (i) more than one guard can address the same process; (ii) when a complement process checks *G-list*(t_l), all guards of t_l which are ready and willing to communicate with the complement process need to be tried; and (iii) a linked list representation is an efficient way to access all available guards (*i.e.*, guards ready and willing to rendezvous with the complement process) compared to a scheme where the *G-list* is indexed by the guard indices.

Tryguard: An IO guard is attempted by executing a *Tryguard* actor. The details of the communication guard are specified by the input c as shown in Fig. 3. Note that the values P_l and c_l are constants derived and associated with the node at compile time. This actor invokes the *Checkguard* procedure to determine whether the corresponding guard is ready to rendezvous. If a matching guard is waiting, then an attempt to capture the *Committed* flags of the respective transactions is made. Successful capturing marks a rendezvous. Failure results in releasing captured transaction, if any. The operational semantics of this actor is presented in Fig. 4.

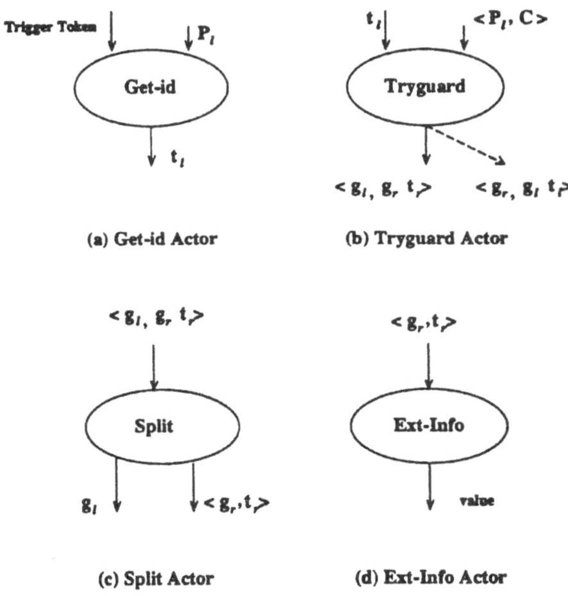

Fig.3 New Data Flow Actors

Split: The *Split* actor receives a triple $\langle g_l, g_r, t_r \rangle$ as its input from a *Tryguard* actor.[2] This actor splits the triple and outputs the g_l value on one output arc and the pair $\langle g_r, t_r \rangle$ on the other output arc. The value of g_l is used to enable the appropriate guarded statement. The $\langle g_r, t_r \rangle$ pair provides the necessary information for the extraction of input data in the IO guards whenever the guard is an Input command.

Ext-Info: When the IO guard is an input command, the corresponding *Tryguard* actor only achieves the synchronization. The actual communication (reception of data) has to be performed with the help of an *Ext-Info* (meaning, extract information) actor. As the necessary synchronization has already been achieved, the data can be input without further synchronization delay.

[2]Or from a *False*-gate when all boolean guards of a transaction are *False*.

The semantics of other data flow actors is same as that presented in [8].

```
PROCEDURE tryguard (t_l, g_l, c);
(* for executing the Tryguard actor the following information are assummed available at the PE
P_l, t_l, g_l : local process id, trans-id, and guard index;
c: communication details of the IO guard; *)
(* P_r, t_r, g_r refer to remote process-id, trans-id, and guard index *)
begin
    Store (P_r, t_l, g_l, c); (* store an entry in G-list (t_l) *)
    t_r := Active-trans (P_r);
    if t_r = 0 then
        skip;
    else
    begin
        g_r := Checkguard (g_l, c);
        if g_r ≠ 0 then
        begin
            t_min := min (t_l, t_r);
            t_max := max (t_l, t_r);
            flag := 0;
            while ((flag=0) AND (NOT Committed (t_min)) AND (NOT Committed (t_max))) do
                flag := fetch-and-increment (Excl (t_min))
                (* capture the transaction with a lower trans-id *)
            if (flag=1) AND (NOT Committed (t_max)) then
            begin release (t_min) := TRUE;
                flag := 0;
                while ((flag=0) AND (NOT Committed (t_max))) do
                    flag := fetch-and-increment (Excl (t_max));
                    (* capture the other transaction *)
                if (flag=1) then
                begin release (t_max) := TRUE;
                    Committed (t_min) := TRUE;
                    Committed (t_max) := TRUE;
                    output-token ((g_r,t_r), Split-Node (t_l));
                    (* output a token to the Split actor of t_l *)
                    output-token ((g_r,t_l), Split-Node (t_r));
                    (* output a token to the Split actor of t_r *)
                end
            end
        end
        if release (t_min) then Excl (t_min):= TRUE; (* release captured transactions, if any *)
        if release (t_max) then Excl (t_max):= TRUE;
    end
end ;
```

Fig. 4. Operational Semantics of *Tryguard* Actor

From Fig.2 and Fig.4, our implementation scheme can be understood as follows. The execution of an alternative command is started by sending a trigger token (see Fig.2) to the *Get-id* node in the data flow graph. The guards of a transaction t_l are attempted concurrently by executing the *Tryguard* actor. This results in writing an entry in *G-list*(t_l, P_r), where P_r is the remote process addressed by g_l. If there is an active transaction t_r in process P_r, then a search in *G-list* (t_r, P_l) is carried out for a matching compatible entry. The execution of the *Tryguard* actor terminates if this search is unsuccessful. Otherwise the PE executing the *Tryguard* actor tries to acquire exclusive access to the *Committed* flags of t_l and t_r in strict order of their *trans-ids*. If one of the *Committed* flags is *True* the execution of the *Tryguard* actor terminates, releasing captured transaction, if any. A rendezvous takes place when both *Committed* (t_l) and *Committed* (t_r) are *False* and have been successfully captured. The success in capturing the transactions results in outputting the triples $\langle g_l, g_r, t_r\rangle$ and $\langle g_r, g_l, t_l\rangle$ on the input arc of the *Split* actors in t_l and t_r respectively. When all the *Tryguard* actors of t_l fail to find a complement transaction willing to rendezvous, the transaction gets into a state similar to the 'waiting' state of [10]. Eventually, some other transaction will find t_l willing to rendezvous, and trigger t_l through a token to its *Split* actor. Following the data flow graph of Fig.2 it can be easily understood how the appropriate guarded statement gets executed. When all the boolean guards fail, the alternative commands results in a runtime error. The data flow actor *UND* is used for this.

It may be observed that more than one guard of a transaction can be executing the respective *Tryguard* actors in parallel; some of them may even succeed in finding their respective partners. However, only one guard can succeed in committing the rendezvous.

3.2 Repetitive and IO Commands

The above scheme can be extended to implement repetitive and simple IO commands. The data flow graph for a repetitive command is similar to the one shown in Fig.2, except for the recirculation of input arguments and trigger token at the end of execution of each iteration. The execution of a repetitive command terminates when all the guards in that command fails. A repetitive command is executed loop sequentially. It may be observed that the semantics of the repetitive command imposes such a loop sequential execution.

Finally, an input command $P_r?x$ generated by a process P_l is implemented by considering it as an alternative command

$$True;\ P_r?x \rightarrow skip.$$

Similarly, an output command $P_r!y$ is translated as:

$$True;\ P_r!y \rightarrow skip.$$

The reason for considering a simple IO command as an alternative command in the implementation is given below. Consider the situation in which a guard of some transaction has its complement as a simple IO command. This guard will never succeed, as the proposed implementation searches the matching compatible guards only in other alternative/repetitive commands. To take care of this situation, the above search (for matching and compatible guard) should be extended to simple IO commands as well. Instead, IO commands are considered as alternative commands to make the implementation simple and uniform.

4 Simulation

It is necessary to ensure that the proposed scheme correctly implements the semantics of guarded commands. This can be established by proving the following: during the potentially infinite execution, all processes (of the application program) and their interplay maintain the invariant properties, namely 'safety' and 'liveness' [10, 16, 19]. The first property safety means any rendezvous that occurs is correct. Liveness ensures two processes which should rendezvous, eventually will, provided either of them does not rendezvous with any other.

Besides safety and liveness, in general, it is desirable to support 'fairness' in implementations involving non-deterministic choices. In particular two kinds of fairness, *weak* and *strong* fairness, have been defined in literature [9, 10]. An implementation of guarded commands is *weakly* fair, if it can be guaranteed that during an infinitely repetitive execution, a guard that remains *continuously* available (*i.e.*, enabled and complement process ready to communicate) will eventually rendezvous. If it can be guaranteed that a guard which is available infinitely often (though not necessarily continuously) will eventually rendezvous, then the implementation is *strongly* fair.

In [12], it has been formally proved that our implementation maintains safety and liveness properties. However, proving fairness (either *strong* or *weak*) is a difficult task, basically due to the absence of an order of execution of the guards. The guards of a single process can be executed in any order, possibly concurrently. Hence no assumption can be made on their execution order, leaving no basis for proving fairness.

In this paper, we establish safety, liveness, and fairness properties empirically using a simulation approach. Another motivation for conducting simulation experiments is to obtain certain performance parameters, concerning efficiency and fairness of our implementation. In this section we highlight the details of the simulation and the results.

4.1 Simulation Details

As mentioned earlier we have chosen the multi-ring data flow machine [5]. The memory modules are connected to the PEs through a multi-stage interconnection network. For simplicity, the number of rings (or PEs) in the data flow machine is made the same as the number of interleaved memory modules. Further buffering of memory requests/responses between the stages of the network has been assumed to avoid contention in a switch. The multi-ring data flow machine is simulated using a discrete event simulation approach.

Constant service times have been assigned to the functional units. This is based on our knowledge of the time taken to process a token by these functional units [5]. Shared memory access cost involves buffering delay, if there is any contention in the interconnection switches, in addition to the logarithmic delay in the multistage network.

Input Parameters

The application program that we run on this simulator is a repetitive command involving m processes and n guards per process. The repetitive commands in the m processes are such that the I/O guards in one have matching and compatible guards in the complement processes. Also, all guards in all processes are enabled; that is, their boolean components always evaluate to *True*. The number of rings in the multi-ring data flow machine (or the number of interleaved memory module) can be varied and is an input parameter for the simulation.

Output Parameters

First, let us concentrate on proving safety, liveness, and fairness properties. We supplement this result with performance parameters which are measures for the efficiency and fairness of the implementation.

In the simulation run we record a log-file for the transactions that successfully rendezvous. Using the log-file, we ensure that each rendezvous is safe. To establish liveness, let only one pair (chosen randomly) of matching guards be available and check whether the rendezvous takes place within a finite time. The implementation is said to be live, if the rendezvous takes place no matter which pair of complement guards is enabled. Further, this should remain true irrespective of the variations in the number of rings. Lastly, to prove *(weak)* fairness of our implementation, we enable all the guards continuously and expect each of them to participate in a rendezvous, at least once, during a finite simulation time.

Apart from the three properties, we define the following six output parameters which are measures for the efficiency and fairness of the implementations.

(i) Average Tries: Average tries is an important performance parameter from efficiency viewpoint. It can be evaluated as the ratio of the sum of the number of guards tried (before a rendezvous) in each transaction to the total number of transactions. The value of average tries for our implementation could be between 2 (corresponding to one guard in each of the complement processes) and 2*n (where n is the number of guards per transaction). The value of average tries can be normalized by dividing it by 2*n.

(ii) Best Case Figure: As mentioned above, in the best case, a transaction can succeed by attempting just two guards, one in each of the complement processes. Best case figure is the ratio of the number of rendezvouses that took place with just 2 tries to the total number of transactions in the simulation time.

(iii) Worst Case Figure: A rendezvous that occurs after all guards in the complement transactions have been tried corresponds to a worst case occurrence. Counting the number of such rendezvouses and normalizing it by the total number of transactions gives the worst case figure.

(iv) Relative Merit Figure: In other implementations [4, 10, 18] an agreement is reached only after all the n guards of one of the transactions have been attempted. So it is meaningful to count the transactions where an agreement is reached by trying fewer than n guards in the complement processes. Averaging this value over the total number of transactions in the simulation period gives the relative merit figure.

(v) Guard Bias Figure: In a fair implementation each guard in a process is equally like to participate in a rendezvous involving that process. The number of times a guard has participated in a rendezvous, called the success count, is measured for each guard. Ideally, the success count for a guard should be equal to

$$\frac{2 * total\, no\, of\ transactions}{number\ of\ guards}$$

The standard deviation of the success counts of guards in a process is a measure of biasing. This figure, referred as the guard bias figure, indicates the extent to which a guard is biased/favored. This value should be very low for a fair implementation.

(vi) Process Bias Figure: The process bias figure is a measure to what extent a process has been biased/favored in all rendezvouses that took place during the simulation period. This figure is similar to the guard bias figure and can be obtained by computing the standard deviation of the success counts of various processes.

4.2 Results

The input program for the simulator was run several times with the following values of m m, the number of processes, and n, the number of guards per process: (i) $m = 5$ and $n = 8$, (ii) $m = 5$ and $n = 16$, and (iii) $m = 9$ and $n = 16$. All three properties, namely safety, liveness, and (weak) fairness have been checked for all of the above cases. The experiments are repeated on different configurations of the data flow machine having 1, 2, 4, 8, 16, or 32 rings. To measure the performance parameters, we ran the application programs long enough for at least 100 rendezvouses to take place.

Average Tries: Table 1 lists the normalized value of average tries (for a rendezvous) for various simulation runs. We observe that this value varies from 0.6 to 0.95 for the different runs. In particular, when the number of guards per process is smaller and the number of rings in the data flow machine is larger, the value of average tries is high. This can be reasoned as follows. When there is low parallelism (*i.e.*, less number of guards to try) and high resource availability, the chances for more number of guards to be tried are high. Further, we have assumed (i) the alternative commands in all the processes are initiated at the same time, and (ii) it takes equal execution time for the all boolean guards. These two assumptions further add to the possibility that the *Tryguard* actors of a process can be executed simultaneously, accounting for the large value of normalized average tries.

Table 1. Normalized Average Tries

Input Programs with		Normalized Average Tries					
		Number of Rings					
Processes	Guards	1	2	4	8	16	32
5	8	0.694	0.690	0.727	0.795	0.822	0.833
5	16	0.600	0.606	0.633	0.662	0.709	0.701
9	16	0.666	0.669	0.639	0.689	0.755	0.750

Best Case, Worst Case, and Relative Merit Figures: We do not observe the occurrence of the best case. As the guards of a process are attempted concurrently and as the *Tryguard* actors of a process are enabled more or less at the same time, it is more likely that at least two guards in a process are tried simultaneously, forbidding the best case occurrence. The worst case and relative merit figures for the various runs are tabulated in Table 2. The worst case figure drops steadily as the parallelism increases. When the parallelism is high, the worst case virtually never occurs. Also, the relative merit figure shows an increasing trend with an increase in the parallelism.

Table 2. Worst Case and Relative Merit Figures

Input Programs with		Parameters	Performance Parameters					
			Number of Rings					
Processes	Guards		1	2	4	8	16	32
			(in %)					
5	8	Worst Case	0.0	0.0	0.0	7.4	10.6	9.7
		Relative Merit	0.0	0.0	6.8	1.6	3.8	0.0
5	16	Worst Case	0.0	0.0	0.0	0.0	0.0	0.0
		Relative Merit	0.0	0.0	6.7	4.9	0.0	0.0
9	16	Worst Case	0.0	0.0	0.0	0.0	4.4	0.9
		Relative Merit	0.0	12.5	21.9	10.0	0.0	0.0

Fairness Measures: The guard bias figures (in a typical process) and the process bias figures for various simulation runs are shown in Table 3. We notice these values are very low, at most 4 transactions out of nearly 100 transactions. Such a low degree of biasing is acceptable to an implementation involving non-determinism.

Table 3. Fairness Measures

Input Programs with			Bias Figures					
			Number of Rings					
Processes	Guards		1	2	4	8	16	32
			(in number of transactions)					
5	8	Guard	1.14	0.97	1.68	2.19	1.93	2.63
		Process	1.67	1.20	1.36	0.97	2.56	2.65
5	16	Guard	0.76	0.78	1.34	1.37	1.74	1.71
		Process	1.35	0.74	2.75	1.16	1.54	3.84
9	16	Guard	0.41	0.64	0.80	1.46	1.21	1.60
		Process	0.49	0.56	1.03	1.07	0.66	1.52

Variable Execution Time for Boolean Guards and Guarded Statements:

As mentioned earlier, in the earlier simulation runs an equal execution time is assumed for all boolean guards and guarded statements (statements following the guards). To study the effect of variable execution times (of boolean guards and guarded statements) on the performance parameters, we use a uniformly distributed execution time. The range of the uniform distribution is from 0 time units to 100 time units. Due to this difference, the *Tryguard* actors may be activated at different time. This is expected to reduce the average tries. The performance results indeed show a decrease in the average tries and a significant increase in the relative merit figure (refer Table 4). The results are for the repetitive command involving 5 processes and 16 guards.

Table 4. Results for Variable Execution Time

Number of Rings	Variable Execution Time for			
	Boolean Guards		Guarded Statements	
	Normalized Average Tries	Relative Merit	Normalized Average Tries	Relative Merit
1	0.60	0.0%	0.58	0.0%
2	0.57	11.1%	0.57	29.1%
4	0.56	33.3%	0.52	59.1%
8	0.54	35.8%	0.51	65.7%
16	0.63	12.6%	0.59	75.8%
32	0.60	23.7%	0.63	67.3%

5 Related Works and Comparisons

Our implementation, for the first time, uses data flow model of computation. As mentioned earlier, data-driven computation facilitates fine grain asynchronous parallelism. In our implementation parallelism exploited is not only among processes but also within a process. A marked difference from other models is the concurrent execution of guards in a process. Further, processors do not busy wait during synchronization.

5.1 Shared Memory Systems

The work reported in [10] is based on shared memory model and resembles our implementation in some aspects. However, our implementation performs better than the former in the following two ways.

In [10], a rendezvous between two transactions takes place only after one of them enters the 'waiting' state. This means that all guards of one transaction have to be attempted before the rendezvous. That is, if there are n guards in each process, then at least $(n+1)$ must be tried for a rendezvous. $2 * n$ tries wil be made in the worst case. The average case results are not reported in [10]. Theoretically, the best and worst case figures for our implementation are 2 and $2 * n$ respectively. Also, the simulation results predict an average of $(0.6) * (2 * n)$ guards are tried for a rendezvous. The relative merit figure shows that our implementation indeed performs better than [10] in many cases.

It is important to realize that the above comparisons are based only on the number of tries and not on the execution time. Given that the guards of a process are attempted concurrently in our implementation, it easily outperforms the other [10] in terms of execution time. The improvement in performance is not without its price. In our implementation, we store the guard index and communication detail for each attempted guard in the *G-list* (t_l); whereas, only the guard details are stored in the *Alt-List* in [10]. However, with the developments in VLSI technology, it can be argued that an improvement in execution speed at the expense of storage is affordable.

5.2 Message Passing Systems

The implementations presented in [2, 3, 4, 7, 18] use message passing architecture. Some of them [2, 7] are based on a two phase algorithm. Our implementation, like the ones presented in [4, 10, 18], involves only one phase. There is no need for a second phase as every *Tryguard* actor terminates with a definite answer. That is, non-committal replies and retries do not occur in our implementation.

It is unfair to evaluate our implementation directly in the light of the six criteria listed in [2, 18] as these are essentially for a loosely-coupled multiprocessor system. However our implementation retains the spirit of these criteria. For instance, five of these six criteria[3] can be appropriately redefined as: (i) the amount of system information stored (in the shared memory) should be minimal; (ii) when both t_l and t_r are ready to select the guards g_l and g_r, respectively, then at least one of the transactions must select a guard (not necessarily g_l or g_r) within a finite time; (iii) the number of attempts made in selecting a guard of a transaction should be bounded; (iv) the time taken by a *Tryguard* actor to determine whether it can establish a rendezvous must be finite; and (v) if a process has a guarded command that is enabled continuously for an infinite time, then it should eventually succeed. It can be easily proved that our implementation satisfies these criteria. In fact it performs better than others [2, 3, 4, 7, 10, 18] with respect to (ii) and (iii), as parallelizing the selection of guards significantly reduces the execution time.

6 Conclusions

In this paper, we have presented a decentralized parallel implementation for the generalized guarded commands of CSP using data flow model of computation. A simulation study has been conducted to measure various performance and fairness parameters. Also, simulation experiments establish safety, liveness, and fairness of our implementation. Finally, a comparison with the existing ones reveals the superiority of our model. We have

[3]Criterion (i) of [18] is not relevant to the discussion.

not addressed the issue of process termination. However, the implementation can be easily extended to take care of process termination.

Acknowledgments

The authors acknowledge Dr.R.A. Nicholl and Dr.V.S.Lakshmanan for their useful comments. The authors are thankful to Bhama for her comments on the initial draft of this paper.

References

[1] Arvind and Gostelow, K.P., "The U Interpreter", *IEEE Computer*, vol.15, no.2, pp.42-49, Feb. 1982.

[2] Back, R.J.R, Ekslund, P., and Kurki-Suonia, R., "A Fair and Efficient Implementation of CSP with Output Guards", Technical Report, Ser. A, No. 38, Abo Akademic, Finland, 1984.

[3] Bagrodia, R., "A Distributed Algorithm to Implement the Generalized Alternative Command in CSP", in *Proc. of the 6th International Conference on Distributed Computing Systems*, pp. 422-427, 1986.

[4] Bagrodia, R., "Synchronization of Asynchronous Processes in CSP", *ACM Transactions on Programming Languages and Systems*, vol.11, no.4, pp.585-597, Oct.1989.

[5] Barahona, P.M.C.C., and Gurd, J.R., "Processor Allocation in a Multi-ring Data Flow Machine," *Journal of Parallel and Distributed Computing*, vol.3, no.3, pp.305-327, 1986.

[6] Bernstein, A.J., "Output Guards and Nondeterminism in Communicating Sequential Processes", *ACM Transactions on Programming Language and Systems*, vol.2, no.2, pp.234-238, April 1980.

[7] Buckley, G.N. and Silberschatz, A., "An Effective Implementation for the Generalized Input-Output Construct of CSP", *ACM Transactions on Programming Languages and Systems*", vol. 5, no. 2, pp. 223-235, 1983.

[8] Davis, A.L. and Keller, R.M., "Data Flow Program Graphs", *IEEE Computer*, vol.15, no.2, pp.26-41, Feb. 1982.

[9] Francez, N., *Fairness*, Springer-Verlag, New York, 1986.

[10] Fujimoto, R.N. and Hwa-chung Feng, "A Shared Memory Algorithm and Proof for the Generalized Alternative Construct in CSP", *International Journal of Parallel Programming*, vol. 16, no. 3, pp. 215-241, 1987.

[11] Gottlieb, A., Grishman, R., Kruskal, C.P., McAuliffe, Rudolph, L., and Snir, M., "The NYU Ultracomputer - Designing an MIMD Shared Memory Parallel Computer", *IEEE Transactions on Computers*, vol.C-32, no.2, pp.175-189, Feb. 1983.

[12] Govindarajan, R. and Yu. S, "Attempting Guards in Parallel: A Data Flow Approach to Execute Generalized Guarded Commands", Technical Report # 273, Department of Computer Science, University of Western Ontario, London, May 1990.

[13] Gurd, J.R., Watson, I., and Kirkham, C.C., "The Manchester Prototype Data Flow Computer", *Communications of the ACM*, vol.28, no.1, pp.34-52, Jan.1985.

[14] Hoare, C.A.R., "Communicating Sequential Processes", *Communications of the ACM*, vol. 21, no. 8, pp. 666-677, Aug. 1978.

[15] Kieburtz, R.B. and Silberschatz, A., "Comments on Communicating Sequential Processes", *ACM Transactions on Programming Language and Systems*, vol.1, no.2, pp.218-225, Oct. 1979.

[16] Owicki, S. and Lamport, L., "Proving Liveness Properties of Concurrent Programs", *ACM Transactions on Programming Languages and Systems*, vol. 6, no. 2, pp. 455-495, July 1982.

[17] Patnaik, L.M. and Basu, J., "Two Tools for Interprocess Communication in Distributed Data Flow Systems", *The Computer Journal*, vol. 29, no. 6, pp. 506-521, Dec. 1986.

[18] Ramesh, S., "A New Implementation of CSP with Output Guards" in *Proc. of the 7th International Conference on Distributed Computing Systems*, pp. 266-273, 1987.

[19] Reed, D.A., Malony, A.D., and McCredie, B.D., "Parallel Discrete Event Simulation: A Shared Memory Approach", in *Proc. of the ACM SIGMETRIC S Conference on Measuring and Modeling Computer Systems*, vol. 15, no.1, pp. 36-38, May 1987.

[20] Silberschatz, A., "Communication and Synchronization in Distributed Systems" *IEEE Transactions on Software Engineering*, vol.SE-5, no. 6, pp.542-546, Nov.1979.

[21] Treleaven, P.C., Brownbridge, D.R., and Hopkins, R.P., "Data-Driven and Demand-Driven Architecture", *Computing Surveys*, vol. 14, no. 1, pp. 93-143, Mar. 1982.

ON THE DESIGN OF DEADLOCK-FREE ADAPTIVE ROUTING ALGORITHMS FOR MULTICOMPUTERS: DESIGN METHODOLOGIES

J. Duato
Dept. de Ingeniería de Sistemas, Computadores y Automática
Facultad de Informática. Universidad Politécnica de Valencia
P.O.B. 22012. 46071 - Valencia, Spain

Abstract

Second generation multicomputers use wormhole routing, allowing a very low channel set-up time and drastically reducing the dependency between network latency and internode distance. Deadlock-free routing strategies have been developed, allowing the implementation of fast hardware routers that reduce the communication bottleneck. Also, adaptive routing algorithms with deadlock-avoidance or deadlock-recovery techniques have been proposed for some topologies, being very effective and outperforming static strategies.

This paper develops the theoretical background for the design of deadlock-free adaptive routing algorithms for wormhole as well as store-and-forward routing. Some basic definitions and four theorems are proposed, developing conditions to verify that an adaptive algorithm is deadlock-free, even when there are cycles in the channel dependency graph. Also, two design methodologies are proposed. The first one supplies algorithms with a high degree of freedom, without increasing the number of physical channels. The second methodology is intended for the design of fault-tolerant algorithms. Some examples are given, showing the application of the methodologies.

1. Introduction

Multicomputers [1] rely on an interconnection network between processors to support the message-passing mechanism. The network latency [1] can be defined as the time from when the head of a message enters the network at the source until the tail emerges at the destination. In first generation multicomputers, a store-and-forward mechanism has been used to route messages. Each time a message reaches a node, it is buffered in local memory, and the processor interrupted to execute the routing algorithm. Accordingly, the network latency is proportional to the distance between the origin and the destination.

However, second generation multicomputers are most distinguished by message routing hardware that makes the topology of the message-passing network practically invisible to the programmer. The message routing hardware uses a routing mechanism known as

This work is partly supported by CICYT grant number TIC 87-0655

wormhole routing [9]. As messages are typically at least a few words long, each message is serialized into a sequence of parallel data units, referred to as flow control units, or flits [8]. The flit at the head of a message governs the route. As the header flit advances along the specified route, the remaining flits follow it in a pipeline fashion. If the header encounters a channel already in use, it is blocked until the channel is freed; the flow control within the network blocks the trailing flits.

This form of routing and flow control has two important advantages over the store-and-forward packet routing used in first generation multicomputers. Firstly, it avoids using storage bandwidth in the nodes through which messages are routed. Secondly, this routing technique makes the message latency largely insensitive to the distance in the message-passing network. Since the flits move through the network in a pipeline fashion, in the absence of channel contention, the network latency equals the sum of two terms:

- T_pD is the time associated with forming the path through the network, where T_p is the delay of the individual routing nodes found on the path, and D is the number of nodes traversed.

- L/B is the time required for a message of length L to pass through a channel of bandwidth B.

In second generation multicomputers, the network latency is dominated by the second term for all but very short messages.

Another improvement in message performance results from selecting the optimal topology for the implementation on printed circuit boards or VLSI chips. As message latency is dominated by the term L/B, more wirable network topologies will increase the bandwidth B at the expense of increasing the network diameter. An analysis [5,7] shows that, under the assumption of constant number of wires through the network bisection, a two dimensional network minimizes latency for typical message lengths for up to 1024 nodes. For larger sizes, a three dimensional network achieves better performance. Among these networks, meshes are preferred because they offer useful edge connectivity, which can be used for I/O controllers. Also, meshes partition into units that are still meshes, simplifying the design of routing algorithms that are independent of the network size, as well as the implementation of space-sharing techniques.

However, deadlocks may appear if the routing algorithms are not carefully designed. A deadlock in the interconnection network of a multicomputer occurs when no message can advance toward its destination because the queues of the message system are full. The size of the queues strongly influences the probability to reach a deadlocked configuration. First generation multicomputers buffer full messages or relatively large packets. By contrary, second generation machines buffer flits, being more deadlock-prone. So, the only practical way to avoid deadlock is to design deadlock-free routing algorithms.

Many deadlock-free routing algorithms have been developed for store-and-forward computer networks [12,14,20]. These algorithms are based on a structured buffer pool. However, with wormhole routing, buffer allocation cannot be restricted, because flits have no routing information. Once the header of a message has been accepted by a channel, the remaining flits must be accepted before the flits of any other message can be accepted. So, routing must be restricted to avoid deadlock.

Dally [9] has proposed a methodology to design static routing algorithms under general assumptions. He defines a channel dependency graph and establishes a total order among channels. Routing is restricted to visit channels in decreasing or increasing order to eliminate cycles in the channel dependency graph. This methodology has been applied to the design of routing chips for multicomputers [8] and multicomputer nodes with integrated communication support [2]. It has also been applied to systolic communication [19,2].

The restriction of routing, although avoids deadlock, can increase traffic jams, specially in heavily loaded networks with long messages. In order to avoid congested regions of the network, an adaptive routing algorithm can be used. Adaptive strategies have been shown to outperform static strategies in store-and-forward routing [3] and in packet-switched communications [18,21]. In general, adaptive routing needs additional hardware support.

Several adaptive algorithms have been developed for wormhole routing. A deadlock-free adaptive algorithm for the hypercube is the Hyperswitch algorithm [4], which is based on backtracking and hardware modification of message headers to avoid congestion and cycles. Another deadlock-free adaptive algorithm has been proposed for the MEGA [13]. This algorithm always routes messages, sending them away from their destination if necessary, like the Connection Machine [15]. If a message arrives to a node without free output channels, deadlocks are avoided by storing the message and removing it from the network. In this respect, it is similar to virtual cut-through [17]. Jesshope [16] has proposed an algorithm for n dimensional meshes, by decomposing them into 2n virtual networks. Inside each virtual network, displacements along a given dimension are always made in the same direction, thus avoiding cycles and deadlock.

An alternative way consists of recovering from deadlock. Reeves et al. [22] have used an abort-and-retry technique to remove messages blocked for longer than a certain threshold from the network. Aborted messages are introduced again into the network after a random delay. In [22] three adaptive routing strategies have been proposed and evaluated for a binary 8-cube.

In [10] we have proposed a very simple methodology to design deadlock-free adaptive routing algorithms for wormhole networks. The routing algorithms obtained from the application of that methodology to 2D and 3D-meshes have been evaluated by simulation.

Also, in [11] we have presented a first version of a theory for the design of deadlock-free adaptive routing algorithms. Some basic definitions and three theorems have been proposed, as well as a design example. However, that theory is only valid for single flit messages, contrary to one of the assumptions stated in the paper, being useful for store-and-forward routing.

This paper develops the theoretical background for the design of deadlock-free adaptive routing algorithms. For wormhole routing, some basic definitions and two theorems are proposed and proved, developing conditions to verify that an adaptive algorithm is deadlock-free, even when there are cycles in the channel dependency graph. For store-and-forward routing, the theory presented in [11] is updated. Also, two design methodologies based on the above mentioned theorems are proposed. The first one supplies adaptive algorithms with a high degree of freedom, without increasing the number of physical channels. The second methodology is intended for the design of fault-tolerant algorithms.

Section 2 develops the new theory for wormhole routing. Section 3 summarizes the theoretical aspects for store-and-forward routing. Section 4 proposes two design methodologies, giving some examples of their application. Finally, some conclusions are drawn.

2. Definitions and theorems for wormhole routing

This section develops the theoretical background for the design of deadlock-free adaptive routing algorithms for networks using wormhole routing.

The basic assumptions are very similar to the ones proposed by Dally [9], except that adaptive routing is allowed. These assumptions are the following:

1) A node can generate messages destined for any other node at any rate.

2) A message arriving at its destination node is eventually consumed.

3) Wormhole routing is used. So, once a queue accepts the first flit of a message, it must accept the remainder of the message before accepting any flits from another message.

4) A node can generate messages of arbitrary length. Packets will generally be longer than a single flit.

5) An available queue may arbitrate between messages that request that queue, but may not choose among waiting messages.

6) A queue cannot contain flits belonging to different messages or packets. After accepting a tail flit, a queue must be emptied before accepting another header flit. Then, when a message or packet is blocked, its header flit will always occupy the head of a queue.

7) The route taken by a message depends on its destination and the status of output channels (free or busy). At a given node, the routing function supplies a set of output channels based on the current and destination nodes. A selection from this set is made based on the status of output channels at the current node. So, *adaptive* routing will be considered.

Before to propose the theorems, some definitions are needed:

Definition 1: An *interconnection network* I is a strongly connected directed multigraph, I = G(N, C). The vertices of the multigraph N represent the set of processing nodes. The edges of the multigraph C represent the set of communication channels. More than a single channel is allowed to connect a given pair of nodes. Each channel c_i has an associated queue denoted queue(c_i) with capacity cap(c_i). The source and destination nodes of channel c_i are denoted s_i and d_i, respectively.

Definition 2: Let B be the set of valid *channel status,* B = {free, busy}. Let T: C $\rightarrow$ B be the status of the output channels in the network.

Definition 3: An *adaptive routing function* R: N x N $\rightarrow$ C^p supplies a set of p alternative output channels to send a message from the current node n_c to the destination node n_d, $R(n_c, n_d) = \{c_1, c_2, \dots c_p\}$. In general, p will be less than the number of output channels per node to restrict routing and obtain deadlock-free algorithms. As a particular case, p = 1 defines a static routing function. Also, the channels in the set supplied by R are not necessarily different. So, p is the maximum number of choices. In particular,

$R(n, n) = \emptyset, \quad \forall\, n \in N.$

Definition 4: A *selection function* $S: C^p \times B^p \rightarrow C$ selects a free output channel (if any) from the set supplied by the routing function. From the definition, S takes into account the status of all the channels belonging to the set supplied by the routing function. The selection can be random or based on static or dynamic priorities. Also, in the same way the result of a static routing function may be a busy channel, if all the output channels are busy, any of them is selected. The decomposition of the adaptive routing into two functions (routing and selection) will be critical while proving the theorems, because only the routing function determines whether a routing algorithm is deadlock-free or not. Then, the selection function will only affect performance. Moreover, it is possible to extend the definition of the selection function by taking into account additional information, either local to the node or remote. We will comment on this in section 4.

Definition 5: A routing function R for a given interconnection network I is *connected* iff

$$\forall\, i, j \in N \quad i \neq j, \quad \exists\, c_1, c_2, \ldots c_k \in C \quad \ni$$

$$c_1 \in R(i, j) \quad \wedge \quad c_{m+1} \in R(d_m, j) \quad \forall\, m \in \{1, k\text{-}1\} \quad \wedge \quad d_k = j$$

In other words, it is possible to establish a path between i and j using channels belonging to the sets supplied by R. Notice that the interconnection network is strongly connected, but it does not imply that the routing function must be connected.

Definition 6: A *routing subfunction* R_1 for a given routing function R and channel subset $C_1 \subseteq C$, is a routing function

$$R_1: N \times N \rightarrow C_1^q, \quad 0 < q \leq p \ \ni$$

$$R_1(i, j) = R(i, j) \cap C_1 \quad \forall\, i, j \in N$$

Definition 7: Given an interconnection network I, a routing function R and a pair of channels $c_i, c_j \in C$, there is a *direct dependency* from c_i to c_j iff

$$c_i \in R(s_i, n) \quad \wedge \quad c_j \in R(d_i, n) \quad \text{for some } n \in N$$

that is, c_j can be used immediately after c_i by messages destined to some node n.

Definition 8: Given an interconnection network I, a routing function R, a channel subset $C_1 \subset C$ which defines a routing subfunction R_1 and a pair of channels $c_i, c_j \in C_1$, there is an *indirect dependency* from c_i to c_j iff

$$\exists\, c_1, c_2, \ldots c_k \in C - C_1 \quad \ni$$

$$c_i \in R_1(s_i, n) \quad \wedge \quad c_1 \in R(d_i, n) \quad \wedge$$

$$c_{m+1} \in R(d_m, n) \quad \forall\, m \in \{1, k\text{-}1\} \quad \wedge$$

$$d_k = s_j \quad \wedge \quad c_j \in R_1(s_j, n) \quad \text{for some } n \in N$$

that is, it is possible to establish a path from s_i to d_j for messages destined to some node n. c_i and c_j are the first and last channels in that path and the only ones belonging to C_1. Then, c_j can be used after c_i by some messages. As c_i and c_j are not adjacent, some other channels belonging to $C - C_1$ are used between them. It must be noticed that, given three channels $c_i, c_k \in C_1$ and $c_j \in C - C_1$, the existence of direct dependencies between c_i, c_j and c_j, c_k, respectively, does not imply the existence of an indirect dependency between c_i, c_k.

Definition 9: A *channel dependency graph* D for a given interconnection network I and routing function R, is a directed graph, D = G(C, E). The vertices of D are the channels of I. The edges of D are the pairs of channels (c_i, c_j) such that there is a direct dependency from c_i to c_j. Notice that there are no 1-cycles in D, because channels are unidirectional.

Definition 10: An *extended channel dependency graph* D_E for a given interconnection network I and routing subfunction R_1 of a routing function R, is a directed graph, $D_E = G(C_1, E_E)$. The vertices of D_E are the channels that define the routing subfunction R_1. The edges of D_E are the pairs of channels (c_i, c_j) such that there is either a direct or an indirect dependency from c_i to c_j.

Definition 11: A *sink* channel for a given interconnection network I and routing function R is a channel $c_i \ni$

$$\forall j \in N, \quad c_i \in R(s_i, j) \quad \Rightarrow \quad j = d_i$$

In other words, all the flits that enter a sink channel reach their destination in a single hop. As a result, there are no outgoing arcs from a sink channel in any channel dependency graph, as can be easily seen from the definitions.

Definition 12: A *configuration* is an assignment of a set of flits to each queue, all of them belonging to the same message or packet (assumption 6). The number of flits in the queue for channel c_i will be denoted size(c_i). The destination node for a flit f_j will be denoted dest(f_j). If the first flit in the queue for channel c_i is a header flit destined for node n_d, then head(c_i) = n_d. If the first flit is not a header and the next channel reserved by its header is c_j, then next(c_i) = c_j, that is, each flit must follow the same path as its header. A configuration is *legal* iff

$$\forall c_i \in C, \quad \text{size}(c_i) \leq \text{cap}(c_i) \quad \wedge$$

$$c_i \in R(s_i, \text{dest}(f_j)) \quad \forall f_j \in \text{queue}(c_i)$$

that is, the queue capacity is not exceeded and all the flits stored in the queue have been sent there by the routing function.

Definition 13: A *deadlocked configuration* for a given interconnection network I and routing function R is a nonempty legal configuration verifying the following conditions:

1) $\forall c_i \in C \ni \text{head}(c_i) \in N \quad \Rightarrow \quad \text{head}(c_i) \neq d_i \wedge \text{size}(c_j) > 0 \quad \forall c_j \in R(d_i, \text{head}(c_i))$

2) $\forall c_i \in C \ni \text{next}(c_i) \in C \quad \Rightarrow \quad \text{size}(\text{next}(c_i)) = \text{cap}(\text{next}(c_i))$

The first condition refers to channels with a header flit at their queue head (head(c_i) $\in$ N). The second condition refers to channels containing a data or tail flit at their queue head, not destined to d_i (next(c_i) $\in$ C). No condition is imposed to empty channels. In a deadlocked configuration there is not any header flit one hop from its destination. Header flits cannot advance because the queues for all the alternative output channels supplied by the routing function are not empty (see assumption 6). As a particular case (for disconnected routing functions), the routing function may not supply any output channel. Data and tail flits cannot advance because the next channel reserved by their message header has a full queue. It must be noticed that a data flit can be blocked at a node even if there are free output channels to reach its destination. Also, in a deadlocked configuration, there is no message whose header flit has already arrived to its destination.

Definition 14: A routing function R for an interconnection network I is *deadlock-free* iff there is not any deadlocked configuration for that routing function on that network.

Two theorems are proposed. The first one is a straightforward extension of Dally's theorem for adaptive routing functions. The second one allows the design of adaptive routing functions with cyclic dependencies in their channel dependency graph. For each theorem, a sketch of the proof as well as the full proof are given.

Theorem 1: A connected and adaptive routing function R for an interconnection network I is deadlock-free if there are no cycles in the channel dependency graph D.

Proof sketch:

$\Leftarrow$ As the channel dependency graph for R is acyclic, it is possible to establish an order between the channels of C. As R is connected, the minimals of that order are also sinks. Suppose that there is a deadlocked configuration for R. Let c_i be a channel of C with a nonempty queue such that there are no channels less than c_i with a nonempty queue. If c_i is a minimal (that is, a sink) then the flits are not blocked and there is no deadlock. Otherwise, using the channels less than c_i, the flit at the queue head of c_i can advance and there is not a deadlock.•

Proof:

$\Leftarrow$ Suppose that there are no cycles in D. Then, one can assign an order to the channels of C so that if $(c_i, c_j) \in E$ then $c_i > c_j$. Consider the channel(s) $c_i \ni$

$$\forall c_j \in C, \quad (c_i, c_j) \notin E$$

Such a channel c_i is a minimal of the order. Let us prove that it is a sink. If it were not a sink, as the routing function is connected, for any legal configuration with a header flit stored in the queue head of c_i

$$d_i \neq \text{head}(c_i) \quad \Rightarrow \quad \exists\, c_k \in C \quad \ni \quad c_k \in R(d_i, \text{head}(c_i))$$

As the configuration is legal then

$$c_i \in R(s_i, \text{head}(c_i)) \quad \Rightarrow \quad (c_i, c_k) \in E$$

contrary to the assumption that c_i is a minimal. So, $d_i = \text{head}(c_i)$ and c_i is a sink of D.

Suppose that there is a deadlocked configuration for R. Let c_i be a channel with a nonempty queue such that there is not any channel less than c_i with a nonempty queue. If c_i is a minimal, it is also a sink and then, all the flits stored in its queue will be destined to d_i and the flit at the head of the queue for c_i is not blocked. If c_i is not a minimal then

$$\text{size}(c_j) = 0 \quad \forall\, c_j \in C \quad \ni \quad c_i > c_j$$

Thus, the flit at the head of the queue for c_i is not blocked, regardless it is a header or a data flit, and there is no deadlock.•

There are some interesting considerations:

1) The theorem gives a sufficient but not necessary condition for an adaptive routing function to be deadlock-free. As will be seen later, the existence of cycles in the channel dependency graph does not imply the existence of a deadlocked configuration.

2) For most networks and routing functions, even for static ones, only a partial order

between channels can be defined, based on the set E. In general, there will be more than a single sink in D.

3) As indicated above, in a legal configuration all the flits stored in a given queue have been sent there by the routing function. Otherwise, the theorem cannot be proved. Consider, for instance, a configuration in which the queues of all the sink channels in D are full of flits destined to nodes not directly connected to those channels.

Theorem 2: A connected and adaptive routing function R for an interconnection network I is deadlock-free if it exists a subset of channels $C_1 \subseteq C$ that defines a routing subfunction R_1 which is connected and has no cycles in its extended channel dependency graph D_E.

Proof sketch:

$\Leftarrow$ The case $C_1 = C$ is trivial. Otherwise $C_1 \subset C$. As the extended channel dependency graph for R_1 is acyclic, it is possible to establish an order between the channels of C_1. As R_1 is connected, the minimals of that order are also sinks. Suppose that there is a deadlocked configuration for R. There are two possible cases:

a) The queues for channels belonging to C_1 are empty. As R_1 is connected and the header flits are at queue heads, they can be routed using channels belonging to C_1 and there is no deadlock.

b) The queues for channels belonging to C_1 are not empty. Let c_i be a channel of C_1 with a nonempty queue such that there are no channels less than c_i with a nonempty queue. Again, there are two possible cases:

b1) If c_i is a minimal (sink) then the flits are not blocked and there is no deadlock.

b2) If c_i is not a minimal, all the channels of C_1 less than c_i will have empty queues, existing three possible cases:

b2.1) If c_i has a header at the queue head, it can be routed because R_1 is connected and there is no deadlock.

b2.2) If there is a data flit at the queue head of c_i and next(c_i) belongs to C_1, that flit can also advance.

b2.3) If next(c_i) belongs to $C - C_1$, we have to use the indirect dependencies in the extended channel dependency graph. Let c_k be the channel containing the header of the data flits contained in c_i. Then, it is possible to find a channel c_j belonging to C_1 to route that header, because R_1 is connected. In that case, there is an indirect dependency from c_i to c_j ($c_i > c_j$), implying that c_j is empty and there is no deadlock.•

Proof:

$\Leftarrow$ Suppose that R_1 is connected and there are no cycles in D_E. If $C_1 = C$ then $D_E = D$, because $C - C_1 = \emptyset$. Thus, there is not any cycle in D and R is deadlock-free by theorem 1. Otherwise $C_1 \subset C$. As there are no cycles in D_E, one can assign an order to the channels of C_1 so that if $(c_i, c_j) \in E_E$ then $c_i > c_j$. Similarly to theorem 1, it can be proved that the minimals of that order are also sinks.

Suppose that there is a deadlocked configuration for R. There are two possible cases:

a) The queues for channels belonging to C_1 are empty. Then, there will be channels belonging to $C - C_1$ with header flits at their queue heads. Let c_i be one of those channels. As R_1 is connected then

$$head(c_i) \neq d_i \Rightarrow \exists\, c_j \in C_1 \ni c_j \in R_1(d_i, head(c_i)) \Rightarrow$$

$$\exists\, c_j \in C \ni c_j \in R(d_i, head(c_i))$$

Also $size(c_j) = 0$ and R does not have a deadlock.

b) The queues for channels belonging to C_1 are not empty. Let c_i be a channel belonging to C_1 with a nonempty queue such that there are no channels less than c_i with a nonempty queue. Again, there are two possible cases:

b1) c_i is a minimal. As shown above, it is also a sink and then, all the flits stored in its queue will be destined to d_i and the flit at the head of the queue for c_i is not blocked.

b2) c_i is not a minimal. Then

$$size(c_j) = 0 \quad \forall\, c_j \in C_1 \ni c_i > c_j$$

existing three possible cases:

b2.1) c_i has a header at the queue head. Taking into account that R_1 is connected

$$head(c_i) \neq d_i \Rightarrow \exists\, c_j \in C_1 \ni c_j \in R_1(d_i, head(c_i)) \Rightarrow$$

$$\exists\, c_j \in C \ni c_j \in R(d_i, head(c_i))$$

Also

$$c_i > c_j \Rightarrow size(c_j) = 0$$

and R does not have a deadlock.

b2.2) c_i has a data flit at the queue head, not destined to d_i and $next(c_i)$ belongs to C_1. Then

$$c_i > next(c_i) \Rightarrow size(next(c_i)) = 0$$

and R does not have a deadlock.

b2.3) c_i has a data flit at the queue head, not destined to d_i and $next(c_i)$ belongs to $C - C_1$. Let $c_1, c_2, \ldots c_k \in C - C_1$ be the set of channels reserved by the message after reserving c_i, c_k containing the message header. Those channels belong to $C - C_1$, because there are no channels less than c_i with a nonempty queue.

$$\exists\, c_1, c_2, \ldots c_k \in C - C_1 \ni$$

$$c_i \in R_1(s_i, head(c_k)) \wedge c_1 \in R(d_i, head(c_k)) \wedge$$

$$c_{m+1} \in R(d_m, head(c_k)) \quad \forall\, m \in \{1, k-1\}$$

As R_1 is connected

$$head(c_k) \neq d_k \Rightarrow \exists\, c_j \in C_1 \ni c_j \in R_1(d_k, head(c_k)) \Rightarrow$$

$$\exists\, c_j \in C \ni c_j \in R(d_k, head(c_k))$$

Thus, there is an indirect dependency from c_i to c_j ($c_i > c_j$), implying that $size(c_j) = 0$. Then, the header at the queue head for c_k is not blocked and R does not have a deadlock.•

Again, there are some interesting considerations:

1) The basic idea behind theorem 2 is that one can have an adaptive routing function with cyclic dependencies between channels, provided that there are alternative paths without cyclic dependencies to send a given flit towards its destination. As messages are several flits long, the extended channel dependency graph must be used to take into account the indirect dependencies.

2) If it were not necessary emptying a queue before accepting the header of another message, then it would be no guarantee that header flits occupy the queue heads and the theorem would not be valid. Consider, for instance, a set of three or more channels with cyclic dependencies between them and a configuration in which the queues of those channels are full, each one containing the tail of a message followed by a fragment of another message destined two nodes away. The rest of that message occupies part of the next channel queue and so on. That configuration is deadlocked because the header flits do not occupy the queue heads and cannot be routed using the alternative paths offered by the routing function.

3) If the routing function were defined as $R: C \times N \rightarrow C^p$, then the theorem would not be valid. Consider, for instance, two subsets of C, namely, C_1 and $C - C_1$, and a routing function defined in such a way that all the messages arriving to a given node through a channel belonging to $C - C_1$ are routed through a channel belonging to the same subset. Suppose that there are cyclic dependencies between the channels belonging to $C - C_1$ and that C_1 defines a routing subfunction which is connected and has no cycles in its channel dependency graph. That routing function is not deadlock-free.

4) The routing subfunction R_1 is not necessarily static. It can be adaptive.

3. Definitions and theorems for store-and-forward routing

This section develops the theoretical background for the design of deadlock-free adaptive routing algorithms for networks using store-and-forward routing. This theory is almost directly derived from the theorems proposed in section 2 for the particular case of messages consisting of a single flit. Thus, this theory avoids deadlock by restricting routing instead of buffer allocation.

The basic assumptions are very similar to the ones proposed in section 2, except that store-and-forward routing is used and then, all the packets have routing information and full packets are stored in each queue buffer. The assumption 6 has no meaning here.

The definitions are also identical, except those ones referring to deadlocked configurations.

Definition 12: A *configuration* is an assignment of a list of nodes to each queue. The number of packets in the queue for channel c_i will be denoted size(c_i). The destination node for a packet f_j will be denoted dest(f_j). If the first packet in the queue for channel c_i is destined for node n_d, then head(c_i) = n_d. A configuration is *legal* iff

$$\forall c_i \in C, \quad \text{size}(c_i) \leq \text{cap}(c_i) \quad \wedge$$
$$c_i \in R(s_i, \text{dest}(f_j)) \quad \forall f_j \in \text{queue}(c_i)$$

that is, the queue capacity is not exceeded and all the packets stored in the queue have been sent there by the routing function.

Definition 13: A *deadlocked configuration* for a given interconnection network I and routing function R is a nonempty legal configuration verifying the following condition:

$$\forall c_i \in C \ni \text{head}(c_i) \in N \Rightarrow \text{head}(c_i) \neq d_i \wedge$$
$$\text{size}(c_j) = \text{cap}(c_j) \quad \forall c_j \in R(d_i, \text{head}(c_i))$$

In a deadlocked configuration there is not any packet one hop from its destination. Packets cannot advance because the queues for all the alternative output channels supplied by the routing function are full. As a particular case (for disconnected routing functions), the routing function may not supply any output channel.

The theorem 1 proposed in section 2 is also valid for store-and-forward routing, the proof being almost identical. In this section, two theorems are proposed. The first one is similar to theorem 2 for wormhole routing, except that the channel dependency graph is used instead of the extended one. The second theorem makes the design of adaptive routing functions more flexible. For each theorem, a sketch of the proof is given.

Theorem 3: A connected and adaptive routing function R for an interconnection network I is deadlock-free if it exists a subset of channels $C_1 \subseteq C$ that defines a routing subfunction R_1 which is connected and has no cycles in its channel dependency graph D_1.

Proof sketch:

$\Leftarrow$ The proof is basically the same as in section 2, changing nonempty queue by full queue and empty queue by nonfull queue. Also, cases b2.2 and b2.3 do not exist. Then, only the channel dependency graph for R_1, instead of the extended one, is required to be acyclic.

Theorem 4: A connected and adaptive routing function R for an interconnection network I is deadlock-free if it exists a subset of channels $C_1 \subseteq C$ that defines a connected and deadlock-free routing subfunction R_1.

Proof sketch:

$\Leftarrow$ If $C_1 = C$ the proof is trivial. Otherwise $C_1 \subset C$. Suppose that there is a deadlocked configuration for R. There are two possible cases:

a) The queues for channels belonging to C_1 are empty. As R_1 is connected, a given packet can be routed using an empty channel belonging to C_1 and there is no deadlock.

b) The queues for channels belonging to C_1 are not empty. As R_1 is deadlock-free, one can find a channel $c_i \in C_1$ such that $c_j \in R_1(d_i, \text{head}(c_i))$ and $\text{size}(c_j) < \text{cap}(c_j)$. As $C_1 \subset C$ and $R_1 \subset R$ then there is not a deadlock, contrary to the initial assumption.

4. Design methodologies

In this section we propose two methodologies for the design of deadlock-free adaptive routing algorithms. Although the same methodologies can be applied for both, wormhole and store-and-forward routing, there are some differences that will be highlighted.

The generation of static deadlock-free routing algorithms requires to restrict routing by removing edges from D to make it acyclic. If it is not possible to make D acyclic without disconnecting the routing function, edges can be added to D by splitting physical channels into a set of virtual channels, each one requiring its own buffer. This technique was introduced by Dally [9] to remove cycles from the channel dependency graph.

However, a physical channel can be split into more virtual channels than the ones strictly necessary to avoid deadlock [6,10]. In such a case, the router can choose among several channels to send a message, reducing channel contention and message delay. Alternatively, more physical channels can be added to each node, increasing the network bandwidth and allowing the design of fault-tolerant adaptive routing algorithms.

A design methodology must supply a way to add channels following a regular pattern, also deriving the new routing function from the old one. A design methodology based on theorem 1 has been presented in [10]. Although the algorithms designed with it behave better than the static ones, a higher degree of freedom can be obtained basing the design either on theorem 2 (wormhole) or on theorems 3 and 4 (store-and-forward). Here we will present some more general methodologies for the design of deadlock-free adaptive routing algorithms.

Methodology 1. This methodology is intended to increase the number of valid alternative paths to send a message towards its destination without increasing the number of physical channels. In general, it will reduce channel contention and message delay but it will not increase fault-tolerance significantly. The steps are the following:

1) Given an interconnection network I_1, define a minimal path connected static routing function R_1 for it, following Dally's methodology and splitting physical channels into virtual ones, if necessary, to guarantee that R_1 is deadlock-free. Alternatively, define a minimal path connected adaptive routing function R_1 and selection function S_1, verifying that R_1 is deadlock-free using either theorem 2 or theorem 3. Let C_1 be the set of channels at this point.

2) Split each physical channel into a set of additional virtual channels. Let C be the set of all the (virtual) channels in the network. Let C_{ij} be the set of output channels from node i belonging to a minimal path from i to j. Define the new routing function R as follows:

$$R(i, j) = R_1(i, j) \cup (C_{ij} \cap (C - C_1)) \quad \forall\, i, j \in N$$

that is, the new routing function can use any of the new channels belonging to a minimal path or, alternatively, the channels supplied by R_1. The selection function can be defined in any way.

3) For wormhole routing only, verify that the extended channel dependency graph for R_1 is acyclic.

Step 1 establishes the starting point. We can use either a static or adaptive routing function as the basic one. Dally's theorem and theorems 2 and 3 can be used to verify that the basic function is deadlock-free. Step 2 indicates how to add more (virtual) channels to the network and how to define a new adaptive routing function from the basic one. Step 3 allows us to verify that the new routing function is deadlock-free. This step is only required for wormhole routing.

For store-and-forward routing, it is easy to see that the proposed methodology directly supplies a deadlock-free routing function R, because

$R_1(i, j) = R(i, j) \cap C_1 \quad \forall\, i, j \in N$

So, it exists a routing subfunction R_1 of R, which is connected and deadlock-free. Then, by theorem 4, R is deadlock-free.

It must be noticed that the methodology can also be applied by adding physical channels instead of virtual ones. The resulting network will be faster and more expensive, but the effective fault-tolerance will not increase. The reason is that the new routing function relies on the set of channels C_1 to guarantee that it is deadlock-free.

Methodology 2. This methodology is intended to increase fault-tolerance in a network. It will add physical channels, instead of splitting channels into virtual ones. Of course, it will also reduce channel contention and message delay. The steps are the following:

1) Given an interconnection network I_1, define a static or adaptive connected routing function R_1 for it, following Dally's methodology, the above proposed methodology or verifying that R_1 is deadlock-free using either theorem 2 or theorem 3. Let C_1 be the set of channels at this point.

2) Duplicate each physical channel. If the original channel was split into several virtual channels, the duplicated channel will also be split into the same number of virtual channels. Let C_2 be the set of duplicated channels and C the set of all the channels. Let R_2 be a routing function identical to R_1, but defined using C_2 instead of C_1. Define the new routing function R as follows:

$R(i, j) = R_1(i, j) \cup R_2(i, j) \quad \forall\, i, j \in N$

that is, the new routing function can use any of the channels supplied by both, R_1 and R_2. Define the selection function giving to the channels belonging to C_1 and C_2 the same probability of use.

Again, step 1 supplies the basic routing function and step 2 adds alternative paths. As can be easily seen, R_2 does not add any cycle neither to the channel dependency graph of R_1 nor to the extended one. Then, R is deadlock-free.

The duplication of channels defines an interconnection network $I_2 = G(N, C_2)$, which is identical to I_1 and shares the same set of nodes N. However, C_1 and C_2 are disjointed sets. R_2 has the same properties as R_1. Also,

$R_1(i, j) = R(i, j) \cap C_1 \quad \forall\, i, j \in N$

$R_2(i, j) = R(i, j) \cap C_2 \quad \forall\, i, j \in N$

So, one can find, at least, two subfunctions of R, which allow us the application of the corresponding theorem to guarantee that R is deadlock-free. Then, the theorems can be applied even if we remove some channels either from C_1 or from C_2.

However, the set of nodes is the same for I_1 and I_2. It seems that the proposed methodology is not tolerant to node faults. But, provided that R_1 and R_2 are adaptive routing functions, in general there will be alternative paths to reach the destination node (assuming that it is not the faulty one). Of course, some mechanism is needed to identify faulty channels, marking them as busy, and faulty nodes, marking all the channels connected to them as busy and avoiding to send messages to them. It must be noticed that if there is not any faulty node, the information about faulty channels is recorded locally.

Finally, step 2 can be applied several times, duplicating each channel as many times as desired.

The proposed methodologies are very simple to apply. They illustrate the power of the theorems. More complex design methodologies can be defined based on the same theoretical background.

As an example, we will present a design based on the above proposed methodologies. Consider a binary n-cube. We will study three cases: a) applying methodology 1; b) applying methodology 2; c) applying methodologies 1 and 2.

a) For the step 1 we can use the conventional static routing algorithm for the binary n-cube. It forwards messages crossing the channels in order of decreasing dimensions. It is well known that this routing function is connected and deadlock-free.

For the step 2, consider that each physical channel c_i has been split into two virtual channels, namely, a_i and b_i. Let C_1 be the set of 'b' channels. The algorithm obtained applying the step 2 can be stated as follows: Route over any useful dimension using 'a' channels. Alternatively, route over the highest useful dimension using 'b' channels. A useful dimension is one that forwards a message nearer to its destination.

It can be seen that the extended channel dependency graph for R_1 is acyclic. Then, R is deadlock-free.

As virtual channels share a single physical channel, the former algorithm effectively allows messages to cross the physical channels corresponding to the n-cube dimensions in any order, increasing considerably the number of alternative paths and decreasing network contention. The simulation results for that algorithm are very promising.

b) Assume that step 1 is applied as in case a), obtaining the conventional static routing algorithm.

For the step 2, consider that each physical channel c_i has been duplicated, obtaining a new channel d_i. The algorithm obtained applying the step 2 can be stated as follows: Route over the highest useful dimension using 'c' channels. Alternatively, route over the highest useful dimension using 'd' channels.

That algorithm increases the tolerance to faulty channels, but it does not take advantage of alternative minimal paths.

c) Assume that we apply the methodology 1, obtaining the algorithm proposed in case a). That algorithm constitutes the step 1.

For the step 2, consider that each physical channel c_i has been duplicated, obtaining a new channel d_i, which is split into two virtual channels, namely, e_i and f_i. The algorithm obtained applying the step 2 can be stated as follows: Route over any useful dimension using either 'a' or 'e' channels. Alternatively, route over the highest useful dimension using either 'b' or 'f' channels.

That algorithm has the advantages of the previous ones at the cost of a slightly more complicated circuitry.

As stated in section 2, the selection function only affects performance. It is not necessary to give a higher priority to the channels in the acyclic dependency subgraph, because when the remaining channels are busy those ones will be used. In general, a higher performance is achieved when the channels in the cyclic dependency subgraph are given a higher priority, because they usually offer a larger number of alternative paths.

Also, the selection function can be extended by including additional information in its domain. For instance, for the algorithm obtained in case a), it is possible to favour the 'a' channel connecting to the neighbour with a higher number of free channels in useful dimensions. This selection function is inspired in an algorithm proposed by Reeves et al. [22], the main difference being that our algorithm does not require a complex mechanism to abort messages because it is deadlock-free.

Another interesting extension of the selection function is taking into account the time a message is waiting in a given node. This information can be used to prevent channel multiplexing when the network traffic is low. That extension gives good results when added to the algorithm obtained in case a). The simulation results will be presented in another paper.

More examples could be presented for other topologies. However, the application of the above proposed methodologies is so easy that we consider that it is not necessary. The simulation under different load conditions will give some insight about the performance of the new family of adaptive algorithms.

5. Conclusions

The theoretical background for the development of deadlock-free adaptive routing algorithms has been proposed for both, store-and-forward and wormhole routing. Firstly, a straightforward extension of Dally's theorem has been presented, allowing the design of adaptive algorithms. However, the absence of cycles in the channel dependency graph is too restrictive.

For wormhole routing, theorem 2 gives a more flexible condition for the development of adaptive algorithms, by allowing the existence of cycles in the channel dependency graph. The only requirement is the existence of a channel subset which defines a connected routing subfunction with no cycles in its extended channel dependency graph.

For store-and-forward routing, theorem 3 develops a sufficient condition similar to theorem 2. Theorem 4 adds more flexibility. It simply requires the existence of a connected and deadlock-free routing subfunction. In turn, that subfunction can be proved to be deadlock-free using theorem 3.

To simplify the application of the theorems, two design methodologies have been proposed. The first one supplies adaptive algorithms with a high degree of freedom. The second one gives a way to design fault-tolerant routing algorithms. Both methodologies can be combined easily.

Finally, an example showing three alternative ways to apply the proposed design methodologies is presented.

References

[1] W.C. Athas and C.L. Seitz, Multicomputers: message-passing concurrent computers, *Computer,* Vol. 21, No. 8, pp. 9-24, August 1988.

[2] S. Borkar et al., iWarp: an integrated solution to high-speed parallel computing, *Supercomputing'88,* Kissimmee, Florida, November 1988.

[3] W. Chou, A.W. Bragg and A.A. Nilsson, The need for adaptive routing in the chaotic and unbalanced traffic environment, *IEEE Trans. Commun.,* Vol. COM-29, No. 4, pp. 481-490, April 1981.

[4] E. Chow, H. Madan, J. Peterson, D. Grunwald and D.A. Reed, Hyperswitch network for the hypercube computer, *Proc. 15th Int. Symp. Computer Architecture,* Honolulu, May-June 1988.

[5] W.J. Dally, *A VLSI architecture for concurrent data structures,* Kluwer Academic Publishers, 1987.

[6] W.J. Dally, Virtual-channel flow control, *Proc. 17th Int. Symp. Computer Architecture,* Seattle, Washington, May 1990.

[7] W.J. Dally, Performance analysis of k-ary n-cube interconnection networks, *IEEE Trans. Computers,* Vol. C-39, No. 6, pp. 775-785, June 1990.

[8] W.J. Dally and C.L. Seitz, The torus routing chip, *Distributed Computing,* Vol. 1, No. 3, pp. 187-196, October 1986.

[9] W.J. Dally and C.L. Seitz, Deadlock-free message routing in multiprocessor interconnection networks, *IEEE Trans. Computers,* Vol. C-36, No. 5, pp. 547-553, May 1987.

[10] J. Duato, Deadlock-free adaptive routing algorithms for multicomputers. Submitted to *Tech. et Sci. Informatiques.*

[11] J. Duato, On the design of deadlock-free adaptive routing algorithms for multicomputers: theoretical aspects, *Proc. 2nd European Distributed Memory Computing Conference,* Munich, April 1991.

[12] D. Gelernter, A DAG-based algorithm for prevention of store-and-forward deadlock in packet networks, *IEEE Trans. Computers,* Vol. C-30, pp. 709-715, October 1981.

[13] C. Germain-Renaud, *Etude des mécanismes de communication pour une machine massivement parallèle: MEGA,* Ph.D. dissertation, Université de Paris-Sud, Centre d'Orsay, 1989.

[14] K.D. Gunther, Prevention of deadlocks in packet-switched data transport systems, *IEEE Trans. Commun.,* Vol. COM-29, pp. 512-524, April 1981.

[15] W.D. Hillis, *The connection machine,* MIT Press, Cambridge, Mass., 1985.

[16] C.R. Jesshope, P.R. Miller and J.T. Yantchev, High performance communications in processor networks, *Proc. 16th Int. Symp. Computer Architecture,* Jerusalem, Israel, May-June 1989.

[17] P. Kermani and L. Kleinrock, Virtual cut-through: a new computer communication switching technique, *Computer Networks,* Vol. 3, pp. 267-286, 1979.

[18] C.K. Kim and D.A. Reed, Adaptive packet routing in a hypercube, *Proc. 3rd Conf. on Hypercube Concurrent Computers & Applications,* Pasadena, California, January 1988.

[19] H.T. Kung, Deadlock avoidance for systolic communication, *Proc. 15th Int. Symp. Computer Architecture,* Honolulu, May-June 1988.

[20] P.M. Merlin and P.J. Schweitzer, Deadlock avoidance in store-and-forward networks - I: Store-and-forward deadlock, *IEEE Trans. Commun.,* Vol. COM-28, pp. 345-354, March 1980.

[21] S. Ragupathy, M.R. Leutze and S.R. Schach, Message routing schemes in a hypercube machine, *Proc. 3rd Conf. on Hypercube Concurrent Computers & Applications,* Pasadena, California, January 1988.

[22] D.S. Reeves, E.F. Gehringer and A. Chandiramani, Adaptive routing and deadlock recovery: a simulation study, *Proc. 4th Conf. on Hypercube Concurrent Computers & Applications,* Monterey, California, March 1989.

A Toolkit for Debugging Parallel Lisp Programs

Hermann Ilmberger, Sabine Thürmel
*EDS Project**
Siemens AG, ZFE IS SOF 22
Otto-Hahn-Ring 6, D-8000 München 83, Germany
e-mail: thuermel@ztivax.uucp, hermann%moony@ztivax.uucp

Abstract

As part of the ESPRIT-II project EDS a toolkit (called Delphi) is under development for the debugging of Lisp programs with explicit parallelism being executed on a homogeneous distributed machine. It assists in the detection of functional and synchronisation errors. It also helps to detect unexpected nondeterminacy and sources of poor program performance. Specific mechanisms allow the user to effectively control several processes in a debugging session. The paper introduces the basic concepts behind the tools and how the user may benefit from them.

1 Motivation and Overview

Many experts consider parallel computation the only promising approach to enhancing the performance of computer systems. This is not only true for numerical algorithms. Significant increases in performance are also expected for data intensive applications using nontrivial algorithms such as the translation of natural languages [EDS89]. This motivates to combine parallel computation and symbolic languages. Such a combination is the goal of the ESPRIT-II project 2025 EDS (European Declarative System). The project will produce the prototype of a homogeneous distributed machine supporting among other things a parallel Lisp dialect (EDS Lisp). In EDS Lisp processes are spawned by the future construct. Communication may be realized through mailboxes. EDS Lisp is intended to support large knowledge based systems, for example for the translation of natural languages or VLSI chip design. As a consequence there is the need for a toolkit that assists users of EDS Lisp in debugging applications consisting of several concurrent processes.

This paper describes Delphi, a toolkit for the debugging and visualisation of EDS Lisp programs. Specifically Delphi addresses the problem of how the user

---

* The EDS-Project (European Declarative System) is partially sponsored by the European Community under ESPRIT-II 2025.

can keep an overview of the parallel program run and effectively control the debugging session.

The paper begins with a short introduction to EDS Lisp and the EDS Machine. Chapter 3 focuses on basic issues in debugging parallel Lisp programs. In chapter 4 an overview of Delphi is presented. Its tools for distributed symbolic debugging are described. It is outlined how Delphi supports user controlled program execution for debugging purposes. We also present a visualisation facility supporting the user in the granularity analysis and the detection of inadequate communication structures. Chapter 5 is devoted to additional, planned features. A comparison to other approaches follows in chapter 6.

2 EDS Lisp and the EDS Machine

EDS Lisp is a parallel Lisp dialect supporting explicit parallelism. It is an extension of Common Lisp [Steele 84]. Common Lisp has been chosen as the base language of EDS Lisp because one goal of the project is to port real life applications to the EDS system.

The following gives a brief overview of the main features of this language (for a detailed introduction see [HamHen90]):
For the spawning of processes EDS Lisp has a single concept, the future construct, known from other parallel Lisp extensions (e.g. [Halst85]). It allows concurrent function evaluation. After spawning a process the newly spawned process immediately returns a placeholder for the result. If there is an access to the result of a spawned process while the return value is not yet computed the spawning process is forced to wait by an implicit synchronisation mechanism.
One may also wait for the result of a process using the wait construct thus causing an explicit synchronisation. This construct can be used if one wants to wait for the sideeffects of a spawned process before resuming evaluation.
In addition EDS Lisp contains two other concepts for explicit synchronisation, critical sections and mailboxes. Critical sections are particularly well suited to synchronise accesses to shared global variables. Processes can communicate via mailboxes which provide asynchronous buffered message passing.

The EDS system is a homogeneous distributed multiprocessor. A prototype will consists of 63 processing elements and a diagnostic node. During the test phase the latter will serve as an interface between the 63 node EDS machine and the user's workstation. The Process and Store Model of EDS Lisp is described in detail in [HamHen90]. We won't refer to it in this paper because our debugging concepts are intentionally independent of the EDS system. So our ideas are also applicable to similar architectures.

3 Basic Issues in Debugging Parallel Lisp Programs

In this paper we want to address some of the basic issues in debugging parallel Lisp programs. They are listed below. The concepts we propose for their solution are introduced briefly. (Chapter 4 is devoted to their detailed description.)

Detecting functional errors is usually necessary for the detection of synchronisation errors

While debugging parallel Lisp programs the detection of functional errors (classically called bugs) cannot be neglected. For example, most synchronisation errors result from incorrect use of a combination of sequential and parallel constructs.
Therefore we propose a straightforward extension to classical symbolic debugging tools. This allows us to extend well known debugging strategies for the parallel case.

Concentrating on debugging one process at a time is not appropriate in debugging parallel programs

Symptoms of a problem may show up on another process rather than on the one where the problem actually occurred. Thus a single debugger examining one process at a time makes debugging a clumsy affair. Tools for distributed symbolic debugging are needed. To provide the user with a detailed view of the program state the tools have to provide a mechanism to compute nonlocal information.

Observing and controlling process states by the user is necessary

For the exact location and analysis of errors the programmer needs assistance in getting an overview of the general program state (e.g. which processes are active, what is their status). In addition the user would like to control the amount of trace information to be displayed (there will be times when the user simply wants to discard trace output). Also it is helpful if broken processes can be continued individually after being inspected. Thus a mechanism for the observation and control of processes is necessary.

Visualising the process history gives a comprehensive overview of the runtime behaviour

An overview focusing on a moment in time is not sufficient for the understanding of the possibly quite complex runtime behaviour of parallel programs. Graphical displays of the program execution have to be supported which allow getting an overview of the total program execution.We propose focusing on two essential aspects: the spawning/joining of processes and the

Supporting user controlled execution enhances the error detection

The debugging tools should enhance the detection of unwanted nondeterminacy. Nondeterminacy is not in itself an error and may be intentional. But careless unsynchronized access to global or dynamic variables may lead to nondeterminate results. Our clique model for the support of user controlled execution helps detecting sources of nondeterminacy.

4 Delphi: A Toolkit for Debugging Parallel Lisp Programs

The EDS Lisp debugger Delphi supports various debugging techniques. It consists of

* basic tools: stepper, tracer and break-mode,
* a mechanism allowing the computation of nonlocal information by communicating sequential debuggers,
* an interactive graphical process observation and control tool,
* a process history visualizer,
* the clique model for user controlled program execution.

Stepper, tracer and break-mode are common for sequential Lisp debuggers. These tools have been extended for use in the parallel environment of EDS Lisp. Special treatment is necessary if stepping/tracing/debugging takes place in several processes. In order not to confuse the programmer, the information is handled in different windows on the terminal, controlled by the user via the graphical representation of the Lisp process tree. Additionally, the debuggers can communicate to compute nonlocal information.

The visualizer offers two complementary views of a program execution, one focusing on the spawning of processes and one concentrating on the communication at mailboxes.

A clique is a group of Lisp processes running quasiparallel. Grouping and scheduling is user controlled. The clique model supports the programmer in detecting communication errors and nondeterminacy and can be used in the testing phase to simulate extreme scheduling behaviour.

4.1 Basic Tools

Stepper

The stepper interactively single-steps through the evaluation of a form. If the stepped form spawns a process, the user is asked whether the new process should be stepped, too. If yes, a new window is opened on the screen, and the user can single-step through both processes, thus controlling their relative progress. As this method is very intrusive, it is only recommended when no timing conditions are involved.

Tracer

The user can specify a set of Lisp functions to be traced. Both Common Lisp functions and the new extensions for parallelism in EDS Lisp can be traced. Each program run logs a minimal trace. There is a separate window for the trace output of each process. To avoid a confusing display, the user can control the window action via a graphical representation of the process tree (ch. 4.3).

Break-Mode - the "Debugger"

According to the semantics of error and break in EDS Lisp, an error in one process of a program causes all processes of the program to be broken. There is a separate debugger for each broken process. Using the graphical representation of the process tree (ch. 4.3), the user can open a window for selected debuggers. Each user-opened debugger communicates with the user via its own window.

4.2 Communicating Sequential Debuggers to Gain Nonlocal Information

Since a program can consist of many processes, it is not convenient to open a window for all the corresponding debuggers. The programmer may want to handle all broken processes using only a few windows. Assume the following scenario:

When a whole EDS Lisp program is broken because of an error, the bug can be located

a) locally in the process A that raised the error, or

b) in one or more other processes which transferred erroneous data to process A.

In case b) the bug lies in other processes, but the error itself shows up in process A. So the programmer will probably first inspect process A, but will sooner or later realize that the real bug lies in some other process. If the process containing the bug is located, the programmer can simply open a window for the debugger of the process and continue debugging there.

If process A is a process that communicates frequently with other processes, it may be more difficult to determine the process(es) containing the bug. For such cases the EDS Lisp debuggers can communicate with each other using the "nonlocal" commands. Each debugger can be remote controlled from any other. The responses from remote debuggers are displayed in the controlling debugger's window. For the nonlocal commands the user can specify from which remote debuggers the information is requested.

Examples of nonlocal commands are:

* list the top-of-stack frames of all broken processes
* continue all processes with name foo
* list the value of the dynamic variable x in all processes (in EDS Lisp each process has its own view of the dynamic binding stack)

The above commands simply broadcast a request to the other debuggers and gather the answers. But there are also more powerful nonlocal commands:

* Which processes are executing the function foo? (requires searching through the backtrace of all processes)
* With which processes did a process communicate? (not directly visible in the program because processes send messages to mailboxes, not to processes)

Thus the nonlocal commands enable the user to get a global view of the intrinsics of a program. How are other existing parallel Lisp systems addressing the problem of debugging many processes?

The parallel Lisp system Mult-T [Kranz89] supports the concept of process groups. If one process in a group raises an error, all processes of this group are stopped. There is one window to debug the whole group. Debugger commands by default refer to the process in which the error occurred. The commands also allow referring to other processes or other stopped groups. But there are no nonlocal commands in our sense; debugger commands influence always only one process.

MultiScheme [Miller87] also stops all processes when an error occurs, but the user can debug only one process at all. Butterfly Lisp [Allen87] opens a separate window for each broken process.

The EDS Lisp Debugger supports this as an option. But often this is not desired. Therefore the opening of windows can be controlled by the user as described in the next section.

4.3 Graphical Process Observation & Control and Visualization of Runtime Behaviour

For locating and analysing errors, the programmer needs assistance in getting an overview of the current program state. But often the current state is not sufficient. An overview of the past program behaviour is also necessary for efficient debugging.

4.3.1 Graphical Process Observation and Control

The graphical observation and control tool gives a global view of the process dependencies and states. It is based on a graphical representation of the program's process tree. The behaviour of trace and debugger windows are controlled via the tree.

Each node in the tree represents a process. The children of a node are the directly spawned future-processes of the parent process, ordered by their time of creation. The root of the tree is the EDS Lisp program's toplevel.

The tree is kept in a separate window on the user's workstation screen. The pictorial representation of a node changes according to the state of the corresponding process. There are the following states:

a) process running
b) process wants to output trace information (but continues to run)
c) process is broken, its debugger is active
d) process completed

In case b) and c) the user can open a trace resp. debugger window by selecting a node. Thus, the user can manage the opening and closing of windows. There is no abundance of tracer and debugger windows to handle. Only the window of the process' debugger where the error occurred is automatically opened. The debuggers for other suspect processes can be opened at will. Processes that are not suspect can be continued, changing to state a).

There is no separate node state for active steppers, because stepper windows are automatically opened by the system. This makes sense since usually only one form is stepped at a time and the user is asked if both processes should be stepped after a future-command.

Opening a debugger or tracer window is one of many commands that may be performed on a node in the process tree display. Other commands are:

* break the process
* restart a broken process
* discard the trace output (if the user does not want to see the trace)
* list basic characteristics of the process (processing element, runtime, ...)
* visualize process history (see below)

Thus the process tree in combination with the node commands is a powerful tool to get a general overview of the current program state and to customise the window management.

4.3.2 Visualisation of Runtime Behaviour

Visualisation of the dynamic behaviour of EDS Lisp programs

* enhances the understanding of the possibly quite complex dynamic

behaviour of EDS Lisp programs,
* assists in finding errors,
* is the basis for granularity analysis performed by the user,
* supports the user in detecting inappropriate communication structures.

Alternative representations of the runtime behaviour include the one favored in literature (e.g. [GeKr86], [Bag89]). It represents the processes that existed during program execution as horizontal bands with arrows between the processes indicating interactions. It seems to be very helpful when one wants to concentrate on certain aspects of an execution. But even when such process graphs are hierarchically organised with adequate operators for hiding processes and their descendants, such systems are limited in the size of an application they can deal with.

Therefore we decided to offer two complementary views of program execution:
The first view focuses on the spawning of processes in an EDS Lisp program execution. It displays a general overview of the possibly quite complex runtime behaviour. It may serve as a basis for granularity analysis. This view is called a process history tree.
The second view depicts the use of EDS Lisp mailboxes. It assists in the detection of inappropriate communication structures in the EDS Lisp program.

Such displays may be generated during execution, displaying the behaviour up to that time, or may be generated post mortem. The process tree displayed in the tool for process observation and control is a snapshot of the program behaviour (i.e. long dead processes are not visible). The process history tree allows the inspection of the runtime behaviour up to the moment when the tree was built.

To produce these views, for every process a trace is generated consisting of basic events (as process creation and termination etc.) and user selected events during execution. At the user's workstation these traces are evaluated. Displays are generated upon request. Since the graphical representations are built offline, the delay of the program execution caused by the visualizer is minimal.

Process History Trees

A process tree for the execution of an EDS Lisp program shows the spawning structure of the EDS Lisp processes generated during the execution.This is exemplified in fig. 1.

To navigate these displays, horizontal and vertical scrollbars are provided allowing depth first and breadth first search. For the hiding/unhiding of processes and their descendents appropriate operators are supported.

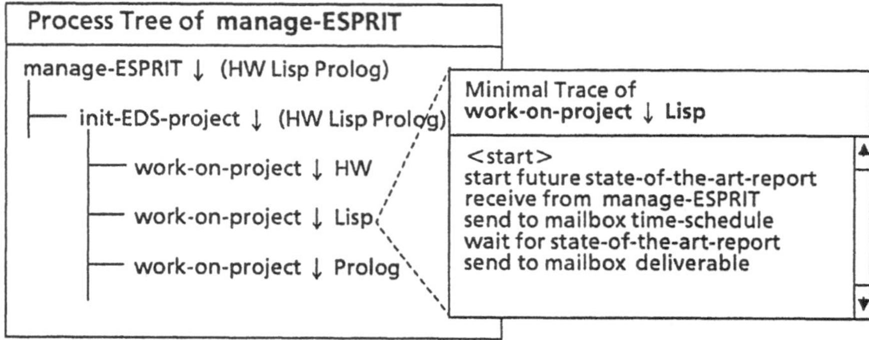

fig. 1: A sample process-tree and minimal trace.
legend: ↓ actual parameters of a process

For each process its process specific trace (process history) as well as some of its basic characteristics (like active and idle time) and number of remote accesses may be displayed upon request.

The trace allows inspecting process specific events. In combination with its basic (system) characteristics, granularity analysis may be performed by the user: An extremely short active time will indicate that the process should be evaluated inline rather than in parallel. Long idle times in combination with a large amount of remote accesses may be a hint to look for inappropriate variable bindings. Thus this display may help the programmer to tune the performance on the EDS Lisp level.

Display of Communication via Mailboxes

The displays of the communication via mailboxes assist the user in finding inappropriate communication structures. Similar to [BuMil88] we want to display all send and receives to and from any user selected mailbox. In concrete terms: The y-axis presents time. The x-axis holds the information on the send and receive operations as demonstrated in figure 2. Every send operation increases the access counter, while a receive operation reduces the access counter. So the region on the negative side of the chart gives an impression on blocked and waiting processes, while the positive side visualizes the number of stored messages. By clicking on a specific access counter value in the display the corresponding message resp. waiting queue is displayed (see fig. 2). For any message the following information is given: who sent the message and who, if any, received it and its identification on source code level (the latter is not represented in the example). Its contents are not displayed because it costs too much time recording it during runtime since any Lisp object may be sent. The waiting queue can be used to identify the processes waiting at that mailbox.

Comparing the waiting queues of different mailboxes when all processes are broken or killed makes deadlock detection feasible. Upon request the representations of the processes involved in the traffic at a specific mailbox will be highlighted in the corresponding process tree. After detecting a disadvantageous communication structure in the mailbox display, the process specific traces of the process tree will help to improve it.

4.4 Support of User Controlled Execution

An EDS Lisp program can spawn several future-processes at runtime. A major problem in debugging parallel programs is that there is more than one process running at the same time. How can the programmer keep the overview and how can s/he handle the parallel processes to find bugs?

One common method is to run the program and wait until it crashes. Using the state of the program (i.e. variable bindings, Lisp stacks, ...) at the moment of the crash, many errors can be detected post mortem. Support for this method was discussed in chapter 4.2 and 4.3.

Sometimes this kind of post mortem detection is not sufficient. If an error occurs in one process of a Lisp program, the semantics of EDS Lisp says that all processes of the program will be broken. Suppose a Lisp program consists of two processes A and B running on different processing elements. If an error occurs in process A, A is broken immediately. Process B can still change its state in the time until the break signal reaches B's processing element. B can for example change the value of a global variable. Post mortem analysis will be difficult if just the former value of this variable caused the error in process A.

We developped the clique model to address this problem.

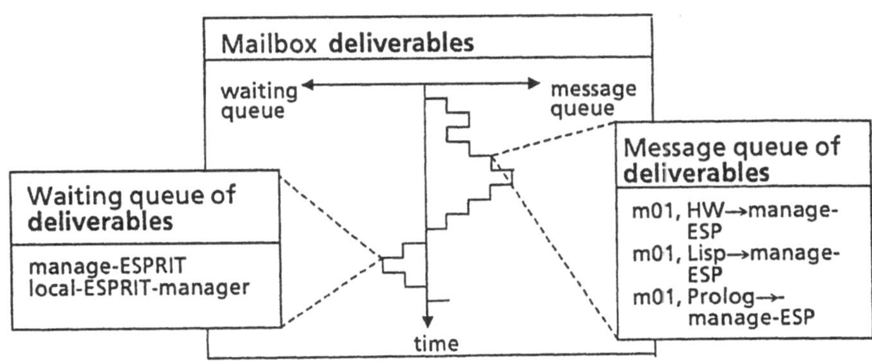

fig. 2: A sample communication at a mailbox

The Clique Model

The EDS Lisp debugger allows the user to control the sequence of execution by grouping processes together (into "cliques") and forcing them to run in a quasiparallel manner with a user defined Lisp level scheduling. If an error occurs in one of the clique processes, the other clique members are not allowed to continue and thus cannot change the program state.

Cliques support not only post mortem analysis but can be used even more profitably in the test phase. They can be used to test the program under unusual scheduling conditions (for example scheduling after each function call), and help programmers familiar with sequential languages get a better understanding how a parallel program works.

Membership

A "clique" is a group of processes which the programmer wants to serialise. Each Lisp process can become a member of (at most) one clique. A process which is not a member of a clique is "clique-free". (for an example see fig. 3)

Scheduling

At most one member of each clique is in the state 'computing' at any moment of time (quasiparallel run). Scheduling within a clique is controlled by the programmer, is visible at the source code level, and is repeatable. Timeslice scheduling is not reasonable, because the exact point of descheduling cannot be controled. Repeatability cannot be guaranteed either.

The straightforward kind of scheduling is function based. The programmer can supply a list of function names, and scheduling takes place before the call or

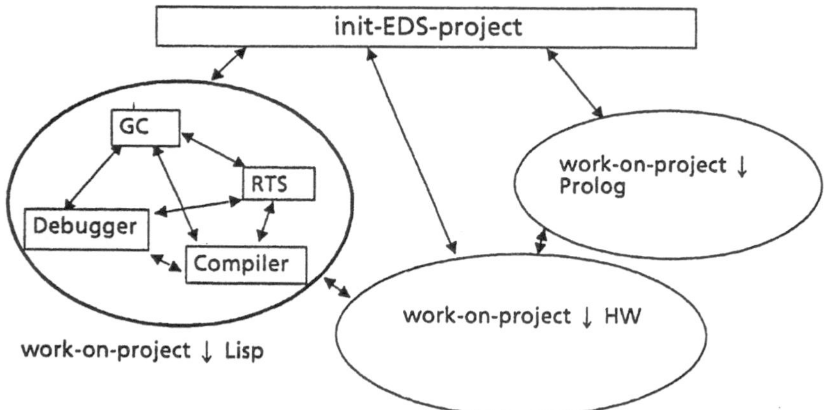

fig. 3: A sample of cliques
legend: ⟷ communication link via a mailbox

after the exit of each of these functions. Changing the list enforces a different scheduling. The scheduling can also be set to 'verbose'. In this mode each (de)scheduling action is printed out at runtime and the user can follow the program.

Application Areas

We see three main applications for cliques:

1. Tracing communication and synchronisation errors: the programmer conjectures that a mailbox communication or a critical section does not work well. Scheduling after the suspected command makes it possible to test whether it cooperates well with the other clique members.

2. Testing unusual scheduling: the programmer can simulate any possible scheduling in order to test highly communicating program parts and to detect unexpected nondeterminacy.

3. Programmers familiar with sequential languages often have difficulties programming parallel applications. Serialization of a program with verbose scheduling may help in better understanding what happens in a program.

Use

What strategy should be used to group processes together?

For programs with only a few processes where the programmer suspects nondeterminacies or communication errors, all processes can be grouped into one clique. There are two possibilities:

1. Shared mode: The clique runs on one processing element. Thus all processes have real shared memory. Errors caused by inconsistent use of distributed memory cannot occur. Furthermore all processes have the same global time, making it possible to totally order the events in the program.

2. Distributed mode: The processes are distributed to the processing elements "as in real life". This makes better use of the EDS machine's storage capacity. Now all the effects of distributed memory can occur. Total ordering is still possible since only one clique member runs at a time.

If the program gets bigger, a bottom up test strategy can be used: In a program there are often groups of processes which cooperate more with each other than with the rest of the program. These processes can form a clique. Several such cliques may be found. These cliques can now be analysed separately, and the other processes are perhaps replaced with stubs. Larger program systems can be tested incrementally. A process is already tested, can be made clique-free and run asynchronously.

5 Additional Features: Replay and Countercheck Sessions

In general, concurrent program runs cannot be repeated exactly, because there may be races between processes. There are two main aspects why a programmer wants to repeat a run:

a) After the occurence of an error it is very useful to be able to replay a program exactly in order to see how the error came into being: "replay".

b) A program shall be tested for nondeterminism. In this case a first run ("initial session") will be compared against a second, third etc. ("countercheck sessions") with same input data, the latter having for example different scheduling strategies and/or different machine load.

Replay

Replay techniques are an ideal approach to reproducing program execution([LeBl87], [Mill88]). Unfortunately they only allow reproducing such program executions where all shared data structures are known at compile time. This is usually not the case for programs written in parallel extensions to Common Lisp. Therefore we do not want to guarantee replay for unsynchronized global variable access (the effect of this is undefined in EDS Lisp anyway). We only want to guarantee replay for mailbox communication (mailboxes are created explicitely) and for access to critical sections (they are created explicitely, too). Inside the critical sections the programmer can access global variables in a well-defined manner. The replay does not need to know which variables are shared as long as all accesses happen within critical sections.

Countercheck Sessions

In a countercheck session the logged events of the initial session are compared to the current run. Any deviation signals a nondeterminism in the program. It is selfevident that countercheck sessions can only help to find nondeterminism, not to prove the absence of nondeterminism. Countercheck sessions seem to come almost for free with replay.

6 Other Approaches to Debugging Parallel Programs

Although there exists a variety of parallel Lisp dialects ([Tar89]) only very few pointers to debugging concepts for these languages can be found. We know of several people working in this field. For example, Bert Halstead is working on visualising Multilisp program execution based on [Bag89]. Later on, other debugging techniques will be integrated into this tool [Halst90].

Apart from work in progress there exists a traditional debugging toolkit

Lisp [TopL89]. It runs in a shared environment and supplies assistance for debugging several processes simultaneously in a traditional manner.

In contrast to the debugging of parallel Lisp dialects there exists a whole variety of debugging techniques for sideeffect free languages or for languages where shared data structures are known at compile time ([Jell90]). Behavioural abstraction techniques, replay mechanisms and static analysis tools are among the most well known.

Behavioural Abstraction ([Bate88], [Baia86]) allows comparing the expected and actual runtime behaviour using specifications based on predefined event classes and operators for their combination. Filter- and clustering techniques make it possible to abstract from unimportant details. But all possibly interesting events have to be specified before program execution.

To reproduce runtime behaviour replay techniques were developed. In [LeBl87] only accesses to shared data are logged. These protocols then control the reexecution. Additional techniques are necessary for the actual debugging. The same is true for [Mill88]. Here a (minimal) program graph is constructed during the (initial) execution. The thus obtained information can by extended by partial reexecution (incremental tracing). The adaption of this concept for EDS Lisp is briefly outlined in chapter 7.

Static analysis techniques were developed to parallelize programs automatically (e.g. [Harr89]). Other approaches allow the detection of potential nondeterminacy in procedural languages ([Emra88], [Bala88], [Call88]). Potential Nondeterminacy means that by solely using the proposed techniques of static analysis it is not decidable whether a certain statement may contribute to a nondeterminate program result. In such cases tracing is necessary. We are investigating how such techniques could be helpful for EDS Lisp. Although static techniques are only of very restricted use in a language such as Common Lisp which is list oriented and which allows dynamic function definition we think such techniques an elaborate pendant to the use e.g. of masterscope ([Inter85]) in sequential Lisp.

7 Conclusion

Delphi is a toolkit for debugging the parallel Lisp dialect EDS Lisp. Although being developed for the distributed EDS system, its debugging strategies are applicable as well to similar distributed systems and shared memory architectures.

Delphi contains tools for distributed symbolic debugging. In addition to classical debugging strategies, the tools allow the computation of nonlocal information. Thus a detailed view of the global state of the parallel program may be obtained in break-mode, whereas the visualizer of the process history and mailbox communication offers a global view of the process dependencies

their communication structure. The visualizer assists in the granularity analysis of parallel Lisp programs and allows the detection of inappropriate communication structures in the EDS Lisp program.Together these tools provide a comprehensive overview of the program.
The debugging session is controlled by the user: The tool for observation and process control makes it possible to inspect the state of the different processes and reducing the amount of information displayed. The developped clique model is a mechanism for user controlled execution. It enhances the detection of unexpected nondeterminacy.

Delphi contains flexible debugging tools that allow the programmer to focus on details as well as to get a general overview of a program execution.

References

[Allen87] D. Allen, S. Steinberg, L. Stabile
Recent developments in Butterfly Lisp, AAAI 87, Seattle, July 1987, pp. 2-6

[Bag89] Laura Bagnell
ParVis: A Program Visualization Tool for Multilisp, S.M. thesis, MIT E.E.C.S. Dept., Cambridge, Ma, Feb. 1989

[Baia86] Fabrizio Baiardi, Nicoletta De Francesco, Gigliola Vaglini
Development of a Debugger for a Concurrent Language, IEEE Transactions on Software Engineering, Vol.SE-12(4), April 1986, pp. 547-553

[Bala88] Vasanth Balasundaram, Donn Bäumgartner, David Callahan, Ken Kennedy, Jaspal Subhlok
PTOOL: A System for Static Analysis of Parallelism in Programs, Rice University, Computer Science Technical Report TR88-71, June, 1988

[Bate88] Peter Bates
Debugging Heterogeneous Distributed Systems Using Event-Based Models of Behavior, Proceedings of the ACM SIGPLAN and SIGOPS Workshop on Parallel and Distributed Debugging, Madison, WI, USA, May 5-6, 1988, pp. 11-22

[BuMil88] Helmar Burkhart, Roland Millen
Performance-Measurement Tools in a Multiprocessor Environment, IEEE Transactions onComputers, Vol.38,No. 5, May 1989

[Call88] David Callahan, Jaspal Subhlok
Static-Analysis of Low-level Synchronization, Proceedings of the

ACM SIGPLAN and SIGOPS Workshop on Parallel and Distributed Debugging, Madison, WI, USA, 1988, pp. 100-111

[EDS89] Carsten Hammer et al.
Volume 5 (Part 2) Language Subsystems The Lisp Subsystem, ESPRITII EP2025, Document: EDS.DD.5S.0001, Dez. 1989

[Emra88] Perry A. Emrath, David A. Padua
Automatic Detection of Nondeterminism in Parallel Programs, Proceedings of the ACM SIGPLAN and SIGOPS Workshop on Parallel and Distributed Debugging, Madison, WI, USA, May 5-6, 1988, pp. 89-99

[Fid88] C. J. Fidge
Partial Orders for Parallel Debugging, Proceedings of the ACM SIGPLAN and SIGOPS Workshop on Parallel and Distributed Debugging, Madison, WI, USA, May 5-6, 1988, pp. 183-194

[GeKr86] Patrick F. McGehearty, Edward J. Krall
Potentials for Parallel Execution of Common Lisp Programs, Proceedings of the 1986 International Conference on Parallel Processing, IEEE, pp.696-702

[Harr89] Williams Ludwell Harrison III
The Interprocedural Analysis and Automatic Parallelization of Scheme Programs, Lisp and Symbolic Computation, Vol.2 No3/4, Okt.1989,pp.185-391

[Halst85] R. Halstead
Multilisp: A Language for Concurrent symbolic Computation, ACM Transactions on Programming Languages and Systems, Okt. 1985

[Halst90] R. Halstead
private communication

[Inter85] Xerox Cooporation
Interlisp-D Reference Manual I-III, Okt.1985

[HamHen90] Carsten Hammer, Thomas Henties
Parallel Lisp for a Distributed memory Machine, Proc. of the EUROPAL workshop on "High Performance and Parallel Computing in Lisp", Nov. 1990, Twickenham, UK

[Jell90] Sylvia Jell
Parallel Debugging - State of the Art Report, ESPRIT-II EP2025, Document: EDS.WP.8S.0002, Mar. 90

[Kranz89] David A. Kranz, Robert H. Halstead Jr., Eric Mohr
Mult-T: A High-Performance Parallel Lisp, SIGPLAN 1989 Sympo-

sium on Programming Language Design and Implementation, Portland, Oregon, June 1989

[LeBl87] Thomas J. LeBlanc, John M. Mellor-Crummey
Debugging Parallel Programs with Instant Replay, IEEE Transactions on Computers, Vol.C-36(4), April 1987, pp. 471-482

[Mill88] Barton P. Miller, Jong-Deok Choi
A Mechanism for Efficient Debugging of Parallel Programs, Proceedings of the SIGPLAN '88 Conference on Programming Language Design and Implementation, Atlanta, Georgia, USA, June 22-24, 1988, pp. 135-144

[Miller87] J. Miller
MultiScheme: A Parallel Processing System Based on MIT Scheme, Ph.D. Thesis, M.I.T. E.E.C.S. Dept., Cambridge, Mass., August 1987

[Sto88] Janice Stone
A graphical representation of concurrent processes, Proceedings of the ACM SIGPLAN and SIGOPS Workshop on Parallel and Distributed Debugging, Madison, WI, USA, May 5-6, 1988, pp. 226-235

[Stee84] Guy Steele
Common LISP: The Language, Digital Press, 1984

[Tar89] Jolan M. Targonski, Parallel Lisp Languages - the State of the Art, ESPRITII EP2025, Document: EDS.WP.5S.0001, Jul.1989

[ToLe90] Top Level, Inc.

Author Index Volume I

Lecture Notes in Computer Science

For information about Vols. 1–411
please contact your bookseller or Springer-Verlag

Vol. 412: L.B. Almeida, C.J. Wellekens (Eds.), Neural Networks. Proceedings, 1990. IX, 276 pages. 1990.

Vol. 413: R. Lenz, Group Theoretical Methods in Image Processing. VIII, 139 pages. 1990.

Vol. 414: A. Kreczmar, A. Salwicki, M. Warpechowski, LOGLAN '88 – Report on the Programming Language. X, 133 pages. 1990.

Vol. 415: C. Choffrut, T. Lengauer (Eds.), STACS 90. Proceedings, 1990. VI, 312 pages. 1990.

Vol. 416: F. Bancilhon, C. Thanos, D. Tsichritzis (Eds.), Advances in Database Technology – EDBT '90. Proceedings, 1990. IX, 452 pages. 1990.

Vol. 417: P. Martin-Löf, G. Mints (Eds.), COLOG-88. International Conference on Computer Logic. Proceedings, 1988. VI, 338 pages. 1990.

Vol. 418: K.H. Bläsius, U. Hedtstück, C.-R. Rollinger (Eds.), Sorts and Types in Artificial Intelligence. Proceedings, 1989. VIII, 307 pages. 1990. (Subseries LNAI).

Vol. 419: K. Weichselberger, S. Pöhlmann, A Methodology for Uncertainty in Knowledge-Based Systems. VIII, 136 pages. 1990 (Subseries LNAI).

Vol. 420: Z. Michalewicz (Ed.), Statistical and Scientific Database Management, V SSDBM. Proceedings, 1990 V, 256 pages. 1990.

Vol. 421: T. Onodera, S. Kawai, A Formal Model of Visualization in Computer Graphics Systems. X, 100 pages. 1990.

Vol. 422: B. Nebel, Reasoning and Revision in Hybrid Representation Systems. XII, 270 pages. 1990 (Subseries LNAI).

Vol. 423: L.E. Deimel (Ed.), Software Engineering Education. Proceedings, 1990. VI, 164 pages. 1990.

Vol. 424: G. Rozenberg (Ed.), Advances in Petri Nets 1989. VI, 524 pages. 1990.

Vol. 425: C.H. Bergman, R.D. Maddux, D.L. Pigozzi (Eds.), Algebraic Logic and Universal Algebra in Computer Science. Proceedings, 1988. XI, 292 pages. 1990.

Vol. 426: N. Houbak, SIL – a Simulation Language. VII, 192 pages. 1990.

Vol. 427: O. Faugeras (Ed.), Computer Vision – ECCV 90. Proceedings, 1990. XII, 619 pages. 1990.

Vol. 428: D. Bjørner, C.A.R. Hoare, H. Langmaack (Eds.), VDM '90. VDM and Z – Formal Methods in Software Development. Proceedings, 1990. XVII, 580 pages. 1990.

Vol. 429: A. Miola (Ed.), Design and Implementation of Symbolic Computation Systems. Proceedings, 1990. XII, 284 pages. 1990.

Vol. 430: J.W. de Bakker, W.-P. de Roever, G. Rozenberg (Eds.), Stepwise Refinement of Distributed Systems. Models, Formalisms, Correctness. Proceedings, 1989. X, 808 pages. 1990.

Vol. 431: A. Arnold (Ed.), CAAP '90. Proceedings, 1990. VI, 285 pages. 1990.

Vol. 432: N. Jones (Ed.), ESOP '90. Proceedings, 1990. IX, 436 pages. 1990.

Vol. 433: W. Schröder-Preikschat, W. Zimmer (Eds.), Progress in Distributed Operating Systems and Distributed Systems Management. Proceedings, 1989. V, 206 pages. 1990.

Vol. 434: J.-J. Quisquater, J. Vandewalle (Eds.), Advances in Cryptology – EUROCRYPT '89. Proceedings, 1989. X, 710 pages. 1990.

Vol. 435: G. Brassard (Ed.), Advances in Cryptology – CRYPTO '89. Proceedings, 1989. XIII, 634 pages. 1990.

Vol. 436: B. Steinholtz, A. Sølvberg, L. Bergman (Eds.), Advanced Information Systems Engineering. Proceedings, 1990. X, 392 pages. 1990.

Vol. 437: D. Kumar (Ed.), Current Trends in SNePS – Semantic Network Processing System. Proceedings, 1989. VII, 162 pages. 1990. (Subseries LNAI).

Vol. 438: D.H. Norrie, H.W. Six (Eds.), Computer Assisted Learning – ICCAL '90. Proceedings, 1990. VII, 467 pages. 1990.

Vol. 439: P. Gorny, M. Tauber (Eds.), Visualization in Human-Computer Interaction. Proceedings, 1988. VI, 274 pages. 1990.

Vol. 440: E. Börger, H. Kleine Büning, M.M. Richter (Eds.), CSL '89. Proceedings, 1989. VI, 437 pages. 1990.

Vol. 441: T. Ito, R.H. Halstead, Jr. (Eds.), Parallel Lisp: Languages and Systems. Proceedings, 1989. XII, 364 pages. 1990.

Vol. 442: M. Main, A. Melton, M. Mislove, D. Schmidt (Eds.), Mathematical Foundations of Programming Semantics. Proceedings, 1989. VI, 439 pages. 1990.

Vol. 443: M.S. Paterson (Ed.), Automata, Languages and Programming. Proceedings, 1990. IX, 781 pages. 1990.

Vol. 444: S. Ramani, R. Chandrasekar, K.S.R. Anjaneyulu (Eds.), Knowledge Based Computer Systems. Proceedings, 1989. X, 546 pages. 1990. (Subseries LNAI).

Vol. 445: A.J.M. van Gasteren, On the Shape of Mathematical Arguments. VIII, 181 pages. 1990.

Vol. 446: L. Plümer, Termination Proofs for Logic Programs. VIII, 142 pages. 1990. (Subseries LNAI).

Vol. 447: J.R. Gilbert, R. Karlsson (Eds.), SWAT '90. 2nd Scandinavian Workshop on Algorithm Theory. Proceedings, 1990. VI, 417 pages. 1990.

Vol. 448: B. Simons, A. Spector (Eds.), Fault Tolerant Distributed Computing. VI, 298 pages. 1990.

Vol. 449: M.E. Stickel (Ed.), 10th International Conference on Automated Deduction. Proceedings, 1990. XVI, 688 pages. 1990. (Subseries LNAI).

Vol. 450: T. Asano, T. Ibaraki, H. Imai, T. Nishizeki (Eds.), Algorithms. Proceedings, 1990. VIII, 479 pages. 1990.

Vol. 451: V. Mařík, O. Stepánková, Z. Zdráhal (Eds.), Artificial Intelligence in Higher Education. Proceedings, 1989. IX, 247 pages. 1990. (Subseries LNAI).

Vol. 452: B. Rovan (Ed.), Mathematical Foundations of Computer Science 1990. Proceedings, 1990. VIII, 544 pages. 1990.

Vol. 453: J. Seberry, J. Pieprzyk (Eds.), Advances in Cryptology – AUSCRYPT '90 Proceedings, 1990. IX. 462 pages. 1990.

Vol. 454: V. Diekert, Combinatorics on Traces. XII, 165 pages. 1990.

Vol. 455: C.A. Floudas, P.M. Pardalos, A Collection of Test Problems for Constrained Global Optimization Algorithms. XIV, 180 pages. 1990.

Vol. 456: P. Deransart, J. Maluszyn'ski (Eds.), Programming Language Implementation and Logic Programming. Proceedings, 1990. VIII, 401 pages. 1990.

Vol. 457: H. Burkhart (Ed.), CONPAR '90 – VAPP IV. Proceedings, 1990. XIV, 900 pages. 1990.

Vol. 458: J.C.M. Baeten, J.W. Klop (Eds.), CONCUR '90. Proceedings, 1990. VII, 537 pages. 1990.

Vol. 459: R. Studer (Ed.), Natural Language and Logic. Proceedings, 1989. VII, 252 pages. 1990. (Subseries LNAI).

Vol. 460: J. Uhl, H.A. Schmid, A Systematic Catalogue of Reusable Abstract Data Types. XII, 344 pages. 1990.

Vol. 461: P. Deransart, M. Jourdan (Eds.), Attribute Grammars and their Applications. Proceedings, 1990. VIII, 358 pages. 1990.

Vol. 462: G. Gottlob, W. Nejdl (Eds.), Expert Systems in Engineering. Proceedings, 1990. IX, 260 pages. 1990. (Subseries LNAI).

Vol. 463: H. Kirchner, W. Wechler (Eds.), Algebraic and Logic Programming. Proceedings, 1990. VII, 386 pages. 1990.

Vol. 464: J. Dassow, J. Kelemen (Eds.), Aspects and Prospects of Theoretical Computer Science. Proceedings, 1990. VI, 298 pages. 1990.

Vol. 465: A. Fuhrmann, M. Morreau (Eds.), The Logic of Theory Change. Proceedings, 1989. X, 334 pages. 1991. (Subseries LNAI).

Vol. 466: A. Blaser (Ed.), Database Systems of the 90s. Proceedings, 1990. VIII, 334 pages. 1990.

Vol. 467: F. Long (Ed.), Software Engineering Environments. Proceedings, 1969. VI, 313 pages. 1990.

Vol. 468: S.G. Akl, F. Fiala, W.W. Koczkodaj (Eds.), Advances in Computing and Information – ICCI '90. Proceedings, 1990. VII, 529 pages. 1990.

Vol. 469: I. Guessarian (Ed.), Semantics of Systeme of Concurrent Processes. Proceedings, 1990. V, 456 pages. 1990.

Vol. 470: S. Abiteboul, P.C. Kanellakis (Eds.), ICDT '90. Proceedings, 1990. VII, 528 pages. 1990.

Vol. 471: B.C. Ooi, Efficient Query Processing in Geographic Information Systems. VIII, 208 pages. 1990.

Vol. 472: K.V. Nori, C.E. Veni Madhavan (Eds.), Foundations of Software Technology and Theoretical Computer Science. Proceedings, 1990. X, 420 pages. 1990.

Vol. 473: I.B. Damgård (Ed.), Advances in Cryptology – EUROCRYPT '90. Proceedings, 1990. VIII, 500 pages. 1991.

Vol. 474: D. Karagiannis (Ed.), Information Syetems and Artificial Intelligence: Integration Aspects. Proceedings, 1990. X, 293 pages. 1991. (Subseries LNAI).

Vol. 475: P. Schroeder-Heister (Ed.), Extensions of Logic Programming. Proceedings, 1989. VIII, 364 pages. 1991. (Subseries LNAI).

Vol. 476: M. Filgueiras, L. Damas, N. Moreira, A.P. Tomás (Eds.), Natural Language Processing. Proceedings, 1990. VII, 253 pages. 1991. (Subseries LNAI).

Vol. 477: D. Hammer (Ed.), Compiler Compilers. Proceedings, 1990. VI, 227 pages. 1991.

Vol. 478: J. van Eijck (Ed.), Logics in AI. Proceedings, 1990. IX, 562 pages. 1991. (Subseries in LNAI).

Vol. 480: C. Choffrut, M. Jantzen (Eds.), STACS 91. Proceedings, 1991. X, 549 pages. 1991.

Vol. 481: E. Lang, K.-U. Carstensen, G. Simmons, Modelling Spatial Knowledge on a Linguistic Basis. IX, 138 pages. 1991. (Subseries LNAI).

Vol. 482: Y. Kodratoff (Ed.), Machine Learning – EWSL-91. Proceedings, 1991. XI, 537 pages. 1991. (Subseries LNAI).

Vol. 483: G. Rozenberg (Ed.), Advances In Petri Nets 1990. VI, 515 pages. 1991.

Vol. 484: R. H. Möhring (Ed.). Graph-Theoretic Concepts In Computer Science. Proceedings, 1990. IX, 360 pages. 1991.

Vol. 485: K. Furukawa, H. Tanaka. T. Fullsaki (Eds.), Logic Programming '89. Proceedings, 1989. IX, 183 pages. 1991. (Subseries LNAI)

Vol. 486: J. van Leeuwen. N. Santoro (Eds.). Distributed Algorithms. Proceedings, 1990. VI, 433 pages. 1991.

Vol. 487: A. Bode (Ed.). Distributed Memory Computing. Proceedings, 1991. XI, 506 pages. 1991

Vol. 488: R. V. Book (Ed.). Rewriting Techniques and Applications. Proceedings, 1991. VII, 458 pages. 1991.

Vol. 489: J. W. de Bakker, W. P. de Roever, G. Rozenberg (Eds.), Foundations of Object-Oriented Languages. Proceedings, 1990. VIII, 442 pages. 1991.

Vol. 490: J. A. Bergstra, L. M. G. Feljs (Eds .), Algebraic Methods 11: Theory, Tools and Applicatlons. VI, 434 pages. 1991.

Vol. 491: A. Yonezawa, T. Ito (Eds.). Concurrency: Theory, Language, and Architecture. Proceedings, 1989. VIII, 339 pages. 1991.

Vol. 492: D. Sriram, R. Logcher, S. Fukuda (Eds.). Computer-Aided Cooperative Product Development. Proceedings, 1989 VII, 630 pages. 1991.

Vol. 493: S. Abramsky, T. S. E. Maibaum (Eds.). TAPSOFT '91. Volume 1. Proceedings, 1991. VIII, 455 pages. 1991.

Vol. 494: S. Abramsky, T. S. E. Maibaum (Eds.). TAPSOFT '91. Volume 2. Proceedings, 1991. VIII, 482 pages. 1991

Vol. 495: 9. Thalheim, J. Demetrovics, H.-D. Gerhardt (Eds.), MFDBS '91. Proceedings, 1991. VI, 395 pages. 1991.

Vol. 497: F. Dehne, F. Fiala. W.W. Koczkodaj (Eds.), Advances in Computing and Intormation - ICCI '91 Proceedings, 1991. VIII, 745 pages. 1991.

Vol. 498: R. Andersen, J. A. Bubenko jr., A. Sølvberg (Eds.), Advanced Information Systems Engineering. Proceedings, 1991. VI, 579 pages. 1991.

Vol. 499: D. Christodoulakis (Ed.), Ada: The Choice for '92. Proceedings, 1991. VI, 411 pages. 1991.

Vol. 500: M. Held, On the Computational Geometry of Pocket Machining. XII, 179 pages. 1991.

Vol. 501: M. Bidoit, H.-J. Kreowski, P. Lescanne, F. Orejas, D. Sannella (Eds.). Algebraic System Specification and Development. VIII, 98 pages. 1991.

Vol. 502: J. Bārzdiņš, D. Bjørner (Eds.), Baltic Computer Science. X, 619 pages. 1991.

Vol. 503: P. America (Ed.), Parallel Database Systems. Proceedings, 1990. VIII, 433 pages. 1991.

Vol. 504: J. W. Schmidt, A. A. Stogny (Eds.), Next Generation Information System Technology. Proceedings, 1990. IX, 450 pages. 1991.

Vol. 505: E. H. L. Aarts, J. van Leeuwen, M. Rem (Eds.), PARLE '91. Parallel Architectures and Languages Europe, Volume I. Proceedings, 1991. XV, 423 pages. 1991.

Vol. 506: E. H. L. Aarts, J. van Leeuwen, M. Rem (Eds.), PARLE '91. Parallel Architectures and Languages Europe, Volume II. Proceedings, 1991. XV, 489 pages. 1991.